# Lecture Notes in Artificial Intelligence 8917

Subseries of Lecture Notes in Computer Science

LNAI Series Editors

Randy Goebel
*University of Alberta, Edmonton, Canada*
Yuzuru Tanaka
*Hokkaido University, Sapporo, Japan*
Wolfgang Wahlster
*DFKI and Saarland University, Saarbrücken, Germany*

LNAI Founding Series Editor

Joerg Siekmann
*DFKI and Saarland University, Saarbrücken, Germany*

Xianmin Zhang   Honghai Liu   Zhong Chen
Nianfeng Wang (Eds.)

# Intelligent Robotics and Applications

7th International Conference, ICIRA 2014
Guangzhou, China, December 17-20, 2014
Proceedings, Part I

 Springer

Volume Editors

Xianmin Zhang
Zhong Chen
Nianfeng Wang
South China University of Technology
School of Mechanical and Automobile Engineering
Wushan Rd. 381, Tianhe District, Guangzhou 510641, China
E-mail: {zhangxm, mezhchen, menfwang}@scut.edu.cn

Honghai Liu
University of Portsmouth
School of Computing
Intelligent Systems and Biomedical Robotics Group
Buckingham Building, Lion Terrace
Portsmouth PO1 3HE, UK
E-mail: honghai.liu@icloud.com

ISSN 0302-9743                       e-ISSN 1611-3349
ISBN 978-3-319-13965-4               e-ISBN 978-3-319-13966-1
DOI 10.1007/978-3-319-13966-1
Springer Cham Heidelberg New York Dordrecht London

Library of Congress Control Number: 2014956249

LNCS Sublibrary: SL 7 – Artificial Intelligence

*Typesetting:* Camera-ready by author, data conversion by Scientific Publishing Services, Chennai, India

Printed on acid-free paper

Springer is part of Springer Science+Business Media (www.springer.com)

# Preface

The Organizing Committee of the 7<sup>th</sup> International Conference on Intelligent Robotics and Applications has been committed to facilitating interactions among those active in the field of intelligent robotics, automation, and mechatronics. Through this conference, the committee intends to enhance the sharing of individual experiences and expertise in intelligent robotics with particular emphasis on technical challenges associated with varied applications such as biomedical application, industrial automations, surveillance, and sustainable mobility.

The 7<sup>th</sup> International Conference on Intelligent Robotics and Applications was most successful in attracting 159 submissions on addressing the state-of-the art developments in robotics, automation, and mechatronics. Owing to the large number of submissions, the committee was faced with the difficult challenge of selecting of the most deserving papers for inclusion in these proceedings. For this purpose, the committee undertook a rigorous review process. Despite the high quality of most of the submissions, only 109 papers (68.5 % acceptance rate) were selected for publication in two volumes of Springer's *Lecture Notes in Artificial Intelligence* as subseries of *Lecture Notes in Computer Science*. The selected papers were presented during the 7<sup>th</sup> International Conference on Intelligent Robotics and Applications held in Guangzhou, China, during December 17 – 20, 2014.

The selected articles represent the contributions of researchers from 19 countries. The contributions of the technical Program Committee and the additional reviewers are deeply appreciated. Most of all, we would like to express our sincere thanks to the authors for submitting their most recent works and the Organizing Committee for their enormous efforts to turn this event into a smoothly running meeting. Special thanks go to South China University of Technology for the generosity and direct support. Our particular thanks are due to Alfred Hofmann and Anna Kramer of Springer for enthusiastically supporting the project.

We sincerely hope that these volumes will prove to be an important resource for the scientific community.

October 2014

Xianmin Zhang
Xiangyang Zhu
Jangmyung Lee
Huosheng Hu

# Organization

## International Advisory Committee

| | |
|---|---|
| Tamio Arai | University of Tokyo, Japan |
| Hegao Cai | Harbin Institute of Technology, China |
| Tianyou Chai | Northeastern University, China |
| Toshio Fukuda | Nagoya University, Japan |
| Sabina Jesehke | RWTH Aachen University, Germany |
| Oussama Khatib | Stanford University, USA |
| Zhongqin Lin | Shanghai Jiao Tong University, China |
| Ming Li | National Natural Science Foundation of China, China |
| Nikhil R. Pal | Indian Statistical Institute, India |
| Grigory Panovko | Russian Academy of Science, Russia |
| Jinping Qu | South China University of Technology, China |
| Clarence De Silva | University of British Columbia, Canada |
| Bruno Siciliano | University of Naples, Italy |
| Shigeki Sugano | Waseda University, Japan |
| Michael Yu Wang | Chinese University of Hong Kong, China |
| Youlun Xiong | Huazhong University of Science and Technology, China |
| Ming Xie | Nanyang Technological University, Singapore |
| Lotfi A. Zadeh | University of California Berkeley, USA |

## General Chair

| | |
|---|---|
| Xianmin Zhang | South China University of Technology, China |

## General Co-chairs

| | |
|---|---|
| Xiangyang Zhu | Shanghai Jiao Tong University, China |
| Jangmyung Lee | Pusan National University, South Korea |
| Huosheng Hu | University of Essex, UK |

## Program Chair

| | |
|---|---|
| Guoli Wang | Sun Yat-Sen University, China |

## Program Co-chairs

| | |
|---|---|
| Youfu Li | City University of Hong Kong, SAR China |
| Honghai Liu | University of Portsmouth, UK |

## Organizing Committee Chairs

Han Ding                    Huazhong University of Science and
                            Technology, China
Feng Gao                    Shanghai Jiao Tong University, China
Zexiang Li                  Hong Kong University of Science and
                            Technology, China
Tian Huang                  Tianjin University, China
Hong Liu                    Harbin Institute of Technology, China
Guobiao Wang                National Natural Science Foundation of China,
                            China
Tianmiao Wang               Beihang University, China
Hong Zhang                  University of Alberta, Canada

## Area Chairs

Suguru Arimoto              Ritsumeikan University, Japan
Jiansheng Dai               King's College of London University, UK
Christian Freksa            Bremen University, Germany
Shuzhi Sam Ge               National University of Singapore, Singapore
Peter B. Luh                Connecticut University, USA

## Awards Chairs

Yangming Li                 University of Macau, SAR China
Wei-Hsin Liao               Chinese University of Hong Kong, SAR China
Caihua Xiong                Huazhong University of Science and
                            Technology, China

## Publication Chairs

Zhong Chen                  South China University of Technology, China
Xinjun Sheng                Shanghai Jiao Tong University, China

## Local Arrangements Chairs

Nianfeng Wang               South China University of Technology, China
Lianghui Huang              South China University of Technology, China

## Finance Chair

Hongxia Yang                South China University of Technology, China
Wei Sun                     South China University of Technology, China

# General Affairs Chair

Zhiwei Tan                          South China University of Technology, China

# Additional Reviewers

We would like to acknowledge the support of the following people who peer reviewed articles for ICIRA 2014.

| | | |
|---|---|---|
| Xianmin Zhang | Guoli Wang | Nianfeng Wang |
| Guoying Gu | Zhong Chen | Tiemin Zhang |
| Yanjiang Huang | Xiangyang Zhu | Xinjun Liu |
| Caihua Xiong | Hegao Cai | Tianyou Chai |
| Tamio Arai | Yan Li | Clarence De Silva |
| Toshio Fukuda | Sabina Jesehke | Michael Yu Wan |
| Zhongqin Lin | Ming Li | Shigeki Sugano |
| Grigory Panovko | Jinping Qu | Jangmyung Lee |
| Bruno Siciliano | Lotfi A. Zadeh | Honghai Liu |
| Huosheng Hu | Youfu, Li | Zexiang Li |
| Han Ding | Feng Gao | Guobiao Wang |
| Tian Huang | Hong Liu | Suguru Arimoto |
| Tianmiao Wang | Hong Zhang | Shuzhi Sam Ge |
| Jiansheng Dai | Christian Freksa | Youlun Xiong |
| Peter B. Luh | Wei-Hsin Liao | |
| Xinjun Sheng | Yangming Li | |

# Table of Contents – Part I

## Recent Advances in Research and Application of Modern Mechanisms

## Rehabilitation Robotics

# Underwater Robotics and Applications

# Agricultural Robot

# Bionic Robotics

## Service Robotics

# Table of Contents – Part II

## Parallel Robotics

## Robot Vision

## Mechatronics

## Industrial Robotics

## System Optimization and Analysis

## Mechanism Design

# Task-Oriented Design Method and Experimental Research of Six-Component Force Sensor

Jiantao Yao[1,2,*], Wenju Li[1], Hongyu Zhang[1], Yundou Xu[1], and Yongsheng Zhao[1,2]

[1]Parallel Robot and Mechatronic System Laboratory of Hebei Province,
Yanshan University, Qinhuangdao 066004, China
[2]Key Laboratory of Advanced Forging and Stamping Technology and Science,
Ministry of Education of China, Yanshan University, Qinhuangdao 066004, China
{jtyao,ydxu,yszhao}@ysu.edu.cn, gu_duxingzhe@yeah.net,
zhyysu@gmail.com

**Abstract.** As the isotropy performance widely used for evaluating parallel six-component force sensor is not suitable for a certain task, the task-oriented design method of the sensor is proposed. The mathematic model of sensor is built with the screw theory; based on the task requirement, the task ellipsoid is established. Then, with ellipsoid method, the design method for the sensor structure is proposed and the structure constraints are deduced. In order to perform peg-in-hole assembly task with the task-oriented sensor, the sensor prototype is designed and manufactured. Based on the assembly platform, the experimental research is carried out. The research results are useful for the design and application of six-component force sensor.

**Keywords:** Parallel six-component force sensor, Design method, Task, Assembly experiment.

## 1    Introduction

With the ability of measuring three force components and three torque components, the six-component force sensor is a necessary tool for achieving six-component force measurement technology. It is widely used to measure the force with changing directions and dimensions, acceleration and realize force and force/position control.

Stewart platform-based six-component force sensor is composed of two platforms and six elastic legs connecting the platforms with spherical joints. Theoretically speaking, each elastic leg of the sensor just sustains tensile strain or compressive strain along its axis with the neglect of the legs' gravity and the frictional moment in the spherical pairs, which can perform the measurement of six-component force/torquewithout stress coupling. So Stewart platform-based six-axis sensor is one of the most widely adopted structures in six-component sensor design and manufacture[1,2,3,4]. The current design studies mainly focus on optimizing the isotropy performance of six-component force sensor with isotropy indices [5,6,7,8], besides, the precondition of the optimal design is

---

* Corresponding author.

X. Zhang et al. (Eds.): ICIRA 2014, Part I, LNAI 8917, pp. 1–12, 2014.

that the task ellipsoid is a ball. However, for practical application of most manipulator tasks, such as contour tracking, peg-in-hole assembly and grasp tasks, the task model is not a ball but an ellipsoid, while the isotropic structure is not optimal[9].Therefore, according to the applications of the six-component force sensors, proposing task-oriented design method has important significance.

The organization of this paper is as follows. Following the introduction, the task model of parallel six-component force sensor is established. With the model and task requirement, the task-oriented design method is proposed and the constraints of the sensor parameter are deduced in the section3. Besides, the in order to perform the peg-in-hole assembly task, the sensor prototype is designed with the proposed method in section 4. Furthermore, based on the assembly platform with sensor prototype, the experimental research is carried out. The paper is concluded at last, summarizing the present work.

## 2    Task Model of Sensor

### 2.1    Mathematic Model

Based on the screw theory [10], the mapping relationships between the elastic legs and the applied force of the parallel sensor can be obtained as

$$F_a = Gf \tag{1}$$

Where $F_a = (F_x\ F_y\ F_z\ M_x\ M_y\ M_z)^{\mathrm{T}}$ is the vector of six-component external force applied on the measuring platform; $f = (f_1\ f_2\ f_3 \cdots f_n)^{\mathrm{T}}$ is the vector composed of the reacting forces of the $n$ legs; $G$ is the first ordered static matrix of affection coefficient which is given by

$$G = \begin{pmatrix} S_1 & S_2 & S_3 & \cdots & S_n \\ S_{01} & S_{02} & S_{03} & \cdots & S_{0n} \end{pmatrix} \tag{2}$$

Where $\$_i = (S_i,\ S_{0i})$ represents the unit line vector along the axis of the $i$ th leg.

Generally, the stiffness of each elastic leg is same, and the reacting forces of the legs can be calculated with the Moore-Penrose generalized inverse as

$$f = G^+ F_a = CF_a \tag{3}$$

Where matrix G+ is the Moore-Penrose generalized inverse of matrix $G$, and we have C=G+.

Considering the dimensional differences of force and moment, the matrix $C$ can expressed as

$$C = \begin{pmatrix} C_F & C_M \end{pmatrix} \tag{4}$$

Where matrix $C_F$ is the first three columns of matrix $C$, which represents the inverse mapping relationship of the three force components; matrix $C_M$ is the last three columns of matrix $C$, which is the inverse mapping relationship of the three moment components.

In addition, matrix $F_a$ can be rewritten as

$$F_a = \begin{pmatrix} F_s & M_s \end{pmatrix}^{\mathrm{T}} \tag{5}$$

Combining Eq. (3), Eq. (4) and Eq. (5), we obtain

$$f = C_F F_s + C_M M_s \tag{6}$$

Let

$$f_F = C_F F_s \tag{7}$$

$$f_M = C_M M_s \tag{8}$$

According to Eq. (7) and Eq. (8), define two balls in the reacting force space of legs with $\| f_F \| = 1$ and $\| f_M \| = 1$; furthermore, mapping the two balls into the space of applied force, the force ellipsoid and the moment ellipsoid can be obtained and expressed as

$$F_s^{\mathrm{T}} C_F^{\mathrm{T}} C_F F_s = 1 \tag{9}$$

$$M_s^{\mathrm{T}} C_M^{\mathrm{T}} C_M M_s = 1 \tag{10}$$

## 2.2    Task Model

Similar to the modeling method of the grasping planning with task ellipsoid[11], the task ellipsoid in the task coordinate system can be obtained as

$$a_1 F_{tx}^2 + a_2 F_{ty}^2 + a3 F_{tz}^2 + a_4 F_{tx} F_{ty} + a_5 F_{ty} F_{tz} + \\ a_6 F_{tz} F_{tx} + a_7 F_{tx} + a_8 F_{ty} + a_9 F_{tz} + a_{10} \leq 0 \tag{11}$$

Where $F_{tx}$, $F_{ty}$, $F_{tz}$ represent three perpendicular force components; $a_i (i=1,2,\cdots,10)$ represents real coefficient.

When the center of the ellipsoid coincides with the coordinate origin, Eq. (11) can be expressed in matrix form as

$$F_t^{\mathrm{T}} C_{tF}^{\mathrm{T}} C_{tF} F_t \leq 1 \tag{12}$$

Where

$$F_t = \begin{pmatrix} F_{tx} & F_{ty} & F_{tz} \end{pmatrix}^{\mathrm{T}};$$

$$C_{tF}^{\mathrm{T}} C_{tF} = -\frac{1}{a_{10}} \begin{pmatrix} a_1 & \dfrac{a_4}{2} & \dfrac{a_6}{2} \\ \dfrac{a_4}{2} & a_2 & \dfrac{a_5}{2} \\ \dfrac{a_6}{2} & \dfrac{a_5}{2} & a_3 \end{pmatrix};$$

As matrix $C_{tF}^{\mathrm{T}}C_{tF}$ is a real symmetric matrix, we have Eq. (13) with the orthogonal diagonal factorization of the matrix.

$$C_{tF}^{\mathrm{T}}C_{tF} = Q_F diag(\lambda_{tFx}, \lambda_{tFy}, \lambda_{tFz})Q_F^{\mathrm{T}} \tag{13}$$

Where $\lambda_{tFx}$, $\lambda_{tFy}$, $\lambda_{tFz}$ represent the eigenvalues of matrix $C_{tF}^{\mathrm{T}}C_{tF}$, and the half lengths of the force ellipsoid's three spindles can be given by

$$a_{tF} = 1/\sqrt{\lambda_{tFx}}, b_{tF} = 1/\sqrt{\lambda_{tFy}}, c_{tF} = 1/\sqrt{\lambda_{tFz}} \tag{14}$$

The three column vectors of matrix $Q_F$ constitute a complete orthonarmal feature vector system of matrix $C_{tF}^{\mathrm{T}}C_{tF}$.

Similarly, the moment ellipsoid of the task can be given by

$$M_t^{\mathrm{T}}C_{tM}^{\mathrm{T}}C_{tM}M_t \leq 1 \tag{15}$$

Furthermore, based on the orthogonal diagonal factorization of matrix $C_{tM}^{\mathrm{T}}C_M$, we have

$$C_{tM}^{\mathrm{T}}C_{tM} = Q_M diag(\lambda_{tMx}, \lambda_{tMy}, \lambda_{tMz})Q_M^{\mathrm{T}} \tag{16}$$

Where $\lambda_{tMx}$, $\lambda_{tMy}$, $\lambda_{tMz}$ represent the eigenvalues of matrix $C_{tM}^{\mathrm{T}}C_{tM}$, and the half lengths of the moment ellipsoid's three spindles can be given by

$$a_{tM} = 1/\sqrt{\lambda_{tMx}}, b_{tM} = 1/\sqrt{\lambda_{tMy}}, c_{tM} = 1/\sqrt{\lambda_{tMz}} \tag{17}$$

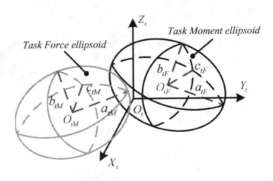

**Fig. 1.** Sketch map of task model

The three column vectors of matrix $Q_M$ constitute a complete orthonarmal feature vector system of matrix $C_{tM}^{\mathrm{T}}C_{tM}$.

Finally, as shown in Fig. 1, the force ellipsoid and the moment ellipsoid of the task are determined in the task coordinate system.

# 3    Task-Oriented Design Method

According to the ellipsoid theory, the force ellipsoid and the moment ellipsoid of sensor should contain the force ellipsoid and the moment ellipsoid of task, respectively. So the length constraints of the ellipsoids' spindles can be given by

$$\begin{cases} \lambda_{sFx} = k_{Fx}\lambda_{tFx} \\ \lambda_{sFy} = k_{Fy}\lambda_{tFy} \\ \lambda_{sFz} = k_{Fz}\lambda_{tFz} \end{cases}$$

$$\begin{cases} \lambda_{sFx} = k_{Fx}\lambda_{tFx} \\ \lambda_{sFy} = k_{Fy}\lambda_{tFy} \\ \lambda_{sFz} = k_{Fz}\lambda_{tFz} \end{cases} \tag{18}$$

Where $\lambda_{sFx}$, $\lambda_{sFy}$ and $\lambda_{sFz}$ represent the eigenvalues of matrix $C_{sF}^{T}C_{sF}$, and the square roots of their reciprocal are half length of the sensor's force ellipsoid spindles, respectively; $k_{Fx}$, $k_{Fy}$ and $k_{Fz}$ are positive real numbers, which are not large than 1.

So the force ellipsoid of sensor in the task coordinatesystem, satisfied the task requirement, can be expressed as

$$^{t}F_{s}^{T}Q_{sF}\,diag\left(\lambda_{sFx},\lambda_{sFy},\lambda_{sFz}\right)Q_{sF}^{T}\,^{t}F_{s} = 1 \tag{19}$$

Where the three column vectors of matrix $Q_{sF}$ represent the directions of the sensor's force ellipsoid spindles.

Similarly, the moment ellipsoid of sensor in the task coordinatesystem, satisfied the task requirement, can be expressed as

$$^{t}M_{s}^{T}Q_{sM}\,diag\left(\lambda_{sMx},\lambda_{sMy},\lambda_{sMz}\right)Q_{sM}^{T}\,^{t}M_{s} = 1 \tag{20}$$

Where the three column vectors of matrix $Q_{sM}$ represent the directions of the sensor's moment ellipsoid spindles; $\lambda_{sMx}$, $\lambda_{sMy}$ and $\lambda_{sMz}$ are the eigenvalues of matrix $C_{sM}^{T}C_{sM}$, and the square roots of their reciprocal are half length of the sensor's moment ellipsoid spindles, respectively, which satisfy the constraints as

$$\begin{cases} \lambda_{sMx} = k_{Mx}\lambda_{tMx} \\ \lambda_{sMy} = k_{My}\lambda_{tMy} \\ \lambda_{sMz} = k_{Mz}\lambda_{tMz} \end{cases} \tag{21}$$

Where $k_{Mx}$, $k_{My}$ and $k_{Mz}$ are positive real numbers which are not large than 1.

As the ellipsoid of sensor described in Eq. (9), Eq. (10) and the ellipsoid of sensor described in Eq. (19), Eq. (20) with the task model are in different coordinate system, the equations should be converted into the same coordinate system; then, make the corresponding terms equal, the relationships between the structural parameters of sensor can be obtained.

## 4      Structural Design of Sensor Prototype

### 4.1    Fully Pre-stressed Structure

Consideringthe shortages of traditional parallel structure six-component force sensor that therelatively bigger contact area in the practical spherical pair leads to the bigger frictional force in the ball and socket joints which results in the bigger stress coupling on measuring legs; the inherent clearance existing in the traditional spherical pair causes the mechanical hysteresis and the damage tothe linearity of the sensor; twelve spherical pairsare required to be adjusted and pre-stressed separately, but it is difficult to make the pre-stressed force uniform, the fully pre-stressed sensor structure is proposed and used as the prototype structure.

**Fig. 2.** Fully pre-stressed parallel six-component force sensor (1) measuring platform; (2) adjust shim; (3) sleeve; (4) measuring leg; (5) pre-stressing platform; (6) pre-stressing bolt; (7) pre-stressing nut; (8) middle platform connected to measuring platform; (9) fixed base.

Fig. 2 illustrates the structure of fully pre-stressed sensor,which mainly consists of a force-measuring platform, a fixed platform, a pre-stressing platform and seven measuring legs divided into two groups and distributed on the two sides of force-measuring platform.

Each measuring leg contains a single-axis force transducer and connects with the two platforms using the cone-shaped spherical pairs with unilateral constraint instead of the traditional spherical pairs. Due to the application of the unilateral constraint, the pre-stressed force is needed to ensure that all the measuring legs are always in compression when subjected to the expected range of external loads. There is a gap or a soft aluminum gasket designed between the pre-stressing platform and fixed base, so by pre-stressing the bolt and nut properly, the platforms can make measuring legs bear an anticipant pre-compression force[12].

### 4.2    Mathematic Model

Based on the fully pre-stressed sensor structure, the structural model can be established as Fig. 3 shown.

The Cartesian coordinate system $O_s\text{-}X_sY_sZ_s$ is set up with its origin located at the center of the gravity of measuring platform. For the generalized fault-tolerant fully pres-stressed six-component force sensor, symbols are defined as follows: $b_i$ and $B_i$ ($i=1,2,\cdots,7$) denote the position vectors of the center of $i$th spherical joint on the force-measuring platform, pre-stressing platform and fixed base with respect to the coordinate system, respectively.

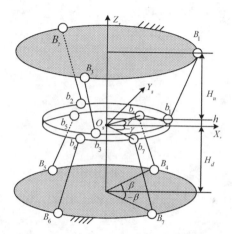

**Fig. 3.** Structure sketch of fully pre-stressed parallel

The first ordered static matrix of affection coefficient can be obtained as

$$G = \begin{bmatrix} \dfrac{b_1 - B_1}{|b_1 - B_1|} & \dfrac{b_2 - B_2}{|b_2 - B_2|} & \dfrac{b_3 - B_3}{|b_3 - B_3|} & \cdots & \dfrac{b_7 - B_7}{|b_7 - B_7|} \\[2mm] \dfrac{B_1 \times b_1}{|b_1 - B_1|} & \dfrac{B_2 \times b_2}{|b_2 - B_2|} & \dfrac{B_3 \times b_3}{|b_3 - B_3|} & \cdots & \dfrac{B_7 \times b_7}{|b_7 - B_7|} \end{bmatrix} \tag{22}$$

Furthermore, combining Eq. (24) and Eq. (3) - Eq. (10), we have

$$F_s^T C_F^T C_F F_s = 1 \tag{23}$$

$$M_s^T C_M^T C_M M_s = 1 \tag{24}$$

Where

$$C_F^T C_F = \begin{bmatrix} \psi_1 & 0 & 0 \\ 0 & \psi_1 & 0 \\ 0 & 0 & \psi_2 \end{bmatrix}; \quad C_M^T C_M = \begin{bmatrix} \psi_3 & 0 & 0 \\ 0 & \psi_3 & 0 \\ 0 & 0 & \psi_4 \end{bmatrix};$$

$\psi_1 = [(3h^2 R^2 - 6rH_u hR - 6rh^2 R + 3r^2 H_u^2 + 6r^2 H_u h + 3r^2 h^2)l_2^2$
$\qquad + 8H_d^2 r^2 l_1^2 \cos^2 \gamma]/12\lambda_1$ ;

$\psi_2 = l_1^2 l_2^2 / (3H_u^2 l_2^2 + 4H_d^2 l_1^2)$ ;

$\psi_3 = [(3R^2 - 6rR + 3r^2)l_2^2 + 8(r^2 \sin^2 \gamma - 2rR \sin \beta \sin \gamma + R^2 \sin \beta)l_1^2]/12\lambda_2$ ;

$$\psi_4 = l_2^2 / \left[ 4r^2 R^2 \sin^2 (\gamma - \beta) \right] ;$$
$$\lambda_1 = ( r^2 \cos \gamma (H_u + H_d + h) - Rr \cos \beta (H_u + h) - rR \cos \gamma (h + H_d) + R^2 h \cos \beta)^2 ;$$
$$\lambda_2 = ( r^2 \sin \gamma (H_d + H_u + h) - rR \sin \gamma (h + H_d) + R^2 h \sin \beta - Rr \sin \beta ( H_u + h))^2 .$$

## 4.3    Design of Prototype Parameters

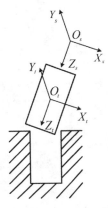

**Fig. 4.** Sketch map of coordinate of sensor and task

The sensor prototype is designed for peg-in-hole assembly, and connected to the shaft with its measuring platform. As Fig. 4 shown, the coordinate system $O_s\text{-}X_sY_sZ_s$ is the reference coordinates of the sensor. The origin $O_s$ of the coordinate system is on the centroid of the lower surface of sensor's measuring platform. Besides, $O_t\text{-}X_tY_tZ_t$ is the task coordinate system. The two coordinate axes are parallel, respectively. The coordinate of the origin $O_s$ in the centroid of shaft is $(0, 0, Z_{t0})$. Assuming the length of the shaft is 180mm and the diameter is 12mm; according to Eq. (12) and Eq. (15), the task force ellipsoid and the task moment force ellipsoid can be expressed as

$$F_t^{\mathrm{T}} C_{tF}^{\mathrm{T}} C_{tF} F_t \leq 1 \tag{25}$$

$$M_t^{\mathrm{T}} C_{tM}^{\mathrm{T}} C_{tM} M_t \leq 1 \tag{26}$$

Where

$$C_{tF}^{\mathrm{T}} C_{tF} = \begin{bmatrix} 16.5 & 0 & 0 \\ 0 & 16.5 & 0 \\ 0 & 0 & 0.29 \end{bmatrix} ;$$

$$C_{tM}^{\mathrm{T}} C_{tM} = \begin{bmatrix} 0.00098 & 0 & 0 \\ 0 & 0.00098 & 0 \\ 0 & 0 & 0.0039 \end{bmatrix} ;$$

Furthermore, converting the sensor ellipsoids from reference coordinate system of sensor to the coordinate system of task and combining Eq. (18), Eq.(21), the parametric constraint equations can be obtained as

$$
\begin{cases}
\{3[(R-r)(h+Z_{t0})-rH_u]^2 + 8[(Z_{t0}+H_d)r\cos\gamma \\
\qquad - Z_{to}R\cos\beta]^2\}l^2/12\lambda_1 = 16.5k_{Fx} \\
\{3[R(h-Z_{t0})+r(H_u+Z_{t0}-h)]^2 - 8[(Z_{t0}+H_d)\times \\
\qquad r\sin\gamma - Z_{to}R\sin\beta]^2\}l^2/12\lambda_2 = 16.5k_{Fy} \\
l^2/(3H_u^2+4H_d^2) = 0.29k_{Fz}
\end{cases}
\tag{27}
$$

$$
\begin{cases}
[3(R-r)^2 + 8(r^2\sin^2\gamma - 2rR\sin\beta\sin\gamma + \\
\qquad R^2\sin\beta)]l^2/12\lambda_2 = 0.00098k_{Mx} \\
[3(R-r)^2 + 8(r^2\cos^2\gamma - 2rR\cos\beta\cos\gamma \\
\qquad + R^2\cos\beta)]l^2/12\lambda_1 = 0.00098k_{My} \\
l^2/[4r^2R^2\sin^2(\gamma-\beta)] = 0.0039k_{Mz}
\end{cases}
\tag{28}
$$

Let

$$
k_{Fx} = k_{Fy} = k_{Fz} = k_{Mx} = k_{My} = k_{Mz} = 0.9
\tag{29}
$$

And

$$
\beta = \pi/6, r = 25mm
\tag{30}
$$

A set of meaningful parameter solution is obtained as

$$
\begin{cases}
\gamma = \pi/4 \\
R = 40mm \\
h = 9.1794mm \\
H_u = 27.1293mm \\
H_d = 25.8428mm
\end{cases}
\tag{31}
$$

# 5 Pose Data Based Identification

## 5.1 Experimental Method

As peg-in-hole assembly is a typical task in the industrial process, and stuck and wedging easily occurred during automatic assembly, that will cause task fault; we installed the sensor prototype on the end effecter of a parallel robot and carried out experimental research of peg-in-hole assembly task.

Fig. 5 illustrates the experimental platform, which is composed of the six-component force sensor prototype, DAQ system and 6-UPS parallel robot. With the design result, the sensor prototype is manufactured and calibrated. The maximal external force and torque magnitudes are ±100N and ±10Nm, and the measurement error of prototype is less than 0.66% after calibration. The 6-UPS parallel robot has six degree-of- freedom, which is worked with position command and adjusted with the force feedback of sensor prototype.

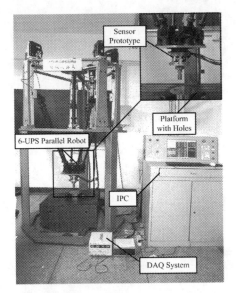

**Fig. 5.** Experimental platform for peg-in-hole assembly task

During the experiment, the shaft and hole is coarse located at first. Then the shaft installed on the end effecter of robot is moved to the assembly area and begin to insert the hole. During the assembly process, the touching force of the shaft and the hole is real-time monitored by the sensor, and the pose of the shaft is adjusted accordingly. At last, when the constraints of touching force and the position conditions are satisfied simultaneously, the assembly task is completed.

## 5.2    Experimental Result

Based on the experiment platform, the peg-in-hole assembly task is carried out. The assembly gap between the shaft and the hole is 0.1mm.

With analysis, it is known that the forces along $X$-axis and $Y$-axis are large and fluctuated obviously during the assembly process with touching force. Besides, the deeper the shaft insert, the more the touching force will reduce.

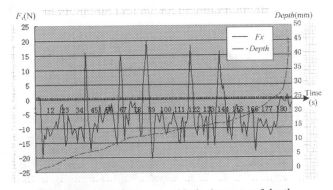

**Fig. 6.** Fx changing curve with the increase of depth

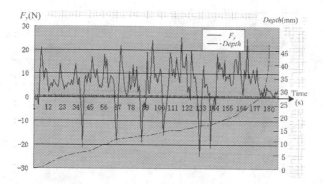

**Fig. 7.** Fy changing curve with the increase of depth

Fig. 6 and Fig. 7 illustrate the measured forces along X-axis and Y-axis during the assembly experiment, respectively. It is obvious that the touching force is reduced with the inserted depth of the shaft increasing; besides, as the axes of shaft and hole tend to overlap, the touching force meets the defined limits at last. The experimental result proves the success of assembly task and the practical value of the task-oriented sensor.

# 6    Conclusions

In order to meeting the requirement in a certain task, the thought of task-oriented design is introduced into the research field of six-component force sensor. With the screw theory, the mathematic model of parallel six-component force sensor is established. With the ellipsoid method, the design method for sensor structure is proposed and the structure constraints are deduced. In order to perform peg-in-hole assembly task with the task-oriented sensor, the sensor prototype is designed and manufactured. Based on the assembly platform contained sensor prototype, the assembly experiment is carried outsuccessfully. The research results demonstrate the feasibility of the task-orienteddesign method, which is useful for the design and applied research of six-component force sensor.

**Acknowledgements.** This research is sponsored by the NSFC (Grant No. 51305383), the financial support of the Major State Basic Research Development Program of China (973 Program)(Grant No. 2013CB733000), and   Specialized Research Fund for the Doctoral Program of Higher Education(SRFDP)(20131333120007).The reviewers are also acknowledged for their critical comments.

# References

1. Kerr, D.R.: Analysis: Properties and Design of a Stewart-Platform Transducer. Mech. Transm. Autom. Design. 1(11), 25–28 (1989)
2. Kang, C.G.: Closed-form force sensing of a 6-axis force transducer based on the Stewart platform. Sens. Actuators A Phys. 90, 31–37 (2001)

3. Dwarakanath, T.A., Dasgupta, B., Mruthyunjaya, T.S.: Design and development of a Stewart platform based force-torque sensor. Mechatronics 11(7), 793–809 (2001)
4. Wang, H., Yao, J.T., Hou, Y.L., et al.: Configuration and isotropy study of a novel fully pre-stressed and double-layer six-component force/torque sensor. In: IEEE Int. Conf. Mechatronics Autom., pp. 3693–3698 (2009)
5. Liu, S.A., Tzo, H.L.: A novel six-component force sensor of good measurement isotropy and sensitivities. Sens. Actuators A Phys. 100, 223–230 (2002)
6. Wang, X.Y., Rui, R.D.: Six-axis force/torque sensor based on Stewart platform. Chinese Journal of Mechanical Engineering 44(12), 118–130 (2008)
7. Yao, J.T., Hou, Y.L., Chen, J., et al.: Theoretical analysis and experiment research of a statically indeterminate pre-stressed six-axis force sensor. Sens. Actuators A Phys. 150, 1–11 (2009)
8. Tong, Z.Z., Jiang, H.Z., He, J.F., et al.: Optimal Design of Isotropy performance of Six-dimensional Force Sensor Based on Standard Stewart Parallel Structure Lying on a Circular Hyperboloid of One Sheet. Acta Aeronautica et Astronautica Sinica 32(12), 232–2334 (2011)
9. Xiong, Y.L.: On isotropy of Robot's Force Sensor. Acta Automatica Sinica 22(1), 10–18 (1996)
10. Huang, Z., Zhao, Y.S., Zhao, T.S.: Advanced Spatial Mechanism. Higher Education Press, Beijing (2006) (in Chinese)
11. Li, Z.X., Sastry, S.: Task Oriented Optimal Grasping by Multifingered Robot Hands. IEEE J. of Robotics and Automation RA2-14(1), 32–44 (1988)
12. Yao, J.T., Zhu, J.L., Wang, Z.J., et al.: Measurement Theory and Experimental Study of Fault-tolerant Fully Pre-stressed Parallel Six-component Force Sensor. IEEE Sensors Journal 13(9), 3472–3482 (2013)

# Mobility Analysis of Two Limited-DOF Parallel Mechanisms Using Geometric Algebra

Xinxue Chai and Qinchuan Li[*]

Mechatronic Institute, Zhejiang Sci-Tech University,
Hangzhou, Zhejiang 310018, China
lqchuan@zstu.edu.cn

**Abstract.** Mobility analysis determines the number of degree of freedom (DOF) and the motion pattern of a mechanism. Geometric algebra is applied to mobility analysis of two limited-DOF parallel mechanisms (PMs). Based on the outer product in geometric algebra, this method has the advantage in terms of geometric interpretation. It also can simplify the calculation because only addition and multiplication are involved during the whole computation.

**Keywords:** Mobility analysis, geometric algebra, limited-DOF parallel mechanism.

## 1    Introduction

Mobility or degree of freedom (DOF) is a basic property of a parallel mechanism (PM). It decided the minimum number of independent coordinates which defines the configuration of a kinematic chain. The purpose of the mobility analysis is to determine the DOF and the motion pattern of a mechanism. The motion pattern includes rotation and translation, or both of them. Lower-mobility PMs is the parallel mechanism which has less than six DOFs. Due to the advantages of reducing cost for fabrication, actuation, control, and maintenance, lower-mobility PMs have attracted more attention in recent years, such as Delta robot [1], Z3 head [2], and Tricept [3] hybrid robot.

Geometric algebra (GA) [4] is a potent computational methodology for geometric applications. Geometric algebra is also called the Clifford algebra, because it was first proposed by William. K. Clifford in 1878 [5-7]. This methodology has obvious advantages of interpretation when dealing with geometric applications. Geometric algebra has already applied in physics, neural computing, robotics, signal and image processing, computer and robot vision [8]. Research of robotics using geometric algebra often focuses on kinematics, dynamics, and robot vision. Aristidou and Lasenby [9] applied geometric algebra to solve the inverse kinematics and proposed an efficient iterative inverse kinematics solver called Fabrik [10]. Zamora and Bayro-Corrochano [11] use geometric algebra to solve the iinverse kinematics of a visually guided robot. However few researchers applied geometric algebra in the study of parallel mechanisms. Tanev [12-13] investigates the singularity of limited-DOF PM using geometric algebra. As far as we know, there is no attempt to systematically analyze the mobility analysis of limited-DOF PMs by geometric algebra.

---

[*] Corresponding author.

X. Zhang et al. (Eds.): ICIRA 2014, Part I, LNAI 8917, pp. 13–22, 2014.
© Springer International Publishing Switzerland 2014

A GA-based method for mobility analysis of limited-DOF PMs is proposed in [14]. In this paper, we applied this method to the mobility analysis of three limited-DOF PMs and further showed its validity.

## 2      Fundamentals of GA-Based Method for Mobility Analysis of Limited-DOF PMs

In this section, some necessary fundamentals of geometric algebra are introduced to do the mobility analysis of limited-DOF PMs. More details can be found in reference [4] which was mentioned in previous section.

For the reader's sake, a brief introduction of the GA-based method for mobility anlalysis of lower-mobility PMs is presented. More details about this method or GA itself can be found in [14-17].

In a $n$-dimension real vector space $V_n$ , there is a set of all multivectors. The set is generated by a new kind of multiplication called geometric product. This set of multivectors is called a geometric algebra $G_n = G(V_n)$ . The geometric product of two vectors $a$ and $b$ can be defined as follows:

$$ab = a \cdot b + a \wedge b \tag{1}$$

where $a \cdot b$ is the inner product of $a$ and $b$, and $a \wedge b$ is the outer product of $a$ and $b$.

The inner product of two vectors geometrically means the perpendicular projection of vector $a$ on vector $b$. Note that the inner product of two vectors is a scalar and it is commutative, namely, $a \cdot b = b \cdot a$ . While the outer product of two vectors means the directed plane segment. It is a parallelogram generated by sweeping vector $b$ along vector $a$ , as shown in Fig. 1. The outer product is anti-commutative, namely, $a \wedge b = -b \wedge a$ . The geometric product also satisfies the associative law and the distributive law, namely, $a(bc)=(ab)c$, $a(b+c)=ab+ac$, $(b+c)a=ba+ca$.

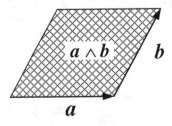

**Fig. 1.** $a \wedge b$

In this paper, we use $G_6$, the geometric algebra of a 6-dimension vector space. The basis of $G_6$ is $\{e_1, e_2, e_3, e_4, e_5, e_6\}$ , which satisfies the following relations:

$$e_i \cdot e_j = \begin{cases} 1 & , \quad i = j \\ 0 & , \quad i \neq j \end{cases} \tag{2}$$

$$e_i \wedge e_i = 0$$

A $k$-blade is built by the outer product of $k$ vectors $a_1, a_2, \ldots, a_k$.

$$\langle A \rangle_k = a_1 \wedge a_2 \wedge \ldots \wedge a_k \tag{3}$$

The geometric interpretation of $k$-blade is a subspace which is spanned by these $k$ vectors. For examples, the outer product of two vectors $a \wedge b$ generates a 2-blade, which denotes a directional plane. The outer product of three vectors $a \wedge b \wedge c$ is a 3-blade, which denotes a directional volume.

The number $k$ is called the grade of the $k$-blade. In geometric algebra, the concept grade is related to dimension. But grade is more likely to describe the new entities which are generated by vector multiplication. In geometric algebra $\mathcal{G}_n$, the dimension $n$ is decided by the maximum grade $n$ element among the non-zero blades.

In geometric algebra $\mathcal{G}_n$, the pseudoscalar $I_n$ is generated by all the $n$ basis, in which $n$ is the maximum grade of $\mathcal{G}_n$.

$$I_n = e_1 e_2 e_3 \cdots e_n \tag{4}$$

A directional line in $\mathcal{G}_6$ can be decided by its direction $u$ and moment $m$ [12]:

$$S = u + r \wedge u + h I_6 u \equiv v_1 e_1 + v_2 e_2 + v_3 e_3 + b_1 e_4 + b_2 e_5 + b_3 e_6 \tag{5}$$

where $r$ is the position vector of any point on the line $l$, $v_i$ (i=1,2,3) and $b_i$ (i=1,2,3) are scalar coefficients, $I_6 = e_1 e_2 e_3 e_4 e_5 e_6$ is the unit pseudoscalar of $\mathcal{G}_6$, $h$ is the pitch of the screw. Note that, in $\mathcal{G}_6$, $e_4$, $e_5$, $e_6$ denotes the moment $m$, $e_4 = e_2 \wedge e_3$, $e_5 = e_3 \wedge e_1$, $e_6 = e_1 \wedge e_2$. Though only 3 basis can totally express the directional line, it is a multivector $s$ which consists of a directional vector of grade 1 and a moment of grade 2 in the geometric algebra of 3-dimension vector space $\mathcal{G}_3$. At this time, the directional line is a vector $S$ in $\mathcal{G}_6$.

Reciprocal transformations [19] of $S$ can be written as

$$\tilde{S} = \Delta S = b_1 e_1 + b_2 e_2 + b_3 e_3 + v_1 e_4 + v_2 e_5 + v_3 e_6 \tag{6}$$

where $\Delta$ is an elliptic polar operator. In fact, equation (6) just exchanges the scalar coefficients $v_i$ and $b_i$ of equation (5).

A parallel mechanism consists of a fixed base, a moving platform and $n$ limbs connecting the base and the moving platform. Each limb can be considered as a serial chain .

Let $S_{ij}$ denotes the $j$th joint twist of the $i$th limb. When $S_{ij}$ denotes a revolute joint, the expression in $\mathcal{G}_6$ is the same as equation (5).

When $S_{ij}$ denotes a translation joint, it can be expressed as follows in $\mathcal{G}_6$:

$$S_{ij} = b_1 e_4 + b_2 e_5 + b_3 e_6 \tag{7}$$

Hence, the outer product of all the $m_i$ twists associated with the joints of the $i$th limb can be written as follows:

$$A_i = S_{i1} \wedge S_{i2} \wedge \cdots \wedge S_{ij} \wedge \ldots \wedge S_{m_i} \tag{8}$$

The $m_i$-blade $A_i$ can be defined as a blade of limb motion (BLM). It denotes the motion subspace which is spanned by all the $m_i$ twists of the $i$th limb.

The dual space of a $k$-blade is the orthogonal complement of the subspace generated by this $k$-blade.

The dual of $A_i$ is defined as

$$D_i = A_i I_6^{-1} = (-1)^{m_i - (6-m_i)} I_6^{-1} A_i \tag{9}$$

where $I_6^{-1}$ is the inverse of the unit psedoscalar in $\mathcal{G}_6$.

A blade of limb constraint(BLC) can be defined as

$$C_i = \tilde{D}_i = \Delta(A_i I_6^{-1}) \tag{10}$$

$\tilde{D}_i$ is a reciprocal transformation in $\mathcal{G}_6$. According to equation (8), $C_i$ is a blade of grade $(6-m_i)$. This $(6-m_i)$-blade denotes a subspace which is spanned by all the constraint wrenches of the $i$th limb. And it represents the constraint space of this limb.

Similarly, the blade of platform constraint (BPC) is defined as

$$A_C = C_1 \wedge \cdots \wedge C_i \wedge \cdots \wedge C_n \tag{11}$$

where $A_C$ is a blade of grade $p$. The $p$-blade means the constraint subspace spanned by all the constraint wrenches of $n$ limbs.

The dual space of the BPC is given by

$$D_C = A_C I_6^{-1} \tag{12}$$

The blades of platform motion (BPM) is then defined as the reciprocal transformation of $D_C$

$$M = \tilde{D}_C = \Delta(A_C I_6^{-1}) \tag{13}$$

Note that $M$ is a blade of grade $(6-p)$. This $(6-p)$-blade means the motion subspace spanned by all the twists of moving platform. In other words, this $(6-p)$-blade can be regarded as the DOF of the parallel mechanism.

## 3    Mobility Analysis of a 2-UPR-RPU PM

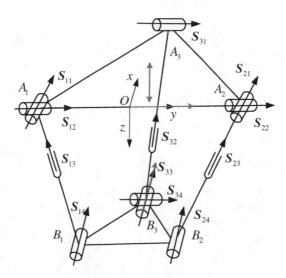

**Fig. 2.** 3-UPR-RPU PM

Fig. 2 shows a 2-UPR-RPU PM in the initial configuration, in which the moving platform is parallel to the base. We choose line $A_1A_2$ as the $y$ axis. The origin $o$ is in the middle of line $A_1A_2$. The $x$ axis passes through the point $A_3$. The coordinates of point $A_1$, $A_2$, $A_3$, $B_1$, $B_2$, $B_3$ are $A_1 = (0, -y_{A_1}, 0)$, $A_2 = (0, -y_{A_1}, 0)$, $A_3 = (x_{A_3}, 0, 0)$, $B_1 = (x_{B_1}, y_{B_1}, z_{B_1})$, $B_2 = (x_{B_2}, y_{B_{21}}, z_{B_2})$, $B_3 = (x_{B_3}, y_{B_3}, z_{B_3})$, respectively.

Each twist of limb 1 can be written as

$$S_{11} = e_1 - y_{A_1}e_6$$
$$S_{12} = e_2$$
$$S_{13} = m_{13}e_5 + n_{13}e_6 \qquad (14)$$
$$S_{14} = e_1 + z_{B_1}e_5 - y_{B_1}e_6$$

The outer product of all the 4 twists yields the BLM of limb 1:

$$A_1 = S_{11} \wedge S_{12} \wedge S_{13} \wedge S_{14}$$
$$= ae_1 \wedge e_2 \wedge e_5 \wedge e_6 \qquad (15)$$

where $a$ is a scalar coefficient.

The BLC of limb 1 can be obtained:

$$C_1 = \Delta(A_1 I_6^{-1}) = ae_1 \wedge e_6 \qquad (16)$$

The BLC of limb 1 is a 2-blade. Thus, the constraint wrenches of limb 1 are $C_{11} = e_1$ and $C_{12} = e_6$, respectively.

Each twist of limb 2 can be written as

$$S_{21} = e_1 + y_{A_1} e_6$$
$$S_{22} = e_2$$
$$S_{23} = m_{23} e_5 + n_{23} e_6 \tag{17}$$
$$S_{24} = e_3 + z_{B_1} e_5 + y_{B_1} e_6$$

The outer product of all the 4 twists yields the BLM of limb 2:

$$A_2 = S_{21} \wedge S_{22} \wedge S_{23} \wedge S_{24}$$
$$= b e_1 \wedge e_2 \wedge e_5 \wedge e_6 \tag{18}$$

where $b$ is a scalar coefficient.

The BLC of limb 2 can be obtained:

$$C_2 = \Delta(A_2 I_6^{-1}) = b e_1 \wedge e_6 \tag{19}$$

The BLC of limb 2 is a 2-blade. Thus, the constraint wrenches of limb 2 are $C_{21} = e_1$ and $C_{22} = e_6$, respectively.

Each twist of limb 3 can be written as

$$S_{31} = e_2 + x_{A_3} e_6$$
$$S_{32} = l_{32} e_4 + n_{32} e_6$$
$$S_{33} = e_1 + z_{B_1} e_5 \tag{20}$$
$$S_{34} = e_2 - z_{B_1} e_4 + x_{B_3} e_6$$

The outer product of all the 4 twists yields the BLM of limb 3:

$$A_3 = S_{31} \wedge S_{32} \wedge S_{33} \wedge S_{34}$$
$$= c(-z_{B_1} e_2 \wedge e_4 \wedge e_5 \wedge e_6 - e_1 \wedge e_2 \wedge e_4 \wedge e_6) \tag{21}$$

where $c$ is a scalar coefficient.

The BLC of limb 3 can be obtained:

$$C_3 = \Delta(A_3 I_6^{-1}) = c(-z_{B_1} e_4 \wedge e_6 + e_2 \wedge e_6) = c(e_2 - z_{B_1} e_4) \wedge e_6 \tag{22}$$

The BLC of limb 3 is 2-blade. Thus, the constraint wrenches of limb 3 are $C_{31} = e_2 - z_{B_1} e_4$ and $C_{32} = e_6$, respectively.

The constraint wrenches of 2-RPU-UPR mechanism are $C_{11} = e_1$, $C_{12} = e_6$, $C_{21} = e_1$, $C_{22} = e_6$, $C_{31} = e_2 - z_{B_1} e_4$, $C_{32} = e_6$, respectively. Since $C_{11} = e_1$ is equal to $C_{21} = e_1$, it means $e_1$ is a redundant constraint. $C_{12} = e_6$, $C_{22} = e_6$, $C_{32} = e_6$ are the same, namely each limb generates a constraint wrench $e_6$. $e_6$ is a common constraint. Thus, the grade of BPC is 3. The BPC can be derived:

$$A_C = C_{11} \wedge C_{12} \wedge C_{31}$$
$$= e_1 \wedge e_2 \wedge e_6 - z_{B_1} e_1 \wedge e_4 \wedge e_6 \tag{23}$$

The BPM of the 2-UPR-RPU parallel mechanism is then obtained

$$M = \tilde{D}_C = -e_1 \wedge e_2 \wedge e_6 + z_{B_1} e_2 \wedge e_5 \wedge e_6 = e_2 \wedge (e_1 + z_{B_1} e_5) \wedge e_6 \quad (24)$$

Equation (24) indicates that the 2-UPR-RPU PM has 3 DOF in the initial configuration. $M_1 = e_1 + z_{B_1} e_5$, $M_2 = e_2$, $M_3 = e_6$.

Geometrically, $M_1$ is a line which passes through point $B_3$ and is parallel to the axis $x$, $M_2$ is a line which coincides with the axis $y$, $M_3$ is a line which coincides with the axis $z$.

Physically, according to equation (5) and equation (7), $M_1$ and $M_2$ denote the rotational DOFs , $M_3$ denotes the translational DOF.

## 4    Mobility Analysis of a 3-PRRR PM

Fig. 3 shows a 3-PRRR PM in the initial configuration in which the moving platform is parallel to the base. We choose the intersection point of ${}^1S_1$ and ${}^2S_1$ as the origin $o$, line $oA_1$ as the $x$ axis, $oA_2$ as the $y$ axis. The coordinates of point $A_1$, $A_2$, $A_3$, $M_1$, $M_2$, $M_3$, $B_1$, $B_2$, $B_3$ are $A_1 = (x_{A_1}, 0, 0)$ , $A_2 = (0, y_{A_2}, 0)$ , $A_3 = (0, 0, z_{A_3})$ , $M_1 = (x_{A_1}, y_{M_1}, z_{M_1})$, $M_2 = (x_{M_2}, y_{A_2}, z_{M_2})$, $M_3 = (x_{M_3}, y_{M_3}, z_{A_3})$, $B_1 = (x_{A_1}, y_{B_1}, z_{A_3})$, $B_2 = (x_{B_2}, y_{A_2}, z_{A_3})$, $B_3 = (x_{B_3}, y_{B_3}, z_{A_3})$, respectively.

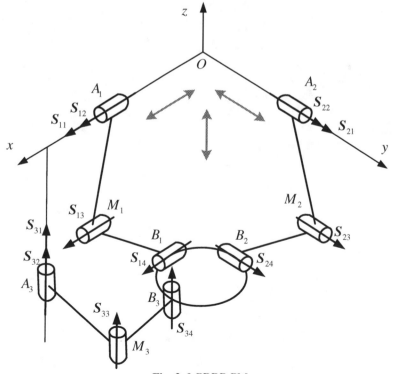

**Fig. 3.** 3-PRRR PM

Each twist of limb 1 can be written as

$$S_{11} = e_4$$
$$S_{12} = e_1$$
$$S_{13} = e_1 + z_{M_1} e_5 - y_{M_1} e_6 \tag{25}$$
$$S_{14} = e_1 + z_{A_3} e_5 - y_{B_1} e_6$$

Generating the outer product of all the 4 twists obtains the BLM of limb 1:

$$A_1 = S_{11} \wedge S_{12} \wedge S_{13} \wedge S_{14}$$
$$= a e_1 \wedge e_4 \wedge e_5 \wedge e_6 \tag{26}$$

where $a$ is a scalar coefficient.

The BLC of limb 1 can be obtained:

$$C_1 = \Delta(A_1 I_6^{-1}) = a e_5 \wedge e_6 \tag{27}$$

The BLC of limb 1 is 2-blade. Thus, the constraint wrenches of limb 1 are $C_{11} = e_5$ and $C_{12} = e_6$, respectively.

Each twist of limb 2 can be written as

$$S_{21} = e_5$$
$$S_{22} = e_2$$
$$S_{23} = e_2 - z_{M_2} e_4 + x_{M_2} e_6 \tag{28}$$
$$S_{24} = e_2 - z_{A_3} e_4 + x_{B_2} e_6$$

Generating the outer product of all the 4 twist obtains the BLM of limb 2:

$$A_2 = S_{21} \wedge S_{22} \wedge S_{23} \wedge S_{24}$$
$$= b e_2 \wedge e_4 \wedge e_5 \wedge e_6 \tag{29}$$

where $b$ is a scalar coefficient.

The BLC of limb 2 can be obtained

$$C_2 = \Delta(A_2 I_6^{-1}) = b e_4 \wedge e_6 \tag{30}$$

The BLC of limb 2 is 2-blade. Thus, the constraint wrenches of limb 2 are $C_{21} = e_4$ and $C_{22} = e_6$, respectively.

Each twist of limb 3 can be written as

$$S_{31} = e_6$$
$$S_{32} = e_3 - L e_5$$
$$S_{33} = e_3 + y_{M_3} e_5 - x_{M_3} e_4 \tag{31}$$
$$S_{34} = e_3 + y_{B_3} e_4 - x_{B_3} e_5$$

Generating the outer product of all the 4 twist obtains the BLM of limb 3:

$$A_3 = S_{31} \wedge S_{32} \wedge S_{33} \wedge S_{34}$$
$$= c e_3 \wedge e_4 \wedge e_5 \wedge e_6 \tag{32}$$

where $c$ is a scalar coefficient.

The BLC of limb 3 can be obtained:

$$C_3 = \Delta(A_3 I_6^{-1}) = c e_4 \wedge e_5 \tag{33}$$

The BLC of limb 3 is 2-blade. Thus, the constraint wrenches on limb 3 are $C_{31} = e_4$ and $C_{32} = e_5$, respectively.

The constraint wrenches of 3-PRRR mechanism are $C_{11} = e_5$, $C_{12} = e_6$, $C_{21} = e_4$, $C_{22} = e_6$, $C_{31} = e_4$, $C_{32} = e_5$, respectively. Redundant constraints appear, since it is the fact that the constraint wrenches $C_{12} = e_6$ and $C_{22} = e_6$ are the same, $C_{11} = e_5$ and $C_{32} = e_5$ are the same, $C_{21} = e_4$ and $C_{31} = e_4$ are the same. The BPC is 3-blade,

$$A_C = C_{21} \wedge C_{11} \wedge C_{12}$$
$$= e_4 \wedge e_5 \wedge e_6 \tag{34}$$

The BPM of the 3-PRRR parallel mechanism is then obtained

$$M = \tilde{D}_C = e_4 \wedge e_5 \wedge e_6 \tag{35}$$

Equation (35) indicates that the 3-PRRR PM has 3 DOF under the initial configuration. $M_1 = e_4$, $M_2 = e_5$, $M_3 = e_6$.

Geometrically, $M_1$ is a line which coincides with the axis $x$, $M_2$ is a line which coincides which the axis $y$, and $M_3$ is a line which coincides the axis $z$.

Physically, according to equation (5) and equation (7), $M_1$, $M_2$, $M_3$ denote the translational axes.

Additionally, in geometric algebra, the 3-blade element $e_4 \wedge e_5 \wedge e_6$ in equation (35) is a volume spanned by the three lines $M_1$, $M_2$, $M_3$. Physically, this volume represents the position of the translational axes. What it means this time is that any axis in this volume can be regard as the translational axis.

## 5  Conclusions

In this paper we analyzed the mobility of two limited-DOF PMs using geometric algebra. Geometric algebra has advantages in terms of geometrical intuition in mobility analysis. The advantage is attributed to the outer product in geometric algebra. With outer product, the motion and constraint subspace can be described clearly. Other advantage is only addition and multiplication are involved in the process of mobility analysis, which simplifies the calculation by avoiding discussion of cases caused by zero denominators. The method can be applied to all kinds of lower-mobility PMs.

**Acknowledgments.** The authors would like to acknowledge the financial support of the Natural Science Foundation of China (NSFC) under Grant 310018 and Natural Science Foundation of Zhejiang Province under Grant LZ14E050005. We also thanks Prof. J. S. Dai for his very valuable suggestions.

# References

1. Clavel, R.: A fast robot with parallel geometry. In: Proc. Int. Symposium on Industrial Robots (1988)
2. Wahl, J.: Articulated tool head. Google Patents (2002)
3. Neumann, K.-E.: Robot. Google Patents (1988)
4. Hestenes, D.: New foundations for classical mechanics. Springer (1999)
5. Clifford, W.K.: On the classification of geometric algebras. Mathematical Papers, 397-401 (1882)
6. Clifford, W.K.: Elements of dynamic: an introduction to the study of motion and rest in solid and fluid bodies. MacMillan and Company (1878)
7. Clifford, W.K.: On the space-theory of matter. In: Proceedings of the Cambridge Philosophical Society (1876)
8. Hitzer, E., Helmstetter, J., Abłamowicz, R.: Square roots of–1 in real Clifford algebras. Quaternion and Clifford Fourier Transforms and Wavelets. Springer (2013)
9. Aristidou, A., Lasenby, J.: Inverse kinematics solutions using conformal geometric algebra. Guide to Geometric Algebra in Practice. Springer (2011)
10. Aristidou, A., Lasenby, J.: FABRIK: a fast, iterative solver for the inverse kinematics problem. Graphical Models 73, 243–260 (2011)
11. Zamora, J., Bayro-Corrochano, E.: Inverse kinematics, fixation and grasping using conformal geometric algebra. In: Proceedings of the 2004 IEEE/RSJ International Conference on Intelligent Robots and Systems (IROS 2004). IEEE (2004)
12. Tanev, T.K.: Singularity analysis of a 4-DOF parallel manipulator using geometric algebra. In: Advances in Robot Kinematics. Springer (2006)
13. Tanev, T.K.: Geometric algebra approach to singularity of parallel manipulators with limited mobility. In: Advances in Robot Kinematics: Analysis and Design. Springer (2008)
14. Li, Q.C., Chai, X.X.: Mobility analysis of limited-DOF parallel mechanisms in the framework of geometric algebra. Submitted to the Journal of Mechanisms and Robotics
15. Hildenbrand, D.: Foundations of Geometric Algebra Computing. Springer (2012)
16. Vince, J. A.: Geometric algebra for computer graphics. Springer (2008)
17. De Sabbata, V., Datta, B.K.: Geometric algebra and applications to physics. CRC Press (2006)
18. Dorst, L., Fontijne, D., Mann, S.: Geometric algebra for computer science (revised edition): An object-oriented approach to geometry. Morgan Kaufmann (2009)
19. Lipkin, H., Duffy, J.: The elliptic polarity of screws. Journal of Mechanical Design 107, 377–386 (1985)

# Design and Kinematic Analysis of a Novel Flight Simulator Mechanism

Sheng Guo[*], Dian Li[**], Huan Chen[***], and Haibo Qu[†]

Beijing Jiaotong University, Beijing 100044, China

**Abstract.** In this paper, a novel flight simulator mechanism is proposed. By using a new parallel manipulator with redundant structure, a mechanism with larger rotational motions is obtained, which makes the flight simulator more feasible to simulate more complicated flight actions. First, the flight simulator mechanism is introduced and the kinematics is studied. Then, the workspace of the mechanism is analyzed. Finally, a model of this flight simulator mechanism is simulated under the virtual environment. It shows the design can well meet the needs of flight action with large angle requirements.

**Keywords:** Flight simulator, Parallel manipulator, Kinematics, Mechanism simulation.

## 1    Introduction

Flight simulator is a kind of simulation equipment which could simulate all kinds of motion gestures of an aircraft on the ground, and it's a representative example used in contour real time simulation system and virtual reality [1]. It's safer, cheaper, more controllable, more reliable than the aviation experiment with real plane, especially when the most dangerous aviation subject trainings proceed on the flight simulator. We could ensure the pilots and the fight equipments' safety. Since the first plane came out, the simulation of aviate training on the ground takes into account [2].

Most kinds of flight simulator's technology went mature now after 70 years' research and development. Full task flight simulator could fulfill most civil aviation's pilots' training demands and airliner's trial demands nowadays, but could hardly fulfills the training demands of fighter pilots. Especially, with the 3rd and the 4th generation fighters' development, the flight simulator needs more maneuverable and agile index, and traditional full task flight simulator couldn't fulfill all the demands any longer. Therefore, a new kind of flight simulator with high motion performance is significant to improve fighter pilots' training efficiency and quality. It's also of great importance to reduce casualties and property losses.

---

[*] Sheng Guo (1972 --), Male, Professor, Research Field: Parallel Robot Mechanisms.
[**] Dian Li (1988 --), Male, Postgraduate, Research Field: Parallel Robot Mechanisms.
[***] Huan Chen (1988 --), Male, Postgraduate, Research Field: Parallel Robot Mechanisms.
[†] Haibo Qu(1983 --), Male, Postdoctor, Research Field: Parallel Robot Mechanisms.

X. Zhang et al. (Eds.): ICIRA 2014, Part I, LNAI 8917, pp. 23–34, 2014.

An integrated flight simulator includes instrument system, inspecting system, motion system, control and controlling load system, sound system, and simulation computer and imitate cockpit et al [3]. The motion system's essence is computer which could control in real time and provides a six degrees of freedom instant overload simulation equipment includes pitching, turning, yawing, go up and down, lengthways and side direction shifting, and it's performance is directly related to the reality of simulator's flight. Traditional flight simulator uses Stewart six degrees of freedom parallel platform [4] as the motion system of the simulator, which could only do some limited movement such as pitching, turning and yawing, usually could only reach plus-minus 35 angle's turning because of the limited of the structure. It could hardly fulfill the demand of real planes which requires large posture and high angle. Nowadays, many scholars are working with a new parallel motion platform which has higher angle [5,6]. Such as a parallel mechanism which uses sphere layout to enlarge the platform's angle limits raised by Tesar. Asada also came up with a sphere mechanism with 3 collinear drives. Gosselin worked out a sphere mechanism with 3 interfaced drives. These new sphere parallel mechanisms have strong three-dimensional turning ability which traditional ones don't, but the load bearing ability is found unsatisfactory because the limits of structure and drive, the motion angle is also under great limit. Zanganeh and Angeles came up with a new redundant parallel manipulator [7], it could improve parallel manipulator's working space effectively. And lots of scholars use cardan, sphere stepping motor et al to enlarge mechanism's angle but could hardly get 180 or above. Most notably, Kim worked out a new fighter flight simulator named "Eclipse II" [8,9] in Seoul National University based on a six degrees of freedom parallel manipulator. The Eclipse II is settled on a roundness lead rail by using three PPRS limbs with six degrees of freedom. The moving platform can achieve a 360 degrees turning and translation motions on 3 directions. Therefore, such parallel mechanism can be used as a moving foundation bed of high-performance flight simulator.

In this paper, a new redundant parallel mechanism is proposed, which can be used for the flight simulator. After sketching the parallel mechanism, the kinematics, the workspace and the simulation of such new mechanism are analyzed. The results show that the proposed parallel mechanism can fulfill the large posture and high angle motion demands.

## 2    Description of the New Flight Simulator Mechanism

### 2.1    Introduction of Redundant Parallel Manipulator

Redundancy is that something exceed the limit of actually need. Based on the different way of redundancy, parallel manipulator's redundancy is mainly divided into two aspects: redundancy of kinematics and redundancy of drive [10]. Redundancy of drive means that joint drive's number is more than offer terminal actuators' need.    To gain redundancy of drive has several ways [11] : first is to add one or more moving branches in non-redundant parallel manipulator, second is to add drive in non-redundant parallel manipulator's passive joint, third is change the design of mechanism to increase its number of initiative joint. Redundancy of drive could improve the performance of parallel manipulator greatly and fix the shortage. As Zanganeh and

Angeles raised a new redundant parallel manipulator, it could improve Stewart parallel manipulator's working space effectively. Kim and Park et al designed a redundancy drive parallel tool named "Eclipse" [12], not only it has six degrees of freedom and keeps the advantages such as high-precision and high-rigidity etc which parallel manipulator has, but also its working space's singularity is much improved.

## 2.2    Assembling Condition of 4PUS-PPPS Redundant Parallel Manipulator

The 4PUS-PPPS redundant parallel manipulator in sketched, as shown in Figure 1, which includes one hexahedron rack, one moving platform and five motion branches (branch 1 to 4 have the same structure). Branch 1 and 2 connects the moving platform by using a spherical pitch point respectively, branch 3 and 4 connects the moving platform using a spherical pitch point respectively, branch 5 using a spherical pitch point to connect with moving platform at the top. Three spherical pitch points settled on the moving platform as an isosceles right triangle. Branch 1-4 connect with fixed base using translation pitch underneath. Branch 5 has 3 translation pitches perpendicularity, so as to make sure branches have three degrees of freedom at three different directions. Seven drive joints and six degrees of freedom make this manipulator as a redundant parallel manipulator. Seven translation pitches are made as the drive joints, and the moving direction is parallel to hexahedron rack's each brim respectively. Establish coordinate system on fixed base and moving platform respectively comes out manipulator's driving parameters $(d_1, d_2, d_3, d_4, x_A, y_A, z_A)$, moving platform's location parameters $(x_A, y_A, z_A, \alpha, \beta, \gamma)$.

## 2.3    Analysis of Degrees of Freedom

In this section, the degrees of freedom of the proposed parallel mechanism are analyzed based on the screw theory [13]. As shown in figure 1, the coordinate system and the moving coordinate system are established. For the PUS branch, such as the branch GC, as shown in figure 2, the Plücker coordinates of each joints can be obtained,

$$\begin{aligned}
&\$_1 : (0 \quad 0 \quad 0; 0 \quad 0 \quad 1); && \$_4 : (1 \quad 0 \quad 0; 0 \quad z_{c2} \quad -y_{c2}) \\
&\$_2 : (1 \quad 0 \quad 0; 0 \quad d_4 \quad -h_1); && \$_5 : (0 \quad 1 \quad 0; -z_{c2} \quad 0 \quad x_{c2}) \quad\quad (1) \\
&\$_3 : (0 \quad m \quad n; -d_4 \cdot m + h_1 \cdot n \quad -L \cdot n \quad L \cdot m); \$_6 : (0 \quad 0 \quad 1; y_{c2} \quad -x_{c2} \quad 0)
\end{aligned}$$

Through the operation of reciprocal product, we know that the PUS branch does not provide any constraint force or moment to the moving platform..
Here are the Plücker joint coordinates of PPPS branch:

$$\begin{aligned}
&\$_1 : (0 \quad 0 \quad 0; 1 \quad 0 \quad 0) && \$_4 : (1 \quad 0 \quad 0; 0 \quad z_a \quad -y_a) \\
&\$_2 : (0 \quad 0 \quad 0; 0 \quad 1 \quad 0) && \$_5 : (0 \quad 1 \quad 0; -z_a \quad 0 \quad x_a) \quad\quad (2) \\
&\$_3 : (0 \quad 0 \quad 0; 0 \quad 0 \quad 1) && \$_6 : (0 \quad 0 \quad 1; y_a \quad -x_a \quad 0)
\end{aligned}$$

Similarly, the PPPS branch does not provide any constraint force or moment to the moving platform..

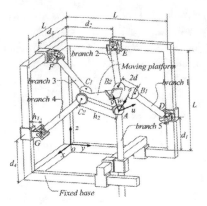
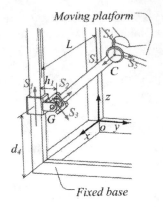

**Fig. 1.** 4PUS-PPPS redundant parallel manipulator    **Fig. 2.** Kinematic screw of one chain

The degrees of freedom can be obtained by using the following modified formula Grübler-Kutzbach [14] criterion.

$$M = d(n - g - 1) + \sum_{i=1}^{g} f_i + v \tag{3}$$

where $M$ stands for the number of degree of freedom; $d$ stands for institutions' order; $d = 6 - \lambda$; $\lambda$ stands for the number of public constraints of institution; $n$ stands for institutions to the total number of component; $g$ means the number of all kinematic pair; $f_i$ means the number of degrees of freedom for the first motion pair; $v$ stands for the number of redundant constraints after removing parallel mechanism's public constraints. we have

$$M = 6(13 - 16 - 1) + 30 + 0 = 6$$

The mechanism has six degrees of freedom, and whether the configuration is, the degrees-of-freedom is not instantaneous because the coordinate system does not change, also, equation (1) and equation (2) do not change. It could fulfill the demands of planes' three direction of translational and turning.

## 3    Kinematic Analysis

The coordinate system of flight simulator's motion mechanism is established, and each parameter is shown in figure 1. The coordinate system $o - xyz$ is settled on the fixed base, and the moving coordinate system $A - uvw$ is attached to the moving platform. In order to describe and simulate flight's movement, the Euler transformation $z(\alpha) - y'(\beta) - x''(\gamma)$ is used to analyze the location. The coordinate transformation is made according to the following ways: moving coordinate $[xa, ya, za]^T$ translation to a new location, rotate angle $\alpha$ to the shaft $z$, then rotate around the new coordinate system's axis $y$ for angle $\beta$, finally rotate around the new shaft $x$ for angle $\gamma$. The rotation matrix can be obtained:

$$_A^oR(\alpha,\beta,\gamma)=\begin{bmatrix} c\alpha c\beta & c\alpha s\beta s\gamma - s\alpha c\gamma & c\alpha s\beta c\gamma + s\alpha s\gamma \\ s\alpha c\beta & s\alpha s\beta s\gamma + c\alpha c\gamma & s\beta s\alpha c\gamma - c\alpha s\gamma \\ -s\beta & c\beta s\gamma & c\beta c\gamma \end{bmatrix} \tag{4}$$

where $c\alpha, c\beta, c\gamma$ stands for $\cos\alpha, \cos\beta$ and $\cos\gamma$ respectively, $s\alpha, s\beta, s\gamma$ stands for $\sin\alpha, \sin\beta, \sin\gamma$ respectively. Here's moving platform's pose transformation matrix:

$$T = \begin{bmatrix} _A^oR & ^oA \\ \mathbf{0} & 1 \end{bmatrix} \tag{5}$$

where $^oA = [x_A, y_A, z_A]^T$ represents for the location of the moving coordinate system's origin vector A. So that each point on the moving platform in the coordinate system can be expressed as $^oB_1 = T.^AB_1, {}^oB_2 = T.^AB_2, {}^oC_1 = T.^AC_1, {}^oC_2 = T.^AC_2$. In figure 1, 4 PUS branches are exactly the same, so, $B_1D=B_2E=C_1F=C_2G= \rho$, where $\rho$ means the length. Therefore, the length constraint equation of the motion mechanism can be obtained:

$$\rho = \left\| \overrightarrow{^oD{}^oB_1} \right\| \quad \rho = \left\| \overrightarrow{^oE{}^oB_2} \right\| \quad \rho = \left\| \overrightarrow{^oF{}^oC_1} \right\| \quad \rho = \left\| \overrightarrow{^oG{}^oC_2} \right\| \tag{6}$$

## 3.1 Inverse Solutions

The kinematics inverse solution is the driving parameter of the moving platform's posture parameter. Make square arithmetic to both sides of the four equations in (6), use substitution method and simplify the mass to each point with triangular parameter, could come to these:

$$d_i^2 + e_{i1}d_i + e_{i2} = 0, (i = 1, \cdots, 4) \tag{7}$$

where $e_{i1}, e_{i2}$ denote the length constraint coefficients, which can be expressed as function of $x_a, y_a, z_a, \alpha, \beta, \gamma$'s

For equation (7), each one contains only one driver parameter, so as to the kinematics inverse solution, when known the position parameters $x_a, y_a, z_a, \alpha, \beta, \gamma$ to the moving platform, drive parameters $d_1, d_2, d_3, d_4$ could be obtained through formula (7). Because of the square, there might be two inverse solution of each drive; we could select the right inverse solution parameters in the process of actual calculation according to the following ways: The two drive solutions are all positive solutions if they are within the scope of the drive. Drive mechanism should be selected according to the initial position, in their shortest trip as the condition, choosing the appropriate solution as the driving parameter. If there is only one real solution in the driving range, it is the only one correct drive parameter. If both real solutions are beyond the scope of drive, it means the driving mechanism is beyond the shifting limit when the moving platform in this posture parameter's range, there's no correct inverse solution in this situation.

## 3.2    Forward Solutions

Find the position parameter of the moving platform after knowing the drive parameter is the way we work out the motion mechanism's positive location solution. As for this flight simulator, drive parameter is $(d_1, d_2, d_3, d_4, x_A, y_A, z_A)$, moving platform's position parameters are $x_a, y_a, z_a, \alpha, \beta, \gamma$. where $x_a, y_a, z_a$ are not only the driving parameters, but also the pose parameters by considering the structural characteristics. When working with the positive solution, $x_a, y_a, z_a$ are always the known parameters, namely, the moving platform's position parameters have been confirmed. So working with the new flight simulator's positive location solution is actually finding the congruent relationship between drive parameters $(d_1, d_2, d_3, d_4)$ and pose parameters $\alpha, \beta, \gamma$ of the moving platform. In the process of actual calculation, generally replace posture parameter $(\alpha, \beta, \gamma)$ in the form of matrix parameter $r_{ij}(i = 1 \cdots 3, j = 1 \cdots 3)$ of the rotation matrix according to the equation (4)'s parameter of the rotation matrix because of difficulty with the trigonometric function's posture parameter calculation. For the parameters of rotation matrix, have the following fixed relationship :

$$r_{12}^{2} + r_{22}^{2} + r_{32}^{2} = 1 \tag{8}$$

$$r_{11}^{2} + r_{21}^{2} + r_{31}^{2} = 1 \tag{9}$$

$$r_{11}r_{12} + r_{21}r_{22} + r_{31}r_{32} = 0 \tag{10}$$

Subtracting equation (8) and equation (7) both sides respectively and combine equation (8) with equation (9), eliminate quadratic term after reduction :

$$c_{11}r_{11} + c_{12}r_{12} + c_{13}r_{21} + c_{14}r_{22} + c_{15}r_{31} + c_{16}r_{32} = 0 \tag{11}$$

Similarly, add equation $(7, i = 3)$ with equation $(7, \ i = 1)$ on both sides respectively, equation $(7, i = 4)$ with equation $(7, \ i = 1)$ on both sides respectively, add(8) with equation (9) and eliminate quadratic term after reduction at the same time:

$$c_{21}r_{11} + c_{22}r_{12} + c_{23}r_{21} + c_{24}r_{22} + c_{25}r_{31} + c_{26}r_{32} = 2e^{2} \tag{12}$$

$$c_{31}r_{11} + c_{32}r_{12} + c_{33}r_{21} + c_{34}r_{22} + c_{35}r_{31} + c_{36}r_{32} = 2e^{2} \tag{13}$$

where $c_{ij}(i = 1, \cdots, 3, j = 1, \cdots, 3)$ stands for the length constraint after simplify the coefficient of mutual transformation, they are the expression of $x_a, y_a, z_a, d_1, d_2, d_3, d_4$ .Then after we work out drive parameter $x_a, y_a, z_a, d_1, d_2, d_3, d_4$ , build simultaneous equation model from equation (8) to (13) come $r_{11}, r_{12}, r_{21}$ and $r_{22}, r_{31}, r_{32}$ . Use equation (14) to work out $r_{13}, r_{23}, r_{33}$ .

$$r_{13} = -r_{22}r_{31} + r_{21}r_{32}; r_{23} = r_{12}r_{31} - r_{11}r_{32}; r_{33} = -r_{12}r_{21} + r_{11}r_{22} \qquad (14)$$

Finally, work out $\alpha, \beta, \gamma$ by equation (15), substitute in to kinematics reverse solution equation, confirm the only group of true value.

$$\beta = ArcTan[-r_{31}, \sqrt{r_{11}{}^\wedge 2 + r_{21}{}^\wedge 2}]$$
$$\alpha = ArcTan[r_{21} / Cos[\beta], r_{11} / Cos[\beta]] \qquad (15)$$
$$\gamma = ArcTan[r_{32} / Cos[\beta], r_{33} / Cos[\beta]]$$

## 4    Workspace Analysis

The workspace of parallel mechanism [15] refers to the reference point on the moving platform which could reach all of the locate points. For flight simulators which use redundant parallel mechanism as the motion mechanism, its working space refers to the work area and the attitude Angle range which the moving platform of the motion simulator could reach, it is an important index to measure the working performance of the flight simulator. The motion parameters of the structure listed in table 1.

**Table 1.** The Architectural parameters of the mechanism

| parameters | Numerical value mm | parameters | Numerical value mm |
|---|---|---|---|
| $h_1$ | 81 | $d$ | 100.5 |
| $h_2$ | 274.5 | $d_1$ | (200,700) |
| $L$ | 867 | $d_2$ | (100,600) |
| $e$ | 472 | $d_3$ | (100,600) |
| $d$ | 100.5 | $d_4$ | (200,700) |

### 4.1    Locational Space

The method of searching three dimensional space was used in solving the locational space. For space analysis of fixed position, a very important position of flight simulator is translation space, namely location space of $\beta = 0, \gamma = 0$ . Flight simulator should slowly return to an initial position and gesture in order to have a bigger position and attitude to move on for the next action, because the position of the flight simulator is very limited compared to the real aircraft. This initial position is usually level flight of planes', so study flight simulator's translation work space is important to flight movements continuous simulation, also have great significance to the training of pilots.

When the moving platform is in position 1 $(\alpha = 3\pi / 4, \beta = 0, \quad \gamma = 0)$ as shown in figure 3, the location of the motion space can be obtained, as shown in figure 4.

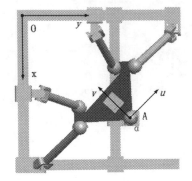

**Fig. 3.** Moving platform of fixing posture 1

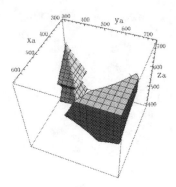

**Fig. 4.** Translational space of fixing posture 1

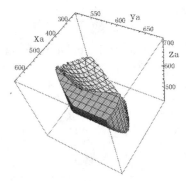

**Fig. 5.** Translational space of posture 1

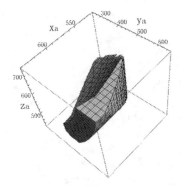

**Fig. 6.** Translational space of posture 1

Similarly, we could get the position location when the institution is in fixed posture 2 $(\alpha = \pi, \beta = 0, \gamma = 0)$ and fixed posture 3 $(\alpha = 2\pi / 4, \beta = 0, \gamma = 0)$, shown in figure 5 and 6.

By calculation and analysis of fixed position's translation space we could know that, when the institution's translation space is $\beta = 0, \gamma = 0$ and $\alpha$ symmetrical about $\alpha = 3\pi / 4$, the result of translation position space also symmetrical about flat $x = y$ which consistent with the characteristics of the institution's configuration, all kinds of offset angle of flight simulator have certain translation space which could greatly improve flight simulator's authenticity and continuity when processing the real flight motion simulation.

## 4.2    Posture Space

Posture space analysis is always the key point of the parallel mechanism's working space analysis. Especially to flight simulator, the size of posture space has a close relationship with some flight movements' implementation degree. Posture space solution uses the method of three-dimensional search which is similar the solution of position space. As shown in figure 7 motion institution's posture space can be obtained in fixed position 1 (500,300,420): $\alpha \in (91°,178°), \beta \in (0,63°), \gamma \in (74°,143°)$.

As shown in figure 8, institution's posture space can be obtained in fixed position 2 (500,600,600) : $\alpha \in (80°,195°), \beta \in (-69°,29°), \gamma \in (29°,120°)$

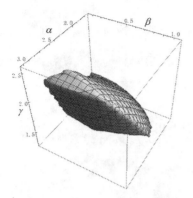

**Fig. 7.** Posture space of position 1          **Fig. 8.** Posture space of   position 2

In order to get the motion mechanism's posture space in the whole space, we could divide the position space and take appropriate feature points in the different area use the principle that around the moving platform's origin posture space also similar. We could measure the posture space of the area by calculating the feature point's posture space on the moving platform, estimate the space position of the moving platform in the whole space by posture spaces in different area. As shown in figure 9, divide the position space of flight simulator with plane which parallels to the coordinate system and get 27 different areas. We could estimate the moving platform's posture space of the whole work space by calculating moving platform's posture space in different areas of point A.

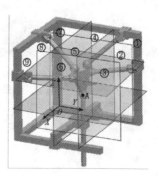

**Fig. 9.** The division of space position

Moving platform's posture space in different position could be obtained by analyze the posture space of characteristics points in different areas of position space which could get the institution's posture space range eventually: $\alpha \in (80°,195°), \beta \in (-69°,29°), \gamma \in (-65°,143°)$ .

Most traditional flight simulators use Stewart platform as the motion mechanism which maximum range of attitude is plus or minus 35 degrees. While new flight simulator's scope of posture space which uses 4PUS-PPPS redundant parallel mechanism is far longer than the traditional one. Among them, the yaw and roll motion space compared to traditional flight simulator has increased nearly doubled, pitch angle and space is 3 times of traditional one.

## 5    Kinematics Simulation

In order to verify the movable ability of the new flight simulator's motion mechanism, the virtual prototype of such motion mechanism is established.

**Fig. 10.** Pugachev Cobra maneuver

As shown in figure 10, Cobra maneuver is a large pitch motion which mainly involves the change of the pitching angle, rolling angle and yawing angle remain the same in the process of movement. The moving platform of flight simulator keeps level flight condition at the beginning and increase the pitch angle gradually, then reduce pitching angle when reaches a certain angle, finally recovers to level flight condition. Traditional flight simulators' angle limit makes them unable to realize the simulation of this kind of action.

**Fig. 11.** The simulation of the large pitch motion

For the virtual prototype, when we put the changing curve of pitching angle as an input parameter, the different configurations of such parallel mechanism can be obtained, as shown in figure 11. We could output the moving platform of flight simulator's displacement parameters curve from each drive through each virtual prototype's sensor, as shown in figure 12. Note that, due to the coordinate system is fixed in the simulation software, the sign of displacement parameters of za, d1, d4 are opposite to kinematics analysis.

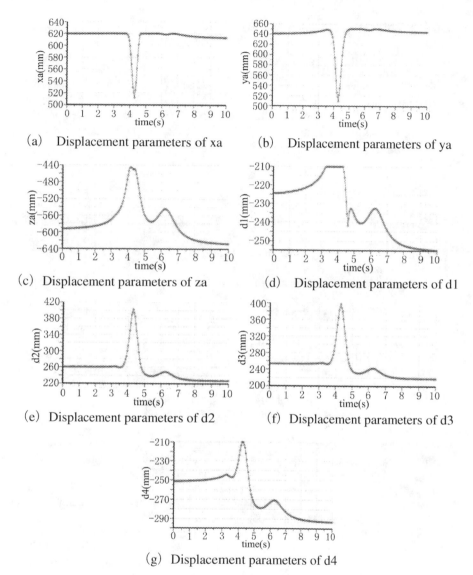

(a)  Displacement parameters of xa

(b)  Displacement parameters of ya

(c)  Displacement parameters of za

(d)  Displacement parameters of d1

(e)  Displacement parameters of d2

(f)  Displacement parameters of d3

(g)  Displacement parameters of d4

**Fig. 12.** Driving parameters in the large pitch motion

The mechanism could achieve great pitching maneuver, and the pitching angle could get $100°$ through kinematics simulation. The mechanism does not produce interference in the process of movement and all drives are moving within the scope of the limit.

## 6    Conclusion

In this paper, a new flight simulator mechanism is proposed, which enlarges the scope of the posture angle of flight simulator.

Compared with the general Stewart flight simulator, the proposed flight simulator mechanism possesses larger rotations of the moving platform, such as the pitching angle $200°$, the rolling angle $130°$, the yawing angle $120°$. Also, the kinematics, the workspace and the simulation of such new mechanism are analyzed.

**Acknowledgments.** This work is supported by National Natural Science Foundation of China (Grant No. 51475035 and No. 51175029), Program for New Century Excellent Talents in University (Grant No. 12-0769), and China Postdoctoral Science Foundation (Grant No. 2014M550601).

# References

[1]  Chongguang, W.U.: Simulation Technology. Chemical Industry Press, Beijing (2000) (in Chinese)

[2]  Zhou, Z., Liu, X., Yu, J., et al.: Modern flight simulation technology. National Defence Industry Press, Beijing (1997) (in Chinese)

[3]  Fennel, N., Hemmens, S.: Engineering Flight Simulator Design for Human in the Loop Interaction. In: SETE 2000 Conference (2000)

[4]  Stewart, D.: A Platform with Six Degree of Freedom. IME (15, pt. I), 371–386 (1965)

[5]  Gosselin, C.M., Vollmer, F., Cote, G., et al.: Synthesis and design of reactionless three-degree-of-freedom parallel mechanisms. IEEE Transactions on Robotics and Automation 20(2), 191–199 (2004)

[6]  Qu, Y.: Research on kinematics behaviors of 2-DOF decoupled spherical parallel mechanisms. Hebei University of Technology (2), 12–14 (2008) (in Chinese)

[7]  Zanganeh, K.E., Angeles, J.: Mobility and Position Analyses of a Novel Redundant Parallel Manipulator. In: IEEE International Conference on Robotics and Automation, pp. 3049–3054 (1994)

[8]  Kim, J., Hwang, J.C., Kim, J.S., Park: Eclipse-II: A New Parallel Mechanism Enabling Continuous 360-degree Spinning Plus Three-axis Translational Motions. In: IEEE International Conference on Robotics and Automation, vol. 18(3), pp. 367–373 (May 2001)

[9]  Kim, J., Kim, S.H.: A New Fighter Simulator Based on a Full Spinning Six Degrees-of-freedom Parallel Mechanism Platform. In: Interservice/Industry Training, Simulation, and Education Conference, I/ITSEC (2005)

[10]  Baron, L., Member, A.J.: The Direct Kinematics of Parallel Manipulators under Joint-Sensor Redundancy. IEEE Transactions on robotics and automation 16(1), 12–19 (2000)

[11]  Yang, J., Yu, Y.: Dynamic analysis of a novel planner 3-DOF redundant parallel manipulator. Machine Design and Research 21(5), 26–28 (in Chinese)

[12]  Kim, J., Park, F.C., Ryu, S.J., et al.: Design and Analysis of a Redundantly Actuated Parallel Mechanism for Rapid Machining. IEEE Transactions on Robotics and Automation 17(4), 423–434 (2001)

[13]  Ball, R.S.: The Theory of Screws. London: Cambridge University Press (1900)

[14]  Huang, Z., Li, Q.C.: General methodology for equation synthesis of lower-mobility symmetrical parallel manipulators and several novel manipulators. Rob Research 21(2), 131–146 (2002)

[15]  Wu, C.: Research of the Workspace of 6-SPS Parallel Robot And Its Optimization Design. HeFei University of Technology, Hefei (2003) (in Chinese)

# Performance Indices for Parallel Robots Considering Motion/Force Transmissibility

Fugui Xie, Xin-Jun Liu[*], and Jie Li

The State Key Laboratory of Tribology and Institute of Manufacturing Engineering,
Department of Mechanical Engineering, Tsinghua University, Beijing 100084, China
xinjunliu@mail.tsinghua.edu.cn

**Abstract.** This paper focuses on the performance evaluation of parallel robots. The existing evaluation methods have been deeply investigated, and we come to the conclusion that some of the existing performance measures are not suitable and there is a need to introduce new ones. Considering the specialties of parallel robots, we think that the performance with respect to motion/force transmission should be evaluated. Consequently, some indices have been defined based on the reciprocal product of screw theory, and suggested as the evaluation criteria for the optimal design of parallel robots. These indices are frame-free and their values fall into a finite range [0 1]. To illustrate their merits and applications, some cases have also been provided.

**Keywords:** Performance evaluation, Parallel robots, Motion/force transmission, Optimal design.

## 1 Introduction

The huge family of robots can be generally classified into two branches, i.e., parallel robots and serial robots. As the counterpart of serial robots, parallel robots are characterized by multi-closed-loop structure. In which, the mobile platform is connected to the base through at least two kinematic chains. Actuating by the input joints mounted on the base, the mobile platform can realize the required motions. These specialties bring the parallel robots lots of merits such as high stiffness (high load to weight ratio), quick response (low inertia), high speed and high acceleration. For such reasons, parallel robots have been extensively used in industry, such as the Metrom [1], Sprint Z3, Tricept and Exechon developed in the field of machine tools, the Delta, H4 [2] used in packaging production lines, and the flight simulator developed on the basis of Gough-Stewart platform.

In the development of parallel robots, optimal design is a fundamental and key step, and is also a very challenging procedure due to the complex kinematics resulted by multi-closed-loop structure [3-5]. In general, there are two issues involved: performance evaluation [6] and dimension synthesis [7].

---

[*] Corresponding author.

X. Zhang et al. (Eds.): ICIRA 2014, Part I, LNAI 8917, pp. 35–43, 2014.

Dimension synthesis is to determine the dimensions (geometric parameters) of the mechanism to be designed. In this area, there are two commonly used methods in classical design, i.e., objective function based method [8] and performance atlas based method [9]. By using the objective function based method, multiple parameters can be simultaneously optimized based on an objective function with specific performance constraints and an optimum algorithm to search the result, and there is no limitation for the number of the parameters to be optimized. But, this method is time consuming, and is difficult to generate the globally optimal solution. This is resulted by the infiniteness of geometric parameters, the selection of initial search point and the antagonism among multiple criteria. Moreover, there is only one solution can be derived by using this method. In case of unpredictable changes occur and designer intervention is required, this is inconvenient for designers to adjust the results according to the task in hand. In contrast, by using the performance atlas based method, the mapping relationship between geometric parameters and the concerned performance can be visually and globally presented in a parameter design space [9], and the antagonism among performance indices can be directly reflected in the generated atlases. Additionally, the result derived by this method is an optimum region, in which all possible solutions are included and designers can adjust the result according to a particular problem. Therefore, this method is flexible. Similarly, there also exists a limitation:  the maximum number of the parameters to be optimized should be less than five. Obviously, the two methods have their own applications. But, they share one common precondition, i.e., the appropriate performance evaluation, which will be discussed in the next section.

## 2    State of the Art

As the precondition of dimension synthesis, performance evaluation [10, 11] is the most important and challenging issue in the field. Lots of indices have been proposed and well defined in the study of serial robots, such as workspace, singularity, stiffness, manipulability, dexterity and accuracy. And these indices have been directly used in the design of parallel robots.

**Are the existing performance measures suitable or is there a need for the introduction of new ones?**

To answer this question, the following problems should be figured out first: what performance should be evaluated for parallel robots and how to evaluate the concerned performance.

For parallel robots, some performances such as workspace, singularity, stiffness [12, 13] and accuracy [14] are still the basic problems which should be fully analyzed. Thus, these performance indices are suitable to be used in the design of parallel robots. Nonetheless, not all the above mentioned indices are suitable to be used in such a way.

Take the well-known index LCI (local conditioning index) [15, 16] as an example. This index is defined as the reciprocal of the condition number of Jacobian matrix.

It is a well-defined index in the field of serial robots and has been extensively used to evaluate the dexterity of serial robots. In view of this, it has been directly used in the evaluation of parallel robots' the accuracy, dexterity and closeness to singularity. However, serious inconsistency (the elements of Jacobian matrix are not homogeneous in terms of units) has been reported when this index is used in parallel robots with both translational and rotational DoFs (degrees of freedom) [15]. Thus, this index is not suitable to be used in parallel robots with mixed DoFs. Moreover, the index LCI has also been used in the elimination of singularity and the near configurations. Usually, this is achieved by defining a good-condition workspace [17] which can be identified by assigning a minimum LCI value. Due to the frame-dependent feature of the LCI, the assigned minimum value is arbitrary and comparative. In other words, the meaning of the given LCI value will be different when the frame is defined in a different way. In addition, an agglomeration phenomenon has been observed when LCI is used in a parallel robot with only translational DoFs [6, 18]. These facts reflect that the physical meaning of LCI is ambiguous. Therefore, the index LCI is not suitable to be used as performance evaluation criterion for the optimal design of parallel robots. The introduction of new indices is necessary.

As is well known, the function of a mechanism is to transmit motion/force between the mobile platform and the base. For serial robots, in general, the transmission of motions is one of the main concerns. So, it is suitable to evaluate the dexterity by using the index LCI. But, for parallel robots, the transmission of forces is a representative distinction/merit when compared with the serial robots. Thus, the motion/force transmission performance instead of the dexterity should be evaluated for parallel robots. Then, how to evaluate this performance?

The classic transmission angle [19], which has been used in the four-bar mechanism to evaluate its force transmission performance, provides an available and possible solution for this question. The four-bar mechanism is a single-closed-loop structure, and a parallel robot is a multi-closed-loop structure. They share some common characteristic in a way. Thus, the concept of transmission angle can be introduced into the motion/force transmissibility evaluation of parallel robots. Inspired by this idea, a local transmission (LTI) has been proposed based on the definitions of forward transmission angle and inverse transmission angle [18]. This index is effective in the evaluation of the motion/force transmission performance for planar or decoupled spatial parallel robots [20]. To evaluate the transmission performance of an arbitrary spatial parallel robot (without decoupled property), a generalized transmission index has been proposed based on the virtual coefficient of screw theory [21]. This index provides an approach to deal with the performance evaluation of an arbitrary non-redundant parallel robot. Based on the definition, however, only the output transmission performance in one limb can be evaluated by this index. Obviously, efforts are still needed in this area. Of note is that, the above mentioned performance evaluation indices are focused only on non-redundant parallel robots, and there is nearly no index for the performance evaluation of redundant parallel robots which have promising potentials and vast application prospects. In view of these, we have done some works in this field, and the details will be presented in the following sections of this paper.

# 3    Performance Indices Considering Motion/Force Transmissibility

To give a comprehensive evaluation on the motion/force transmission performance, the concept of virtual coefficient or reciprocal product [22] between wrench screw and twist screw has also been used. On this basis, a local transmission index [23] for the performance evaluation of non-redundant parallel robots has been defined as follows.

For an $n$-DoF non-redundant parallel robot, the input number should be $n$. Assuming that, the input in the $i$-th limb can be represented by a unit input twist screw $\$_{1i}$ ($i$=1, 2, …, $n$), and this motion will be transmitted to the mobile platform through a unit transmission wrench screw $\$_{Ti}$. When other ($n$-1) inputs (i.e., except the $i$-th input) are fixed, only the transmission wrench screw $\$_{Ti}$ can contribute to the movement of the mobile platform. The generated motions of the mobile platform can be represented by a unit output twist screw $\$_{Oi}$. Then, by using the power coefficient used in [23], an input transmission index of the $i$-th limb can be defined as

$$\lambda_i = \frac{\left|\$_{Ti} \circ \$_{1i}\right|}{\left|\$_{Ti} \circ \$_{1i}\right|_{\max}}. \tag{1}$$

where, $\left|\$_{Ti} \circ \$_{1i}\right|$ represents the reciprocal product of the input twist screw and the transmission wrench screw in the $i$-th limb; and $\left|\$_{Ti} \circ \$_{1i}\right|_{\max}$ represents the potential maximum of the reciprocal product.

Similarly, an output transmission index with respect to the $i$-th input can be defined as the power coefficient of the transmission wrench screw and the output twist screw. This index can be expressed as

$$\eta_i = \frac{\left|\$_{Ti} \circ \$_{Oi}\right|}{\left|\$_{Ti} \circ \$_{Oi}\right|_{\max}}. \tag{2}$$

where, $\left|\$_{Ti} \circ \$_{Oi}\right|$ represents the reciprocal product of the output twist screw and the transmission wrench screw; and $\left|\$_{Ti} \circ \$_{Oi}\right|_{\max}$ represents the potential maximum of the reciprocal product.

Obviously, a larger value of $\lambda_i$ or $\eta_i$ indicates better input or output transmission performance for the $i$-th limb of the discussed mechanism. If the value of $\lambda_i$ or $\eta_i$ is equal to zero, the input or output transmission singularity will occur. Thus, the specific value of $\lambda_i$ or $\eta_i$ can be used to evaluate the closeness to singularity. If all the transmission indices $\lambda_i$ and $\eta_i$ are large enough, the discussed mechanism will have good motion/force transmission performance and will be far away from singularity. Then, an index can be defined as

$$\gamma = \min\{\lambda_i, \ \eta_i\}, \quad (i = 1, 2, ..., n). \tag{3}$$

For a different position and orientation, the $\gamma$ will be different. All the points in the concerned workspace can be evaluated by $\gamma$ one by one. For such a reason, this index is referred to as the local transmission index (LTI), which can be used to evaluate the motion/force transmission performance of all non-redundant parallel robots.  Of note is that, when this index is applied to evaluate the performance of planar parallel robots, the derived results are consistent with that generated by using the index proposed in [18] which is based on the concept of transmission angle.

Due to the fact that the reciprocal product of the wrench screw and the twist screw has no relationship with the definition of the coordinate system, that is to say the values of $\lambda_i$, $\eta_i$ and $\gamma$ are frame-free. Additionally, based on the concept of power coefficient, all these indices' values should fall into a finite range, i.e., $[0, 1]$. These features make the LTI superior to the previously mentioned index LCI. Benefiting from these features, the LTI provides an available and effective performance evaluation criterion for the comparison of different non-redundant parallel robots. For such a reason, the index LTI can also be used in the structure optimization and selection of non-redundant parallel robots.

The motion/force performance of non-redundant parallel robots can be evaluated by the index LTI, then how to evaluate the performance of redundant parallel robots? Generally, by appropriately introducing redundancy, a redundant parallel robot can outperform the non-redundant one. This is why the redundant parallel robots are becoming popular in the field. Consequently, the need to evaluate their performance and lay the foundation for their optimal design is increasing. Based on the definition of LTI, we have done some works in this area. The transmission indices for redundant parallel robots can be summarized as follows.

For a parallel robot with actuation redundancy, there exists mutual interference (assuming the number of the redundant inputs is $r$) among the $k$ inputs. By removing the $r$ redundant inputs, a non-redundant robot can be derived. Obviously, there are $q$ such robots, and $q = C_k^r$. Based on the definition of LTI presented in Eq. (3), for an arbitrary given position and orientation, the LTI value for each robot can be derived and denoted by $\kappa_i$ $(i = 1, 2, ..., q)$. Among the $q$ non-redundant robots, there is a non-redundant robot which can transmit motion/force better than the others. Here, we define the LTI value of this non-redundant robot as the local minimized transmission index (LMTI) of the discussed redundant robot. The LMTI can be represented by

$$\mathbb{M} = \max\left\{\kappa_1, \kappa_2, ..., \kappa_q\right\}, \quad q = C_k^r. \tag{4}$$

The value of LMTI can be used to reflect the minimum motion/force transmission ability of a parallel robot with actuation-redundancy.

For a parallel robot with kinematic redundancy, there is no unique solution for its inverse kinematics. The results can be represented by a set $G$. For an arbitrary result in set $G$, the value of the LTI defined in Eq. (3) will be different. When the result of the inverse kinematics varies within the set $G$, a maximal value $\kappa_g$ $(g \in G)$ can be derived. Here, this maximal value $\kappa_g$ is defined as the local optimal-transmission index (LOTI) of a parallel robot with kinematic redundancy. That is

$$\Theta = \kappa_g, \quad g \in G. \tag{5}$$

The LOTI can be used to evaluate the best motion/force transmission ability of a parallel robot with kinematic redundancy. Additionally, LOTI can also be as a criterion for the identification of optimal inverse kinematics for such a robot.

## 4     Optimal Design and Application Cases

In order to illustrate the application and the effect of the proposed indices, some cases are summarized and presented in this section.

To carry out the structure optimization and selection of two similar parallel robots, i.e., the 2PRU-1PRS (Fig. 1) and 2PRU-1PUR (Fig. 2) robots, the LTI in Eq. (3) is used to be as the criterion. With respect to the same geometric parameters and $\gamma = \sin 40°$, the good transmission workspaces (GTWs) for the two robots have been identified in Fig. 3 [24]. In which, GTW1 indicates the GTW of the robot in Fig. 1, and GTW2 indicates the GTW of the robot in Fig. 2. Obviously, GTW1 is larger than GTW2. Thus, it can be concluded that the structure of the robot in Fig. 1 is better. Therefore, the robot in Fig. 1 is used in our practical application.

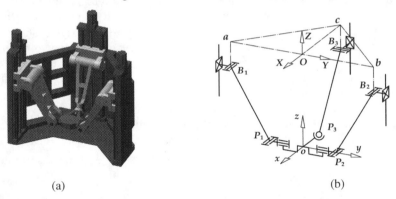

(a)                                      (b)

**Fig. 1.** A 2PRU-1PRS parallel robot: (a) CAD model; (b) kinematic scheme

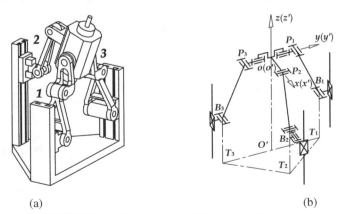

(a)                                      (b)

**Fig. 2.** A 2PRU-1PUR parallel robot: (a) CAD model; (b) kinematic scheme

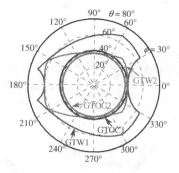

**Fig. 3.** Good transmission workspaces for the two parallel robots

To carry out the optimal design of a 4-DoF parallel robot as presented in Fig. 4, the index defined in Eq. (2) has been used. To make the robot far away from singularity and have high rotational capability, $\eta_i = 0.05$ ($i = 1, 2, 3, 4$) is taken as the criterion. On this basis, the rotational capability of the robot has been investigated and the result is provided in Fig. 5. The areas $\theta_{ABS} \geq 90°$ (which means the rotational capability can reach more than $\pm 90°$) and $\theta_{ABS} \geq 80°$ have been identified, respectively. These results are very helpful to the design and application of the robot.

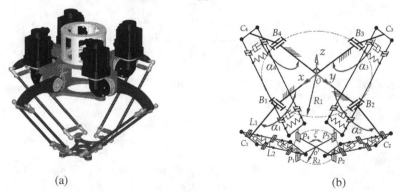

(a)                                                        (b)

**Fig. 4.** A 4-DoF parallel robot: (a) CAD model; (b) kinematic scheme

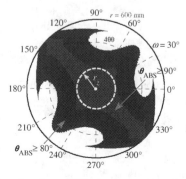

**Fig. 5.** Rotational capability in the concerned workspace

# 5    Conclusions

This paper talks about the indices for performance evaluation of parallel robots. Based on the investigation of the existing indices, such as the well-known LCI etc., some new indices considering the motion/force transmissibility of parallel robots have been presented instead. For non-redundant parallel robots, a local transmission index (LTI) is suggested as the evaluation criterion. While, a local minimized transmission index (LMTI) is defined as the performance evaluation criterion for parallel robots with actuation redundancy and a local optimal-transmission index (LOTI) is defined as the corresponding criterion for parallel robots with kinematic redundancy. All of these indices are based on the reciprocal product of screw theory and have clear physical meaning. What is more, they are frame-free and their values fall into [0 1]. These indices have been well applied in the optimal design of different kinds of parallel robots, and some of the cases have also been included in this paper to show their merits.

**Acknowledgment.** This work was supported in part by the National Natural Science Foundation of China under Grants 51305222 and 51375251.

# References

1. Schwaar, C., Neugebauer, R., Schwaar, M.: Device for the displacement and/or positioning of an object in five axes. US Patent, Patent No.: US 7104746 B2 (2006)
2. Pierrot, F., Company, O.: H4: a new family of 4-dof parallel robots. In: 1999 IEEE/ASME International Conference on Advanced Intelligent Mechatronics, pp. 508–513 (1999)
3. Kim, S.M., Kim, W., Yi, B.J.: Kinematic analysis and optimal design of a 3T1R type parallel mechanism. In: IEEE International Conference on Robotics and Automation, Kobe, Japan (2009)
4. Sun, T., Song, Y.M., Dong, G., Lian, B.B., Liu, J.P.: Optimal design of a parallel mechanism with three rotational degrees of freedom. Robot. Cim-Int. Manuf. 28(4), 500–508 (2012)
5. Pierrot, F., Nabat, V., Company, O., Krut, S., Poignet, P.: Optimal design of a 4-dof parallel manipulator: from academia to industry. IEEE Transactions on Robotics 25(2), 213–224 (2009)
6. Liu, X.J., Wu, C., Wang, J.S.: A new index for the performance evaluation of parallel manipulators: a study on planar parallel manipulators. In: Proceedings of the 7th World Congress on Intelligent Control and Automation, Chongqing, China (2008)
7. Chu, J.K., Sun, J.W.: A new approach to dimension synthesis of spatial four-bar linkage through numerical atlas method. Journal of Mechanisms and Robotics-Transactions of the ASME 2, 041004-1, 14 (2010)
8. Menon, C., Vertechy, R., Markot, M.C., Parenti-Castelli, V.: Geometrical optimization of parallel mechanisms based on natural frequency evaluation: application to a spherical mechanism for future space applications. IEEE Transactions on Robotics 25(1), 12–24 (2009)
9. Liu, X.J., Wang, J.S.: A new methodology for optimal kinematic design of parallel mechanisms. Mech. Mach. Theory 42(9), 1210–1224 (2007)
10. Kim, Y.S., Lee, J.H., Yoo, H.S., Lee, J.B., Jung, U.S.: A performance evaluation of a Stewart platform based hume concrete pipe manipulator. Automat. Constr. 18(5), 665–676 (2009)

11. Kucuk, S., Bingul, Z.: Comparative study of performance indices for fundamental robot manipulators. Robot. Auton. Syst. 54(7), 567–573 (2006)
12. Xu, Q.S., Li, Y.M.: Stiffness optimization of a 3-dof parallel kinematic machine using particle swarm optimization. In: 2006 IEEE International Conference on Robotics and Biomimetics, Kunming, China (2006)
13. Pashkevich, A., Wenger, P., Chablat, D.: Kinematic and stiffness analysis of the Orthoglide, A PKM with simple, regular workspace and homogeneous performances. In: Proceedings of the 2007 IEEE International Conference on Robotics and Automation, Roma, Italy (2007)
14. Briot, S., Bonev, I.A.: Accuracy analysis of 3-dof planar parallel robots. Mech. Mach. Theory 43(4), 445–458 (2008)
15. Merlet, J.P.: Jacobian, manipulability, condition number, and accuracy of parallel robots. J. Mech. Design 128(1), 199–206 (2006)
16. Liu, X.J., Jin, Z., Gao, F.: Optimum design of 3-dof spherical parallel manipulators with respect to the conditioning and stiffness indices. Mech. Mach. Theory 35, 1257–1267 (2000)
17. Liu, X.J., Wang, J., Zheng, H.J.: Optimum design of the 5R symmetrical parallel manipulator with a surrounded and good-condition workspace. Robot. Auton. Syst. 54(3), 221–233 (2006)
18. Wang, J.S., Liu, X.J., Wu, C.: Optimal design of a new spatial 3-dof parallel robot with respect to a frame-free index. Science in China Series E: Technological Sciences 52(4), 986–999 (2009)
19. Balli, S.S., Chand, S.: Transmission angle in mechanisms (triangle in mech). Mech. Mach. Theory 37(2), 175–195 (2002)
20. Xie, F.G., Liu, X.J., Wang, L.P., Wang, J.S.: Optimal design and development of a decoupled A/B-axis tool head with parallel kinematics. Advances in Mechanical Engineering 2010(474602), 14 pages (2010)
21. Chen, C., Angeles, J.: Generalized transmission index and transmission quality for spatial linkages. Mech. Mach. Theory 42(9), 1225–1237 (2007)
22. Ball, R.S.: A treatise on the theory of screws. Cambridge University Press (1900)
23. Wang, J.S., Wu, C., Liu, X.J.: Performance evaluation of parallel manipulators: motion/force transmissibility and its index. Mech. Mach. Theory 45(10), 1462–1476 (2010)
24. Xie, F.G., Liu, X.J., Li, T.M.: A comparison study on the orientation capability and parasitic motions of two novel articulated tool heads with parallel kinematics. Advances in Mechanical Engineering 2013(249103), 11 pages (2013)

# Instantaneous Motion of a 2-RCR Mechanism with Variable Mobility

Xiang Liu, Jingshan Zhao[*], and Zhijing Feng

State Key Laboratory Triboliogy, Department of Mechanical Engineering,
Tsinghua University, Beijing, China
jingshanzhao@mail.tsinghua.edu.cn

**Abstract.** Mobility is a very important parameter for mechanisms, and many methods for calculating the mobility of mechanisms have been proposed till now since it came to be drawn attention in the middle of 19th century. The CKG formula is widely used in the textbook, manuals and applications. However, it has been proved repeatedly to fail to deal with many classical linkages and modern spatial mechanisms as well. On the other hand, although many modifications or extensions of CKG formulas have been proposed, all of them aim at calculating the number of mobility but ignoring other mobility information, such as type, direction and location of motion. Compared with the existing CKG formulas, the analytical method is regarded as a more general and reliable method which could obtain the full information of mobility. By using this method, this paper investigated the instantaneous motion of a 2-RCR mechanism that the number of its mobility is invariable but the type is variable corresponding to different configurations.

**Keywords:** Reciprocal screw theory, mobility, analytical method, mechanisms.

## 1 Introduction

As the foundation and support structure of a robot, mechanism determines the capacity of a robot directly. Mobility is a primary parameter of mechanisms. It is responsible to what and how the robot can move. As a result, mobility analysis is essential important for robot design.

Mobility is defined by the IFToMM (International Federation for the Promotion of Mechanism and Machine Science) as the number of independent coordinates needed to define the configuration of a kinematic chain or mechanism [1]. However, a rigorous concept of mobility should include not only the number but also the type, such as translation, rotation and helical motion [2, 3]. A large number of formulas or methods based on the CKG formulas have been proposed till now. Such as Hunt [4], Hervé [5], Duffy [6, 7], Dai [8], Huang [9], Kong [10], Rico [11, 12], Gogu [13], Zhang [14], Shukla [15] and Yang [16] have made great effort on finding a general formula to calculate the mobility of mechanisms. Almost all of them are focusing on

---

[*] Corresponding author.

X. Zhang et al. (Eds.): ICIRA 2014, Part I, LNAI 8917, pp. 44–55, 2014.

calculating the number of mobility and these formulas have difficulty in analyzing the mobility of mechanisms some of which have invariable numbers but variable types.

An analytical method proposed by Zhao [2, 3] has the capacity to obtain the full information of mobility, including number, type, direction and location of the instantaneous motion. As a result, it is a general and efficient method for mobility analysis. Accordingly, it is adopted here to investigate the instantaneous motion of a 2-RCR mechanism.

## 2    Analytical Method for Mobility Analysis

A screw, $S$, is defined by a straight line associated with a pitch. It can be represented by

$$S = \begin{cases} \begin{bmatrix} s \\ s_0 \end{bmatrix} & \text{if } h \text{ is finite} \\[2em] \begin{bmatrix} 0 \\ s \end{bmatrix} & \text{if } h \text{ is infinite} \end{cases}$$

where $s$ is a unit vector along the direction of screw axis and $s_0 = r \times s + hs$, $r$ is the position vector of any point on the screw axis with respect to the reference coordinate system origin and $h$ denotes the pitch.

If $h$ is finite, $S$ represents an helical joint or constraint helical force with direction $s$ and pitch $h$. Especially when $h = 0$, it degenerates into a revolute joint or constraint force. If $h$ is an infinite, $S$ indicates a prismatic joint or constraint moment with direction $s$. Hence, a unit screw is also termed as a kinematic screw or constraint screw.

Suppose that the kinematic screw or constraint screw $S$ is known, there will be $s \cdot s_0 = s(r \times s + hs) = h\|s\|^2$. Considering $\|s\|^2 = 1$, the pitch can be expressed as

$$h = s \cdot s_0 \tag{1}$$

Since $s \times (s_0 - hs) = s \times (r \times s) = (s \cdot s)r - (r \cdot s)s = r$, the position vector is expressed as

$$r = s \times [s_0 - (s \cdot s_0)s] = s \times s_0 \tag{2}$$

It should be pointed out that the position vector is not unique. Two screws, $S$ and $S^r$, are called reciprocal when they satisfy the reciprocal equation

$$S^T \Delta S^r = 0 \tag{3}$$

where $\Delta = \begin{bmatrix} 0 & I_{3\times3} \\ I_{3\times3} & 0 \end{bmatrix}$ and $I_{3\times3}$ is an identity matrix. The physical meaning of equation (3) is that the work done at any instant by the external force, represented by $S^r$, to a stable rigid body should always be zero. Accordingly, $S^r$ could represent a constraint force or moment.

Assume that a kinematic screw is expressed as $^L S$ and $S$ in the local coordinate system and the absolute one, respectively. Then the transformation between the two screws is

$$S = \mathbf{R}\,^L S \tag{4}$$

where $\mathbf{R}$ is the transformation matrix from the local coordinate system to the absolute one and $\mathbf{R} = \begin{bmatrix} \mathbf{T} & \mathbf{0}_{3\times3} \\ \hat{\rho}\mathbf{T} & \mathbf{T} \end{bmatrix}$, $\mathbf{T}$ is a transformation matrix from the local coordinate system to the absolute one and $\rho$ is the vector of reference origin with respect to the absolute coordinate system. Assume $\rho = \begin{bmatrix} x_o & y_o & z_o \end{bmatrix}^T$, then

$$\hat{\rho} = \begin{bmatrix} 0 & -z_o & y_o \\ z_o & 0 & -x_o \\ -y_o & x_o & 0 \end{bmatrix}.$$

A mechanism with multi kinematic chains is shown in Figure 1. $B_i P_i (i = 1, 2, \cdots, n)$ represents the $i$th kinematic chain. $B_1 B_2 B_i B_n$ and $P_1 P_2 P_i P_n$ represent the fixed base and end effector, respectively. For the $i$th kinematic chain, a local coordinate system $o_i x_i y_i z_i$ is established with the principle that makes the expressions of local kinematic screws as simple as possible. In general, the origin point of local coordinate system, $o_i$, is superposed with the first connected joint, $B_i$. Assume the location of the local origin point, $o_i$, is $\rho_i$ in the absolute coordinate system, $oxyz$, and the transformation matrix from $o_i x_i y_i z_i$ to $oxyz$ is $\mathbf{T}_i$. Then the transformation associated with equation (4) is expressed as $\mathbf{R}_i = \begin{bmatrix} \mathbf{T}_i & \mathbf{0}_{3\times3} \\ \hat{\rho}\mathbf{T}_{ii} & \mathbf{T}_i \end{bmatrix}$.

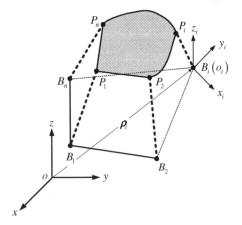

**Fig. 1.** Sketch of a multi-loop mechanism

A general analysis procedure of the analytical method is summarized as follows [2].

(1) Write the local kinematic screws of $i$th kinematic chain in the local coordinate system, $o_i x_i y_i z_i$, and obtain the corresponding local kinematic screw matrix ${}^i\mathbf{S}_i$.

(2) Solve the local constraint screw matrix of $i$th kinematic chain according to equation (3), denoted as ${}^i\mathbf{S}_i^r$.

(3) Associated with equation (4) and obtain the expression of constraint screw matrix in the absolute coordinate system, $\mathbf{S}_i^r = \mathbf{R}_i\,{}^i\mathbf{S}_i^r$.

(4) Repeat steps 1-3 and obtain the constraint screw matrix of the end effector,
$$\mathbf{S}^r = \begin{bmatrix} \mathbf{S}_1^r & \mathbf{S}_2^r & \cdots & \mathbf{S}_n^r \end{bmatrix}.$$

(5) Substitute $\mathbf{S}^r$ into the reciprocal equation, $\begin{bmatrix} \mathbf{S}^r \end{bmatrix}^T \Delta\mathbf{S} = 0$, and solve the kinematic screw matrix of the end effector, $\mathbf{S}$.

Accordingly, the rank of $\mathbf{S}$ presents the number of mobility, and each column of $\mathbf{S}$ is a kinematic screw which shows the type, direction and location of the mobility.

## 3     Instantaneous Motion of the 2-RCR Mechanism

As Figure 2 shows, a 2-RCR mechanism consists of two identical RCR (revolute-cylinder-revolute) kinematic chains, $ABC$ and $DEF$, which locate at $\pi_1$- plane and $\pi_2$-plane, respectively. The unit direction vectors of different joints are $n_A$, $n_B$, $n_C$, $n_D$, $n_E$ and $n_F$. The axes of two revolute joints of each kinematic chain are both perpendicular to the axis of corresponding cylinder joint. The subtended angle between $\pi_1$-plane and $\pi_2$-plane is $\theta = \pi/2$, so do $n_C$ and $n_D$. $\alpha_1$ and $\alpha_2$ are

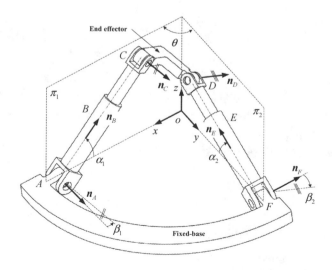

**Fig. 2.** Geometry of the 2-RCR mechanism

the twist angles between the axes of two cylinder joints and $xoy$-plane. $\beta_1$ and $\beta_2$ are the twist angles between $n_A$ and $n_C$, $n_D$ and $n_F$, respectively. Suppose that the revolute joints $A$ and $F$ locate on a circle whose radius equals $r$ and the lengths of two kinematic chains are $l_1$ and $l_2$, respectively.

From Figure 2, the unit direction vectors of kinematic chain $ABC$ and $DEF$ are

$$\begin{cases} n_A = \begin{bmatrix} 0 & 1 & 0 \end{bmatrix}^T \\ n_B = \begin{bmatrix} -\cos\alpha_1 & 0 & \sin\alpha_1 \end{bmatrix}^T \\ n_C = \begin{bmatrix} \sin\alpha_1\sin\beta_1 & \cos\beta_1 & \cos\alpha_1\sin\beta_1 \end{bmatrix}^T \end{cases}, \begin{cases} n_F = \begin{bmatrix} -1 & 0 & 0 \end{bmatrix}^T \\ n_E = \begin{bmatrix} 0 & -\cos\alpha_2 & \sin\alpha_2 \end{bmatrix}^T \\ n_D = \begin{bmatrix} -\cos\beta_2 & \sin\alpha_2\sin\beta_2 & \cos\alpha_2\sin\beta_2 \end{bmatrix}^T \end{cases}$$

The position vectors of the revolute joints are

$$\begin{cases} r_A = \begin{bmatrix} r & 0 & 0 \end{bmatrix}^T, r_C = \begin{bmatrix} r-l_1\cos\alpha_1 & 0 & l_1\sin\alpha_1 \end{bmatrix}^T \\ r_D = \begin{bmatrix} 0 & r-l_2\cos\alpha_1 & l_2\sin\alpha_2 \end{bmatrix}^T, r_F = \begin{bmatrix} 0 & r & 0 \end{bmatrix}^T \end{cases}$$

Accordingly, the kinematic screw matrix of kinematic chain $ABC$ is

$$S_{ABC} = \begin{bmatrix} S_A & S_B & S_C \end{bmatrix}$$

where $S_A = \begin{bmatrix} 0 \\ 1 \\ 0 \\ 0 \\ 0 \\ r \end{bmatrix}, S_B = \begin{bmatrix} -\cos\alpha_1 & 0 \\ 0 & 0 \\ \sin\alpha_1 & 0 \\ 0 & -\cos\alpha_1 \\ -r\sin\alpha_1 & 0 \\ 0 & \sin\alpha_1 \end{bmatrix}, S_C = \begin{bmatrix} \sin\alpha_1\sin\beta_1 \\ \cos\beta_1 \\ \cos\alpha_1\sin\beta_1 \\ -l_1\sin\alpha_1\cos\beta_1 \\ (l_1-r\cos\alpha_1)\sin\beta_1 \\ (r-l_1\cos\alpha_1)\cos\beta_1 \end{bmatrix}.$

Substituting $S_{ABC}$ into the reciprocal equation (3), the constraint screw matrix of the kinematic chain $ABC$ can be solved as

$$S_{ABC}^r = \begin{bmatrix} S_{ABC}^{r1} & S_{ABC}^{r2} \end{bmatrix}$$

When $\beta_1 \neq 0$ holds, there will be

$$S_{ABC}^{r1} = \begin{bmatrix} n_A \\ r_C \times n_A \end{bmatrix}, \quad S_{ABC}^{r2} = \begin{bmatrix} n_C \\ r_A \times n_C \end{bmatrix}$$

where $S_{ABC}^{r1}$ indicates a constraint force along $n_A$ and passing through point $C$, $S_{ABC}^{r2}$ indicates a constraint force along $n_C$ and passing through point $A$, as Figure 3 shows. The notation '——▷' stands for a constraint force.

When $\beta_1 = 0$ holds, there will be

$$S_{ABC}^{r1} = \begin{bmatrix} n_A \\ r_C \times n_A \end{bmatrix}, \quad S_{ABC}^{r2} = \begin{bmatrix} 0 \\ n_{AB} \end{bmatrix}$$

where $S_{ABC}^{r1}$ indicates a constraint force along $n_A$ and passing point $C$, $S_{ABC}^{r2}$ indicates a constraint moment about the axis perpendicular to both $n_A$ and $n_B$, as Figure 4 shows. Namely, $n_{AB} = \begin{bmatrix} \sin\alpha_1 & 0 & \cos\alpha_1 \end{bmatrix}^T$. The notation '⟶▷▷' indicates a constraint moment.

Similarly, we can obtain the constraint screw matrix of kinematic chain *DEF* as following.

$$\boldsymbol{S}_{DEF}^{r} = \begin{bmatrix} \boldsymbol{S}_{DEF}^{r1} & \boldsymbol{S}_{DEF}^{r2} \end{bmatrix}$$

When $\beta_2 \neq 0$ holds, there will be

$$\boldsymbol{S}_{DEF}^{r1} = \begin{bmatrix} \boldsymbol{n}_F \\ \boldsymbol{r}_D \times \boldsymbol{n}_F \end{bmatrix}, \quad \boldsymbol{S}_{DEF}^{r2} = \begin{bmatrix} \boldsymbol{n}_D \\ \boldsymbol{r}_F \times \boldsymbol{n}_D \end{bmatrix}$$

When $\beta_2 = 0$ holds, there will be

$$\boldsymbol{S}_{DEF}^{r1} = \begin{bmatrix} \boldsymbol{n}_F \\ \boldsymbol{r}_D \times \boldsymbol{n}_F \end{bmatrix}, \quad \boldsymbol{S}_{DEF}^{r2} = \begin{bmatrix} \boldsymbol{0} \\ \boldsymbol{n}_{FD} \end{bmatrix}$$

where $\boldsymbol{n}_{FD} = \begin{bmatrix} 0 & \sin\alpha_2 & \cos\alpha_2 \end{bmatrix}^T$.

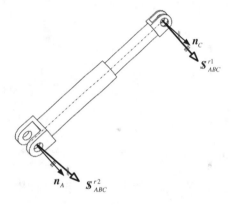

**Fig. 3.** Terminal constraints of RCR kinematic chain when $\beta_1 \neq 0$

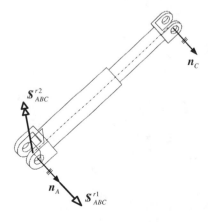

**Fig. 4.** Terminal constraints of RCR kinematic chain when $\beta_1 = 0$

Therefore, the constraint screw matrix of the end effector is expressed as

$$\mathbf{S}_{CD}^{r} = \begin{bmatrix} \mathbf{S}_{ABC}^{r} & \mathbf{S}_{DEF}^{r} \end{bmatrix}$$

Considering the arbitrariness of $\beta_1$, $\beta_2$, there are three different cases that should be discussed, namely 1) $\beta_1 = \beta_2 = 0$  2) $\beta_1 = 0, \beta_2 \neq 0$  or  $\beta_1 \neq 0, \beta_2 = 0$  3) $\beta_1, \beta_2 \neq 0$.

1) When $\beta_1 = \beta_2 = 0$. In this case, there are two constraint forces, whose directions are $\mathbf{n}_A$ and $\mathbf{n}_F$, and two constraint moments, whose directions are $\mathbf{n}_{AB}$ and $\mathbf{n}_{EF}$, exerting on the end effector. Therefore, the instantaneous mobility of the end effector must be a translation along the direction which is perpendicular to both $\mathbf{n}_A$ and $\mathbf{n}_F$, and a rotation about the axis which is perpendicular to both two constraint moments. Substituting the corresponding constraint screw matrix into reciprocal equation yields

$$\mathbf{S}_{CD} = \begin{bmatrix} \mathbf{S}_{CD}^{1} & \mathbf{S}_{CD}^{2} \end{bmatrix}$$

where $\mathbf{S}_{CD}^{1} = \begin{bmatrix} 0 \\ 0 \\ 0 \\ 0 \\ 0 \\ 1 \end{bmatrix}$, $\mathbf{S}_{CD}^{2} = \begin{bmatrix} -\cos\alpha_1 \sin\alpha_2 \\ -\sin\alpha_1 \cos\alpha_2 \\ \sin\alpha_1 \sin\alpha_2 \\ r\sin\alpha_1 \sin\alpha_2 \\ -r\sin\alpha_1 \sin\alpha_2 \\ r\left(\cos\alpha_1 \sin\alpha_2 - \sin\alpha_1 \cos\alpha_2\right) \end{bmatrix}$.

According to equation (1) and (2), the $\mathbf{S}_{CD}^{1}$ and $\mathbf{S}_{CD}^{2}$ can be rewritten as

$$\mathbf{S}_{CD}^{1} = \begin{bmatrix} \mathbf{0} \\ \mathbf{n}_{CD}^{1} \end{bmatrix} \text{ and } \mathbf{S}_{CD}^{2} = \begin{bmatrix} \mathbf{n}_{CD}^{2} \\ \mathbf{r}_{CD}^{2} \times \mathbf{n}_{CD}^{2} \end{bmatrix}, \text{ where } \mathbf{r}_{CD}^{2} = \begin{bmatrix} r & r & 0 \end{bmatrix}^{T}.$$

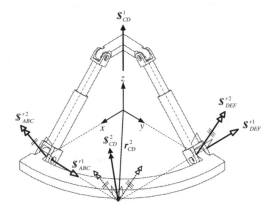

**Fig. 5.** Mobility of the 2-RCR mechanism- $\beta_1 = \beta_2 = 0$

Therefore, $S_{CD}^1$ indicates a translation along $z$-axis, $S_{CD}^2$ indicates a rotation about the axis $n_{CD}^2$ and passing through $r_{CD}^2$ which are shown in Figure 5. The translation and rotation are represented by the notations '——▶' and '——▶▶', respectively. Assume that the end effector move along $z$-axis, the end effector is always parallel to the $xoy$-plane, that is to say, the condition $\beta_1 = \beta_2 = 0$ always holds. Hence, the translation along $z$-axis is a full-cycle mobility in this case. However, once the rotation about the axis $n_{CD}^2$ occurs, the end effector will be not parallel to the $xoy$-plane any more. Namely, the condition $\beta_1 = \beta_2 = 0$ does not hold any more. Therefore, the rotation about the axis $n_{CD}^2$ is an instantaneous mobility in this case.

2) When $\beta_1 \neq 0$ and $\beta_2 = 0$. The terminal constraints are three constraint forces along $n_A$, $n_C$ and $n_F$, and one constraint moment perpendicular to $n_E$. Considering that the three constraint forces are not coplanar, the instantaneous mobility of the end effector must be two rotations or helical motions.

Since $\beta_2 = 0$ holds, the end effector could rotate about the $x$-axis. Assuming that the rotation angle of joint $D$ with respect to the $y$-axis is $\varphi$, the direction vector of joint $C$ can be obtained as $n_C' = [0 \quad \cos\varphi \quad \sin\varphi]^T$. At the same time, the direction vector of joint $C$ satisfies to $n_C = [\sin\alpha_1 \sin\beta_1 \quad \cos\beta_1 \quad \cos\alpha_1 \sin\beta_1]^T$ when $\beta_1 \neq 0$ holds. Therefore, there must be $\sin\alpha_1 \sin\beta_1 = 0$, that is $\alpha_1 = 0$.

Accordingly, the kinematic screw matrix of the end effector can be solved by reciprocal equation.

$$S_{CD} = \begin{bmatrix} S_{CD}^1 & S_{CD}^2 \end{bmatrix}$$

where $S_{CD}^1 = \begin{bmatrix} 1 \\ 0 \\ 0 \\ 0 \\ 0 \\ 0 \end{bmatrix}$ and $S_{CD}^2 = \begin{bmatrix} 0 \\ \cos\alpha_2 \\ -\sin\alpha_2 \\ -r\sin\alpha_2 \\ -\sin\alpha_2(l_1 - r) \\ r\cos\alpha_2 + l_1 \cot\beta_1 \sin\alpha_2 \end{bmatrix}$.

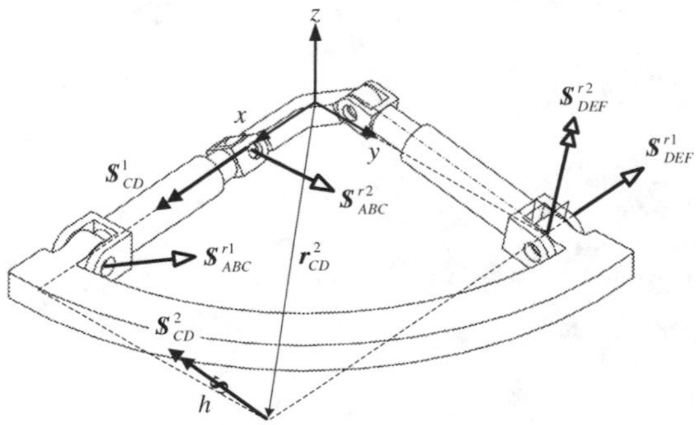

**Fig. 6.** Mobility of the 2-RCR mechanism- $\beta_1 \neq 0, \beta_2 = 0$

According to equation (1) and (2), the kinematic screws can be rewritten as

$$\boldsymbol{S}_{CD}^1 = \begin{bmatrix} \boldsymbol{n}_{CD}^1 \\ \boldsymbol{r}_{CD}^1 \times \boldsymbol{n}_{CD}^1 \end{bmatrix}, \boldsymbol{S}_{CD}^2 = \begin{bmatrix} \boldsymbol{n}_{CD}^2 \\ h_2 \boldsymbol{n}_{CD}^2 + \boldsymbol{r}_{CD}^2 \times \boldsymbol{n}_{CD}^2 \end{bmatrix}$$

where $\boldsymbol{n}_{CD}^1 = \begin{bmatrix} 1 & 0 & 0 \end{bmatrix}^T, \boldsymbol{r}_{CD}^1 = \begin{bmatrix} 0 & 0 & 0 \end{bmatrix}^T, \boldsymbol{n}_{CD}^2 = \begin{bmatrix} 0 & \cos\alpha_2 & -\sin\alpha_2 \end{bmatrix}^T$,
$h_2 = -l_1 \sin\alpha_2 \left[ \cos\alpha_2 + \cot\beta_1 \sin\alpha_2 \right]$,
$\boldsymbol{r}_{CD}^2 = \begin{bmatrix} r + l_1 \sin\alpha_2 (\cot\beta_1 \cos\alpha_2 - \sin\alpha_2) & r & 0 \end{bmatrix}^T$.

As Figure 6 shown, $\boldsymbol{S}_{CD}^1$ indicates a rotation about $x$-axis and passing through $\boldsymbol{r}_{CD}^1$, $\boldsymbol{S}_{CD}^2$ denotes a helical motion, which is represented by a notation '⟶', parallel to $\boldsymbol{n}_E$ and passing through $\boldsymbol{r}_{CD}^2$ with a pitch $h_2$.

Since the rotation $\boldsymbol{S}_{CD}^1$ is parallel to the axes $\boldsymbol{n}_D$ and $\boldsymbol{n}_F$, it will not change the geometry condition that $\beta_2 = 0$. Therefore, the rotation $\boldsymbol{S}_{CD}^1$ is a full-cycle mobility in this case. But the rotation $\boldsymbol{S}_{CD}^2$ is an instantaneous mobility as the geometry condition $\beta_2 = 0$ will change once it occurs.

3) A general case is that $\beta_1 \neq 0$ and $\beta_2 \neq 0$. The terminal constraints exerted on the end effector are four constraint forces along $\boldsymbol{n}_A$, $\boldsymbol{n}_C$, $\boldsymbol{n}_D$ and $\boldsymbol{n}_F$, respectively. Considering they are neither coplanar nor intersection, the rank of the constraint screw matrix is 4 according to the Grassmann line geometry. Therefore, the reciprocal screws must be a two-order system with two rotations or helical motions. The constraint screw matrix can be solved using the reciprocal equation.

$$\boldsymbol{S}_{CD}^r = \begin{bmatrix} \boldsymbol{S}_{CD}^1 & \boldsymbol{S}_{CD}^2 \end{bmatrix}$$

where "s" and "c" stand for sine and cosine, respectively.

$$\boldsymbol{S}_{CD}^1 = \frac{\sigma_1}{\sqrt{\sigma_1^2 + \sigma_{12}^2}} \begin{bmatrix} \dfrac{\sigma_{12}}{\sigma_1} \\ 1 \\ 0 \\ -l_2 \, \mathrm{s}\,\alpha_2 \\ \dfrac{l_1 \, \mathrm{s}\,\alpha_1 \sigma_{12}}{\sigma_1} \\ K_1 \end{bmatrix}, \boldsymbol{S}_{CD}^2 = \frac{\sigma_1}{\sqrt{\sigma_1^2 + \sigma_{22}^2}} \begin{bmatrix} -\dfrac{\sigma_{22}}{\sigma_1} \\ 0 \\ 1 \\ r - l_2 \, \mathrm{c}\,\alpha_2 \\ K_{21} \\ K_{22} \end{bmatrix},$$

and $\sigma_1 = l_1 \, \mathrm{s}\,\alpha_1 \, \mathrm{s}\,\beta_2 (\mathrm{c}\,\beta_1 \, \mathrm{c}\,\alpha_2 - \mathrm{c}\,\alpha_1 \, \mathrm{s}\,\beta_1 \, \mathrm{s}\,\alpha_2) - r \mathrm{c}\,\alpha_1 \, \mathrm{s}\,\beta_1 \, \mathrm{c}\,\alpha_2 \, \mathrm{s}\,\beta_2$,
$\sigma_{12} = l_2 \, \mathrm{s}\,\beta_1 \, \mathrm{s}\,\alpha_2 (\mathrm{c}\,\alpha_1 \, \mathrm{c}\,\beta_2 + \mathrm{s}\,\alpha_1 \, \mathrm{c}\,\alpha_2 \, \mathrm{s}\,\beta_2) + r \mathrm{c}\,\alpha_1 \mathrm{s}\,\beta_1 \mathrm{c}\,\alpha_2 \mathrm{s}\,\beta_2$,
$\sigma_{22} = l_1 \, \mathrm{c}\,\alpha_1 \, \mathrm{s}\,\beta_2 (\mathrm{c}\,\beta_1 \, \mathrm{c}\,\alpha_2 - \mathrm{c}\,\alpha_1 \, \mathrm{s}\,\beta_1 \, \mathrm{s}\,\alpha_2) - l_2 \, \mathrm{s}\,\beta_1 \, \mathrm{c}\,\alpha_2 (\mathrm{c}\,\alpha_1 \, \mathrm{c}\,\beta_2 - \mathrm{s}\,\alpha_1 \, \mathrm{c}\,\alpha_2 \, \mathrm{s}\,\beta_2) + r \mathrm{s}\,\beta_1 \, \mathrm{s}\,\beta_2 \, \mathrm{s}(\alpha_1 + \alpha_2)$,

$$K_1 = \frac{-l_1 \, \mathrm{s}\,\alpha_1 \, \mathrm{c}\,\beta_1 \dfrac{\sigma_{12}}{\sigma_1} + l_2 \, \mathrm{s}\,\alpha_1 \, \mathrm{s}\,\beta_1 \, \mathrm{s}\,\alpha_2 + r \mathrm{c}\,\alpha_1 \, \mathrm{s}\,\beta_1}{\mathrm{c}\,\alpha_1 \, \mathrm{s}\,\beta_1}, K_{21} = l_1 \, \mathrm{c}\,\alpha_1 - r - \frac{l_1 \, \mathrm{s}\,\alpha_1 \sigma_{22}}{\sigma_1},$$

$$K_{22} = \frac{l_1\left(s\alpha_1\,c\,\beta_1\,\dfrac{\sigma_{22}}{\sigma_1} - c\,\alpha_1 c\beta_1\right) + l_2\,s\,\alpha_1\,s\,\beta_1\,c\,\alpha_2 - r s\alpha_1\,s\,\beta_1}{c\,\alpha_1\,s\,\beta_1}.$$

Therefore, $\boldsymbol{S}^1_{CD}$ denotes a helical motion about the axis parallel to $xoy$-plane and its pitch and location are

$$h_1 = \frac{\sigma_1\sigma_{12}}{\sigma_1^2 + \sigma_{12}^2}\left(l_1\,s\,\alpha_1 - l_2\,s\,\alpha_2\right),$$

$$\boldsymbol{r}^1_{CD} = \frac{\sigma_1^2}{\sigma_1^2 + \sigma_{12}^2}\left[K_1 \quad -\frac{K_1\sigma_{12}}{\sigma_1} \quad \frac{l_1\,s\,\alpha_1\sigma_{12}^2}{\sigma_1^2} + l_2\,s\,\alpha_2\right]^T.$$

$\boldsymbol{S}^2_{CD}$ indicates a helical motion about the axis parallel to $xoz$-plane and its pitch and location are

$$h_2 = -\frac{\sigma_1\sigma_{22}}{\sigma_1^2 + \sigma_{22}^2}\left(r - l_2\,c\,\alpha_2\right) + \frac{\sigma_1^2 K}{\sigma_1^2 + \sigma_{22}^2},$$

$$\boldsymbol{r}^2_{CD} = \frac{\sigma_1^2}{\sigma_1^2 + \sigma_{22}^2}\left[-K_{21} \quad r - l_2\,c\,\alpha_2 + \frac{K_{22}\sigma_{22}}{\sigma_1} \quad -\frac{K_{21}\sigma_{22}}{\sigma_1}\right]^T.$$

And they are full-cycle mobility because the geometry condition could be invariable when any one of them occurs. But it should be pointed out that the mobility type might change suddenly if the geometry condition changes during the movement of the end effector.

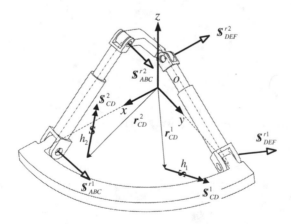

**Fig. 7.** Mobility of the 2-RCR mechanism- $\beta_1 \neq 0, \beta_2 \neq 0$

From the analysis we can find that, the full-cycle mobility is a relative concept for the 2-RCR mechanism as the geometry condition might change between three different configurations which correspond to three different forms even the number is invariable. The 2-RCR mechanism is a special mechanism which has only 2 DOFs but can achieve three types of full-cycle motions, namely translation, rotation and helical motion. This study not only shows the advantage of the analytical method, but also indicates the necessity to study the full mobility information.

## 4     Conclusions

In the past 160 years, many scholars have made great efforts to find a general formula or method to calculate the mobility of any mechanisms. But the fact is that either the "general formula" fails to deal with some special mechanisms, or the analysis process is too complex to be understood by an ordinary engineer. In addition, almost all of the proposed formulas and methods have paid much attention to calculate the number of mobility but ignoring the information of type, direction and location of the instantaneous motions. In this paper, a general analytical method for mobility analysis is discussed based on the previous work on the screw theory. It is the primary principle of the method to use the reciprocal equation twice to obtain the analytical expression of the kinematic screw matrix of the end-effecter, which shows the detail information of the mobility, including the number, type, direction and position. By using the analytical method, the mobility of the 2-RCR mechanism is investigated. The analysis results show that the number of the mobility equals two. However, the type of the mobility is variable corresponding to different configurations. Compared with the metamorphic mechanisms, even though the topological structure of the 2-RCR mechanism is not changed, the mobility could be changed between translation, rotation and helical motion. This paper indicates the necessity to study the full mobility information and paves a way to study the exact mobility of every type of mechanisms.

## References

1. Ionescu, T.G.: Terminology for Mechanisms and Machine Science. Mech. Mach. Theory 38, 774 (2003)
2. Zhao, J.S., Feng, Z.J., Ma, N., Chu, F.L.: Advanced Theory of Constraint and Motion Analysis for Robot Mechanisms. Elsevier (2014)
3. Zhao J. S., Feng Z. J., Ma N., Chu F. L.: Design of Special Planar Linkages. Springer (2013)
4. Hunt, K.H.: Kinematic Geometry of Mechanisms. Oxford University Press, Oxford (1978)
5. Hervé, J.M.: Intrinsic Formulation of Problems of Geometry and Kinematics of Mechanisms. Mech. Mach. Theory 17(3), 179–184 (1982)
6. Sugimoto, K., Duffy, J.: Application of Linear Algebra to Screw Systems. Mech. Mach. Theory 17(1), 73–83 (1982)
7. Mohamed, M.G., Duffy, J.: Direct Determination of the Instantaneous Kinematics of Fully Parallel Robot Manipulators. J. Mech. Trans. Autom. in Des. 107(2), 226–229 (1985)
8. Dai, J.S., Huang, Z., Lipkin, H.: Mobility of Overconstrained Parallel Mechanisms. J. Mech. Des 128, 220–229 (2006)
9. Huang, Z.: Theory of Parallel Mechanisms. Springer (2013)
10. Kong, X.W., Gosselin, C.M.: Type Synthesis of Parallel Mechanisms. Springer (2007)

11. Rico, J.M., Ravani, B.: On Mobility Analysis of Linkages Using Group Theory. J. Mech. Des. 125(1), 70–80 (2003)
12. Rico, J.M., Aguilera, L.D., et al.: A More General Mobility Criterion for Parallel Platforms. J. Mech. Des. 128(1), 207–219 (2005)
13. Gogu, G.: Mobility of Mechanisms: ACritical Review. Mech. Mach. Theory 40, 1068–1097 (2005)
14. Zhang, Y.T., Mu, D.J.: New Concept and New Theory of Mobility Calculation for Multi-loop Mechanisms. Sci. China Technol. Sci. 53(6), 1598–1604 (2010)
15. Shukla, G., Whitney, D.E.: The Path Method for Analyzing Mobility and Constraint of Mechanisms and Assemblies. IEEE Trans. Autom. Sci. Eng 2(2), 184–192 (2005)
16. Yang, T.L., Sun, D.J.: A General Degree of Freedom Formula for Parallel Mechanisms and MultiloopSpatial Mechanisms. J. Mech. Robot. 4, 11001–1, 17 (2012)

# Mechatronic Design of an Upper Limb Prosthesis with a Hand

Lei He, Caihua Xiong, and Keke Zhang

School of Mechanical Science and Engineeing,
Huazhong University of Science and Technology, Wuhan, China
{d201277111,chxiong}@hust.edu.cn, zhangkeke@saicmotor.com

**Abstract.** This paper presents an upper limb prosthesis called IRR, which consists of a 7 DOF fully-actuated arm and a 15-DOF underactuated hand. The prosthesis is aimed to grasp a wide range of objects and perfectly reproduce the movement of the human upper limb. The main development is focused on the requirements of small size, light weight, high functionality and good cosmetic appearance. First, mechanisms of the prosthesis have been designed. Then, highly reliable embedded electronics, effective control architecture are devised to manage complexity of the system. Finally, multi-mode grasping and drinking water experiments have been done to test the ability of the prosthesis to reproduce human activities of daily lives.

**Keywords:** upper limb prosthesis, mechatronics, underactuation, rehabilitation.

## 1    Introduction

Human upper limb is a very powerful tool, whose loss would result in severe physical, psychological and vocational consequences. According to a survey made by the National Bureau of Statistics of the People's Republic of China, up to 2006, the disabled account for 6.34% of the total people in china[1], 29.07% of whom are physically disabled. Strong motivation to help them to return to normal life leads to the development of upper limb prostheses.

The upper limb prosthesis has been an active research topic for several decades. However, in spite of many multifunctional prostheses available in laboratories and companies, a lot of amputees would prefer to use passive prostheses. The main drawbacks of the active prostheses are summarized as follows: 1) lack of anthropomorphic size, weight and cosmetic appearance, 2) low functionalities and 3) control complexity.

Natural movement of the human arm is of great complexity, reproduction of which by prostheses with few DOFs is visually artificial. Our goal of the prosthetic arm design is to mimic human upper limb in size and weight, and to be dexterous enough to complete activities of daily living (ADLs). Commercial prosthetic arms, for example, Boston Elbow and Utah Arm have few degrees of freedom (DOFs)[2]. Although working well in really life, they have low functionality and less anthropomorphic movement. The DLR hand arm system[3] has been designed by the Germany Aerospace Center to improve the robustness, dynamic performance and dexterity of the

X. Zhang et al. (Eds.): ICIRA 2014, Part I, LNAI 8917, pp. 56–66, 2014.
© Springer International Publishing Switzerland 2014

anthropomorphic hand arm system, however, 52 motors are included to achieve the goals, which implies complex control. Recently developed experimental prosthesis like Revolutionizing Prosthetics[4] has made many breakthroughs, however there is still a long way from practical stage.

Additionally, many robotic hands have been developed. Pioneer dexterous robotic hands, like Utah/MIT hand[5] and Salisbury hand[6], adopt the design philosophy of full actuation. Those two hands are useful to serve as a research platform to test theory of grasp and manipulation, but are embarrassing to act as a prosthetic hand owing to the control complexity and great size and weight. To avoid these drawbacks, underactuation is a generally accepted solution. Underactuation is a method of the robot design in which the number of actuators is less than that of degree of freedoms (DOFs), i.e. couplings exist between the joints. Considering the feasibility, two common approaches are used to realize such coupling, i.e. the linkage and the tendon. In Asada's hand[7], coupling is achieved by tendon based mechanism. It utilizes the principal component analysis method, which reduce the number of actuators to two. However, the hand has two main shortcomings: 1) it could not comply to the shape of the objects to be grasped. 2) It is too bulky for use in amputees. The TBM hand[8] introduces adaptation between fingers through springs but fixed couplings in each finger make it less anthropomorphic.

The objective of this paper is to develop a prosthetic limb with a hand to restore motion capability of upper extremity amputee patients. The prosthesis should mimic the human upper limb in size, weight, functionality and cosmetic appearance, and its movement ought to be anthropomorphic, all of which are main concerns of amputees. The IRR prosthesis developed is shown in Fig. 1. However, considering the state of the art of current technology, dilemma exists between the technical implementation and multifunctionality requirement. This gives rise to our design of the prosthesis involving a 15 DOF underactuated hand and a 7 DOF fully actuated arm.

An outline of this paper is as follows. First, mechanical structure of the limb is investigated, which consists of a fully actuated 7 DOF arm and an underactuated 15 DOF hand. Then, highly reliable electronics and control architecture is presented. Finally, the experiment validation is made to demonstrate the functionality and robustness through the drinking water activity.

**Fig. 1.** A comparison between the IRR prosthetic arm-hand system and a male limb

## 2     Mechanical Design

The prosthesis is dimensioned according to anthropometric information available, and main specifications of the hand arm system designed are shown in Table 1.

**Table 1.** Main features of the IRR prosthesis

| Specification | Value |
| --- | --- |
| Whole Size (maximum Height*Width*Depth) | $71*12.3*12.3 \text{ mm}^3$ |
| Weight of the arm | 4 kg |
| Number of DOFs of the arm | 7 |
| Number of actuators of the arm | 7 |
| Weight of the hand | 0.45kg |
| Number of DOFs of the hand | 15 |
| Number of actuators of the hand | 3 |

### 2.1     Arm Mechanics

To perform various activities as anthropomorphic as possible, two aspects must be considered: 1) the mechanical structure of a prosthetic arm should possess the similar distribution of arm joints and lengths, as shown in Fig. 2. 2) dexterity must be guaranteed to achieve various ADLs. Considering the above two concerns, a physiological based method is adopted and a brief analysis of the anatomy of the human upper limb reveals the joint configuration, type and corresponding number of DOF, which are shown in Table 2 in detail. In order to attain high dexterity, a fully actuated 7-DOF prosthetic arm is designed, shown in 错误!未找到引用源。.

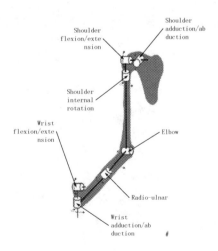

**Fig. 2.** Anatomy of the human upper limb

To realize such a structure, an easily conceivable idea is to implement each joint with a rotary pair, where actuators, reducers and shafts are mounted together around the joint. But this would result in large joints, which is incongruous with other parts of the arm.

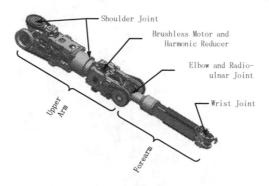

**Fig. 3.** The 3D model of the IRR prosthetic arm

The innovation of the IRR prosthetic arm is the effective utilization of differential bevel gears as the main joint type and the motors and reducers are placed inside the upper-arm and fore-arm, far away from the joints. The benefit of such strategy is the minimization of the whole arm size and weight, obtaining a more anthropomorphic characteristics. The differential gear mechanism mainly has two advantages over other mechanisms. One is that two DOFs could be implemented within a compact space, greatly reducing the size of the joint. The other is that it allows us to utilize the torque of both motors rather than one, which results in smaller motor and reducer size, thus further space saving. In order to reduce total space of the arm joints and, in the meantime, provide enough power, high power density brushless DC motors (BLDCM) in series with harmonic reducers are used.

**Table 2.** Distribution of joints and mechanisms

| Joint Name | Joint Type | DOF | Implementation |
|---|---|---|---|
| Shoulder Joint | The Ball and Socket Joint | 3 | Differential Gear and Rotary Pair |
| Elbow Joint | Hinge Joint | 1 | Differential Gear |
| Radio-ulnar Joint | Pivot Joint | 1 | |
| Wrist Joint | Ellipsoid Joint | 2 | Differential Gear |

## 2.2    Hand Mechanics

The IRR hand has 15 DOFs and 5 passively adaptive fingers which are actuated by 3 motors: one to control the thumb and other two to control two fingers each. To achieve this purpose, underactuation is implemented within each finger and between the fingers.

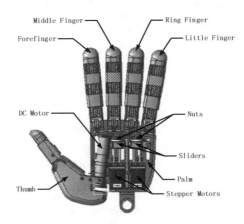

**Fig. 4.** The 3D model of the IRR prosthetic hand

Each finger has three phalanges (proximal, intermediate and distal) and 3 joints (metacarpo-phalangeal joint, MCP, proximal-interphalangeal, PIP and distal-interphalangeal, DIP), much like the human hand. Transmission mechanism in each finger is tendon-based because of the advantages of compact space, constant arm of force, high efficiency compared with another typical choice of linkage mechanism[9]. This scheme is highly adaptive. When the fingers come into contact with a rigid object, the configuration is determined by the object geometry. By incorporating a torsional spring in parallel with the pulley, movement of a finger in free space could be determined by the radius of joint pulley, thus a more anthropomorphic behavior could be obtained through appropriate pulley radius distribution.

In order to provide adaptation between fingers, a sliding pulley based inter-finger underactuated mechanism is designed. Totally, two sliding pulleys are utilized to drive four fingers (one for the forefinger and the middle finger, the other for the ring finger and the little finger). This mechanism allows the force provided by the nut to be distributed among two output drive tendon. The actuation unit is linear stepper motor with lead screw output, which provide self-locking. By means of sliding pulley, adaptation between two fingers can be obtained.

In order to minimize the space and weight of the thumb, an innovative thumb using an external geneva and crank-slider mechanism is designed to realize flexion/extension, abduction/adduction movements with a single motor.

## 3      Electronics and Control

The control of a 22-DOF upper limb prosthesis is a challenging task, for huge control electronics is needed. To manage the complexity of the control system, a hierarchical architecture is presented, as shown in Fig. 5. The architecture consists of four levels: a user interface (a personal computer with an application GUI for diagnosis or simply buttons for amputees), a motion controller, a driver array and a motor array.

In prosthetic limb, position and force are two main variables to be controlled. In order to facilitate feedback control, the system incorporates rich sensor resources. Position sensors are in the form of digital hall effect sensors inside the brushless DC motors. Torque information in the arm is measured indirectly by the current sensing resistor in the three phase inverter of the servo driver, utilizing the fact that in BLDC motors current is proportional to armature current. Force sensing in the hand are achieved by means of force sensing resistors to provide force feedback.

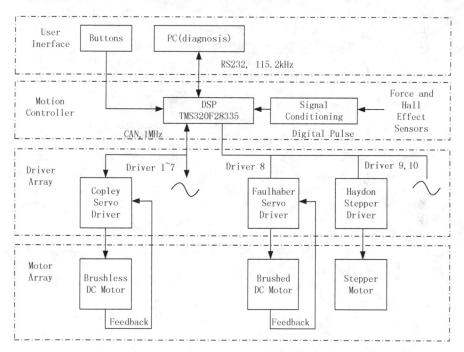

**Fig. 5.** The hardware architecture of the prosthetic limb

### 3.1      Prosthetic Electronics

The PC (personal computer) provides the user with a friendly GUI (Graphical User Interface), an application routine written in C++, including the control and supervision of the whole prosthesis. This interface allows the developers to diagnose the prosthesis and prosthetists to make adjustments easily, and helps with the fitting and training process. RS232 communication is used to perform data transmission, including control and status information, between the PC and the motion controller.

The motion controller is responsible for high level control, whose main function is to coordinate the movement of the hand and arm to perform a specific task. The motion controller and motor drivers in the arm are linked through the CAN (controller area network) bus, which is a high speed bidirectional communication. And, the motion controller communicate with the hand motor drivers through digital pulse. The reason we choose the CAN bus is to reduce electric wires and facilitate control and status information transmission. In order to implement highly reliable, real time control, a CAN-based high level application layer protocol CANopen is used. Various communication mechanisms including NMT (Network Management), SDO (Service Data Object) and PDO (Process Data Object) ensure the reliability the entire system. In addition, the tendon tension sensors, limit switches, initial position switches and buttons are all connected to the motion controller.

To achieve robust position control, servo drive electronics is added to improve the dynamics behavior of the motors. ACK-055-10 (Copley control Inc.) is a very compact digital servo driver, which consists of control and inverter electronics. The PWM frequency is 30 kHz, which minimize the current ripple of the inverter.

Digital hall effect sensors are also employed to implement initialization and protection of the hand, much like the functionality of digital hall effect sensors mounted in the arm. Digital limit switches are mounted to detect the mechanical ends of every joint in case of exceeding the range of motion. The switch is in the form of a unipolar digital hall effect position sensor (SS441A, Honeywell Inc.) and a rare earth pressed bar magnet (103MG5, Honeywell Inc.). Switches of this form, called initial position switches, are also used to facilitate initialization when the system is powered up. Grip force is utilized as the feedback to prevent the grasping force from exceeding the normal range. Each of the actuator in the hand is equipped with a force sensing resistor FSR400 (Interlink Electronics) located between the slider and the nut and the slider to measure the tendon tension. When the slider moves along the rail, it will exert a force on the sensor, inducing a change of resistance. Voltage dividers are built to convert the force into the voltage, and then goes through several signal conditioning circuits, including voltage buffers to reduce output impedence, RC filters to attenuate noise, clamp diodes to prevent the voltage from exceeding the range of the integrated analog-digital converter in the DSP.

Reliability is another important issue, which would be severely decreased if little attention is paid to noise and electromagnetic interference of the control electronics. Noise of the system mainly comes from parasitic impedence of long wires and large power devices. In order to eliminate these effects, bypassing and decoupling are carefully designed and isolation must be used between control and power electronics. Furthermore, high speed communication is rather susceptible to electromagnetic interference, which is mainly caused by high power devices like 220V domestic electricity, switching power supplies and servo drivers. Thus, it is crucial that reasonable layout of the electronic devices, routing, effective grounding and shielding.

### 3.2    Prosthetic Control Architecture

For a human body, the arm is usually used to reach the desired position and orientation as accurately as possible, while the hand is used to grasp the object with the ability to adapt to various objects. Different tasks determine different control strategy in each of them. The total control architecture of the prosthesis is shown in Fig. 6.

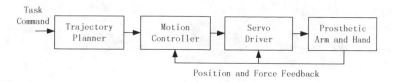

**Fig. 6.** The prosthetic arm control architecture

Instead of handling the complex dynamics of the prosthetic arm, we deal with the required torques to cause the motion as the disturbance torque of the individual servo control system. Varying disturbance torques would cause steady state tracking error, so PI (Proportional- Integral) controllers are adopted to improve the system perfor-mance. Each of the arm joint is equipped with a servo driver, which has the cascaded control architecture shown in Fig. 7.

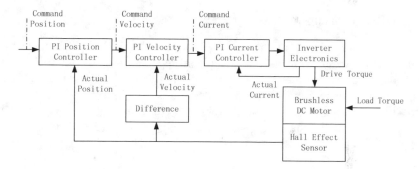

**Fig. 7.** Individual Servo Driver Control Architecture

Due to highly adaptive mechanical structure, control of the hand is significantly simplified. On-off control is needed to grasp and force feedback is used to avoid too large force applied on objects.

# 4     Experiment Validation

The capability of the IRR prosthesis to perform ADLs is characterized in terms of multi-mode hand grasping and the drinking water activity.

## 4.1     Multi-mode Hand Grasping

The capability of the hand is investigated by grasping objects with different sizes and shapes. Five postures of the grasp taxonomy described by Cutkosky[10], i.e. a disk grasp (e.g., for grasping a cap), a cylinder grasp (e.g., for grasping a bottle), a lateral pinch (e.g., for grasping a key), a tripod grasp (e.g., for grasping a tape) and a hook grasp(e.g., for grasping a bucket), are chosen to test the performance, shown in Fig. 8.

**Fig. 8.** IRR hand grasping experiment

The result shows that five grasping modes have been achieved successfully. With compliant mechanism, object with different shape can be grasped.

## 4.2    Drinking Water Activity

In drinking water experiment, the objective is to test the ability of the prosthetic arm-hand system to restore motion of the human upper limb. Frist, the VICON F20-6 motion capture system (Oxford Metrics Ltd.) is used to measure the real drinking trajectory of the upper limb of an experimenter, as shown in Fig. 9. Then, filtering and curve fitting algorithms are designed to obtain a more smooth curve, and joint angles are calculated. Finally, the data is imported into the motion controller, and trajectory is replicated by the prosthesis.

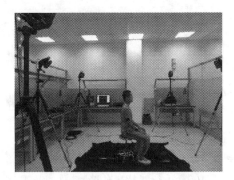

**Fig. 9.** Motion Capture Experiment

The overall trajectory consists of 247 joint angle values, and is executed in 10Hz, i.e. the whole activity costs 24.7s. A sequence of drinking water activity (first 10 seconds) is shown in Fig. 10. The prosthesis first grasps a bottle of water, and then feed it to an imaginary amputee successfully.

**Fig. 10.** Sequence of the IRR prosthetic limb drinking a bottle of water

## 5    Conclusion and Future Work

This paper presents a highly anthropomorphic prototype arm-hand prosthesis. Its design is based on physiological considerations and needs of emputees, e.g. small size, light weight, high functionality and good cosmetic appearance. A comparison between the developed IRR prosthesis and a male limb is shown in Fig. 1. Multi-mode grasping and drinking water experiment was made to validate the performance. The results show that the prosthesis could perform them conveniently and firmly.

Future work mainly includes two aspects: 1) further integration of the electronics into the prosthesis and 2) Introduction of myoelectric control, and vision-based hand-eye coordination.

**Acknowledgment.** The authors would like to thank Dazhu Xiong, Ligui Liu, Wenrui Chen, Shijie Yan and Heng Liu for their contributions to IRR prosthesis.

This work is partly funded by supported by the National Basic Research Program of China (973 Program, Granted No. 2011CB013301).

## References

1. China Disabled Persons' Federation: Communique on major statistics of the Second China National Sample Survey on Disability (2006)
2. Toledo, C., Leija, L., Munoz, R., Vera, A., Ramirez, A.: Upper limb prostheses for amputations above elbow: A review. In: Health Care Exchanges, PAHCE 2009, Pan American, pp. 104–108 (2009)

3. Grebenstein, M., Albu-Schäffer, A., Bahls, T., Chalon, M., Eiberger, O., Friedl, W., et al.: The DLR hand arm system. In: 2011 IEEE International Conference on Robotics and Automation (ICRA), pp. 3175–3182 (2011)
4. Adee, S.: The revolution will be prosthetized. IEEE Spectrum 46, 44–48 (2009)
5. Jacobsen, S., Iversen, E., Knutti, D., Johnson, R., Biggers, K.: Design of the Utah/MIT dextrous hand. In: Proceedings of the 1986 IEEE International Conference on Robotics and Automation, pp. 1520–1532 (1986)
6. Salisbury, J.K., Craig, J.J.: Articulated hands force control and kinematic Issues. The International Journal of Robotics Research 1, 4–17 (1982)
7. Brown, C.Y., Asada, H.H.: Inter-finger coordination and postural synergies in robot hands via mechanical implementation of principal components analysis. In: IEEE/RSJ International Conference on Intelligent Robots and Systems, IROS 2007, pp. 2877–2882 (2007)
8. Dechev, N., Cleghorn, W., Naumann, S.: Multiple finger passive adaptive grasp prosthetic hand, Mechanism and machine theory 36, 1157–1173 (2001)
9. Massa, B., Roccella, S., Carrozza, M.C., Dario, P.: Design and development of an underactuated prosthetic hand. In: Proceedings of the IEEE International Conference on Robotics and Automation, ICRA 2002, pp. 3374–3379 (2002)
10. Cutkosky, M.R.: On grasp choice, grasp models, and the design of hands for manufacturing tasks. IEEE Transactions on Robotics and Automation 5, 269–279 (1989)

# Synergistic Characteristic of Human Hand during Grasping Tasks in Daily Life

Mingjin Liu and Caihua Xiong

School of Mechanical Science and Engineering,
Huazhong University of Science and Technology, Wuhan, China
{mjliu,chxiong}@hust.edu.cn

**Abstract.** It is amazing for human to control highly complex hand with many degrees of freedom. To explore the mystery of hand, we use correlation analysis on human hand movement dataset, which is recorded from 33 kinds of grasping tasks in daily life, and obtain correlation relationships of all joints by hierarchical cluster analysis. The correlation relationships imply the feature of human hand movement. Thumb move relatively independently and other fingers move relatively synergistically during all grasping tasks. Moreover, DIP and PIP joints of all four fingers connect closer together than MCP joints. Before that work in this paper, we try to use dimensional reduction method, which is the main technique, to study the synergistic characteristic. It also supports the conclusion by the considerable inhomogeneity of index of RREV, which is raised to assess the error of each joint variable.

**Keywords:** Synergistic Characteristic, Human Hand, Rehabilitation.

## 1 Introduction

Human hand can grasp a variety of different objects and complete operational tasks excellently. It is a complex and hyper-redundant biomechanical system consisting of 27 bones, 31 muscles, and over 25 degrees of freedom (DOF). Merely considering the sign of the motion at each of the DOFs would yield millions possibilities combinations and it seems quite difficult to control [1]. Amazing, it is known that human beings are able to simply and flexibly control the highly complex hand. The dexterous characteristic of human hand reveals the mysteries of nature and has been a topic of interests to understand how the humans control their hand. In addition, understanding the characteristic of human hand was also the basic work for some applications such as biomechanics, hand surgery and rehabilitation because doctors and researchers could act appropriately to the situation only when known the intrinsic feature of human hand.

The notion of synergy is raised to interpret the great flexibility motor control of human. The typical work is the results of Santello et al. [2], who applied dimension reduction method on a variety of human hand postures data from everyday hand tasks. Their results showed that only 2 variables contained as much as 80% of hand posture information, suggesting that the human hand postures data was redundant and the

X. Zhang et al. (Eds.): ICIRA 2014, Part I, LNAI 8917, pp. 67–76, 2014.
© Springer International Publishing Switzerland 2014

movement of each finger was not independent but coordinate during grasp task. Thus, the motor control of hand posture can be described by a small number of synergies, which represent a collection of relatively independent degrees of freedom that behave as a single functional unit [3]. Although many researchers have done a lot of work on synergy, reviewed in [4], the main method to study the synergy is always dimensional reduction of all joints. But this method is based on the hypothesis that all joints of human hand exist pairwise synergistic relationship during grasping tasks. It is a pity that many researchers ignore the rationality of the assumption. In this paper, we will discuss this by analysis on the motion data of human hand in two aspects and obtain synergistic characteristic of human hand.

On the one hand, we try to follow the common way, which ignores the degree of synergistic relationship between all joints, to explore the synergistic of human hand. It is generally considered that each synergy covers all joints, and only fewer synergies are needed to remain because they account for most posture information. We consider that the evaluation criterion is not very exact because it is an overall evaluation containing all of the joint variables without specific composition by each joint variable respectively. As a result, it may not guarantee the functional realization of grasp tasks even though it account for most information of original data. So we raise a more detailed evaluation index to assess the effect of functional realization of using the common method which is based on all joints to analyze the synergy. According to result of our index, it is necessary to make clear the synergistic characteristic of human hand.

On the other hand, from perspective of biological characteristics of human hand, a synergy does not always mean the synchronous movement of all joints. This also requires us to make clear whether there exit synchronous movement between joints.

Therefore, this paper is composed as follow. Section 2 describes the experimental protocol of the motion dataset of human hand during grasping tasks in daily life. A more detailed evaluation index is raised in section 3 to assess the effect of functional realization. After the discussing about the result of the index, we discover it is required to study intrinsic relationship between joints. Section 4 conducts the research using correlation analysis and hierarchical cluster analysis on hand movement dataset and makes clear the synergistic characteristic of human hand.

## 2    Experimental Protocol

Motion data analysis of human hand was a feasible and widely used method to investigate the characteristic of human hand, because it was a noninvasive and repeatable way in healthy human subjects, compared to the other research methods like anatomy. Establishing human hand motion dataset was the foundation work. It should be concerned to ensure the movement dataset to be able to represent the daily tasks. We chose Feix taxonomy [5] as grasping tasks as it was a much more comprehensive and detailed organization of human grasps. Feix taxonomy (Fig. 1) had 33 different grasp types and allowed a better description of the human grasping capabilities compared to the common classification of Cutkosky [6].

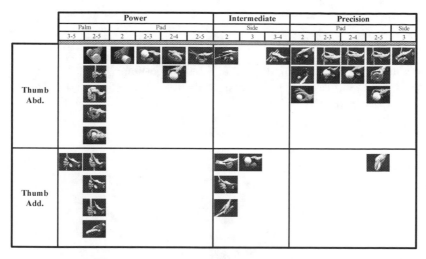

**Fig. 1.** Modified Feix grasp taxonomy with hand wearing glove

Subjects were instructed to shape the right hand to perform 33 different grasp tasks according to Feix grasp taxonomy, and every task contained 3 times with different size or shape objects. Common object were chose to assess the consequent modulation of hand posture. The objects contained cylinders, disks, spheres and cards with a large range of sizes.

Before each trial, subjects put hand near the object in a preset initial position (fingers extension in natural). Upon hearing the "go" command (in Chinese language), the subjects started to grasp the object at a comfortable speed and held the position for two seconds after completing the grasp task. Every trial was presented in pseudo-randomized way to avoid the systematic effects.

Twelve asymptomatic subjects (8 males and 4 females, 23 years old on average) participated in the experiment. All were right-handed.

A right-handed glove (Cyberglove, Virtual Technologies, Palo Alto, CA) with embedded sensors was used to measure the joint angles of the hand during grasping movement. Though CyberGlove could record 22 joint angles, we remained only 16 joints (as shown in Fig. 2), which were important for grasping function, and consisted of the metacarpal-phalangeal (MCP), proximal interphalangeal (PIP) and distal interphalangeal (DIP) joint angles for the four fingers (Index: I, Middle: M, Ring: R, Little: L), the carpo-metacarpal (CMC), metacarpal-phalangeal (MCP), interphalangeal (IP) and the abduction (ABD) between the thumb and the palm of the hand joint angles for the thumb (T). The output of the transducers was sampled every 0.02 second. The data acquisition system started to record the data on the "go" command, and it determined the onset of the movement. Due to the subject hold the hand for a moment after completing the task, the recorded data were not all movement data. We determine the termination of the movement where the sample data did not change in each trial.

All joint angular values at a moment specified hand posture

$$\mathbf{q}_j = \begin{bmatrix} q_{1j} & \cdots & q_{ij} & \cdots & q_{16,j} \end{bmatrix}^T \in \mathfrak{R}^{16 \times 1} \tag{1}$$

where $q_{ij}$ is the angular value of $i$-th joint in $j$-th hand posture.

For each subject, all of 33 grasp tasks formed a human hand grasping movement dataset matrix in daily life

$$\mathbf{Q} = \begin{bmatrix} \mathbf{q}_1 & \cdots & \mathbf{q}_j & \cdots & \mathbf{q}_n \end{bmatrix} \in \mathfrak{R}^{16 \times n} \tag{2}$$

where $\mathbf{q}_j$ is the angular value vector of 16 joints in $j$-th hand posture, n is the number of samples.

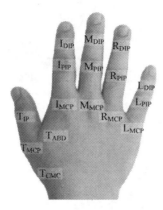

**Fig. 2.** The distribution of 16 joints

## 3    Assessing the Effect of Functional Realization

The concept of synergy is often quantified and defined through dimensionality reduction method. Although there are a number of dimensionality reduction techniques [7], principal component analysis (PCA) is a widely used tool because it's simple and understandable. Specifically, PCA method is aim to find, among linear combinations of the data variables, a sequence of orthogonal principal components (PCs) that most efficiently explain the variance of the original set of angular data[8]. Moreover, each PC represents simultaneous motion of the joint and indicates a basis movement function. Thus, the movement can be characterized by a compilation of PCs. Since the variance, which is explained by each successive PC, diminishes sharply, it is generally considered that only a little dominant PCs are needed to reconstruct the hand movement with closely approximate to the original hand postures. However, we sense that the conclusion may be not accurate.

As we all know, the evaluation index of the approximation between the reconstruction data and original data is the cumulative contribution rate of the reserved PCs, and the error can be expressed as the contributions of the abandon PCs. In addition, the error index is also equal to Relative Reconstruction Error (RRE), which represents the error of the whole joint variables. Generally speaking, to ensure satisfactory approximation, RRE should be lower than some threshold. Thus, RRE is also used to determine the number of PCs, which are needed to be reserved to reconstruct the most information of hand motion. RRE is defined by

$$RRE = \frac{\sum_{j=1}^{n}\left\|\tilde{\mathbf{q}}_j - \mathbf{q}_j\right\|^2}{\sum_{j=1}^{n}\left\|\mathbf{q}_j - \bar{\mathbf{q}}\right\|^2} \tag{3}$$

where $\bar{\mathbf{q}} = \begin{bmatrix} \bar{q}_1 & \cdots & \bar{q}_i & \cdots & \bar{q}_{16} \end{bmatrix}^T \in \Re^{16 \times 1}$ , $\bar{q}_i = \frac{1}{n}\sum_{j=1}^{n} q_{ij}$ , $\tilde{\mathbf{q}}_j$ is the reconstructed

angular value vector of 16 joints in $j$-th hand posture.

We use PCA method on our experimental dataset and obtain the RRE of reserving different number of PCs, as show in Fig. 3 (Subjects mean $\pm$ SD).

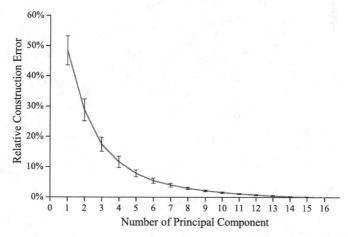

**Fig. 3.** Relative Reconstruction Error. It falls with the increase of the number of PCs. Error bars indicate the standard deviation across subjects.

In our results, when reserving the first two PCs, the RRE is less than 30%, slightly more than the results of Santello et al. discussed above. From this result merely, we may draw a conclusion that just two or a little more PCs are needed to be reserved to reproduce overwhelming information of hand posture and each PC can represent a fixed synergistic relationship. However, the index of RRE is an overall evaluation containing all of the joint variables without specific composition by each joint variable respectively. Therefore, we want to discuss the error of each joint variable, and give the individual evaluation index, Relative Reconstruction Error of Variable (RREV). It is displaced by

$$RREV = \frac{\sum_{j=1}^{n}\left\|\tilde{q}_{ij} - q_{ij}\right\|^{2}}{\sum_{j=1}^{n}\left\|q_{ij} - \overline{q}_{i}\right\|^{2}} \tag{4}$$

where $\tilde{q}_{ij}$ is the reconstructed angular value of $i$-th joint in $j$-th hand posture.

In condition of reserving the first two PCs, the RRE is so small that we use the individual evaluation index to extend the RRE to get the composition of each joint variable, and the result of RREV is showed in Fig. 4.

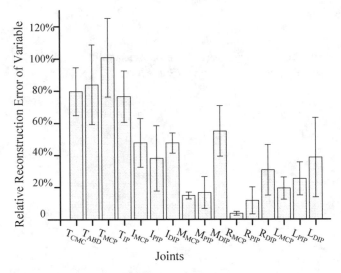

**Fig. 4.** Bars diagram representing Relative Reconstruction Error of each joint variable averaged over subjects. Error bars indicate the standard deviation across subjects.

The bar graph shows that the RREV of each joint is quite different. While the errors of some joints are very small, such as RMCP with only 3.66%, others are quite large, especially, errors of thumb joints of TCMC, TABD, TMCP and TIP reach up to 80%, or even 100%, which are unacceptable because thumb is very important of function realization in clinical studies. Such large errors indicate that reserving the first two PCs cannot reproduce human hand function although the overall error (RRE) is small. Therefore, both overall error and individual errors are needed to be considered to evaluate the effect of reproducing the human hand movement. More important is that the considerable inhomogeneity of RREV of each joint also implies that there exist some intrinsic features between all joint variables. In addition, when quantitatively define the synergies of all joints by dimensional reduction method, it does not qualitatively study whether some joints may be synchronous movement. In a word, it should make clear the synergistic characteristic of human hand.

# 4     Correlation Analysis of Human Hand Movement

A synergy means simultaneous movement between joints. If two joint variables move synchronously, we consider they are correlated and the degree of correlation could be quantitatively described by correlation coefficient on the math. The larger value of correlation coefficient between two joint angular values represents the more simultaneous movement of the two joints.

Ingram et al. [9] had calculated correlation coefficient to analyze the correlations for heterogeneous joint pairs of the fingers, and found strong correlation between PIP and DIP joints and weak between the MCP and PIP joins and MCP and DIP joints. We expand the correlation coefficient to all joints. Our dataset of human hand movement have 16 joint variables and get a correlation coefficient matrix $\mathbf{R} \in \mathfrak{R}^{16 \times 16}$ by every two joints. A possible investigational approach is to extend to study all the correlation coefficients. However, these studies are limited to the relationship of every two joint variables, and can't give a clear and intuitive structural description of correlation relationship of all variables.

Hierarchical cluster analysis is applied to obtain the structural description of all joint variables. Cluster analysis is an exploratory statistical method which aims to classify several data objects into some groups (clusters) according to similarities between them. Correlation coefficient is taken as similarity between two joint variables and the clustering is based on variables. Cluster analysis has different algorithms based on different tasks. We employ hierarchical cluster analysis as it's a well technique for data interpretation. Hierarchical cluster can construct a tree structure (dendrogram) of joint variables by agglomerative approach. Each joint variable starts out as a separate cluster. In successive step of the agglomerative algorithm, the two similar clusters are merged together based on similarity measures in subsequent steps and the total number of clusters is decreased by one. These steps can be repeated until one cluster remained. In each step, similarity measure between two clusters should be defined which is also needed to recalculated. Popular choices are known as single-linkage method to measure similarity between two clusters, which is represented of the similarity between the closest pair of joint variable belonging to different clusters. Thus, similarity measure between cluster U and V can be given by

$$r_{UV} = \max\left\{ r_{ij} \, \middle| \, i \in U, j \in V \right\} \tag{5}$$

where $r_{ij}$ is the correlation coefficient between $i$-th element of cluster U and $j$-th element of cluster V.

The result of hierarchical cluster can be represented using a dendrogram, such as Fig. 4. The top row of nodes is a cluster which includes all other clusters, the remaining nodes represent the clusters to which the data belong and the bottom row are single element clusters representing joint variables. Vertical axis indicates the similarity between clusters. The similarity between merged clusters is decreasing with the level of the merger, and the height of each node in the plot is proportional to the value of the intergroup dissimilarity between its 2 branches. In other words, if the height of a node is lower, the joint variables of its two branches are more similarity.

Correlation coefficient is regarded as the similarity measure between two joint variables in the clustering. Since correlation coefficient also measures the degree of simultaneous movement between two joints, the plot (Fig. 5) can represent graphically not only the structural description of the degree of correlation relationship between all of joint variables, but also the movement feature of joint variables of human hand.

Fig. 5 represents the average result across subject. This plot shows that the four joints of thumb (TCMC, TMCP, TABD, TIP) are separated from all the joints of four fingers (index, middle, ring, little) obviously. This result also verifies the independent of thumb by statistical data analysis on human hand movement. The conclusion is similar to the works of Häger-Ross [10] and Ingram [9], but we give a novel way, which is based on all of the joint variables, to analysis the correlations and receive an intuitive description. Moreover, it is just because there are low correlated between thumb joints and other four fingers, the reconstructed posture information by 2 variables (as showed in Fig. 4) is more representative of most correlated joints, namely, the joints of four fingers. As a result, the errors of thumb joints are very high. Therefore, it is not necessary to put all the joints together to study synergy. In other words, using dimensional reduction method to study the synergies of all joints is not appropriate. On the contrary, the joints of thumb should be separated from other joints to analyze the synergistic relationship.

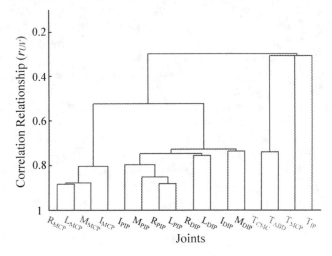

**Fig. 5.** Dendrogram of the clustering of joints. It graphically shows the degree of correlation relationship among joints. The lower of the branch points of the tree indicates the more correlated between the 2 branches.

In addition, an interesting conclusion from Fig. 5 is that all joints of four fingers can be separated 2 groups: MCP joints, PIP and DIP joints. It means that there are strong correlation between the PIP and DIP joints and weak between the MCP and PIP joins or MCP and DIP joints. Compared to the similar work of Ingram [9], our conclusion is based on all fingers and the discovery represents more universal rule of human hand movement. Recognizing the correlation between DIP and PIP joint of each finger, Liu et al. [11] couple middle phalanx and distal phalanx in design each finger of DLR/HIT

II. However, using our findings that all the DIP and PIP joints of four fingers are correlated and can be classified as a group, all DIP and PIP joints of four fingers could be coupled so that only one actuator is needed. Thus, exploring the synergistic characteristic of human hand has not only the significant of science, but also more perfect application value.

## 5    Conclusion

In this paper, we use correlation analysis and hierarchical cluster analysis on human hand movement dataset to get the degree of correlation relationships of all joints. And such correlation relationships imply the feature of human hand movement, thumb move relatively independently and other fingers (Index, Middle, Ring, Little) move relatively synergistically during all grasping tasks. Moreover, DIP and PIP joints of all four fingers connect closer together than MCP joints. The characteristics can explain considerable inhomogeneity of RREV and make researchers concern more local relationship between joints. Moreover, it promotes the awareness of human hand by the dendrogram of the clustering of joints, which provides an intuitive description of relationship of all joints.

**Acknowledgements.** This work was partly supported by the National Basic Research Program of China (973 Program, Grant No. 2011CB013301), National Natural Science Foundation of China (Grant No. 51335004) and National Science Fund for Distinguished Young Scholars of China (Grant No. 51025518).

## References

1. Soechting, J.F., Flanders, M.: Flexibility and repeatability of finger movements during typing: Analysis of multiple degrees of freedom. Journal of Computational Neuroscience 4, 29–46 (1997)
2. Santello, M., Flanders, M., Soechting, J.F.: Postural hand synergies for tool use. The Journal of Neuroscience 18, 10105–10115 (1998)
3. Turvey, M.T.: Action and perception at the level of synergies. Human Movement Science 26, 657–697 (2007)
4. Santello, M., Baud-Bovy, G., Jörntell, H.: Neural bases of hand synergies. Frontiers in Computational Neuroscience 7 (2013)
5. Feix, T., Pawlik, R., Schmiedmayer, H., Romero, J., Kragic, D.: A comprehensive grasp taxonomy. In: Robotics, Science and Systems: Workshop on Understanding the Human Hand for Advancing Robotic Manipulation, pp. 2–3 (2009)
6. Cutkosky, M.R.: On grasp choice, grasp models, and the design of hands for manufacturing tasks. IEEE Transactions on Robotics and Automation 5, 269–279 (1989)
7. Tresch, M.C., Cheung, V.C., D'Avella, A.: Matrix factorization algorithms for the identification of muscle synergies: evaluation on simulated and experimental data sets. Journal of Neurophysiology 95, 2199–2212 (2006)

8. Ciocarlie, M., Goldfeder, C., Allen, P.: Dimensionality reduction for hand-independent dexterous robotic grasping. In: IEEE/RSJ International Conference on Intelligent Robots and Systems, IROS 2007, San Diego, CA, pp. 3270–3275 (2007)
9. Ingram, J.N., Körding, K.P., Howard, I.S., Wolpert, D.M.: The statistics of natural hand movements. Experimental Brain Research 188, 223–236 (2008)
10. Häger-Ross, C., Schieber, M.H.: Quantifying the independence of human finger movements: comparisons of digits, hands, and movement frequencies. The Journal of Neuroscience 20, 8542–8550 (2000)
11. Liu, H., Wu, K., Meusel, P., Seitz, N., Hirzinger, G., Jin, M.H., Liu, Y.W., Fan, S.W., Lan, T., Chen, Z.P.: Multisensory five-finger dexterous hand: The DLR/HIT Hand II. In: IEEE/RSJ International Conference on Intelligent Robots and Systems, IROS 2008, pp. 3692–3697 (2008)

# UKF-SLAM Based Gravity Gradient Aided Navigation

Meng Wu[1] and Ying Weng[2,*]

[1] Institute of Pattern Recognition & Artificial Intelligence,
Huazhong University of Science & Technology, P.R. China
[2] School of Computer Science, Bangor University, United Kingdom
wumenghust@163.com, y.weng@bangor.ac.uk

**Abstract.** Considering the two characteristics: (1) simultaneous localization and mapping (SLAM) is a popular algorithm for autonomous underwater robot, but visual SLAM is significantly influenced by weak illumination; (2) geomagnetism-aided navigation and gravity-aided navigation are equally important methods in the field of robot navigation, but both are affected heavily by time-varying noises and terrain fluctuations; however, gravity gradient vector can avoid the influence of time-varying noises, and is less affected by terrain fluctuations. To the end, we proposes a UKF-SLAM based gravity gradient aided navigation in this paper with the following advantages: (1) the UKF-SLAM is an efficient way to avoid linearization errors compared with the EKF-SLAM; (2) it improves the accuracy of navigation system without the help of any geophysical reference map; (3) it is suitable for a robot to navigate itself under the environment of weak illumination and time-varying disturbances. Experimental results also show that our proposed method has a less localization error than the SLAM-based geomagnetic aided navigation.

**Keywords:** UKF-SLAM, gravity gradient aided navigation, geophysical reference map, geophysical navigation.

## 1 Introduction

The simultaneous localization and mapping (SLAM) algorithm was first proposed by Smith and Cheeseman in 1988 to provide localization and map building for mobile robots, and first used for an unmanned underwater vehicle (UUV) navigation in September 1997 in a collaborative project between the Naval undersea Warfare Center (NUWC) and Groupe d'Etudes Sous-Marines de l'Atlantique (GESMA). The objective of using SLAM was to get a UUV starting in an unknown location and without previous knowledge of the environment to build a map via its onboard sensors and then use the same map to compute the robot's location. In general, SLAM is widely used in visual navigation because vision is the richest source of information from our environment. The visual SLAM techniques can be classified into using Stereo camera and Monocular camera. It is clear that visual SLAM has its disadvantages in mobile robot, which is affected greatly by illumination and camera

---

\* Corresponding Author.

X. Zhang et al. (Eds.): ICIRA 2014, Part I, LNAI 8917, pp. 77–88, 2014.

pose [1]. If a camera is in an environment with poor illumination, visual signals cannot be exactly obtained to be used as landmarks in visual SLAM, and a large measurement bias from camera cannot guarantee that the unscented Kalman filter (UKF)-SLAM has a good convergence. Finally, the estimated location cannot be accurate enough. Especially, in the underwater environment, visual signals from stereo cameras are easily distracted by weak illumination, which brings large measurement errors into UKF-SLAM. Comparing with the visual SLAM method, geophysical navigation approach is less affected by weak illumination, and is more suitable for the autonomous underwater vehicle to acquire the navigation updates by surfacing or nearly surfacing [2][3]. Therefore, SLAM based on geophysical information is an alterative way in cases of visual signal drop-off. Liu et al. introduced a kind of SLAM-based geomagnetic aided inertial navigation method to improve the accuracy of navigation and localization system [4]. Wang et al. considered a fact that it is difficult for a geomagnetism matching guidance system to acquire a high precision geomagnetic map under the environment of various time-varying and electromagnetic disturbances so combines geomagnetic navigation algorithm with SLAM together to realize an autonomous navigation system [5]. However, time-varying disturbances have influenced a lot to the measurement errors of geomagnetic navigation system. Such errors lead to the divergence of UKF-SLAM based on geomagnetic information from fluxgate sensor. In particular, electromagnetic interference is a major disturbance in the process of geomagnetic navigation, which brings noises into measurement values of fluxgate sensor and UKF-SLAM cannot have a good convergent character.

Considering various kinds of time-varying noises and terrain fluctuations, and especially the variations in gravity or geomagnetism are insufficient in some areas, we add gravity gradient information into UKF-SLAM to propose a SLAM-based gravity gradient aided inertial navigation method. In our method, the gravity gradient values of landmarks are measured by gradiometer along the trajectory of an underwater robot. When the robot is moving along a definite trajectory, the gradiometer measures gravity gradient values of landmarks around the trajectory according to gravity gradient values, and combing a method from [3], it is easy to obtain localizations of such landmarks. Then localization information of landmarks is combined with the localization and velocity state of underwater robot in order to let the state matrix of UKF-SLAM to be an augment state matrix. Finally, the augment state input matrix is as the input state matrix of UKF-SLAM. The prediction and update of UKF-SLAM is to find out a series of optimal estimations for the underwater robot in different discrete time. According to the localizations of landmarks and estimated localizations of UKF-SLAM, the robot could build a map by such information. Thus, without the help of any reference maps, the underwater robot can navigate itself by UKF-SLAM.

This paper is organized as follows. In Section 2, the state and measurement models of a navigation system are introduced and analyzed. In section 3, the UKF-SLAM algorithm is presented. In Section 4, experiment and stimulation results are discussed. Conclusions and future work are summarized in Section 5.

# 2   Process Model and Measurement Model in the UKF-SLAM Algorithm

## 2.1   Process Model

The process and observation models of the autonomous underwater vehicle system exhibit significant nonlinearities. When the vehicle is in a state of motion, the values of the gravity gradient are measured at discrete points simultaneously using gravity gradiometer fixed on underwater robots. $x_t$ is the system state of UKF-SLAM

$$\mathbf{X}_t = [r_{xt}, r_{yt}, v_{xt}, v_{yt}, \mathbf{p}_{it}]^T \tag{1}$$

where $r_{xt}$ and $r_{yt}$ are the coordinates of the underwater robot's position in the Cartesian reference frame, respectively; $v_{xt}$ and $v_{yt}$ are the velocity components of the underwater robot in the $x$ and $y$ directions, respectively; and $\mathbf{p}_{it}$ denotes the i-th landmark's position information at the time $t$

$$\mathbf{p}_{it} = \begin{bmatrix} \mathbf{x}_{it} \\ \mathbf{y}_{it} \end{bmatrix} \tag{2}$$

where $\mathbf{x}_{it}$ and $\mathbf{y}_{it}$ stand for the i-th landmark's position information at the time t in the direction of x and y , respectively. In the underwater environment, the method in [3] is a good way to obtain positions of such landmarks by a gravity gradiometer. After calculating the gravity gradient vectors by gradiometer, the position $\mathbf{p}_{it}$ of a landmark is easy to be calculated. In this paper, the position of a static landmark varies little in the different time $t$

$$p_{i(t+1)} = p_{it} = p_i, i = 1, 2, \cdots\cdots, N \tag{3}$$

The state process model is illustrated as

$$x_{t+1} = \Phi(t)x_t + n_k \tag{4}$$

where $\Phi(t) = \begin{pmatrix} 1 & 0 & 0 & \Delta t & 0 \\ 0 & 1 & 0 & 0 & \Delta t \\ 0 & 0 & 1 & 0 & 0 \\ 0 & 0 & 0 & 1 & 0 \\ 0 & 0 & 0 & 0 & 1 \end{pmatrix}$ is a state transition matrix, the process noise is $n_k$

which is an additive, zero-mean Gaussian noise with covariance $R_k$ , i.e., $\varepsilon_k \sim N(0, R_k)$, and $\Delta t$ is the time-step.

## 2.2    Measurement Models

Using gradiometer, the gravity gradient values in different landmarks can be obtained individually

$$Z_t = h(x_t) + \delta_k \tag{5}$$

where $\delta_k$ is an additive, zero-mean Gaussian noise with covariance $Q_k$, i.e., $\delta_k \sim N(0, Q_k)$. The observation $Z_t$ is a function h of the current state corrupted by additive Gaussian noise $\delta_k$ with covariance $Q_k$.

In general, gravity gradient could be used to detect abnormal objects in underwater environment. Gravity gradient anomalies of partial area, which are caused by a landmark, can be measured by a gravity gradiometer. Then anomalies can be inversed with a gravity gradient inversion algorithm. Finally, the mass and position of an object can be estimated [3].

According to the principle mentioned in [3], the position information of a static landmark along a trajectory of underwater robot can be calculated

$$\begin{cases} \theta(x,y,z) = \arccos(\dfrac{\Delta z(x,y,z)}{R(x,y,z)}) = \arccos\sqrt{\dfrac{1}{(\dfrac{\Gamma_{xy}(x,y,z)}{\Gamma_{yz}(x,y,z)})^2 + (\dfrac{\Gamma_{xy}(x,y,z)}{\Gamma_{xz}(x,y,z)})^2 + 1}} \\[2em] \varphi(x,y,z) = \arctan(\dfrac{\Delta y(x,y,z)}{\Delta z(x,y,z)}) = \arctan(\dfrac{\Gamma_{yz}(x,y,z)}{\Gamma_{xz}(x,y,z)}) \\[2em] R(x,y,z) = \sqrt[3]{\dfrac{GM}{\Gamma_{xx}(x,y,z) + \Gamma_{yy}(x,y,z)}(1 - \dfrac{3}{(\dfrac{\Gamma_{xy}(x,y,z)}{\Gamma_{yz}(x,y,z)})^2 + (\dfrac{\Gamma_{xy}(x,y,z)}{\Gamma_{xz}(x,y,z)})^2 + 1})} \end{cases} \tag{6}$$

where $\Gamma_{ij}(x,y,z)$ denotes the gravity gradient value in a landmark with the coordinate x, y and z, respectively; $\theta(x,y,z)$ and $\varphi(x,y,z)$ stand for the orientation of a landmark individually; $R(x,y,z)$ is a distance between the gradiometer and a landmark which is around the trajectory of an underwater robot; $\Delta y(x,y,z)$ and $\Delta z(x,y,z)$ are the distances between a landmark and a gravity gradiometer in the y and z directions, respectively; $M$ is the mass of the landmark. G is the gravitation constant; $\Gamma_{xx}(x,y,z)$, $\Gamma_{xy}(x,y,z)$, $\Gamma_{xz}(x,y,z)$, $\Gamma_{yz}(x,y,z)$ and $\Gamma_{yy}(x,y,z)$ are gravity gradient tensor components which are caused by a real landmark. The five tensor components are measured by gravity gradiometer. Assuming the position of a gradiometer onboard is known, according to different gravity gradient vectors such as $\Gamma_{xy}$, $\Gamma_{yz}$, the position of a landmark is easy to be obtained by (6). In general, landmarks around a trajectory of underwater robot is static, the varying gravity gradient tensor responses such as $\Gamma_{xy}$ at the position of the gravity gradiometer could be measured synchronously. Substituting these responses into (6), the orientation and position information of the landmarks can be calculated with the mass of landmarks.

Assuming that we do not know the exact position of a landmark but know the accurate position information of gravity gradiometer onboard in advance, using (6), it is easy to calculate relative position between gradiometer and a landmark. After obtaining position information of different landmarks along the route of underwater robot and combing the position information of underwater robot, the robot can navigate itself by UKF-SLAM and update the estimated state.

## 3    UKF-SLAM Algorithm Based on Gravity Gradient Measurement Information

A block diagram of UKF-SLAM based on gravity gradient algorithm is shown in Fig. 1. Gradiometer measures gravity gradient values along the route of underwater vehicle and obtains different gravity gradient information from different landmarks. According to the method mentioned in [3], the positions of landmarks around the trajectory of underwater robot can be calculated by gradiometer onboard at the same time. Gyroscope and Accelerometer provide the state information of underwater robot. Based on the state and measurement information of the robot and different landmarks around the trajectory, UKF-SLAM is used to update the state of this underwater robot navigation system. Then the underwater robot adopts positions of landmarks and state of its movement along a definite trajectory to locate itself and build a real environmental map simultaneously and realize autonomous navigation.

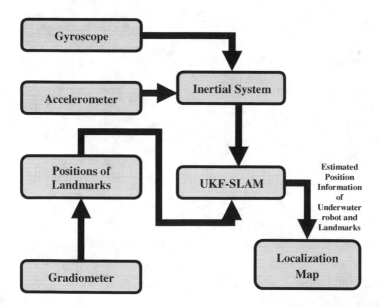

**Fig. 1.** The block diagram of gravity gradient navigation system based on UKF- SLAM

A solution to the SLAM problem using UKF, with many interesting theoretical advantages, is extensively described in the research literature. Compared with extended Kalman filter (EKF)-SLAM, UKF-SLAM is more accurate than EKF-SLAM for estimation of nonlinear systems. EKF for linearization of nonlinear system only uses the first term of Taylor series. Therefore, in the cases where the system is fully nonlinear, higher terms in the Taylor series are fully nonlinear, and higher terms in the Taylor series are important, linearization model by this method is not a good approximation of the real system, which leads to lack of accuracy in the estimation of states or parameters of these types of systems. In UKF-SLAM, there is no need to calculate the Jacobian matrix. The UKF [6] is a nonlinear filter based on the unscented transform (UT). For nonlinear systems, the UKF avoids linearization of the state and measurement equations. Additionally, the UKF principle is simple and easy to implement as it does not require the calculation of Jacobian matrix at each time step, and the UKF is accurate up to second order moments in the probability distribution function propagation whereas the EKF is accurate up to first order moment. UKF-SLAM has two steps, i.e., prediction and update. The measured sensor data of an underwater robot is from gravity gradiometer.

The fundamental UKF algorithm is expressed as below:

(a) Initialization

$$\begin{cases} \hat{x}_0 = E[x_0] \\ P_0 = E[(x_0 - \hat{x}_0)(x_0 - \hat{x}_0)^T] \end{cases} \tag{7}$$

where $x_0$ is the initial system state at time 0, $P_0$ is the initial state covariance matrix, and $\hat{x}_0$ is the mean of $x_0$.

(b) Calculate sigma points with corresponding weights

$$\begin{cases} (\chi_t)_0 = \hat{x}_t \\ (\chi_t)_i = \hat{x}_t + (\sqrt{(L+\lambda)P_t})_i & i = 1,...,L \\ (\chi_t)_i = \hat{x}_t - (\sqrt{(L+\lambda)P_t})_{i-L} & i = L+1,...,2L \\ W_0^{(m)} = \lambda/(L+\lambda) \\ W_0^{(c)} = \lambda/(L+\lambda)+(1-\alpha^2+\beta) \\ W_i^{(m)} = W_i^{(c)} = 1/\{2(L+\lambda)\} & i = 1,...,2L \\ \lambda = \alpha^2(L+\beta)-L \end{cases} \tag{8}$$

Equation (8) is a kind of UT for calculating the statistics of a random variable which undergoes a nonlinear transformation. $\hat{x}_t$ is the mean of the system state, and $P_t$ is the covariance matrix of system state. An L-dimensional random variable can be approximated by a set of 2L+1 weighted sigma points $(\chi_t)_i$ given by (8). In general, the UT performs this approximation by extracting so-called sigma points $(\chi_t)_i$ from the Gaussian estimate and passing them through (4). $\alpha$ and $\beta$ are the coefficients that with their adjustment, the estimation error can be minimized and their values influence on the error rate resulting from higher terms in the Taylor series. $(\sqrt{(L+\lambda)P_t}\,)_i$ is the i-th column of the matrix square root, $W_i^{(m)}$ is a weight parameter associated with state equation and $W_i^{(c)}$ is a weight parameter associated with measurement equation. $W_0^{(m)}$ is the initial parameter of $W_i^{(m)}$ on the condition that $i = 0$. $W_0^{(c)}$ is the initial parameter of $W_i^{(c)}$ on the condition that $i = 0$. $\lambda$ is a scaling parameter. $(\chi_t)_i$ is the i-th sigma point in time $t$.

(c) Time update

$$
\begin{cases}
(\chi_{t+1|t})_i = \Phi[(\chi_t)_i] \\
\hat{x}_{t+1|t} = \sum_{i=0}^{2L} W_i^{(m)}(\chi_{t+1|t})_i \\
P_{t+1|t} = \sum_{i=0}^{2L} W_i^{(c)}[(\chi_{t+1|t})_i - \hat{x}_{t+1|t}][(\chi_{t+1|t})_i - \hat{x}_{t+1|t}]^T \\
(Z_{t+1|t})_i = H[(\chi_{T+1|t})_i] \\
\hat{Z}_{t+1|t} = \sum_{i=0}^{2L} W_i^{(m)}(Z_{t+1|t})_i
\end{cases}
\tag{9}
$$

where each sigma point $(\chi_t)_i$ is propagated through (4) in order to obtain the predicted mean $\hat{x}_{t+1|t}$ and covariance matrix $P_{t+1|t}$. A new set of sigma points $(\chi_{t+1|t})_i$ is also calculated by (9). At the same time, a new set of measurement sigma points $(Z_{t+1|t})_i$ is calculated by (5). $(Z_{t+1|t})_i$ is used to obtain the observation mean matrix $\hat{Z}_{t+1|t}$.

(d) Measurement update

$$
\begin{cases}
P_{z_{t+1|t}z_{t+1|t}} = \sum_{i=0}^{2L} W_i^{(c)}[(z_{t+1|t})_i - \hat{z}_{t+1|t}][(z_{t+1|t})_i - \hat{z}_{t+1|t}]^T \\
P_{x_{t+1|t}z_{t+1|t}} = \sum_{i=0}^{2L} W_i^{(c)}[(\chi_{t+1|t})_i - \hat{x}_{t+1|t}][(\chi_{t+1|t})_i - \hat{x}_{t+1|t}]^T \\
\kappa_{t+1} = P_{x_{t+1|t}z_{t+1|t}} P_{z_{t+1|t}z_{t+1|t}}^{-1} \\
\hat{x}_{t+1} = \hat{x}_{t+1|t} + \kappa_{t+1}(z_{t+1} - \hat{z}_{t+1|t}) \\
P_{t+1} = P_{t+1|t} - \kappa_{t+1} P_{z_{t+1|t}z_{t+1|t}} \kappa_{t+1}^T
\end{cases}
\tag{10}
$$

where the innovation covariance matrix is $P_{y_{t+1|t}y_{t+1|t}}$, the cross correlation matrix is $P_{x_{t+1|t}y_{t+1|t}}$, and the Kalman gain is $\kappa_{t+1}$. The final estimated state is $x_{t+1}$ and the estimated covariance matrix is $P_{t+1}$. $z_{t+1}$ is a real measurement value from gravity gradiometer which can be calculated by (5).

# 4     Simulation and Experimental Results

To improve effectiveness of the proposed algorithm, the simulation is implemented with constraints on velocity, system noise, observation noise, etc. The gravity gradient data arise from some real measurement data measured by a gravity gradiometer in the Sea of Japan. The details of the data are shown in Fig.2 and Table1.

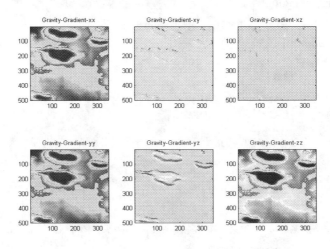

**Fig. 2.** Gravity-gradient data in the Sea of Japan

**Table 1.** Parameters of Gravity Gradient Data in the Sea of Japan

| Range of Longitude | Range of Latitude |
|---|---|
| [140.85°~ 141.1325°] | [39.70°~ 39.9825°] |

**Table 2.** Stimulation Parameters of UKF-SLAM

| Parameters of Stimulation | Values |
|---|---|
| $r_x$ ( $m$ ) | $100 \sim 200$ |
| $r_y$ ( $m$ ) | 300 |
| $v_x$ ( $m/s$ ) | 15 |
| $v_y$ ( $m/s$ ) | 0 |
| $R_k$ ( $E^2$ ) | $10^{-4}$ |
| $Q_k$ ( $E^2$ ) | $10^{-2}$ |

In the simulation, we assume that the state covariance $R_k$ and measurement covariance $Q_k$ are Gaussian White noises. In practice, gravity gradiometer is a kind of high-accuracy measurement sensor, and its measurement covariance $Q_k$ is not necessary to be set as a big value. The movement state of underwater robot is stable so the state covariance $R_k$ is also a small value in order to guarantee that the velocity of underwater robot is with a stable state.

Fig.3 clearly shows the effectiveness of UKF-SLAM combined with gravity gradient information. The blue curve is the real trajectory of underwater robot, and green trajectory is the estimated localization result of UKF-SLAM. The blue stars stand for real localizations of landmarks, and red stars stand for the predicted localizations of such landmarks using UKF-SLAM. It is clear that there are some biases between the blue and red stars, which arise from measurement errors of gravity gradiometer and fluctuations of terrain in underwater environment. In fact, the fluctuations of underwater terrain have an impact on the gradiometer. How to avoid the negative influence from fluctuations of terrain is research task in the future work.

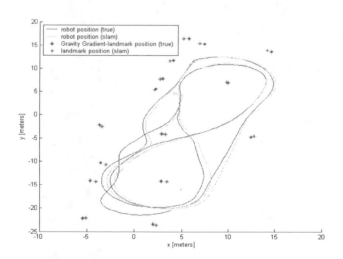

**Fig. 3.** Localization results using UKF-SLAM based on gravity gradient information

As an underwater robot is moving along its definite trajectory, the gradiometer starts to measure gravity gradient values of landmarks along such trajectory. After obtaining gravity gradient values of these landmarks, the measurement equation (10) is used to get localization information of these landmarks. The localization values of landmarks are aggregated with the movement state of underwater robot to form an augment state equation (1). Then UKF-SLAM combines the state equation (1) and measurement equation (9) to locate the underwater robot itself without the help of any reference map.

In Fig.4, the red and green curves denote X and Y localization errors in discrete time, respectively. The blue curve is the average of X and Y localization errors. The fluctuation trends of the three curves are related to disturbances from gradiometers. In general, the noises arising from measurement of gradiometer bring some outliers into UKF-SLAM, which leads to an uncertainty of measurement model. The outliers in gravity gradiometer are usually considered as Gaussian Noises. Although the outliers exist in the system of UKF-SLAM based on gravity gradient, Fig.4 shows that the average localization error is within a controllable level. Therefore, UKF-SLAM is an efficient way to deal with uncertainty of measurement model.

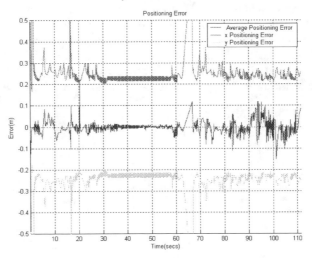

**Fig. 4.** Localization errors of UKF-SLAM based on gravity gradient information.

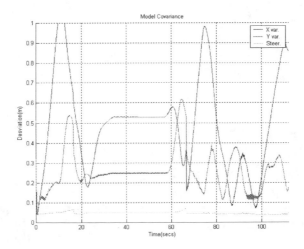

**Fig. 5.** The variety of state model in the movement of underwater robot using UKF-SLAM based on gravity gradient information

Fig.5 illustrates that some uncertainty of state model arising from noises in underwater environment bring several fluctuations into the state model of underwater robot. When the underwater robot is moving with increased time t, the localization information of underwater robot is updated by UKF-SLAM, and gradiometer measures gravity gradient values along the trajectory in order to get new localization information in different landmarks, which adds complexity into the state model. As a result, the red and blue curves show a period of fluctuation in x and y direction of state model, respectively. The green curve denotes the attitude information of underwater robot. It is obvious that the attitude information changes little because the attitude of underwater robot is mostly dependent on the stability of Inertial System, and is less affected by accuracy of gravity gradiometer and UKF-SLAM.

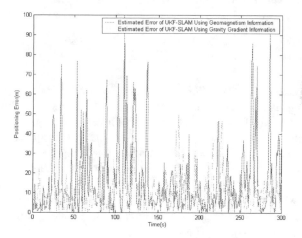

**Fig. 6.** Comparison between two UKF-SLAM Using different geophysical landmarks

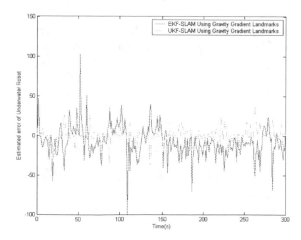

**Fig. 7.** Comparison between EKF-SLAM and UKF-SLAM based on gravity gradient information

From Fig.6, it is obviously seen that gravity gradient landmarks are more suitable to be used in UKF-SLAM compared with geomagnetic landmarks. In general, geomagnetic signals are affected heavily by time-varying noises while Gravity Gradient landmarks are keeping stable under time-varying disturbances. Based on such consideration, gravity gradient landmarks are more effective in UKF-SLAM.

Generally, EKF is less effective in dealing with uncertainty of state model and measurement equation in the process of underwater navigation. Considering various noises and disturbances from underwater environment, EKF-SLAM loses convergence due to uncertainty of measurement equations. Especially, the state covariance matrix loses positive definite character under big measurement noises. Compared with EKF-SLAM, UKF-SLAM is a better way to deal with uncertainty in measurement equation and more robust against different time-vary noises. The Fig. 7 shows a good convergence of UKF-SLAM.

## 5      Conclusions

This paper proposes a kind of UKF-SLAM method and combines gravity gradient values in different landmarks along the trajectory of underwater robot to locate the robot itself. Based on the experimental results, it proves that UKF-SLAM with gravity gradient information is feasible under harsh and real-time underwater condition. Especially, compared with geomagnetism navigation using SLAM, gravity gradient vectors are less affected by time-varying disturbances, and UKF is a better way to tackle uncertainty of state and measurement models.

In the future work, one important issue is how to improve the robustness of UKF-SLAM against fluctuations of underwater terrain, the other is how to improve the measuring accuracy of gravity gradiometer.

## References

1. Bailey, T., Durrant-Whyte, H.: Simultaneous localization and mapping (SLAM): part II. IEEE Robotics and Automation Magazine 13(3), 108–117 (2006)
2. Zheng, H., Wang, H., Wu, L., Chai, H., Wang, Y.: Simulation Research on Gravity-Geomagnetism Combined Aided Underwater Navigation. Royal Institute of Navigation 66(1), 83–98 (2013)
3. Wu, L., Tian, X., Ma, H., Tian, J.W.: Underwater Object Detection Based on Gravity Gradient. IEEE Geoscience and Remote Sensing Letters 7(2), 362–365 (2010)
4. Ming, L., Hai-Jun, W., Yan-Song, J.: System Modeling of SLAM-based Geomagnetic Aided Inertial Navigation. Aviation Precision Manufacturing Technology 47(6), 13–16 (2011)
5. Wang, S.-C., Sun, D.-W., Zhang, J.-S., Chen, L.-H.: Research on Geomagnetism Navigation and Localization Based on SLAM. Fire Control & Command Control 35(12), 35–37 (2010)
6. Julier, S.J., Uhlmann, J.K.: Unscented filtering and nonlinear estimation. Proceedings of the IEEE 92(3), 401–422 (2004)

# Fuzzy Entropy-Based Muscle Onset Detection Using Electromyography (EMG)

Ming Lyu, Caihua Xiong*, Qin Zhang, and Lei He

Institute of Rehabilitation and Medical Robotics, State Key Lab of Digital Manufacturing Equipment and Technology, School of Mechanical Science and Engineering, Huazhong University of Science and Technology, Wuhan, Hubei 430074, China
chxiong@hust.edu.cn

**Abstract.** Muscle onset detection using Electromyography (EMG) plays an important role in movement pattern recognition, motor and posture control, and assessment of neuromuscular and psychomotor diseases. When signal-to-noise ratio (SNR) of EMG signals is low owing to Gaussian noise and Electrocardiography (ECG), conventional detection methods are not effective. In this paper, we propose an automatic and user-independent detection method based on fuzzy entropy (FuzzyEn), which has advantages of high accuracy, light computational load and simple parameter selection. Besides, the processing time of the proposed method was short enough for rehabilitation robotic control.

**Keywords:** Electromyography (EMG), onset detection, fuzzy entropy (FuzzyEn).

## 1    Introduction

The onset detection of muscle activity plays an important role in motor and posture control [1-2], movement pattern recognition [3-5], assessment of neuromuscular and psychomotor diseases [6-7]. Visual onset detection by EMG is heavily dependent on experience of examiners and is incapable of automatic processing. Consequently, computer-based auto onset and offset detection of muscle activity by EMG is important.

Several methods have been used for muscle onset detection [8-14]. The amplitude-based methods were dependent on the choice of threshold very much, and became even inapplicable when SNR was low. Double-threshold detector was proposed by Bonato et al. [15] as a refinement of the single-threshold method. They allowed the user to select the couple false alarm-detection probability freely by setting three parameters, namely, false-alarm probability, detection probability, and time resolution, and thus achieved higher accuracy. However, the method required whitening process of EMG and complex threshold selection procedure.

---

\* Corresponding author.

X. Zhang et al. (Eds.): ICIRA 2014, Part I, LNAI 8917, pp. 89–98, 2014.

Teager–Kaiser energy (TKE) operator was another onset detection method [16, 17], which was relevant to instantaneous amplitude and frequency of the signal at the same time. The method could achieve improved performance in condition of low SNRs. But its advantage was mainly limited to the Gaussian background noise [18].

Zhang et al. [18, 19] introduced sample entropy (SampEn) to the EMG-based onset time detection field. It performed well in many poor conditions such as low SNRs and ECG artifacts. While according to the SampEn theory and experiment results, the SampEn value was stable and credible only when the length of data template in the computation process was above 100, which meant that it would need long computation time and might not suitable for real-time online applications.

In this paper, we propose an approach for muscle onset detection. The FuzzyEn-based method is a refinement of SampEn by introducing fuzzy concept, thus obtaining better relative consistency, less calculating time and higher accuracy for smaller dataset size. It is an automatic and user-independent method, which is not dependent on parameter selection very much. The effect of the presented method was evaluated for three subjects in different conditions.

## 2 Methods

### 2.1 Subjects

Three male subjects participated in the study. They were all right-handed. The study was approved by Tongji Medical College of Huazhong University of Science and Technology (TJMC, HUST).

### 2.2 Data Acquisition

In this study, EMG signals were obtained mainly from four muscles of subjects: the biceps brachii muscle (BI), the triceps brachii muscle (TR), the pectoralis major muscle (PM), and the latissimus dorsi muscle (LD). The selected muscles and their main functions were shown in Table 1. Considering the under-actuated characteristics of the rehabilitation robot used in the study, there were some alternative muscles to select for a certain motion pattern. For example, we could capture EMG signals from the anterior deltoid muscle (AD) instead of BI for movement pattern recognition of touching right ear with right hand, because the movement involved SF and EF at the same time. The flexible selectivity of muscles was vital for people whose EMG signals were not satisfactory due to muscle weakness.

We captured 4-channel EMG signals from the above muscles using ME6000 bipolar EMG system (Mega Electronics Ltd., Kuopio, Finland) and disposable Ag/AgCl surface electrodes with 3 cm in diameter. The electrodes were located along the muscle belly with 3 cm length between centers of electrode pair. Raw EMG signals were recorded at the sampling frequency of 2 KHz. Filters were not applied here.

**Table 1.** Selected muscles and corresponding main functions

| Channel number | Muscle | Main functions |
| --- | --- | --- |
| 1 | BI | Elbow flexion (EF) |
| 2 | TR | Elbow extension (EE) |
| 3 | PM | Shoulder flexion (SF) and shoulder adduction (SAD) |
| 4 | LD | Shoulder abduction (SAB) |

## 2.3    Experimental Protocol

This study was designed to comprise two experiments. The subjects were asked to perform the same motions in each experiment. Considering that all subjects were right-handed, all the movements mentioned below were completed with their right hands except for situation of ECG acquisition.

In Experiment 1, the subject sat on a chair with arms in neutral position. The trunk movement was forbidden with his back against chair. The subject was instructed to make EF and EE movements, SF and SE movements and SAD and SAB movements randomly. There was a relaxation time of about 3 s between movements. Experiment 1 consisted of 20 repetitions for each movement pattern. EMG data were recorded from BI, TR, PM, LD and AD using ME6000 bipolar EMG system.

In Experiment 2, the subject sat on a chair with arms in neutral position. The trunk movement was forbidden with back against chair. The subject was instructed to make touching left ear (TLE) movements, touching forehead (TFH) movements and touching right ear (TRE) movements respectively. When each movement was completed, the arm maintained for 2 s and retrieving (RET) movement was required to return to the initial position. Each movement was performed for three durations (4, 6 and 10 s). Experiment 2 consisted of 20 repetitions for each movement pattern. EMG data were recorded from BI, TR, PM and LD.

## 2.4    Fuzzy Entropy

The conventional Shannon entropy is not suitable for bio-signal processing, because its performance decreases significantly for short time data and noisy data. Zadeh et al. introduced the fuzzy concept into entropy, thus dealing with vagueness and ambiguous uncertainties instead of probabilistic uncertainties [20, 21]. Based on SampEn [22], FuzzyEn was implemented mainly by replacing conventional Heaviside function with fuzzy function. The fuzzy calculation also decreased computation time by abandoning comparison operations. In recent years, FuzzyEn has been successfully applied in bio-signal processing filed such as feature extraction [23, 24].

For a time series $\{u(i), 1 \leq i \leq N\}$, a delayed m-dimensional vector $\mathbf{x}(i)$ is defined as

$$\mathbf{x}(i) = \{u(i), u(i+1), ..., u(i+m-1)\} \in \mathfrak{R}^m, 1 \leq i \leq N - m \qquad (1)$$

where $m$ is embedding dimension for phase space reconstruction.

The distance between $\mathbf{x}(i)$ and $\mathbf{x}(j)$ is defined as

$$d_{ij} = d[\mathbf{x}(i), \mathbf{x}(j)] = \|\mathbf{x}(i) - \mathbf{x}(j)\|_{\infty}, j \neq i. \tag{2}$$

Define the similarity degree $D_{ij}$ of $\mathbf{x}(i)$ and $\mathbf{x}(j)$ using fuzzy function $\mu$ as

$$D_{ij}(n, r) = \mu(d_{ij}, n, r). \tag{3}$$

$C_m(n, r)$ is thus the probability that vector $\mathbf{x}(j)$ is within similarity tolerance $r$ of the template $\mathbf{x}(i)$, calculated by

$$C_m(n, r) = \frac{1}{N-m} \sum_{i=1}^{N-m} (\frac{1}{N-m-1} \sum_{j=1, j \neq i}^{N-m} D_{ij}). \tag{4}$$

Finally, FuzzyEn is defined as natural logarithm of $C_m(n, r)$ and $C_{m+1}(n, r)$:

$$FuzzyEn = \ln(C_m(n, r) / C_{m+1}(n, r)). \tag{5}$$

## 2.5    Performance Evaluation

To compare with the proposed FuzzyEn method, this paper gives a brief introduction of some onset and offset detection methods and their parameters.

The classic detection method is based on amplitude such ad root mean squares (RMS) of EMG signals.

The performance of RMS method is heavily affected by low SNRs. The threshold $Th$ is usually set to $3 \times$ SD of baseline, which is the background noise of EMG data lasting for 500 ms.

Li et al. [25] introduced TKE operator to the onset detection field. For a time series $x(i)$, the discrete TKE operator is defined as:

$$TKE[i] = x^2(i) - x(i+1) \cdot x(i-1). \tag{6}$$

The $Th$ value is set to $8 \times$ SD and a sliding window may be applied to improve performance.

To evaluate performance of different methods on onset detection of muscle activity, the latency of time $\Delta t$ was defined as:

$$\Delta t = |t_P - t_T| \tag{7}$$

where $t_P$ was the predicted onset time and $t_T$ was the true onset time. The one-way ANOVA was undertaken on EMG data with MATLAB (The Mathworks, Inc., 1993). A level of statistical significance was set at the 0.05 level of confidence.

# 3    Results

## 3.1    Parameter Selection of FuzzyEn

We compared the performance of FuzzyEn with different parameters on EMG data in Experiment 1. EMG signals were synthesized by juxtaposing three EMG segmentations, i.e., relaxed state for 500 ms, contractive state for 1000 ms and relaxed state for 500 ms respectively.

The first parameter to be determined was $m$. In our study, $m$ was set to 2, because large value increased computational difficulty but not improve performance much [23].

The exponential function $\mu(d_{ij}, n, r) = \exp(-(d_{ij})^n / r)$ was chosen as the fuzzy function in FuzzyEn. The parameter $r$ was typically set to a small value below one (typically 0.25) and $n$ was set to a small value (typically 2).

The performance of FuzzyEn was not dependent on parameters ($r$ and $n$) much [23, 24]. The parameter $N$ (window length) was the most important among all parameters for affecting both computation accuracy and computation time. The result of performance of FuzzyEn with different values of $N$ was shown in Table 2. Considering both computational complexity and accuracy, the optimal parameters were chosen as follows: $m = 2$, $r = 0.25$, $n = 2$ and $N = 64$.

**Table 2.** Performance of the parameter $N$ (window length) on onset detection of synthetic EMG signals for FuzzyEn

| Window Length (ms) | 16 | 32 | 64 | 128 |
|---|---|---|---|---|
| Latency (ms) | 123±42 | 18±6 | 76±4 | 136±8 |

## 3.2    Performance with Synthetic EMG Signals

To evaluate the performance of onset and offset detection using different methods, EMG sequences with exactly known onset and offset time were required. In this paper, we employed two ways to address the problem, i.e., synthetic EMG signals, real EMG signals with onset time determined by joint angular data.

We could add Gaussian distribution as simulated stochastic noise to synthetic EMG signals, getting data in condition of different SNRs.

We changed EMG segmentations and Gaussian noises, and obtained 100 synthetic EMG signals with SNR from 4 dB to 20 dB. Fig.1 showed an example of onset and offset detection on synthetic EMG signal at 12 dB SNR. The experiment results could be seen in Table 3. Though the performances of the three methods were similar at high SNRs, the FuzzyEn-based method achieved best accuracy and robustness at low SNRs.

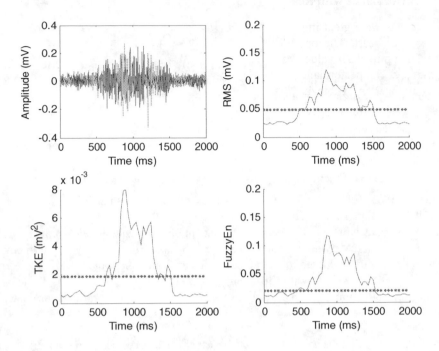

**Fig. 1.** Onset detection with different methods on synthetic EMG signal with true onset time at 500 ms. The red dashed lines represented thresholds for each method.

**Table 3.** Performance of the three algorithms evaluated by mean error (ME) and standard error of the mean (SEM) in ms for onset detection of synthetic EMG signals with different SNRs

| SNR (dB) | RMS | TKE | FuzzyEn | $P$ value |
|----------|-----|-----|---------|-----------|
| 20 | 39±10 | 31±12 | 38±11 | $p1^a = 0.26$<br>$p2^b < 0.05$ |
| 16 | 44±8 | 43±8 | 40±10 | $p1<0.05$<br>$p2<0.05$ |
| 12 | 47±12 | 43±12 | 42±11 | $p1<0.05$<br>$p2= 0.81$ |
| 8 | 128±46 | 103±41 | 56±23 | $p1<0.05$<br>$p2< 0.05$ |
| 4 | 170±20 | 161±15 | 146±19 | $p1<0.05$<br>$p2<0.05$ |

[a] The p value calculated with the latency from RMS and FuzzyEn.
[b] The p value calculated with the latency from TKE and FuzzyEn.

### 3.3    Performance with Real EMG Signals with ECG

ECG artifacts were great impediments to onset time detection for EMG signals from muscles near heart. The optimal *Th* value of TKE was not available (NA) in our study. The optimal *Th* values of TKE and FuzzyEn were settled to $10 \times$ SD and 0.65 respectively. The performance of FuzzyEn-based method against ECG artifacts could be seen in Fig.2 and Table 4.

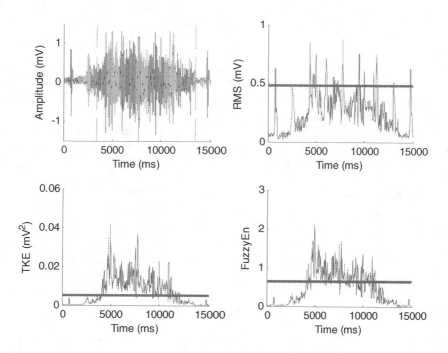

**Fig. 2.** Onset detection with different methods on real EMG signal contaminated with ECG from PM of Group 1

**Table 4.** Performance evaluated by ME and SEM in ms for onset detection of EMG signals contaminated with ECG

| RMS | TKE | FuzzyEn | *P* value |
|-----|-----|---------|-----------|
| NA[a] | 567±88 | 113±21 | p2<0.05 |

[a] The onset time was not available for RMS-based method.

### 3.4    EMG-Based Upper-Extremity Rehabilitation Robotic Control

The core step of a control scheme of EMG-based under-actuated exoskeleton robotic control for upper-extremity rehabilitation is the onset detection. Regardless of traditional on-off control or advanced pattern recognition control, onset detection is very important. Considering that the whole process time of the proposed method was less than 300 ms, it is feasible for on-line real time rehabilitation robotic control.

## 4     Discussion

Onset detection of muscle activity is very important, especially in the field of rehabilitation robotic control. Though voice can be chosen as signal source and speech recognition can thus be applied to control rehabilitation robot, the mapping from speeches to joint movements is not natural and direct contrasted to EMG signals which are produced by muscle contractions. EMG-based on-off control strategy of rehabilitation robot is chosen naturally. Hence, the onset detection of muscle activity becomes a key issue.

FuzzyEn is chosen to measure of the EMG system complexity because of better relative consistency, less calculation time and higher accuracy for smaller dataset length compared with SampEn. The FuzzyEn-based onset detection method is automatic, real-time and user-independent. Compared with conventional methods such as visual detection method, amplitude-based methods and TKE method, it achieves higher accuracy and stronger robustness under most circumstances. It performs especially well in poor conditions such as low SNRs and ECG artifacts. Besides, the effect of FuzzyEn is not dependent on parameter choices very much, which means you need not focus on complicated parameter selection techniques and not adjust parameters for different users and different environments. These advantages make it important in real-time bio-signal analysis and process.

In condition of high SNRs ($\geq$ 12 dB), the FuzzyEn method performed as well as the other two methods, but it performed better in condition of low SNRs (< 12 dB), which meant that it could work well in poor conditions such as movement artifacts of the electrode–skin interface and muscle weakness. The complexities of EMG signals caused by muscle voluntary contractions and involuntary background noise were different, so the FuzzyEn method performed better than that in poor conditions such as ECG.

Though the effect of the proposed method has been tested and verified on healthy subjects, its effect on people with stroke has not been tested. For people with severe stroke, the onset detection of muscle activity is more important because of muscle weakness, spasticity, muscle tremor and abnormal muscle co-activation patterns. Besides, the proposed method will be applied to an upper-extremity rehabilitation robotic control.

## 5     Conclusion

Onset and offset detection of muscle activity by EMG is very significant in bio-signal analysis and process, which especially plays an important role in movement pattern recognition and rehabilitation robotic control. In this paper, we proposed a FuzzyEn-based onset detection method, which achieved high accuracy, short computation time and strong robustness in both good and poor conditions such as different SNRs and ECG artifacts. The presented technique could be applied to on-line real time control of upper-extremity rehabilitation robotic.

**Acknowledgments.** This work was partly supported by the National Natural Science Foundation of China (Grant No. 51335004 and No. 51305148), National Basic Research Program of China (973 Program, Grant No. 2011CB013301), and National Science Fund for Distinguished Young Scholars of China (Grant No. 51025518).

# References

1. Latash, M.L., Aruin, A.S.I., Neyman, N.J.: Anticipatory Postural Adjustments During Self Inflicted and Predictable Perturbations in Parkinson's Disease. Neurol. Neurosur. Ps. 58, 326–334 (1995)
2. Wentink, E.C., Schut, V.G.H.E., Prinsen, C., Rietman, J.S., Veltink, P.H.: Detection of The Onset of Gait Initiation Using Kinematic Sensors and EMG in Transfemoral Amputees. Gait. Posture. 39, 391–396 (2014)
3. Yang, D., Zhao, J., Jiang, L.I., Liu, H.: Dynamic Hand Motion Recognition Based on Transient and Steady-State EMG Signals. Int. J. Hum. Robot. 9, 1250007–1250018 (2012)
4. Dipietro, L., Ferraro, M., Palazzolo, J.J., Krebs, H.I., Volpe, B.T., Hogan, N.: Customized Interactive Robotic Treatment for Stroke: EMG-Triggered Therapy. IEEE. T. Neur. Sys. Reh. 13, 325–334 (2005)
5. Lucas, L., DiCicco, M., Matsuoka, Y.: An EMG-Controlled Hand Exoskeleton for Natural Pinching. J. Robotic. Mec. 16, 482–488 (2004)
6. Karst, G.M., Willett, G.M.: Onset Timing of Electromyographic Activity in The Vastus Medialis Oblique and Vastus Lateralis Muscles in Subjects With and Without Patellofemoral Pain Syndrome. Phys. Ther. 75, 813–823 (1995)
7. Traub, M., Rothwell, J., Marsden, C.: Anticipatory Postural Reflexes in Parkinson's Disease and Other Akinetic-Rigid Syndromes and in Cerebellar Ataxia. Brain. A. J. Neur. 103, 393–412 (1980)
8. Vaisman, L., Zariffa, J., Popovic, M.R.: Application of Singular Spectrum-Based Change-Point Analysis to EMG-Onset Detection. J. Electromyogr. Kines. 20, 750–760 (2010)
9. Micera, S., Vannozzi, G.A., Sabatini, D.P.: Improving Detection of Muscle Activation Intervals. IEEE. Eng. Med. Biol. 20, 38–46 (2001)
10. Rasool, G., Iqbal, K., White, G.A.: Myoelectric Activity Detection During A Sit-To-Stand Movement Using Threshold Methods. Comput. Math. Appl. 64, 1473–1483 (2012)
11. Severini, G., Conforto, S., Schmid, M., D'Alessio, T.: Novel Formulation of A Double Threshold Algorithm for The Estimation of Muscle Activation Intervals Designed for Variable SNR Environments. J. Electromyogr. Kines. 22, 878–885 (2012)
12. Staude, G., Wolf, W.: Objective Motor Response Onset Detection in Surface Myoelectric Signals. Med. Eng. Phys. 21, 449–467 (1999)
13. Solnik, S., Rider, P., Steinweg, K., DeVita, P., Hortobágyi, T.: Teager–Kaiser Energy Operator Signal Conditioning Improves EMG Onset Detection. Eur. J. Appl. Physiol. 110, 489–498 (2010)
14. Allison, G.: Trunk Muscle Onset Detection Technique for EMG Signals With ECG Artefact. J. Electromyogr. Kines. 13, 209–216 (2003)
15. Bonato, P., D'Alessio, T., Knaflitz, M.: A Statistical Method for The Measurement of Muscle Activation Intervals from Surface Myoelectric Signal During Gait. IEEE. T. Bio-Med Eng. 45, 287–299 (1998)
16. Li, X., Zhou, P., Aruin, A.S.: Teager–Kaiser Energy Operation of Surface EMG Improves Muscle Activity Onset Detection. Ann. Biomed. Eng. 35, 1532–1538 (2007)

17. Solnik, S., DeVita, P., Rider, P., Long, B., Hortobágyi, T.: Teager–Kaiser Operator Improves The Accuracy of EMG Onset Detection Independent of Signal-To-Noise Ratio. Acta. Bioeng. Biomech. 10, 65–68 (2008)
18. Zhang, X., Zhou, P.: Sample Entropy Analysis Of Surface EMG for Improved Muscle Activity Onset Detection Against Spurious Background Spikes. J. Electromyogr. Kines. 22, 901–907 (2012)
19. Zhou, P., Zhang, X.: A Novel Technique for Muscle Onset Detection Using Surface EMG Signals Without Removal of ECG Artifacts. Physiol. Meas. 35, 45–54 (2014)
20. Zadeh, L.: Fuzzy set, Information and Control. 8338–353 (1965)
21. Al-sharhan, S., Karray, F., Gueaieb, W., Basir, O.: Fuzzy Entropy: a Brief Survey. In: 2001 IEEE International Fuzzy Systems Conference, pp. 1135–1139 (2001)
22. Richman, J.S., Moorman, J.R.: Physiological Time-Series Analysis Using Approximate Entropy and Sample Entropy. Am. J. Physiol-Heart. C. 278, H2039–H2049 (2000)
23. Chen, W., Wang, Z., Xie, H., Yu, W.: Characterization of Surface EMG Signal Based on Fuzzy Entropy. IEEE. T. Neur. Sys. Reh. 15, 266–272 (2007)
24. Chen, W., Zhuang, J., Yu, W., Wang, Z.: Measuring Complexity Using FuzzyEn, ApEn, and SampEn. Med. Eng. Phys. 31, 61–68 (2009)
25. Langhorne, P., Coupar, F., Pollock, A.: Motor Recovery After Stroke: A Systematic Review. The Lancet Neurology 8, 741–754 (2009)

# Experimental Study on Cutter Deflection in Multi-axis NC Machining

Xianyin Duan[1], Fangyu Peng[2,*], Rong Yan[1], Zerun Zhu[1], and Bin Li[2]

[1] National NC System Engineering Research Center,
School of Mechanical Science and Engineering,
Huazhong University of Science and Technology, Wuhan 430074, China
[2] State Key Laboratory of Digital Manufacturing Equipment and Technology,
School of Mechanical Science and Engineering,
Huazhong University of Science and Technology, Wuhan 430074, China
pengfy@hust.edu.cn

**Abstract.** In five-axis sculptured surface machining, the effect of cutter deflection on tool orientation planning is important. This paper studied the method for online measurement of cutter deflections along two axes simulation. The measurement equipment was designed and implemented to acquire the displacements of cutter under cutting force online. Acquired data were processed to static values and then compensated by geometric analysis. The cutter deflection conditions were analyzed and divided into different types. The corresponding geometrical equations of the relationship of deflections of measured values and actual values were built. The inter-coupling values were decoupled by solving the geometrical equations. The changing regulations of cutter deflection with tool orientations were analyzed, which could provide support for the study of tool orientation planning. The effectiveness of measurement error compensation was verified by the difference between measured values and actual values of cutter deflections under various tool-workpiece inclination angles. This work could be further employed to optimize tool orientations for suppressing the surface errors due to cutter deflections and achieving higher machining accuracy.

**Keywords:** experimental study, cutter deflection, multi-axis machining, online measurement.

## 1 Introduction

Five-axis machining provides machine tool with two rotation axes to enlarge accessible space of the cutter, which meets the needs of machining sculptured surface

---

*Corresponding author at: B418, Advanced manufacturing building, Huazhong University of Science & Technology (HUST), 1037 Luoyu Road, Hongshan District, Wuhan, Hubei Province, PR China. Area code: 430074. Tel.: +86 13986168308. Fax: +86 27 87540024.

X. Zhang et al. (Eds.): ICIRA 2014, Part I, LNAI 8917, pp. 99–109, 2014.

such as aeronautical components. By planning tool orientations, it is probable to avoid interference among cutter, workpiece and other parts, raise the contact order between tool envelope surface and design surface of workpiece to increase machining efficiency, and so forth [1]. Most crucial parts of aeronautical components, such as compressor impeller, landing gear, and rocket engine shell, possess not only complicated surface but also ultra-high strength material. That can engender great cutting force acting on the multi-axis machining system which contains long kinematic chains. Meantime, the cutting tool usually works in an abnormal posture relative to the normal vector of the surface at the cutter contact (CC) point. These conditions all probably produce noteworthy cutter deflection that may make serious consequences. Generally, cutter deflection is the more major issue of sources suppressing the machining precision compared with tool wear, cutter run-out, and chatter vibration [2, 3]. Large cutter deflection can bring about unacceptable machining surface errors and restrict the improvement of production efficiency, or even destroy the machining system.

Series of research have been carried out on the prediction of cutting force induced cutter deflection as follows. Landon predicted the cutter deflection without using the cutting force model, in which a data block was created for each machine/mill/material from experiments [4]. That method only aimed to concrete cases for three-axis milling applications. Dow et al. [5] calculated the deflection of small ball-end mill, and also only considered the flexibilities of the tool and spindle. Chanal et al. [6] computed the cutter deflection due to static structure deflection based on machine structure and cutting load, in which one-tooth flat-end mill and drill were adopted and both considered as a solid body. Besides, the structure model aimed to parallel kinematics machine of tricept legs and was not depicted concretely, and the cutting pressures were identified experimentally. Dépincé et al. [7] dealt with calculation of tool deflection in flat-end milling in which only the cutting force modeling was proposed. Wang et al. [8] described the modeling of robot deformation caused by the external process forces from the machining applications which was only a conceptual model. Soori et al. [9] presented a virtual machining system in order to enforce tool deflection in three-axis milling operations in which only the flexibility of the cutter was computed. Rodríguez et al. [10] developed a tool deflection model based on the tool geometry and elasticity theory of the material, which was used for two- and three-axis micro-milling processes.

As mentioned above, some took the cutting tool as rigid body, another took the spindle and the handle (or tool-holder) as rigid body, and most did not take the transmission axes into account. Furthermore, precise experimental verification of the model was lacking. Most models could only be used for three-axis machining which did not take advantage of five-axis machining fully, or did not connect it with lead and tilt angles for tool orientation planning. As the continuation of our preliminary studies on geometrical error analysis and machine tool characteristic [11, 12], the main purpose of this paper was to present a method, which could precisely measure the cutter deflections and study the variety discipline of cutter deflection under variable tool orientations in five-axis end milling. Then tool orientation planning based on cutter deflection model could be implemented easily.

The remainder of this paper was structured as follows. In section 2, the online measurement equipment for obtaining cutter deflections was described. That was followed by Analysis of measurement errors in section 3.1. Compensation of measurement errors was studied in section 3.2. In section 3.3, comparison and analysis of uncompensated and compensated values of cutter deflection was explained. The whole paper was concluded in section 4.

## 2    Online Measurement of Cutter Deflections

The equipment to measure cutter deflections online was shown in Fig. 1. There were four modules, namely machining system, clamping system, laser displacement measurement system, and data acquisition system. The cutting tool was the end of the machining system and the component under measurement. The milling center was a five-axis machine tool whose type was Mikron UCP 800 Duro. The two rotation axes of the milling center were at the end of the workbench. So tool orientation towards the machine tool of this structure was actually tool-workpiece relative orientation. The function of the clamping system was to fix the heads to the spindle nose (translates and does not rotate with the spindle) in the way that laser emission direction of one head passed through the tool axis along x axis and that of the other along y axis. In addition, the heads holder of the clamping system could be adjusted along tool axis to make the laser point was right on the lowest part of the shank, and the distance between the emission point reflection point was always within the measurement range of the heads. The laser displacement measurement system was a CCD ultra-high accuracy non-contact laser displacement device and contained two heads (Keyence LK-H020), two head-to-controller cable (Keyence CB-A5), one controller (Keyence LK-G5001V), and a power supply (Keyence MS2-H50). The screen of the heads was protected from high-speed chips by fixing ultrathin gorilla screens to maintain high transparency. The data acquisition system was made up by a computer, a high-bandwidth backplane (NI PXIe-1082), a dynamic signal acquisition card (NI PXIe-4499), and LabVIEW software.

The workpiece material was 300M steel which possessed ultra-high strength and was widely adopted in the aviation industry. The purpose was to make the cutter deflections as big as possible. The machining experiments were performed as shown in Fig. 1. The main geometrical and process conditions in experiments were listed as follows. In the experiments, the diameter of shank of the cutting tool was 10mm;

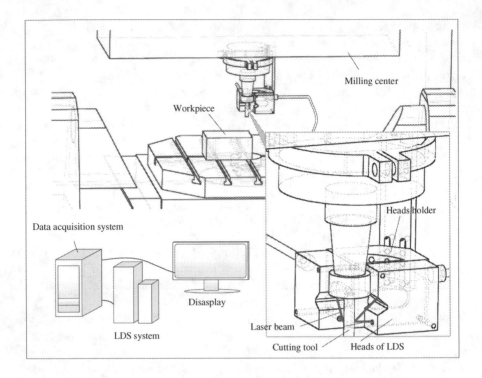

**Fig. 1.** Cutter deflection online measurement equipment

the radius of arc surface of the cutter was 2mm; the number of flutes of the cutter was 4; the nominal helix angle of the cutter was -50 reg; the length of the tool was 100mm; the length of the flute was 30mm; the feed rate per tooth was 0.04mm; the spindle speed was 1000rev/min; the cutting depth 1.4mm; the milling mode was down milling, and no cutting fluid was used.

To study the regularity of the tool-workpiece inclination angles and cutter deflections, the experiment was designed into several groups with different lead angles and tilt angles. Concretely, all groups were divided into five sets, and each set was divided into four groups. The detailed grouping situation was as follows.

The lead angle, $\alpha$, was set within {0, 10, 20, 30, 40}, and for each of these lead angles, the tilt angle, $\beta$, was varied within {0, 15, 30, 45, 60}. Visibly, except the first group, there were four lead angles and five tilt angles, and 20 groups were combined altogether.

The surface of the workpiece was divided into five regions. The lead angles were the same within each region and different between each pair of region. Further, each region was divided into several paths, and the tilt angles were different between each pair path within a region.

# 3    Analysis and Compensation of Measurement Error

## 3.1    Analysis of Measurement Errors

The shank of the cutting tool was cylindrical surface, which made deflections along x axis and y axis were inter-coupling. In detail, deflection along x or y axis could brought about change of measured value along y or x axis, and the changed value was just measurement error which should be calculated and compensated. Here equations of geometrical relationship between measured values and actual values were built and solved to compensate the measurement errors.

During machining process, there were nine possibilities of cutter deflection directions combination along $x^t$ and $y^t$ axes. The nine conditions were shown in Fig. 2 which were divided by the positive and negative of cutter deflections. Necessarily, an assumption should be pointed that the values of cutter deflections all smaller that the radius of the cutter, which was general condition in the actual processing. In Fig. 2, the circles represented cross sections along $z^t$ axis at the optical reflection point of the cutter, and the points represented intersection of the cross section and the tool axis. The blue circle represented cross section of the cutter before deviating from initial position, and the green circle represented that after deviating from initial position. The blue point represented the initial position of the cutter, and the green point represented the deviated position. In the fifth condition, the deviated position overlapped with the initial position, which meant the cutter did not deflected and was an ideal case. The rest eight conditions could be divided into two types by if there existed cutter deflection equaling to zero along $x^t$ or $y^t$ axis.

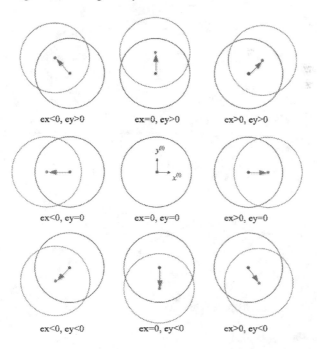

Fig. 2. Nine conditions that the cutter deflects

The first type contained the second, fourth, sixth, and eighth conditions, and an example was shown in Fig. 3, in which there were only cutter deflection along x axis. At first, we set that the directions of lasers emitted were along positive y axis and negative x axis respectively in all conditions. Then concrete analysis for the condition in Fig. 3 was given as follows.

For condition of Fig. 3, $O_1$ represented the initial position of the cutter and $O_2$ represented the deflected position of that. The distance of $O_1$ and $O_2$ was just the actual deflection along x axis ($e_x$). The distance of rightmost points on both the two circle was the measured deflection along x axis ($e_{xm}$). Obviously, the measured value in x direction ($e_{xm}$) equaled actual value ($e_x$). There was no measurement error along x axis in this condition ($e_x'=0$). In y direction, we denoted the initial reflection point by $B$, and the reflection point after the cutter deflected by $A$. Then the length of $AB$ was both the measured deflection ($e_{ym}$) and measurement error ($e_y'$) along y axis. The actual value in y direction ($e_y$) equaled zero, and the measured value in y direction ($e_{ym}$) was measurement error. In practice, it was unknown what kind the condition was, so the measured value in y direction was used to determine that. According to the Pythagorean Theorem and considering the uncertain environmental error, the equation could be given by

$$\left| (\rho - e_{ym})^2 + e_{xm}^2 - \rho^2 \right| < m_0 \tag{1}$$

where $m_0$ was an custom small quantity.

By determination equation mentioned above, when deflection in one direction was small enough, we could treat that condition this type, or it should be handled as followed type.

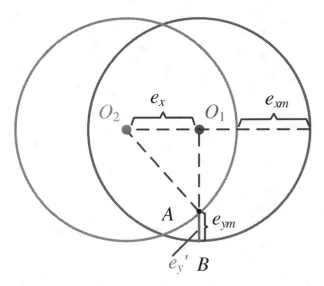

**Fig. 3.** condition when the cutter deflects along only one axis

The second type contained the first, third, fifth, and seventh conditions, and an example was shown in Fig. 4. There were cutter deflections along both x axis and y axis. Likewise, the lasers were set in directions along positive y axis and negative x axis respectively in all conditions. Then geometrical equations between measured and actual distances could be established as follows.

For condition of Fig. 4, $e_x$ was the length of $AO_2$; $e_y$ was the length of $CO_2$; $e_{xm}$ was the length of $DE$; and $e_{ym}$ was the length of $BF$. By geometrical analysis of them, two equations could be obtained in $\mathrm{RT}\,\Delta BAO_2$ and $\mathrm{RT}\,\Delta DCO_2$ respectively as

$$\begin{cases} (\rho - e_{ym} + e_y)^2 + e_x^2 = \rho^2 \ (\mathrm{RT}\,\Delta BAO_2) \\ (\rho + e_x - e_{xm})^2 + e_y^2 = \rho^2 \ (\mathrm{RT}\,\Delta DCO_2) \end{cases}$$

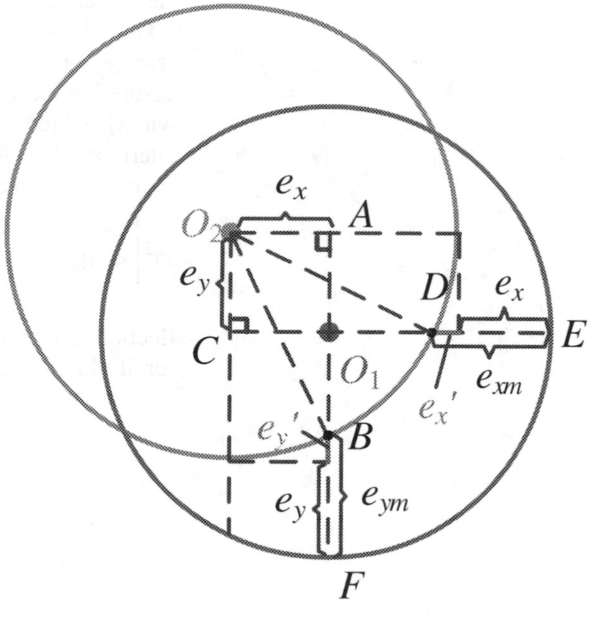

**Fig. 4.** Condition when the cutter deflects along both two axes

## 3.2    Compensation of Measurement Errors

To reveal the solution of the geometrical equation for measurement errors compensation, the solutions were given as

$$\begin{cases} e_x = \dfrac{EX_1 \pm \sqrt{EX_2}}{2EX_3}, \\[3mm] e_y = \dfrac{EY_1 \pm \sqrt{EY_2}}{2EY_3}, \end{cases} \tag{2}$$

where $EX_1$, $EX_2$, $EX_3$, $EY_1$, $EY_2$, and $EY_3$ were intermediate variables and given as

$$EX_1 = 2\rho^3 + 4\rho^2 e_{xm} + 2\rho^2 e_{ym} + 3\rho e_{xm}^2 + \rho e_{ym}^2 + 2\rho e_{xm} e_{ym} + e_{xm}^3 + e_{xm} e_{ym}^2,$$

$$EX_2 = 4\rho^6 + 8\rho^5 e_{ym} - 4\rho^4 e_{xm}^2 - 8\rho^4 e_{xm} e_{ym} - 4\rho^3 e_{xm}^3 - 12\rho^3 e_{xm}^2 e_{ym} - 20\rho^3 e_{xm} e_{ym}^2$$
$$-12\rho^3 e_{ym}^3 - \rho^2 e_{xm}^4 - 8\rho^2 e_{xm}^3 e_{ym} - 14\rho^2 e_{xm}^2 e_{ym}^2 - 16\rho^2 e_{xm} e_{ym}^3 - 13\rho^2 e_{ym}^4$$
$$-2\rho e_{xm}^4 e_{ym} - 4\rho e_{xm}^3 e_{ym}^2 - 8\rho e_{xm}^2 e_{ym}^3 - 4\rho e_{xm} e_{ym}^4 - 6\rho e_{ym}^5 - e_{xm}^4 e_{ym}^2$$
$$-2e_{xm}^2 e_{ym}^4 - e_{ym}^6,$$

$$EX_3 = 2\rho^2 + 2\rho e_{xm} + 2\rho e_{ym} + e_{xm}^2 + e_{ym}^2,$$

$$EY_1 = 2\rho^3 + 4\rho^2 e_{ym} + 2\rho^2 e_{xm} + 3\rho e_{ym}^2 + \rho e_{xm}^2 + 2\rho e_{xm} e_{ym} + e_{ym}^3 + e_{ym} e_{xm}^2,$$

$$EY_2 = 4\rho^6 + 8\rho^5 e_{xm} - 4\rho^4 e_{ym}^2 - 8\rho^4 e_{xm} e_{ym} - 4\rho^3 e_{ym}^3 - 12\rho^3 e_{ym}^2 e_{xm} - 20\rho^3 e_{ym} e_{xm}^2$$
$$-12\rho^3 e_{xm}^3 - \rho^2 e_{ym}^4 - 8\rho^2 e_{ym}^3 e_{xm} - 14\rho^2 e_{ym}^2 e_{ym}^2 - 16\rho^2 e_{ym} e_{xm}^3 - 13\rho^2 e_{xm}^4$$
$$-2\rho e_{ym}^4 e_{xm} - 4\rho e_{ym}^3 e_{xm}^2 - 8\rho e_{ym}^2 e_{xm}^3 - 4\rho e_{ym} e_{xm}^4 - 6\rho e_{xm}^5 - e_{ym}^4 e_{xm}^2$$
$$-2e_{ym}^2 e_{xm}^4 - e_{xm}^6,$$

$$EY_3 = 2\rho^2 + 2\rho e_{ym} + 2\rho e_{xm} + e_{ym}^2 + e_{xm}^2,$$

Then it could be reached to calculate the error compensation values along $x$ and $y$ axis $e_x{}'$ and $e_y{}'$ as

$$\begin{cases} e_x{}' = e_{xm} - e_x, \\ e_y{}' = e_{ym} - e_y. \end{cases}$$

### 3.3     Comparison of Uncompensated and Compensated Values

As shown in Fig. 5, uncompensated values of cutter deflections were indicated by solid lines, and compensated values of them were indicated by dotted lines. Differences between the uncompensated and compensated values varied with the tool-workpiece inclination angles. To express these differences and the relationship with inclination angles more clearly, figures of measurement error and measurement error ratio were also given as follows. The changing regulations of the uncompensated and compensated values with the lead angles and tilt angles were roughly same. The compensated deflections were taken as examples to analyze the general variation trend. From these figures, $e_x$ varied from 0.1mm to 0.35mm in general. And $e_y$ varied from 0.2mm to 0.35mm except groups of conditions when $\alpha=0$. It indicated that $e_x$ increased with $\alpha$ in the range of test parameters when $\beta$ was given from the figure.

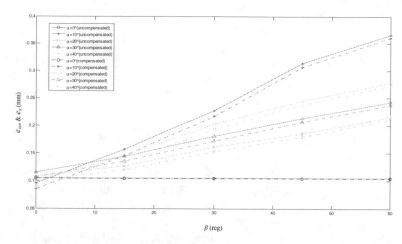

**Fig. 5.** Comparison of uncompensated and compensated cutter deflections along x axis

The changing trends of uncompensated and compensated values along y axis were shown in Fig. 6. Being different from those along x axis, compensated values were bigger than uncompensated values along y axis. Likewise, differences between the uncompensated and compensated values varied with the lead angles and tilt angles. The changing regulations of the uncompensated and compensated values with the lead angles and tilt angles were roughly same. The compensated deflections were taken as examples to analyze the general variation trend.

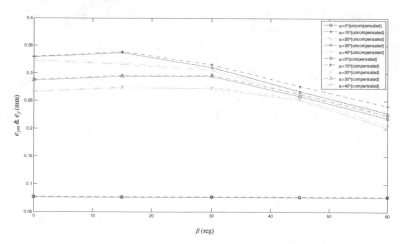

**Fig. 6.** Comparison of uncompensated and compensated cutter deflections along y axis

# 4     Conclusions

In this paper, a method of measuring cutter deflections online involving tool-workpiece inclination angles in five-axis machining was proposed. Study of measurement errors compensation was done to analysis the effect of them.

From the measurement during machining experiments, the cutter deflections in two directions were both the minimum when the tool orientation overlapped with the normal vector at the cutter contact point, but that condition could not meet the actual needs especially in multi-axis machining. Within the scope of experimental parameters, cutter deflection in x direction would decrease with increasing lead angle when the tilt angle was constant, and increase with increasing tilt angle when the lead angle was constant; cutter deflection in y direction would also decrease with increasing lead angle when the tilt angle was constant, and increase with decreasing tilt angle when the lead angle was constant..

From the measurement error compensation and analysis, the changing regulations of measurement errors along x axis basically conformed to those of cutter deflections along y axis, and the general trends of measurement errors along y axis were symmetrical along horizontal direction with those of cutter deflections along x axis. The changing regulations of the uncompensated and compensated values with the lead angles and tilt angles were roughly same. Differences between the uncompensated and compensated values varied with the lead angles and tilt angles.

Based on the proposed method of cutter deflection measurement involving lead and tilt angle, next research will be tool orientation planning for multi-axis NC machining of sculptured surface. An optimized tool orientation space in which the tool orientations can be selected arbitrarily and the cutter deflection is within acceptable limits will be generated. It also provides support for the study of predicting the surface errors due to cutter deflections accurately.

**Acknowledgements.** This work was financially supported by the Major Science and Technology Innovation Program of Hubei Province (2013AAA008), 973 the National Basic Research Program of China (2011CB706803) and the National Natural Science Foundation of China (NSFC) under Grant (51121002).

# References

1. Fard, M.J.B., Feng, H.Y.: Effective Determination of Feed Direction and Tool Orientation in Five-Axis Flat-End Milling. ASME J. Manuf. Sci. Eng. 132(6), 61011 (2010)
2. Yang, M.Y., Choi, J.G.: A Tool Deflection Compensation System for End Milling Accuracy Improvement. ASME J. Manuf. Sci. Eng. 120(2), 222–222 (1998)
3. Sutherland, J.W., Devor, R.E.: An Improved Method for Cutting Force and Surface Error Prediction in Flexible End Milling Systems. ASME J. Manuf. Sci. Eng. 108(4), 269–279 (1986)
4. Landon, Y., Segonds, S., Lascoumes, P., Lagarrigue, P.: Tool Positioning Error (TPE) Characterisation in Milling. Int. J. Mach. Tools Manuf. 44(5), 457–464 (2004)

5. Dow, T.A., Miller, E.L., Garrard, K.: Tool Force and Deflection Compensa-tion for Small Milling Tools. Precis. Eng. 28, 31–45 (2004)
6. Chanal, H., Duc, E., Ray, P.: A Study of the Impact of Machine Tool Structure on Machining Processes. Int. J. Mach. Tools Manuf. 46(2), 98–106 (2006)
7. Dépincé, P., Hascoët, J.Y.: Active Integration of Tool Deflection Effects in End Milling. Part 1. Prediction of Milled Surfaces. Int. J. Mach. Tools Manuf. 46(9), 937–944 (2006)
8. Wang, J.J., Zhang, H., Fuhlbrigge, T.: Improving Machining Accuracy with Robot Deformation Compensation. In: The 2009 IEEE/RSJ International Conference on Intelligent Robots and Systems, pp. 3826–3831 (2009)
9. Soori, M., Arezoo, B., Habibi, M.: Virtual machining considering dimen-sional, geometrical and tool deflection errors in three-axis CNC milling machines. J. Manuf. Syst. (in Press, 2014), doi:10.1016/j.jmsy.2014.04.007
10. Rodríguez, P., Labarga, J.E.: Tool Deflection Model for Micromilling Processes. Int. J. Adv. Manuf. Technol. (2014), doi:10.1007/s00170-014-5890-8
11. Lai, X.D., Zhou, Y.F., Zhou, J., Peng, F.Y., Yan, S.J.: Geometrical Error Analysis and Control for 5-axis Machining of Large sculptured surfaces. Int. J. Adv. Manuf. Technol. 21, 110–118 (2003)
12. Peng, F.Y., Rong, Y., Yang, J., Yang, J.Z., Li, B.: Anisotropic Force Ellipsoid Based Multi-axis Motion Optimization of Machine Tool. Chinese Journal of Mechanical Engineering 25(5), 960–967 (2012)

# A Visual Measurement of Fish Locomotion Based on Deformable Models

Chunlei Xia[1,2], Yan Li[3], and Jang-Myung Lee[2]

[1] Yantai Institute of Coastal Zone Research, Chinese Academy of Sciences,
Yantai 264003, P.R. China
[2] School of Electrical Engineering, Pusan National University,
Busan (Pusan) 609-735, Republic of Korea
[3] Shenyang Institute of Automation, Chinese Academy of Sciences,
Shenyang 110016, P.R. China
c.xia2009@gmail.com

**Abstract.** Measurement of fish locomotion is an essential issue not only for biological studies but also valuable for robotics researchers. In this study, an automatic marker less method was proposed for recording fish locomotion by using digital camera. A fish observation system was presented to capture fish motion from top view. And the active shape model was utilized to construct the deformable fish model for tracking fish locomotion and acquiring the precise fish posture. Subsequently, the fish model was applied to tracking the movement of a single fish. The skeleton of fish body was further calculated from the deformable fish model. The two-dimensional posture of fish body was described by the 20 points on the skeleton. Experimental results demonstrated that the proposed locomotion tracking method was efficient to measure the shape variation of the fish body.

**Keywords:** Fish Tracking, Model based Tracking, Active Shape Model, Posture Measurement, Swimming Modes.

## 1 Introduction

Through millions of years of evolution fish achieved remarkable swimming ability by natural selection, especially, in hovering, turning in intricate water currents and dexterous manipulation under floating conditions [1]. The perfect body mechanisms and swimming modes of fish could inspire innovative designs to improve the ability of locomotion of aquatic robots [2]. Biomimetic robotics has attracted an increasing number of attentions from the researchers. Recently, the fish-like propulsion mechanism, the fin material, and the mechanical structures have been focused on in research works.

Most of the previous works studied the fish locomotion patterns and swim modes by the mathematical models and computer simulations. The artificial fish model was studied for producing the animation of fish school for computer graphic [3]. As the demand of bio-inspired robotics the swimming pattern of fish was studied and a

X. Zhang et al. (Eds.): ICIRA 2014, Part I, LNAI 8917, pp. 110–116, 2014.
© Springer International Publishing Switzerland 2014

number of fish model has been proposed and applied to control locomotion of robotic fish [4, 5, 7].

In the last decade, modeling of fish swimming had a great progress and was successfully applied to robotic fish control. Fish locomotion should not only be studied by the mathematical models but also need to be verified by the actual fish locomotion data. The fish swimming modes were measured by video camera and colored markers attached to the fish body [6]. Five markers were manually attached to fish body from tail to head. The body motion data was obtained by the 5-point data. In order to track the body motion, markers should have high contrast with the fish body in gray scale. Attaching markers to fish body needs a lot of manual operations and it requires skilled operator to minimize the positioning error of markers. Consequently, the marker less automatic observation method is necessary to track and measure the fish locomotion, such as model based posture analysis [8].

In this paper, a deformable fish model is constructed based on the active shape model (ASM). Variation of fish body is modeled in two-dimension. Top view of fish body is modeled by using ASM for measuring precise fish posture. And the fish posture is simplified as a skeleton line for further analysis.

## 2 Structure of Fish Observation System

In this study, the fish motion data was captured and recorded by a digital camera. The structure of fish observation system is presented in Fig. 1. This observation system consisted of a small aquarium, a digital camera, a light source and a desktop computer. Top view observation was preferred to measure and analyze the fish locomotion. The digital camera was placed over the aquarium that capturing top view images of the aquarium. Since the shape of fish body would not change much in the top view image, it is stable and reliable to analyze the posture of fish and track fish movements. The field of view of the camera was required to cover the whole movement area of the fish. The field of view of the camera was optimized by adjusting the distance from camera to the water surface in the arena. According to the experimental conditions, the light source was placed below the aquarium in order to acquire the maximal contrast between the fish and the background. Locomotion of a single fish was observed in this study.

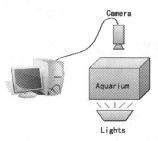

**Fig. 1.** Structure of fish measurement equipment

# 3     Visual Measurement of Fish Locomotion

## 3.1     Extraction of Individual Fish Image

Fig. 2 shows a captured image of a single Zebrafish (*Danio rerio*) moving in a round arena. Since the observation was conducted under stabilized light condition and background of aquarium was clean which providing the optimized contrast for extracting fish images. The fish showed a strong contrast to the white background which could be easily extracted by the thresholding method. Comparing with the widely used background subtraction method, adaptive thresholding is more robust for fish segmentation against the illumination changes. Usually, background subtraction requires calculating mean image of background, such as MOG (Mixture of Gaussians), to deal with the light changes. However, fish could not continuously move all the time when a fish stay for a certain period in the image it would be counted as the background in the process of calculating mean background over time. Therefore, the adaptive local thresholding method was utilized to extract the individual from the background for enhance the robustness of fish segmentation. The fish body was presented in a white shape in the binary image after segmentation. Further analysis was conducted based on the binary images which minimized the computational complexity and processing time.

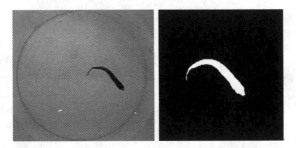

**Fig. 2.** Top view of fish body and its segmented image

## 3.2     Construction of Deformable Fish Model

Since the fish body changes its shape and posture flexibly, a statistical deformable model, active shape model, was utilized to describe the two-dimensional fish body. Active shape model is built on a priori knowledge of the object which is robust to shape changes in the complicated background [9]. In active shape model, fish body is represented by a shape vector xi with n landmarks:

$$\mathbf{x}_i = (x_{i,0}, y_{i,0}, x_{i,1}, y_{i,1}, \cdots, x_{i,n-1}, y_{i,n-1})^T, i = 1, \cdots, N \qquad (1)$$

where $(x_{i,j}, y_{i,j})$ are the coordinates of the $j$th landmark of the $i$th shape in the training sets, $N$ is the number of images in the training set, and $T$ is the transpose operator. The ASM learns a priori knowledge of the shape changes from a set of shape and image samples which is called training set.

To build a deformable fish model, the typical fish postures were manually selected from the video clips recorded by the observation system. And these fish images were manually marked with landmarks along their contours.

According to the authors' experiences and the image size of fish, 42 landmarks in total were determined to describe the fish shapes in this work. An example of fish shape is presented in Fig. 3, landmarks were linked by lines to show the overall shape of a fish. The starting point (0th) indicated the tip of fish head and the 20th point represented the fish tail. And the x and y axis described the fish size in pixels. The sequence and position of landmarks should be kept consistent throughout the training.

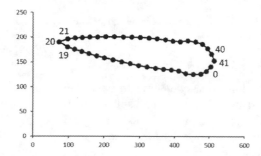

**Fig. 3.** A mean shape of fish calculated from training samples

In this work, 20 images of individual fish were selected as the training set. In the training phase of ASM, shape variations were obtained from the training samples. Any shape in the training set could be represented approximately by the mean shape and weighted modes of variations:

$$\mathbf{x}_i = \overline{\mathbf{x}} + \mathbf{P}\mathbf{b} \tag{2}$$

where $P=[p_1,p_2,...,p_t]$ is the matrix of the first $t$ eigenvectors and $b$ is the vector for weights. The model could be deformed from the mean (i.e., $\overline{\mathbf{x}}$) shape to fit the new data by changing the weight vector $b$. The detailed algorithm of modeling shape variation is given in [9].

## 3.3 Locomotion Measurement Using Fish Model

The constructed fish model was subsequently utilized to track the fish locomotion in a video clip. The fish model initially was manually placed in the arena and the fish model iteratively matching to the fish image by searching the optimized boundary of fish images. Consequently, the fish model continuously follows the fish movement. The detailed matching algorithm of ASM model is described in [9]. In this study, binary image of fish was obtained in advance to locomotion measurement and the matching process was conducted on the binary fish image (Fig. 2). Measuring fish posture from binary image could highly decrease the computational complexity and could improve the measurement speed. Fig. 4 shows a fish image fitted by the proposed fish model from the video clip.

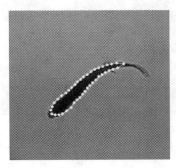

**Fig. 4.** An actual fish image represented by fish model

### 3.4    Calculating Skeleton of Fish

Since the fish body showing a symmetric structure in shape the fish posture could be simplify represented by calculating the skeleton of fish for further analysis. In the previous steps, fish locomotion was modeled into 2D shape. The skeleton of fish could be easily obtained from the shape model. The skeleton is described as a series of points which calculated from the corresponding pair of landmarks on the left and right side of fish body. Each of the points of skeleton was the mean values of the corresponding pair landmarks.

## 4    Experiments and Results

The proposed locomotion tracking method was tested with a single Zebrafish in the laboratory condition. A round aquarium with diameter of 20cm was selected for the experiments, and the water depth was 5 cm. Five video clips of fish were recorded by using Logitech C905 webcam and each video clip recorded 2 minutes of fish movement. The resolution of video clips was 1280*720 pixels. The fish image was approximate 500*80 pixels. The proposed method was implemented by Microsoft Visual C++ 10.0 and an open source computer vision library, OpenCV, on a personal computer (Intel® Core™ 2 Duo CPU E4500@ 2.20GHz).

Fig. 5 shows various patterns of fish locomotion which described by the deformable fish model (yellow contour). In addition, the fish skeleton was presented in green points and lines. These results demonstrated that the proposed fish model could accurately fit the actual fish image and measure the fish posture. The measurement data could describe the precise body motion that could be used for studying the mathematical fish locomotion models or for behavioral fish from the biological aspects.

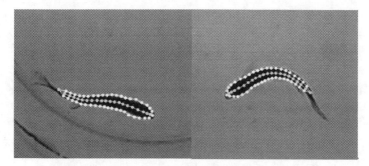

**Fig. 5.** Individual fishes represented by the fish model (yellow contour) and their skeletons (green line)

One of the advantages of the ASM is to segment objects from complicated background. Due to the background was rather simple in this study binary image of fish was adopted for shape analysis. However, in the case of long term observation, many unexpected objects would be occurred in bottom of the aquarium, such as discharges of fish, which would contribute severe noise to the background and increase the segmentation error in binary fish image. Computational function for describing the boundary feature of fish body, such as Mahalanobis distance [9] or neural network [10], would be an efficient way to segment the accurate fish body. Furthermore, since the ASM could identify the occluded objects by *a priori* knowledge of shapes the proposed model could be also utilized to recognize and segment multiple fishes from occlusions.

## 5    Conclusion

A model based fish tracking is presented for measuring fish posture in this study. The fish model was constructed based on the active shape model and shape variations were modeled from the sample data of fish images. The fish model represented precise fish posture and the skeleton of fish was further calculated from the model. The skeleton of fish described the posture and bending of fish body. The performance of the proposed method was demonstrated through experiments. The proposed method could provide accurate fish posture data for more depth studies, such as investigating hydrodynamics models for controlling robotic fish.

**Acknowledgements.** This project is supported by National Natural Science Foundation of China (Grant No.41406112).

## References

1. Wu, T.Y.: Fish swimming and bird/insect flight. Annual Review of Fluid Mechanics 43, 25–58 (2011)

2. Sfakiotakis, M., Lane, D.M., Davies, J.B.C.: Review of fish swimming modes for aquatic locomotion. IEEE Journal of Oceanic Engineering 24(2), 237–252 (1999)
3. Tu, X.: Artificial Animals for Computer Animation. LNCS, vol. 1635. Springer, Heidelberg (1999)
4. Zhou, C., et al.: The design and implementation of a biomimetic robot fish. International Journal of Advanced Robotic Systems 5(2), 185–192 (2008)
5. Hu, T., Low, K.H., Shen, L., Xu, X.: Effective Phase Tracking for Bioinspired Undulations of Robotic Fish Models: A Learning Control Approach. IEEE/ASME Transactions on Mechatronics 19(1), 191–200 (2014)
6. Yan, H., Su, Y.-M., Yang, L.: Experimentation of fish swimming based on tracking locomotion locus. Journal of Bionic Engineering 5(3), 258–263 (2008)
7. Hu, H., et al.: Design of 3D swim patterns for autonomous robotic fish. In: IEEE/RSJ International Conference on Intelligent Robots and Systems. IEEE (2006)
8. Lai, C.-L., Tsai, S.-T., Chiu, Y.-T.: Analysis and comparison of fish posture by image processing. In: 2010 International Conference on Machine Learning and Cybernetics (ICMLC), vol. 5. IEEE (2010)
9. Cootes, T.F., Taylor, C.J., Cooper, D.H., Graham, J.: Active shape models-their training and application. Computer Vision and Image Understanding 61(1), 38–59 (1995)
10. Xia, C., Lee, J.M., Li, Y., Song, Y.H., Chung, B.K., Chon, T.S.: Plant leaf detection using modified active shape models. Biosystems Engineering 116(1), 23–35 (2013)

# Design and Pressure Experiments of a Deep-Sea Hydraulic Manipulator System

Zhang Qifeng[1], Zhang Yunxiu[1,2], Huo Liangqing[1], Kong Fandong[1],
Du Linsen[1], Cui Shengguo[1], and Zhao Yang[1]

[1] Key Laboratory of RoboticsShenyang Institute of Automation, CAS Shenyang, China
[2] ChinaUniversity of Chinese Academy of Sciences, Beijing, China
{zqf,zhangyunxiu}@sia.cn

**Abstract.** The design and realization of a 7-Function master-slave deep-sea hydraulic manipulator can be used in 7000 meters depth is proposed. Linear actuator, rotary actuator and cycloid motor are the three basic modules of the slave arm. To achieve smooth control result, PI control algorithm with variable gains is applied on the control of the slave arm. The control algorithm is also certified effective in alleviating the movement delay and overshoot jitter of slave-arm. Based on the reliable running experiments on land, a pressure experiment system to test the manipulator's index of 7000m depth is proposed. The pressure experimenting process is described in detail. Finally, actually application testing in the sea is presented. During the process of application testing, the manipulator movement function was normal and successfully completed the goal underwater fetching tasks. The whole result of experiment shows that the hydraulic manipulator satisfies 7000m design depth index.

**Keywords:** Deep-sea hydraulic manipulator, PI control algorithm, 7000m pressure experiment, Application testing in the sea.

## 1 Introduction

It has been decades since underwater vehicles applied in underwater exploring and research, and have been considered as efficient tools in the field of marine science and engineering. However an underwater vehicle can only take survey tasks unless equipped with an underwater operation tool. Underwater manipulator is a typical underwater operation tool which is widely used for performing underwater tasks [1]. Recently, autonomous manipulation has been a hot research field in underwater-vehicle manipulator system (UVMS) [2], but master-slave manipulators mounted on remotely-operated vehicles and human occupied vehicles are still important tools, because of their irreplaceable performance in unstructured and complicated environment [3].

A main difference between underwater and land is that pressure increases with water depth at a rate of one bar per 10 meters, which is a crucial factor of hindering development of deep-sea underwater vehicles and manipulators. From materials, manufacture, power supply to communication, the design of deep-sea manipulator is

X. Zhang et al. (Eds.): ICIRA 2014, Part I, LNAI 8917, pp. 117–128, 2014.

far different from that operated on land, and more difficult. Some industrial companies such as Schilling Robotics [4], Hydro-Lek [5] and ISE [6], have already realized manipulator productization, the normally designed depth rating is 6000 meters or less. However oceans deeper than 6000 meters account for the bottom 45 percent of the full ocean depth range, where we barely know about. It is clear that deep-sea manipulator system is an important research field. Published literatures of manipulator mainly concentrate their studies on controller design [7-10] or dynamic analysis [11, 12]. Reference [13] describes a 7-Function hydraulic underwater manipulator which working depth is 3000 meters, including the system configuration, main components, control system and fault and error diagnosis scheme.

As other underwater equipment especially deep-sea equipment, pressure experiment is a very important and necessary step in the manipulator's development. This paper briefly presents the whole system of 7000 meters depth-rating hydraulic manipulator and the 7000m pressure experiment, and finally, actually application testing in the sea is presented. The organization of the paper is as follows. The system design is described in section 2, where mechanical, electronic, slave arm control contrast experiment, algorithm design and simulation are separately represented. 7000m experiment system construction is introduced in section 3. In section 4, the process of 7-Function hydraulic manipulator to carry out integrated application testing is presented. In section 5, we give the conclusion about the design, 7000m pressure experiment and application test in the sea of the manipulator system. There is acknowledgement in section 6.

## 2    The 7-Function Manipulator System Design

### 2.1    System Overview

The 7-function manipulator system could be mounted on remotely-operated vehicles or human occupied vehicles to perform underwater tasks, and normally works in a master-slave mode. The system is divided into surface part and underwater part, the former is composed of a surface control box and a power supply junction box, while the latter is composed of a hydraulic control valve pack, a slave arm, a hydraulic power unit and a pressure compensator, and linked to the surface part with a cable. The whole system configuration is shown in Fig. 1. The slave arm is the system's actual actuator which is powered by hydraulic with 6-DOFs movement joints and a gripper, each joint equipped with a displacement sensor to achieve servo control. The slave arm is controlled by the master arm, besides automatic extension and recovery functions are included [14, 15].

Surface control box is the control terminal of the system, consisting of a master controller, a control panel and a master arm. Operators can choose control mode and set parameters of the system through the function keys on the control panel, an LED monitor is used for visualized human-machine interaction. The master arm is a small scaled model of the slave arm.

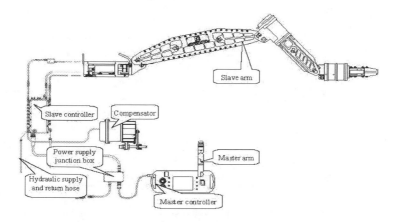

**Fig. 1.** System configuration of the 7-Function manipulator

Hydraulic control valve pack is an underwater sealed container, and filled with compensation oil in order to work stably without variation of water depth. A slave controller and seven servo valve blocks are contained inside the valve pack. The slave controller receives the command from the master controller to drive the motion of the slave arm, and returns the real-time work state of the underwater part. The servo valve block is integrated electro-hydraulic servo valve, relief valve and locking valve.

The power supply junction box is connected to the 220VAC power source and links the surface part with underwater part through a cable. The hydraulic supply and return hoses transport hydraulic oil between the hydraulic power unit and other hydraulic components. A pressure compensator is used to compensate the variation of pressure caused by water depth for the valve pack and other housing sealed components so as to prevent seawater from entering the system.

## 2.2    Mechanical Design

The 7-function hydraulic slave arm is a system's actual actuator to perform underwater tasks, in addition to completing the appropriate action, having sufficient strength and stiffness, but also to facilitate the installation, testing, maintenance and repair, so mechanism design of the slave arm is the emphasis and difficulty of the entire system realization.

The seven functions of the slave arm are composed of 6-DOFs motion and a grip function. The hydraulically powered 7-function slave arm is driven by linear cylinder, rotary actuator and cycloid motor. Displacement sensors integrated inside movement joints and electro-hydraulic servo valve are used to drive the actuator. The main performance specification of the slave arm is shown in Table 1. The joint function configuration, actuator type and nominal range of the slave arm are shown in Table 2.

**Table 1.** Performance specification of the slave arm

| Item | Value |
|---|---|
| Depth rating | 7000 m |
| Weight in air | 85 kg |
| Lift at full extension | 65 kg |
| Maximum reach | 1.9 m |
| Working pressure | 21 MPa |
| Elbow torque | 340 Nm |
| Wrist torque | 290 Nm |
| Wrist rotate | 7~32 rpm |

**Table 2.** Joint function configuration, actuator type and nominal range

| Joint function | Actuator type | Motion range |
|---|---|---|
| Shoulder yaw | Linear | 120 degree |
| Shoulder pitch | Linear | 120 degree |
| Elbow pitch | Linear | 120 degree |
| Elbow yaw | Rotary | 180 degree |
| Wrist pitch | Linear | 120 degree |
| Wrist rotate | Gerotor | 120 degree |
| Gripper | Linear | 140 mm |

Linear motion of cylinder's piston could be converted into rotation through joint transmission, thus is used to drive the motion of shoulder yaw and pitch, elbow pitch, and wrist pitch. Since angular displacement of joint motion can be measured directly or by calculating conversion from piston displacement, integrating the displacement sensor inside linear cylinder would simplify structural design and improve the system stability and modularity. A short linear cylinder driving a linkage mechanism is designed to realize gripper close and open. The diagram of gripper driven mechanism is shown in Fig. 2.

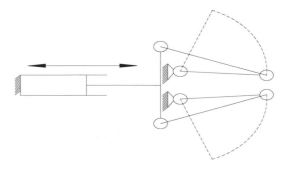

**Fig. 2.** Diagram of the gripper drive mechanism

Rotary actuator which typically used to drive elbow yaw is a compact module to achieve rotary motion of the7-Function slave arm, and is the most difficult module to realize because of high torque-to-volume ratio and multi-path oil passing through mechanism. A double-screw-pair swing rotary actuator is designed to realize rotary driving, multi-path oil and electric cable passing through and rotary angle measuring. As shown in Fig. 3 is the internal mechanism design of multi-path oil and electric cable passing through.

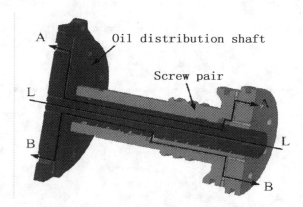

**Fig. 3.** Internal structure diagram of the swing rotary actuator, path A-A and B-B is an example of oil passing through, path L-L is an electric cable passing through

Cycloid motor is a core component to achieve continuous rotation. Restructuring has been carried out to minimize the dimension of wrist and gripper module, which using a cycloid motor directly drives the rotation of the short linear cylinder.

## 2.3    Electronic Design

Electronic system contains surface part and underwater part, which is the master controller and slave controller of the 7-function manipulator separately. The whole architecture diagram of the electronic system is shown in Fig. 4.

The surface part adopts industrial micro-computer as main controller. An LED monitor is connected to the motherboard through video terminal interface, which achieved friendly human-machine interaction. A data acquisition board is used to acquire displacement sensor signal of the master arm, which connected to the motherboard through PCI bus. Since the displacement sensor signal is analog, thus an analog to digital module is possessed on the data acquisition board.

A small and stable controller based SMC technology is used as the slave controller in underwater part, communicating with the surface part through RS-232 serial interface. A data acquisition board with an analog to digital module and a digital to analog module is connected to the controller. The analog to digital module acquires the displacement sensor signal of slave arm, while the digital to analog module converts control signal to analog one exported to servo amplifier. Servo amplifier is a key component of an electro-hydraulic servo control system which drives and controls electro-mechanical conversion device. The underwater electronic subsystem is designed to be directly submerged in compensating oil.

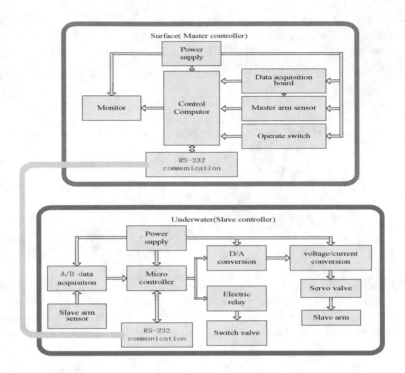

**Fig. 4.** Architecture diagram of electronic design

## 2.4    Algorithm Design and Simulation

Variability of underwater environment, floating behavior of fixed platform, kinematic and dynamic coupling between vehicle and manipulator make the control problem of underwater manipulator more challenging. The 7-Function hydraulic manipulator system works in a master-slave control mode. Slave arm follows the motion of master arm, thus the accuracy of following motion determines the system performance and task execution efficiency.

PID control algorithm is widely used in robotic control schemes for its simplicity and practicality. However control parameters of traditional PID algorithm remain constant, resulting the contradiction between control accuracy and stability difficult to balance. To improve the performance of traditional PID algorithm, a PI control algorithm with variable gains is presented in this section, which achieving a smooth control result with high accuracy and strong stability.

Assuming e is control error, while ė is the differential of control error. PI control algorithm with variable gains is described as follows:

$$\text{Rule 1: if } e \cdot \dot{e} < 0 \text{, then } U(k) = k_p \cdot e \text{ ;}$$

$$\text{Rule 2: if } e \cdot \dot{e} \geq 0 \text{, then } U(k) = k_p \cdot e + k_i \int e \text{ ;}$$

where $k_p$ is proportional coefficient, $k_p = k_1 + k_2 \cdot |e|$, $k_1$ and $k_2$ is a small positive number, $k_i$ is integral coefficient.

The proportional coefficient kp is a function of e. It approaches k1 when e approaching zero, thus proportional control action reduced. As a result, controlled object keeps rising, which helpful to reduce overshoot without increasing rise-time. If the system output pass over a given value, leading to a larger control error, so the integral part is activated that the system returns to a steady state quickly.

In order to verify the effectiveness of the proposed algorithm, motion control experiment of shoulder pitch which is driven by a linear cylinder and bears the largest load variation is conducted. The servo control model of this joint is described in [16]. The open-loop transfer function of piston output displacement relative to servo amplifier input voltage of shoulder pitch becomes (1):

$$\frac{Y(s)}{U(s)} = \frac{\omega_h^2 K_a K_{sv} K_x / A_m}{s^3 + 2x_h \omega_h s^2 + \omega_h^2 s} \tag{1}$$

where $w_h$=352Hz, is hydraulic natural frequency; $x_h$=0.76, is hydraulic damping ratio; $K_a$=1, is transfer function of servo amplifier; $K_{sv}$=5.18×10$^{-3}$, is transfer function of servo valve; $K_x$=2.5, is gain coefficient of servo valve flow; $A_m$=1.264×10$^{-3}$, is average area of piston;

Applying the value into (1), yields (2):

$$\frac{Y(s)}{U(s)} = \frac{1.27 \times 10^6}{s^3 + 535s^2 + 123904s} \tag{2}$$

Simulation results of proposed PI algorithm with variable gains based on Matlab simulation platform is presented in this section. The contrast diagram of step response is shown in Fig. 5.

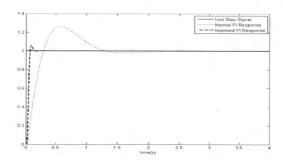

**Fig. 5.** Contrast diagram of step response

In traditional PI control algorithm, integral part is activated to eliminate the steady-state error, but a large overshoot and long rise-time occurs. The PI algorithm with variable gains only introduces integral part when output deviating from the target value, and increases the proportional coefficient to reduce settling time, thus improving the system real-time performance.

## 2.5 Slave Arm Control Contrast Experiment

The proposed algorithm is applied in the motion control of shoulder pitch of actual 7-function hydraulic manipulator system. Working hydraulic pressure is set at 10 MPa, and control period time is 0.2 s. The experiment result is shown in Fig. 6.

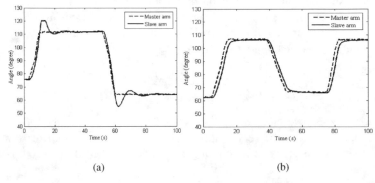

          (a)                                        (b)

**Fig. 6.** Contrast experiment of (a) traditional PI algorithm and (b) PI algorithm with variable gains

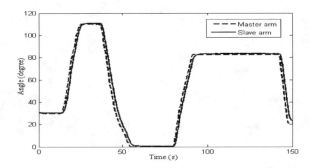

**Fig. 7.** PI algorithm with variable gains of increased coefficients

    In traditional PI algorithm, $k_p$=0.4, $k_i$=0.1; while $k_p$=0.4+0.03×|e|, $k_i$=0.1 in proposed algorithm. The result shows that the proposed algorithm doesn't have overshoot, but a little hysteresis occurs. This is because the control is weak when output close to the target value, and integral part is not obvious. To improve the control performance, a larger $k_p$ and $k_i$ is applied to the system. The experiment result is shown in Fig. 7.

# 3     7000M Experiment System Design

## 3.1     Experiment System Overview

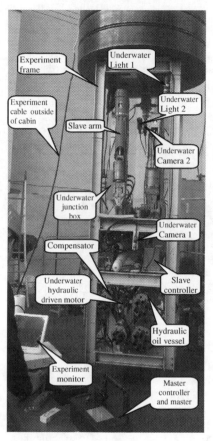

| name | function | Location |
|------|----------|----------|
| Master controller and arm | Send control command | Outside pressure cabin |
| Experiment monitor and video recorder | Show and record the two camera's video | |
| Experiment frame | Underwater equipment fixed base | inside pressure cabin |
| Slave arm | Experiment object | |
| Slave controller | | |
| Underwater junction box | Equipment cable junction | |
| Light 1 and 2 | Supply light | Fixed on experiment frame, inside pressure cabin |
| Camera 1 | Compensator oil compressing status monitor | |
| Camera 2 | Slave arm motion monitor | |
| Compensator | Junction box, salve controller housing oil compensation | |
| Hydraulic system | hydraulic supply for slave arm | |

**Fig. 8.** The 7000m pressure whole experiment system configuration location and function of equipments in the experiment

The whole 7000m pressure experiment system configuration and the location and function of equipments in the experiment are shown as Fig. 8. Besides the manipulator system, assistant equipments (underwater light, camera etc.) and hydraulic system (AC motor, pump, compensator and hydraulic oil in vessel) are included.

## 3.2     Experiment Process Design

The experiment is to validate（1）the slave controller and salve arm could endure hydraulic pressure correspond with 0~7000m ocean pressure,（2）no structure and component  of the slave controller and salve arm should be destroyed and  （3） the 6-DOFs movement joints and gripper of the salve arm should work well.

In the experiment, every 10MPa added of the cabin water pressure from 0Mpa to 71.5MPa （in freshwater, 71.5MPa is   equivalent to pressure arising in 7000m seawater）, the master arm would be operated to monitor the slave arm joint and gripper movement in small range. The master arm operating and the slave arm monitor scenes are shown as Fig. 9. To validate the manipulator underwater subsystem's structure intensity, we choose 78.65MPa as the max experiment pressure(in this case, the manipulator would not be operated), which is 1.1 times of the max working pressure.

**Fig. 9.** Master arm operating and slave arm experiment scene

### 3.3    Problems in the Experiment

In the experiment, some phenomenon that didn't occur in pool or low pressure environment arise when the experiment pressure is above 50MPa,  the most critical problem is that the double-screw-pair swing rotary actuator stop rotating. The reason is that some gaps are not filled with compensating oil, so great pressure difference between a sealing plane make the friction large enough to stop the actuator's rotating. Some holes are drilled to fill the cavity with compensating oil. Complementary experiment aiming at the double-screw-pair swing rotary actuator in high pressure environment are carried out, the actuator rotating normally in pressure from 0MPa to 71.5MPa in the cabin.

## 4     Application Test in the Sea

In 2014, during from May 8 to May 25，one type of ROV be equipped with the 7-function hydraulic manipulator to carry out integrated application testing (as shown in Fig. 10) .In the application testing, the power of surface control box and master arm comes from the sea surface, while the slave arm and hydraulic control valve pack from ROV. The distance of fiber-optic communications nearly has 1700m between surface from underwater controller.

**Fig. 10.** ROV equipment with7-Function manipulator and deep-sea operation

The 7-Function hydraulic manipulator was in good condition during carry out integrated application testing. The test showed that underwater movement function is normal and could successfully complete the underwater fetching tasks. The user gave a high evaluation about the 7-function hydraulic manipulator technology state and surface friendly human-machine interaction function.

## 5     Conclusion

The design, 7000m pressure experiment and application test in the sea of a deep-sea hydraulic manipulator are proposed in this paper. The system mainly consists of surface part including surface control box and a power supply junction box, and underwater part including a hydraulic control valve pack, a slave arm, a hydraulic power unit and a pressure compensator. A detailed describes of system design including mechanical structure, electronic subsystem, algorithm and slave arm control contrast experiment are represented subsequently. Some phenomena never occurred in running on land arose in 30MPa and 40MPa cabin hydraulic pressure, the reasons are analyzed and complementary experiments are carried on later. The whole system's and complementary experiment show that the manipulator designs satisfy the 7000m design depth index. During application testing, 7-function hydraulic manipulator underwater movement function was normal and successfully completed the underwater fetching tasks.

**Acknowledgment.** The authors would graciously acknowledge the funding provided by the China national 863 plan (2012AA091101), and also acknowledge the support for this work provided by the Dept. of Ocean Technology R&D, Shenyang Institute of Automation, thank for Qingmei Wang for her software support about the manipulator system.

## References

1. Dunnigan, M.W., Lan, D.M., Clegg, A.C., Edwards, I.: Hybrid position/force control of a hydraulic underwater manipulator. IEEE Proceedings: Control Theory and Applications 143, 145–151 (1996)

2.  Marani, G., Choi, S.K., Yuh, J.: Underwater autonomous manipulation forintervention missions AUVs. Ocean Engineering 36, 15–23 (2009)
3.  Shim, H., Jun, B.H., Lee, P.M., Baek, H., Lee, J.: Workspace control system of underwater tele-operated manipulators on an ROV. Ocean Engineering 37, 1036–1047 (2010)
4.  Schilling robotics [EB/OL] (January 4, 2011), http://www.schilling.com/
5.  Hydro-Lek innovative design engineering [EB/OL] (January 4, 2011), http://www.hydro-lek.com
6.  International Submarine Engineering [EB/OL] (January 4, 2011), http://www.ise.bc.ca/
7.  Shim, H., Jun, B.H., Lee, P.M., Kim, B.: Dynamic workspace control method for underwater manipulator of floating ROV. International Journal of Precision Engineering and Manufacturing 14, 387–396 (2013)
8.  Shen, X., Xu, G., Yu, K., Tang, G., Xu, X.: Development of a deep ocean master-slave electric manipulator control system. In: Jeschke, S., Liu, H., Schilberg, D. (eds.) ICIRA 2011, Part II. LNCS, vol. 7102, pp. 412–419. Springer, Heidelberg (2011)
9.  Yao, J.J., Wang, C.J.: Model reference adaptive control for a hydraulic underwater manipulator 18, 893–902 (2012)
10. Ryu, J.H., Kwon, D.S., Lee, P.M.: Control of underwater manipulators mounted on an rov using base force information. In: Proceedings of the 2001 IEEE International Conference on Robotics and Automation, vol. 4, pp. 3238–3243 (2001)
11. Levesque, B., Richard, M.: Dynamic Analysis of a Manipulator in a Fluid Environment. International Journal of Robotics Research 13, 221–231 (1994)
12. Kawamura, S., Sakagami, N.: Analysis on dynamics of underwater robot manipulator basing on iterative learning control and time-scale transformation. In: Proceedings of the 2002 IEEE International Conference on Robotics and Automation, Washington, pp. 1088–1094 (2002)
13. Yao, J.J., Wang, L.Q., Jia, P., Wang, Z.: Development of a 7-Function hydraulic underwater manipulator system. In: Proceedings of the 2009 IEEE International Conference on Mechatronics and Automation, Changchun, China, pp. 1202–1206 (2009)
14. Zhang, Q., Chen, J., Huo, L., Sun, B., Zhao, Y., Shengguo, C.: Design and Experiments of a Deep-sea Hydraulic Manipulator System, Oceans, San Diego (2013)
15. Huo, L., Zhang, Q., Zhang, Z.: A method of inverse kinematics of a 7-Function underwater hydraulic manipulator, Oceans, San Diego (2013)
16. Li, L., Sun, B., Zhang, Q.: Model and Simulation Analysis of Asymmetrical Hydraulic Cylinder Controlled by Servo-valve. Coal Mine Machinery 32(10), 89–91 (2011)

# Optimal Sensors Deployment for Tracking Level Curve Based on Posterior Cramér-Rao Lower Bound in Scalar Field

Wentao Zhao[1,2], Jiancheng Yu[1], Aiqun Zhang[1], and Yan Li[1]

[1] State Key Laboratory of Robotics, Shenyang Institute of Automation,
Chinese Academy of Sciences, 110016, Shen Yang, China
[2] University of Chinese Academy of Sciences, Beijing 100049, China
{yjc,zhaowt}@sia.cn

**Abstract.** This paper focuses on discussing the space distance of gliders in a group for level curve tracking task. A developed adaptive space distance algorithm for glider formation based on Posterior Cramér-Rao Lower Bound (PCRLB) is proposed. For a feature-tracking application with scalar sensors, gliders are adopted to track a level curve in 2D space. In this work, the white noise from the measurement process and oceanic background is taken into account, as well as the effect of omitting the higher order terms in the Taylor series and roughly estimated Hessian Matrix. Since the PCRLB is an effective criterion to quantify the performance of all unbiased nonlinear estimators of the target state, our adaptive space distance algorithm for gliders may be functional when implemented with many kinds of nonlinear filters together. Finally, the performance of the proposed algorithm in this study is evaluated on simulated platforms by applying it with the Extended Kalman Filter(EKF) and Particle Filter.

**Keywords:** Adaptive space distance algorithm, Posterior Cramér-Rao Lower Bound (PCRLB), Gliders cooperation, Feature tracking, Level curve.

## 1    Introduction

Underwater gliders as energy efficient vehicles have been used in a number of shallow and deeper-water missions such as ocean sampling and surveillance. Since the maneuverability of gliders increased, also the cost of implementing gliders, sensors and communications decreased, cooperation of gliders become realized[1, 2]. Leonard and Fiorelli present a framework for coordinate and distributed control of multiple autonomous vehicles using artificial potentials and virtual leaders[3]. And then, Leonard et al. address the design of mobile sensor networks with gliders formation[4]. Smith present algorithms that determine paths for AUVs to track evolving of interest in the ocean by considering the output of predictive ocean models, with the intent to increase the skill of future predictions in the local region [5].

X. Zhang et al. (Eds.): ICIRA 2014, Part I, LNAI 8917, pp. 129–141, 2014.

For a feature tracking mission a cooperative group of mobile sensor gliders are expected to be superior to a single glider. The space distance between the center of glider formation and individual in the team is an important parameter for gliders team. Derek, Zhang and Leonard design and demonstrate an automated control system that performs feedback control at the level of the fleet, in 2008[6]. By extending the adaptive scheme previously[2], Zhang and Leonard present a kind of cooperative EKF with adaptive formation method to minimize the estimation error based on instantaneous measurements[7]. To detect and track one of the lines of curvature on a desired level surface, Wu and Zhang study the problem of controlling the formation[8], and investigate the problem of using multiple mobile sensing agents to track one moving target[9].

Posterior Cramér-Rao Lower Bound (PCRLB) has been used in optimal placement of sensors in various fields. Hernandez, Kirubarajan and Bar-Shalom use the PCRLB to optimize multisensory resource deployment in target tracking[10, 11]. Zhong, Premkumar and Madhukumar employ Posterior Cramér-Rao Bound with Particle Filter together for 2D direction of arrival tracking utilizing an acoustic vector sensor[12].

While exploring ocean, scientist often expect to use a fleet of platforms to track the level curve in scalar fields such as a temperature or a salinity field. Through study we are aware of that the space distance of the formation can impact the accuracy of the estimation of the scalar value and gradient at the center of the glider formation. As such, this paper proposes a method to use the PCRLB to adapt the space distance between the agents during this kind of task. In real ocean situation, the field usually contains background noise and the measurements noise taken by sensors which is inevitable. And the Taylor series is the common way to approximatively model the scalar field. However, because the data obtained from the gliders is limited, the Hessian Matrix is always roughly estimated by using the history data, also we must ignore the high order items of the Taylor series in the model of state dynamics and measurement. Based on the PCRLB technology, we will take these influences into account to investigate an adaptive algorithm for space distance between the center of gliders formation and individual in the team.

In this paper we consider a task to track a level curvature in a 2D scalar field using glider formation. The glider in the field can measure the scale value and feedback to the controller. The controller will calculate the optimal space distance of the formation for the next sample step, using our adaptive algorithm. Since the PCRLB is a useful criterion to quantify the performance of all unbiased estimators of the target state, thus we can use the PCRLB to optimize the space distance in the team. The PCRLB is an evaluation criterion independent from the filter, therefore the algorithm we present in this paper is not bounded with a certain kind of filter. This makes the choice of filter more flexible for different feature tracking task.

This paper is organized as follows. In section 2, we illustrate models of the state and measurement used in this paper. In section 3, the PCRLB is applied to adapt the space distance of the formation. Firstly, we review the recursive equation giving the sequence of PCRLB. Subsequently, we use the model in section 2 to get the simplified PCRLB recursion. In section 4, the simulation results are presented to show the performance of our formation space distance adaptive algorithm. Besides, we give out the results by applying our algorithm with different kind of filters to show the universality of our algorithm. Future works in section 6 show some effects that can be done to make the algorithm work more robustly and effectively.

## 2    Model of State Dynamics and Measurement

We are interested in the scalar value and gradient vector at the center of the glider formation, when the task is to track a level curve in a scalar field. Thus we define state variables vector in step k as $x_k = \left[ Z_{c,k}, \nabla Z_{c,k}^T \right]^T$. In the state variables vector, $Z_{c,k}$ is the scalar value in the center and $\nabla Z_{c,k}^T$ represents the gradient at this position. Assuming that in the formation there are N gliders, we let $r_{c,k} = \left( r_{c,k,x}, r_{c,k,y} \right)$ , $r_{i,k} = \left( r_{i,k,x}, r_{i,k,y} \right), i = 1 \dots N$ on behalf of the position vector of formation center and of corresponding agent respectively. The relationship between $r_{c,k}$ and $r_{i,k}$ is $r_{c,k} = \dfrac{1}{N} \sum_{i=1}^{N} r_{i,k}$ .

### 2.1    State Model

Slightly modifying the state dynamics model addressed by Ogren[2] and Zhang[7], then we give out the state model as    by changing the covariance matrix. The elements in this equation have the form expressed in .

$$x_{k+1} = A_k^s x_k + h_k + \varepsilon_p$$
$$\varepsilon_p \in N(0, \Sigma) \tag{2.1}$$

$$A_k^s = \begin{bmatrix} 1 & \left( r_{c,k+1} - r_{c,k} \right)^T \\ 0 & I_{2\times2} \end{bmatrix}$$
$$h_k = \left[ 0, E\left[ H_{c,k} \left( r_{c,k+1} - r_{c,k} \right) \right] \right]^T \tag{2.2}$$
$$H_{c,k} = \begin{bmatrix} H_{c,k,11} & H_{c,k,12} \\ H_{c,k,21} & H_{c,k,22} \end{bmatrix}$$

$$\Sigma = \text{diag}\begin{pmatrix} \left(\frac{1}{2}E\left[\left(r_{c,k+1}-r_{c,k}\right)\otimes\left(r_{c,k+1}-r_{c,k}\right)^{T}\bar{H}_{c,k}\right]\sigma_{pH}\right)^{2}+\sigma_{pe}^{2}, \\ \left(E\left[H_{c,k}\left(r_{c,k+1}-r_{c,k}\right)\right]_{1}\sigma_{pH}\right)^{2}+\sigma_{pe}^{2}, \\ \left(E\left[H_{c,k}\left(r_{c,k+1}-r_{c,k}\right)\right]_{2}\sigma_{pH}\right)^{2}+\sigma_{pe}^{2} \end{pmatrix} \qquad (2.3)$$

Especially, since we just can roughly estimate the Hessian Matrix, and must omit the high order terms in Taylor series, we must consider the effect of these in $\Sigma$. In the equation of $\Sigma$, we have $E\left[H_{c,k}\left(r_{c,k+1}-r_{c,k}\right)\right]_{1}$ and $E\left[H_{c,k}\left(r_{c,k+1}-r_{c,k}\right)\right]_{2}$ represent the first and second row of $E\left[H_{c,k}\left(r_{c,k+1}-r_{c,k}\right)\right]$ respectively. And $\sigma_{pH}$ is a factor that can be altered following error caused by rough Hessian estimation and omitting high order terms in the Taylor series. Similarly, $\sigma_{pe}^{2}$ reflect the effect of the process noise and positioning errors.

Though the equation, we can get the possibility density function (pdf) of the state dynamics expressed as equation .

$$p\left(x_{k+1} \mid x_{k}\right) = \text{N}\left(A_{k}^{s}x_{k}+h_{k},\Sigma\right) \qquad (2.4)$$

## 2.2   Measurement Model

Similar to the state dynamics model, we modify the model used by Ogren[2] and Zhang [7]. And $y_{k} = \left[y_{1},\cdots,y_{i},\cdots,y_{N}\right]^{T}$ is used to represent the measurement vector, then the Measurement Model can be shown as the following equation .

$$y_{k} = C_{k}x_{k}+D_{k}\bar{H}_{c,k}+\varepsilon_{H}+\varepsilon_{M} \qquad (2.5)$$

The elements in the upper equation have the form expressed in equation . In this equation, $\otimes$ represent the Kronecker product; $\bar{H}_{c,k}$ is the vector form of Hessian Matrix. $\varepsilon_{H}$ is to reflect error caused by rough Hessian estimation and omitting high order terms in the Taylor series, and $\varepsilon_{M}$ is an element determined by the noise caused by measurement and background. We assume that $\varepsilon_{H,i}$ and $\varepsilon_{M,i}$ are i.i.d. Gaussian noises.

$$C_k = \begin{bmatrix} 1 & \left(r_{1,k} - r_{c,k}\right)^T \\ \vdots & \vdots \\ 1 & \left(r_{N,k} - r_{c,k}\right)^T \end{bmatrix}$$

$$D_k = \begin{bmatrix} \frac{1}{2}\left(r_{1,k} - r_{c,k}\right) \otimes \left(r_{1,k} - r_{c,k}\right)^T \\ \vdots \\ \frac{1}{2}\left(r_{N,k} - r_{c,k}\right) \otimes \left(r_{N,k} - r_{c,k}\right)^T \end{bmatrix}$$

$$\vec{H}_{c,k} = \left[ H_{c,k,11}, H_{c,k,12}, H_{c,k,21}, H_{c,k,22} \right]^T$$

$$\varepsilon_H = \left[ \varepsilon_{H,1}, \cdots, \varepsilon_{H,i}, \cdots, \varepsilon_{H,N} \right]^T \tag{2.6}$$

$$\varepsilon_{H,i} \sim N\left( 0, \left( \frac{1}{2}\left(r_{i,k} - r_{c,k}\right) \otimes \left(r_{i,k} - r_{c,k}\right)^T \vec{H}_{c,k}\sigma_H \right)^2 \right)$$

$$\varepsilon_M = \left[ \varepsilon_{M,1}, \cdots \varepsilon_{M,i} \cdots, \varepsilon_{M,N} \right]^T$$

$$\varepsilon_{M,i} \sim N\left(0, \sigma_M^2\right)$$

Eventually, we get the pdf of measurement model represented by equation . In this pdf, Similar with the first diagonal element in $\Sigma$, we use $\Sigma_y$ to synthesize the effect of $\varepsilon_H$ and $\varepsilon_M$ in expression . As the covariance matrix , $\Sigma_y$ has the form shown in equation .

$$p\left(y_k \mid x_k\right) = N\left(\left(C_k x_k + D_k \vec{H}_{c,k}\right), \Sigma_y\right) \tag{2.7}$$

$$\Sigma_y = diag \begin{bmatrix} \left(\frac{1}{2}\left(r_{1,k} - r_{c,k}\right) \otimes \left(r_{1,k} - r_{c,k}\right)^T \vec{H}_{c,k}\sigma_H\right)^2 + \sigma_M^2 \\ \vdots \\ \left(\frac{1}{2}\left(r_{i,k} - r_{c,k}\right) \otimes \left(r_{i,k} - r_{c,k}\right)^T \vec{H}_{c,k}\sigma_H\right)^2 + \sigma_M^2 \\ \vdots \\ \left(\frac{1}{2}\left(r_{N,k} - r_{c,k}\right) \otimes \left(r_{N,k} - r_{c,k}\right)^T \vec{H}_{c,k}\sigma_H\right)^2 + \sigma_M^2 \end{bmatrix} \tag{2.8}$$

## 3    PCRLB for Adaptive Space Distance

It is always difficult thing to quantify the performance of a nonlinear filter. The PCRLB is the inverse of the Fisher Information Matrix and it gives the lower bound for the performance of unbiased estimator[13]. We let $\hat{x}_{k+1}$ be the unbiased estimator of $x_{k+1}$ based on the measurements $y_k$. Then we have the inequality equation as . In this inequality $J^{-1}$ represents the PRCLB matrix.

$$E\left[ \left( \hat{x}_k - x_k \right) \left( \hat{x}_k - x_k \right)^{\mathrm{T}} \right] \geq J^{-1} \tag{3.1}$$

As a important evaluation criterion for nonlinear estimator, Šimandl provide the recursion for PCRLB as equation.

$$F_{k+1|k} = K_{k+1}^{k+1} - K_{k+1}^{k+1,k} \left( K_{k|k}^{k} + F_{k|k} \right)^{-1} K_{k+1}^{k,k+1}$$

$$F_{k|k} = F_{k|k-1} + L_k^k$$

$$K_{k+1}^{k} = E\left\{ -\nabla_{x_k} \left[ \nabla_{x_k} \ln p\left( x_{k+1} \mid x_k \right) \right]^{\mathrm{T}} \right\}$$

$$K_{k+1}^{k,k+1} = E\left\{ -\nabla_{x_{k+1}} \left[ \nabla_{x_k} \ln p\left( x_{k+1} \mid x_k \right) \right]^{\mathrm{T}} \right\} = \left( K_{k+1}^{k+1,k} \right)^{\mathrm{T}}$$

$$K_{k+1}^{k+1} = E\left\{ -\nabla_{x_{k+1}} \left[ \nabla_{x_{k+1}} \ln p\left( x_{k+1} \mid x_k \right) \right]^{\mathrm{T}} \right\} \tag{3.2}$$

$$L_k^k = E\left\{ -\nabla_{x_k} \left[ \nabla_{x_k} \ln p\left( y_k \mid x_k \right) \right]^{\mathrm{T}} \right\}$$

$$K_0^0 = E\left\{ -\nabla_{x_0} \left[ \nabla_{x_0} \ln p\left( x_0 \right) \right]^{\mathrm{T}} \right\}$$

$$F_{0|-1} = K_0^0$$

In the recursion , there are some factors we should evaluate to use it under our model for the state dynamics and measurement. Thus we give the following equations for these factors. In these equation, $\left( r_{c,k+1} - r_{c,k} \right)_x$, $\left( r_{c,k+1} - r_{c,k} \right)_x$ represent the x-axis and y-axis value of $\left( r_{c,k+1} - r_{c,k} \right)$, respectively. $\Sigma_i^{-1}, i = 1,2,3$ represents the i-th diagonal element of $\Sigma^{-1}$, so as $\Sigma_{y,i}^{-1}$.

$$K_{k+1}^{k} = E\left\{ \begin{bmatrix} \Sigma_1^{-1} & \Sigma_1^{-1}\left( r_{c,k+1} - r_{c,k} \right)_x & \Sigma_1^{-1}\left( r_{c,k+1} - r_{c,k} \right)_y \\ \Sigma_1^{-1}\left( r_{c,k+1} - r_{c,k} \right)_x & \Sigma_1^{-1}\left( r_{c,k+1} - r_{c,k} \right)_x^2 + \Sigma_2^{-1} & \Sigma_1^{-1}\left( r_{c,k+1} - r_{c,k} \right)_x \left( r_{c,k+1} - r_{c,k} \right)_y \\ \Sigma_1^{-1}\left( r_{c,k+1} - r_{c,k} \right)_y & \Sigma_1^{-1}\left( r_{c,k+1} - r_{c,k} \right)_x \left( r_{c,k+1} - r_{c,k} \right)_y & \Sigma_1^{-1}\left( r_{c,k+1} - r_{c,k} \right)_y^2 + \Sigma_3^{-1} \end{bmatrix} \right\} \tag{3.3}$$

$$K_{k+1}^{k,k+1} = \left(K_{k+1}^{k+1,k}\right)^{\mathrm{T}} = E\left\{-1\begin{bmatrix} \Sigma_1^{-1} & \Sigma_1^{-1}\left(r_{c,k+1}-r_{c,k}\right)_x & \Sigma_1^{-1}\left(r_{c,k+1}-r_{c,k}\right)_y \\ 0 & \Sigma_2^{-1} & 0 \\ 0 & 0 & \Sigma_3^{-1} \end{bmatrix}\right\} \tag{3.4}$$

$$K_{k+1}^{k+1} = E\left\{\frac{1}{2}\begin{bmatrix} 2\Sigma_1^{-1} & 0 & 0 \\ 0 & 2\Sigma_2^{-1} & 0 \\ 0 & 0 & 2\Sigma_3^{-1} \end{bmatrix}\right\} = E\left\{\begin{bmatrix} \Sigma_1^{-1} & 0 & 0 \\ 0 & \Sigma_2^{-1} & 0 \\ 0 & 0 & \Sigma_3^{-1} \end{bmatrix}\right\} \tag{3.5}$$

$$L_k^k = E\left\{\begin{bmatrix} \Sigma_{y,1}^{-1} & \cdots & \Sigma_{y,i}^{-1} & \cdots & \Sigma_{y,N}^{-1} \\ \Sigma_{y,1}^{-1}\left(r_{1,k}-r_{c,k}\right)_x^{\mathrm{T}} & \cdots & \Sigma_{y,i}^{-1}\left(r_{1,k}-r_{c,k}\right)_x^{\mathrm{T}} & \cdots & \Sigma_{y,N}^{-1}\left(r_{N,k}-r_{c,k}\right)_x^{\mathrm{T}} \\ \Sigma_{y,1}^{-1}\left(r_{1,k}-r_{c,k}\right)_y^{\mathrm{T}} & \cdots & \Sigma_{y,i}^{-1}\left(r_{1,k}-r_{c,k}\right)_y^{\mathrm{T}} & \cdots & \Sigma_{y,N}^{-1}\left(r_{N,k}-r_{c,k}\right)_y^{\mathrm{T}} \end{bmatrix}\begin{bmatrix} 1 & \left(r_{1,k}-r_{c,k}\right)_x^{\mathrm{T}} & \left(r_{1,k}-r_{c,k}\right)_y^{\mathrm{T}} \\ \vdots & \vdots & \vdots \\ 1 & \left(r_{1,k}-r_{c,k}\right)_x^{\mathrm{T}} & \left(r_{1,k}-r_{c,k}\right)_y^{\mathrm{T}} \\ \vdots & \vdots & \vdots \\ 1 & \left(r_{N,k}-r_{c,k}\right)_x^{\mathrm{T}} & \left(r_{N,k}-r_{c,k}\right)_y^{\mathrm{T}} \end{bmatrix}\right\} \tag{3.6}$$

After getting $F_{k|k}$, in step k we can easily evaluate the $F_{k+1|k}$. Because in the ocean, scalar field is relative steadily, thus the Hessian matrix change slowly. So, when predict the $L_{k+1}^{k+1}$, denoted by $L_{k+1,P}^{k+1}$, we can use the Hessian matrix that we got in step k. Through the $L_{k+1,P}^{k+1}$, we evaluate the $F_{k+1|k+1,P}$ eventually. With the control algorithm addressed by Zhang[14], we assume that we can control the formation hold the equilateral polygon shape. In equilateral polygon shape formation, we get the equation for $d_f$ and $r_{i,k} - r_{c,k}$ expressed by equation , and in this equation $\psi$ represents the heading angle of glider formation. Thus, the space distance between dual gliders in the formation could be determined once space distance from the center of the formation to every agent in the team, denoted by $d_f$ is ascertained. Denoted by $\psi$ in Fig. 1, heading angle of the fleet is controlled by the algorithm addressed by Zhang[14], as expressed by equation . In this equation, $Z_{\mathrm{Tag}} - Z_{Cur}$ means the difference between the current estimated formation center scalar value and value on the target level curve. Besides, $\angle\dfrac{\nabla Z_{Cur}}{\|\nabla Z_{Cur}\|}$ represents the direction of the current estimated gradient vector; $\angle\dfrac{\nabla Z_{Cur}^{\perp}}{\|\nabla Z_{Cur}^{\perp}\|}$ represents the normal direction of the current level curve, pointing to the target level curve. $w$ is a factor determining the intensity of the heading angle control.

Maximizing the determinant of $F_{k+1|k+1,P}$, we get the optimized space distance $d_{f\_op,k+1}$ for the next sample step as shown in equation .

$$\left(r_{i,k} - r_{c,k}\right)^{\mathrm{T}} = \left(d_f \times \cos(\psi + \frac{i-1}{N}), d_f \times \sin(\psi + \frac{i-1}{N})\right) \qquad (3.7)$$

$$\psi = \frac{w\psi_{\nabla \mathbf{Z}_{Cur}^{\perp}} + (1-w)\psi_{\nabla \mathbf{Z}_{Cur}}}{\left\| w\psi_{\nabla \mathbf{Z}_{Cur}^{\perp}} + (1-w)\psi_{\nabla \mathbf{Z}_{Cur}} \right\|}$$

$$\psi_{\nabla \mathbf{Z}_{Cur}} = \mathrm{atan}\left(\left\| Z_{\mathrm{Tag}} - Z_{Cur} \right\|\right) \angle \frac{\nabla \mathbf{Z}_{Cur}}{\left\| \nabla \mathbf{Z}_{Cur} \right\|} \qquad (3.8)$$

$$\psi_{\nabla \mathbf{Z}_{Cur}^{\perp}} = \mathrm{atan}\left(\frac{1}{\left\| Z_{\mathrm{Tag}} - Z_{Cur} \right\|}\right) \angle \frac{\nabla \mathbf{Z}_{Cur}^{\perp}}{\left\| \nabla \mathbf{Z}_{Cur}^{\perp} \right\|}$$

$$\mathbf{F}_{k+1|k+1,\mathrm{P}} = \mathbf{F}_{k+1|k} + \mathbf{L}_{k+1,\mathrm{P}}^{k+1}$$

$$d_{f\_op,k+1} = \max_{d_f \in \mathrm{R}} \left[ \det(\mathbf{F}_{k+1|k+1,\mathrm{P}}) \right] \qquad (3.9)$$

# 4    Simulation Results

As shown in the state dynamics model ( or measurement model) in this paper , we treat the effect of Hessian Matrix and the high order terms in Taylor series as noise generator. And add effects of them to the covariance matrix of the pdf directly, through multiplying $\sigma_{pH}$ (or $\sigma_H$). This will make our algorithm sensitive to the changing circumstance. Also compared with the adaptive formation algorithm addressed by Zhang[7], our algorithm can implement with different nonlinear filters. We will show the universality of our algorithm by applying it with EKF and particle filter.

In the simulation, we have 3 gliders in the fleet, and the sketch map of the glider formation is shown in Fig. 1. The expression for the scalar field is $T = \frac{(x^2 - 25)^2}{500} + \frac{y^2}{8}$ , and we track the level curve $T_{Con} = 3$. During the simulation, we let $\sigma_{pH} = \sigma_H = 1$ . And at the simulation beginning, $p(x_0)$ and $d_{f\_op,1}$ could be set arbitrary by just roughly detecting the circumstance in the first step, and it just influence the results slightly.

Several results implementing our algorithm to optimize the space distance in the glider formation are given out. From the upper left to the low right, figures are about the position of formation center; optimized space distance; error of estimated center scalar value(EECSV) and square error of estimated center gradient vector(SEECGV).

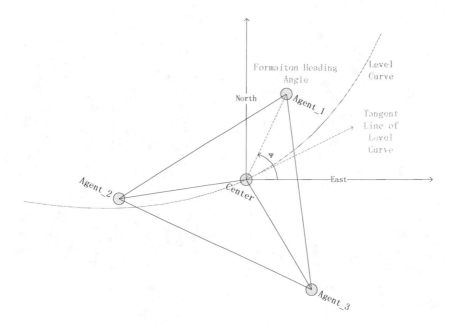

**Fig. 1.** Sketch map of the glider formation

Firstly, the process noise rate $\sigma_{pe}$ and observation noise rate $\sigma_M$ are set to be 0.1. We present the result applying our algorithm together with particle filter in Fig. 2.

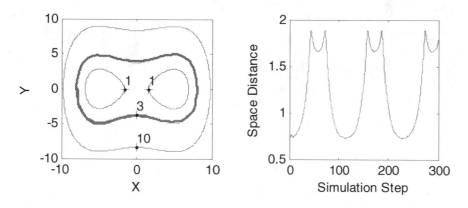

**Fig. 2.** Results of Particle Filter

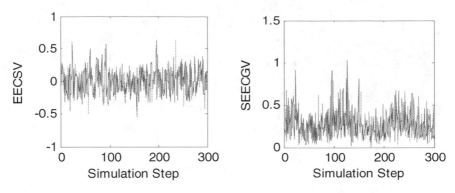

**Fig. 2.** (*Continued*)

Subsequently, we utilize our algorithm together with EKF under the same value of $\sigma_{pe}$ and $\sigma_M$. Results are shown in Fig. 3.

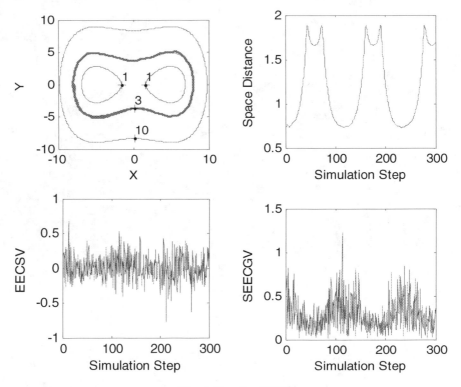

**Fig. 3.** Results of EKF

Thirdly, the process noise rate $\sigma_{pe}$ and observation noise rate $\sigma_M$ are set to be 0.1 at beginning. But we double the observation noise during the last half simulation process, i.e., $\sigma_M = 0.2$. And our algorithm is used together with particle filter in this time. Results is shown in Fig. 4.

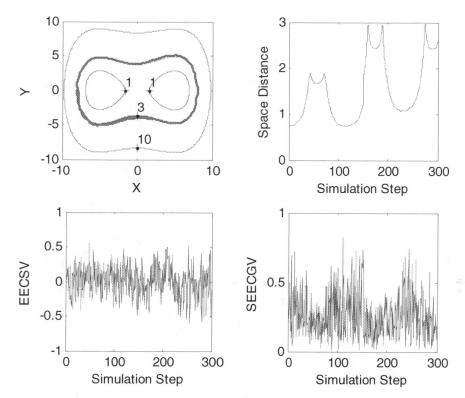

**Fig. 4.** Results of Particle Filter
Double the observation noise during the last half simulation process

At last, the observation noise rate is doubled at the last half simulation process too. However, the EKF is employed together with our algorithm. The results are shown in Fig. 5.

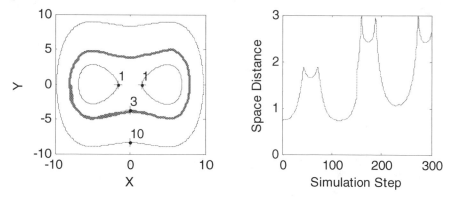

**Fig. 5.** Results of EKF
Double the observation noise during the last half simulation process

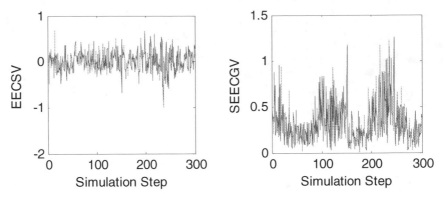

**Fig. 5.** (*Continued*)

Through upper results, we know that our algorithm can be integrated with different nonlinear filters, thus we can employ different filters with different feature tracking tasks. This will be a great advantage for the adaptive space distance algorithm based on PCRLB. Through analyzing results in Fig. 2 and Fig. 3, we know that at different position of the scalar field, our algorithm can adapt the space distance of the formation effectively. Besides, results in Fig. 4 and Fig. 5 show us that our algorithm can adjust the space distance in the formation to adapt with the changing measurement noise quickly.

Also, comparing Fig. 2 with Fig. 4 or comparing Fig. 3 with Fig. 5, we are aware that although observation noise is doubled but the EECSV and SEECGV are not doubled. It is because that after the observation noise is doubled, our algorithm make the optimized space distance larger than before, thus the EECSV and SEECGV are restrained relatively.

## 5    Future Works

Since the glider could be asynchronous when back to the sea level, thus we should develop distributed/self-organized adaptive algorithm for the teamwork and space distance of the formation.

The factor $\sigma_{pH}$ (**or** $\sigma_H$ ) could be adaptive during the tracking process. And the Hessian estimation could be more accurate using more history data or we could use other nonlinear model to represent the state transmission. Also, in real ocean circumstance the background noise in measurement model is not i.i.d. Gaussian noise and it should be colored noise. Therefore, in the future work, we will consider and discuss these in our model.

**Acknowledgment.** The work of this article is in part supported by National Natural Science foundation under grant 61233013.

# References

1. Fiorelli, E., Leonard, N.E., Bhatta, P., Paley, D., Bachmayer, R., Fratantoni, D.M.: Multi-AUV Control and Adaptive Sampling in Monterey Bay. IEEE Journal of Oceanic Engineering 31(4), 935–948 (2006)
2. Ogren, P., Fiorelli, E., Leonard, N.E.: Cooperative control of mobile sensor networks: Adaptive gradient climbing in a distributed environment. IEEE Transactions on Automatic Control 49(8), 1292–1302 (2004)
3. Leonard, N.E., Fiorelli, E.: Virtual leaders, artificial potentials and coordinated control of groups. In: Proceedings of the 40th IEEE Conference on Decision and Control, pp. 2968–2973. IEEE (2001)
4. Leonard, N.E., Paley, D.A., Lekien, F., Sepulchre, R., Fratantoni, D.M., Davis, R.E.: Collective Motion, Sensor Networks, and Ocean Sampling. Proceedings of the IEEE 95(1), 48–74 (2007)
5. Smith, R.N., Yi, C., Li, P.P., Caron, D.A., Jones, B.H., Sukhatme, G.S.: Planning and Implementing Trajectories for Autonomous Underwater Vehicles to Track Evolving Ocean Processes Based on Predictions from a Regional Ocean Model. The International Journal of Robotics Research 29(12), 1475–1497 (2010)
6. Paley, D.A., Zhang, F., Leonard, N.E.: Cooperative Control for Ocean Sampling: The Glider Coordinated Control System. IEEE Transactions on Control Systems Technology 16, 735–744 (2008)
7. Zhang, F., Leonard, N.E.: Cooperative Filters and Control for Cooperative Exploration. IEEE Transactions on Automatic Control 55(3), 650–663 (2010)
8. Wu, W., Zhang, F.: Cooperative exploration of level surfaces of three dimensional scalar fields. Automatica 47(9), 2044–2051 (2011)
9. Wu, W., Zhang, F.: A Switching Strategy for Target Tracking by Mobile Sensing Agents. Journal of Communications 8, 47–54 (2013)
10. Hernandez, M.L., Kirubarajan, T., Bar-Shalom, Y.: Multisensor resource deployment using posterior Cramer-Rao bounds. IEEE Transactions on Aerospace and Electronic Systems 40, 399–416 (2004)
11. Tharmarasa, R., Kirubarajan, T., Hernandez, M., Sinha, A.: PCRLB-based multisensor array management for multitarget tracking. IEEE Transactions on Aerospace and Electronic Systems 43, 539–555 (2007)
12. Zhong, X., Premkumar, A.B., Madhukumar, A.S.: Particle Filtering and Posterior Cramér-Rao Bound for 2-D Direction of Arrival Tracking Using an Acoustic Vector Sensor. Sensors Journal 12(2), 363–377 (2012)
13. Van Trees, H.L.: Detection, Estimation, and Modulation Theory. Wiley, New York (1968)
14. Zhang, S., Yu, J., Zhang, A., Zhao, W.: Tracking strategy analysis with multi underwater vehicles for ocean feature. Chin. Sci. Bull (Chin. Ver.), 1–5 (2014)

# Path Planning Method of Underwater Glider Based on Energy Consumption Model in Current Environment

Yaojian Zhou[1,2], Jiancheng Yu[1], and Xiaohui Wang[1]

[1] Shenyang Institute of Automation, CAS
Shenyang, China
[2] University of Chinese Academy of Sciences
Beijing, China
Zhouyj@sia.cn

**Abstract.** It is generally considered that the speed of underwater glider is the function of buoyancy and gliding angle. However, the buoyancy and gliding angle are adjustable, which makes the speed of underwater glider within an adjustable range, however, it is usually taken as a constant in current documentations. Considering the path planning in ocean currents, if the maximum speed of a glider can find a path that connects the start point and the target point, then it can decrease its consumption by adjusting its speed in current field of some regions to fit the favorable currents and overcome the influence of adverse currents. Based on the above facts, the paper presents a new path planning method of adjustable speed glider in currents, and the simulation result is shown. According to the result: compared with the path planning method in constant speed, the adjustable speed glider can utilize the current in a better way and save the consumption of energy in a further way.

**Keywords:** optimal energy consumption, adjustable speed, iteration, underwater glider, path planning.

## 1 Introduction

Underwater gliders are a class of Autonomous Underwater Vehicles (AUVs) that are buoyancy-driven, deploying with lower operating cost and long duration in the ocean deployments. They go especially well with marine observation and have been used extensively in oceanography of late[1][2][3]. Their speeds are typically lower than motor-driven AUVs, reaching a maximum speed of around 1knot.Due to the low surge speed, they are susceptible to ocean currents.

Path planning of AUVs have been discussed in many literatures. At the beginning, the only consideration was the obstacles in ocean environment without currents [4][5][6].Till 2005,Garau et al. [7] transformed obstacles of A* graph search algorithm into unreachable grid points in ocean currents, then the path planning in currents field had been widely researched. Based on[7],the author of [8]built Rapidly-Exploring Random Trees(RRTs) that connect the start point and the target point, then utilized A* algorithm to search a path with lower energy consumption, but the

X. Zhang et al. (Eds.): ICIRA 2014, Part I, LNAI 8917, pp. 142–152, 2014.

energy-consuming model they used was extremely simplified. Constraining the vehicle to move in an 8-connected grid also means that optimality is compromised in the graph discretisation alone, a continuous technique is desirable. Continuous planning techniques have been explored for both AUVs and gliders alike in[8][10][11]and[12] .Without the constraining of 8-connective,the problem had been considered further in [9]and[13]. The gliding motion control of underwater gliders can be described by considering a typical diving and surfacing cycle, but we can only control the glider at the surfacing cycle for the unavailable underwater communication. Meanwhile, in a typical cycle, compared with diving ,the time of surfacing can be neglected. Based on the above mentioned facts, authors of [9]and[13] proposed methods to determine optimal paths that account for the influence from ocean currents .

In the above literatures, speeds of AUVs are considered to be constant generally. For a underwater glider, once the structure is fixed, the water-referenced speed of underwater glider is the function of buoyancy and gliding angle[14][15][16],and we can adjust the water-referenced speed in a certain range by controlling the driven-buoyancy or pitching angle. Compared with vehicle of constant speed, the vehicle of adjustable speed can utilize currents far more efficient.

Based on a high efficiency path planning searching algorithm in ocean currents field, this paper describes and evaluates a new algorithm for glider path planning, which contains a precise energy consumption model. Once the path exists, the proposed scheme would choose the most appropriate speed of glider automatically to fit the currents, which actualized by a iterative optimization process, and it can save energy consumption further.

## 2    The Description of Current Field of Adjustable Speed Glider Path Planning

The information of ocean currents can usually be obtained from marine environment numerical prediction. And it can assume that local currents remain unchanged over a period of time. Once given the start point and the target point, if the maximum speed of glider can find a path that connects the start point and the target point, the optimization of energy consumption of the glider can be considered further. Assuming that the glider moving to a certain domain, if currents of the area are beneficial for the glider to drive to its target, the speed of glider can be considered to decrease to save the energy consumption. On the contrary, if the currents are unfavorable, glider is supposed to increase speed properly to overcome the influence of the currents. In that case, it can across the unfavorable region as soon as possible.

Assuming a domain $D$ , global currents of the environment $V_c$ , setting a start point $p_1$ and a target point $p_N$ , the glider begin to perform its deployment from the start point. According to reference [17], the water-referenced speed of gliders has the following description in horizontal direction

$$v_{glider} = \cos \gamma \sqrt{\frac{B \cos \gamma}{K_{L0} + K_L \alpha(\gamma)}} \tag{1}$$

$$\alpha(\gamma) = \frac{K_L}{2K_D} \tan \gamma (-1 + \sqrt{1 - 4\frac{K_D}{K_L^2} \cot \gamma (K_{D0} \cot \gamma + K_{L0})}) \tag{2}$$

The $K_{L0}$, $K_L$, $K_{D0}$, $K_D$ are lift and drag coefficients. By equation (1), $v_{glider}$ is a function of the net buoyancy $B$ and gliding angle $\gamma$ .In this paper, the gliding angle is fixed as $\gamma = 20°$, so $v_{glider}$ is only the function of the buoyancy $B$ . Setting the range of the net buoyancy $B$ into $[B_{min} \ B_{max}]$ and substituting into equation (1), the adjustable speed range of glider satisfied $v_{glider} \in [v_{r\,min} \quad v_{r\,max}]$

The goal of the path planning is to find a path with least energy consumption from the start point to the target point. The path of the glider is consist of gliding cycles one by one, but it can only be controlled when raising to the surface. Owing to this, there is no guarantee that the glider can arrive the target point precisely, so we set a target domain with center $p_N$ . When the glider surfacing at a point located in the target domain, we regard the glider as to approach the target point. Therefore, all the way-points and the trajectory under the influence of currents between the each pair way-points composed a path $P = \{p_1, \cdots p_i, p_{i+1} \cdots p_N'\}$

Glider can be controlled at the location of way-points. The speed of glider is controlled by adjusting the net buoyancy of the glider. The bearing of the glider is controlled by adjusting the heading. Each pair of two adjacent way-points meet the relationship

$$p_{i+1} = p_i + \int_{T(i)}^{T(i)+t_s(i)} v(x)dt \tag{3}$$

$t_s(i)$ is a gliding cycle time cost ,is the function of diving depth $h$ , gliding angle $\gamma$ and speed $v_{glider}$

$$t_s(i) = \frac{2h}{v_{glider,i} \tan \gamma}, v_{glider,i} \in [v_{r\,min} \quad v_{r\,max}] \tag{4}$$

Where  $v(x)$ is the glider speed referenced to ground, it is the vector sum of the current speed and water-referenced speed of the glider.

$$v(x) = v_c(x) + v_{glider}(x) \tag{5}$$

The segment  $p_i p_{i+1}$  corresponds energy consumption $E_{i,i+1}$ . According to the reference [18]. The energy consumption of the underwater glider a gliding cycle can be divided into two parts: one is related to time, which is the power consumption of sensor and control unit, and the other part is the power consumption of buoyancy device and the position device, which has nothing to do with the time. Addressed as follow.

$$E_{(i,i+1)t} = P_t \frac{2h}{\tan \gamma} \frac{1}{v_{glider,i}} \tag{6}$$

$$E_{(i,i+1)c} = \frac{2}{\rho g}(P_c + \frac{kh}{q_p})\left|K_{L0} + K_L \alpha(\gamma)\right| v_{glider,i}^2 \cos \gamma + K \tan \gamma \tag{7}$$

Among (7)

$$K = \frac{4P_m M \Delta h}{m v_m} \tag{8}$$

Therefore, the total energy consumption of a complete gliding cycle is

$$E_{i,i+1} = E_{(i,i+1)t} + E_{(i,i+1)c} \tag{9}$$

The problem of glider path planning in the current environment can be described as below:

Given: The global information current field  $V_c$  The adjustable range of glider net buoyancy is $B \in [B_{min} \ B_{max}]$ , The glider dynamic model, the start point  $p_1$ ,  the target point $p_N$ .

Required:   path  $P = \{p_1, \cdots p_i, p_{i+1} \cdots p_N'\}$    ,the   corresponding   energy consumption , time consumption and the glider control law of every way-point in the path  $p_i \rightarrow (v_{glider,i}, \theta_i), i = 1, 2 \cdots N$

Planning objectives and constraint conditions are as follows:

$$\min E = \sum_{i=1}^{i=N} E_{i,i+1} = P_t \frac{2h\cos\gamma}{\sin\gamma} \sum_{i=1}^{i=N} \frac{1}{v_{glider,i}} + \frac{2}{\rho g}(P_c + \frac{kh}{q_p}) \left|\frac{K_{L0} + K_L\alpha(\gamma)}{\cos^3\gamma}\right| \sum_{i=1}^{i=N} v^2_{glider,i}$$

$$+ nK\tan\gamma$$

s.t.

$$v_{glider} = \cos\gamma \sqrt{\frac{B\cos\gamma}{K_{L0} + K_L\alpha(\gamma)}}$$

$$B \in [B_{min} \ B_{max}], \theta \in [0 \ 2\pi)$$

$$\left|p_N - p_N'\right| \leq C$$

(10)

## 3    Path Planner Introduction

There exist plenty of constant speed AUVs path planning algorithms in ocean currents. The method CTS-A* is a variant of the classic A* where the time between two consecutive surfacings is kept constant. This method discretizes the bearings that can be commanded on each surface. Since it is a high efficiency path planning method for underwater gliders, this paper adopt this method and improve it slightly. A glider energy consumption model and a novel heuristic function are integrated in CTS-A*,and then iterative optimization process is implemented. Ultimately, we gain a minimum energy consumption path.

CTS-A* includes a notable modification to the original A* algorithm. The main difference between the two is the process of generating successors:

1)For the glider at node $p_i$, the water-referenced speed is $v_{gliderC}$,which is kept constant. We select $K$ sample bearings in $0 \sim 360°$ to represent all bearings .While we neglect the vertical currents, the time of a gliding cycle can be calculated as

$t_s = \dfrac{2h}{v_{gliderC}\tan\gamma}$ . For each bearing $\theta_k$ we integrate the glider trajectory for the

surfacing time $t_s$ .i.e. we compute the trajectory followed by a glider that keeps a bearing $\theta_k$ under the influence of instantaneous ocean currents, which is described in(3).

2) The final location $p_{i+1}$ of each trajectory is stored in the Nearest Neighbor node, which is shown is Fig.1. The Nearest Neighbor nodes are regarded as the generating successors in CTS-A*, rather than node $p_{i+1}$ .

3) If two locations fall into the same node, we take the one with lower cost. If both have the same cost, we take the closest to the node.

4) Although the Nearest Neighbor node has been stored, node $p_{i+1}$ should be selected as the current node in the next turn of generating successors (from $p_{i+1}$ to $p_{i+2}$),rather than the Nearest Neighbor node.

For classic A* and variants of the classic A*, choosing a appropriate heuristic function may cause a significant influence of the algorithm efficiency. It is generally believed that if the cost of heuristic function is less than the real cost, then the optimal solution can be got. Corresponding with the energy consumption mode, we can divide the cost of heuristic function into two parts, one part of which is associated with time, and the other part has no relationship with time. The time in the first part should be minimized. Since the time is the product of distance and the reciprocal of resultant speed, we should maximize the resultant speed and minimize the distance from current node to the goal. The distance from $p_i$ to $p_N$ can be described as below:

$$d = |p_i - p_N| \tag{11}$$

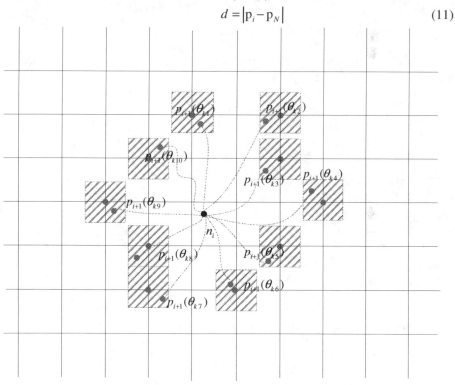

**Fig. 1.** Schematic diagram of the successors generated in CTS-A*

While the resultant speed should be maximized, considering the constant glider speed, we choose the maximum current speed in the domain as the estimated current speed, then add their absolute value together as the resultant speed. In that case, the time $t_p$ can be expressed as follows:

$$t_p = \frac{d}{\left|v_{gliderC}\right| + \left|v_{c\max}\right|} \tag{12}$$

When considering the term that has no association with time, we found that the number of gliding cycles is the only factor that should be taken into account. In calm water, the horizontal distance is fixed with $2D_h = \dfrac{2h}{\tan\gamma}$ ,so the minimum number of total cycles is $\left\lceil \dfrac{d}{2D_h} \right\rceil = \left\lceil \dfrac{d\tan\gamma}{2h} \right\rceil$ ,in which, the symbol $\lceil \bullet \rceil$ means ceiling. Adding the two energy consumption terms ,it is obviously that the heuristic function can be written as below:

$$h(x) = P_t t_p + E_{(i,i+1)c} \left\lceil \frac{d}{2D_h} \right\rceil \tag{13}$$

Substituting(7) and(12)to(13), we obtain the energy consumption heuristic function as

$$h(x) = P_t \frac{d}{\left|v_{gliderC}\right| + \left|v_{c\max}\right|}$$
$$+ \left\lceil \frac{d\tan\gamma}{2h} \right\rceil (\frac{2}{\rho g}\left|(P_c + \frac{kh}{q_p})(K_{L0} + K_L\alpha(\gamma))\cos\gamma\right|v_{gliderC}^2 + K.\tan\gamma) \tag{14}$$

This paper uses CTS-A* with heuristic function (14) as the basic search algorithm, then repeat the iteration with different water-referenced speeds, to realize the further energy consumption optimization between the two adjacent way-points. The final goal is to make sure the less consumption globally compared with the constant velocity situation. In section 2, we obtain the range of glider speed $v_{glider} \in \left[v_{r\min}\quad v_{r\max}\right]$ within the range $B$ allowed, then we select $n$ samples of it to represent all values, so $v_{glider} \in \left\{v_{r1}, v_{r2} \cdots v_{rn}\right\}$ .

In currents environment, given start point $p_1$ and target point $p_N$ ,we utilize CTS-A* to search path by different $v_{glider}$ ,if the path exist, then the  optimal energy consumption can be got. Assuming that the global energy consumption is $E_1$ and the corresponding constant speed $v1 \in \left\{v_{r1}, v_{r2} \cdots v_{rn}\right\}$ ,the path can be written as $P_1 = \left\{p_1, x_{1,1}, \cdots x_{1,n}, x_{1,n+1} \cdots p_{N1}\right\}$ . $E_1$ can be divided into two parts: one part is the energy cost from $p_1$ to $x_{1,1}$ ,and the other part is the energy consumption from $x_{1,1}$ to $p_{N1}$ .

$$E_1 = E_{p_1,x_{1,1}} + E_{x_{1,1},p_{N1}} \tag{15}$$

Saving the first way-point $p_1$ ,second way-point $x_{1,1}$ and the corresponding energy consumption $E_{p_1,x_{1,1}}$ between the two adjacent way-points of path $P_1$ , $x_{1,1}$ is regarded as the new start point and target point is kept the same. If we reuse CTS-A* to search paths from $x_{1,1}$ to $p_N$, the path $\{x_{1,1}, \cdots x_{1,n}, x_{1,n+1} \cdots p_{N1}\}$ would be found, and the corresponding energy cost is $E_{x_{1,1},p_{d1}}$. Generally, it may found less energy consumption path $P_2 = \{x_{1,1}, x_{2,1} \cdots x_{2,n}, x_{2,n+1} \cdots p_{N2}\}$, which corresponding energy cost $E_{x_{1,1},p_{N2}}$, and $E_{x_{1,1},p_{N2}} \le E_{x_{1,1},p_{N1}}$, then we save the first way-point $x_{1,1}$, second way-point $x_{2,1}$ and the corresponding energy consumption $E_{x_{1,1},x_{2,1}}$. Repeating the iterative process until we get path $P_k = \{x_{k-1,1}, p_{Nk}\}$, at this time, $p_{Nk}$ is just located in the domain of target point $p_N$. We found all the saved way-points constitute a new path $P = \{p_1, x_{1,1}, x_{2,1}, x_{3,1}, \cdots x_{k-1,1}, p_{Nk}\}$. If we let $x_{i,1} = p_{i+1}, p_{dk} = p_N$, the path can be written as $P = \{p_1, \cdots p_i, p_{i+1} \cdots p_N\}$, and the total energy consumption of $P$ is the sum of each segment.

$$E = E_{p_1,x_{1,1}} + E_{x_{1,1},x_{2,1}} + \cdots E_{x_{k-1,1},p_{Nk}} \tag{16}$$

From above mentioned, apparently, the inequation $E \le E_1$ could be satisfied.

# 4      Simulation and Analysis

With hydrodynamic parameters testing, the hydrodynamic parameters of glider model it the part 2 can be listed as follows:

$$K_{L0} = -0.421, K_L = 488.7837, K_{D0} = 6.7143, K_D = 435.052$$

$$P_i = 3W, P_{p0} = 20.618W, k = 0.0459W/m, P_v = 12W$$

$$q_p = 4.0 \times 10^{-3} L/s, q_v = 2.6 \times 10^{-3} L/s, M = 65kg$$

$$m = 11.278kg, Pm = 16W, v_m = 3.5 \times 10^{-3} m/s, \rho = 1.025kg/L$$

$$\Delta h = 0.05m$$

In addition, we set the gravitational acceleration $g = 9.8m/s^2$, and fixed the maximum depth of glider every cycle as $h = 200m$ within $100km \times 100km$ domain. The environment is divided into squares $1km \times 1km$. Assume that the underwater glider net buoyancy $B \in [0\ 8]N$, we select five samples to represent all values of $B$, $B = [0.2465\quad 0.9861\quad 2.2187\quad 3.9443\quad 6.1630]N$, which corresponds to $v_{glider} = [0.1\quad 0.2\quad 0.3\quad 0.4\quad 0.5]m/s$, it is convenient to calculate. In CTS-A*, we set 20 glider bearings in every generating successors process, and the 20 glider bearings distributed uniformly in $\left[0\quad 360°\right)$. As the underwater glider can only be controlled when surfacing, this paper argues that the glider diving into the target domain （1.1kilometer, one horizontal distance of a gliding cycle in calm water approximately） can be regarded as the target is arrived. Maximum depth of every diving cycle is fixed $h = 200m$ .gliding fixed angle is $\gamma = 20°$ ,the horizontal speed of the glider is only associated with buoyancy $B$ . The start point is （15.1 ,40.2） ,the target point is (70,70). The current field is described as a moment of hyperbolic currents. Hyperbolic currents and the parameters are defined as follows:

$$x = \frac{xx}{100}, y = \frac{yy}{100}, \quad xx = 1:100, yy = 1:100$$

$$u = -\pi A \sin(\pi f(x))\cos(\pi y)$$

$$v = \pi A \cos(\pi f(x))\sin(\pi y)\frac{df}{dx}$$

Among then $f(x,t) = a(t)x^2 + b(t)x, a(t) = \varepsilon \sin(\omega t), b(t) = 1 - 2\varepsilon \sin(\omega t)$ , $A$ 、 $\varepsilon$ are related to the amplitude of the current field. $\omega$ represents frequency. Setting $A = 0.4, \varepsilon = 0.25, \omega = 0.2\pi$ ,and setting t=6s, Fig.2. shows the simulation result. Given in table 1, speed of 0.3 m/s under the constant velocity of CTS-A *algorithm and adjustable speed obtained respectively under the iterative solution of energy consumption and time .Fig. 3. shows the iterative algorithm to get the path to the specific control scheme.

**Table 1.** hyperbolic currents, constant CTS - A * (0.3 m/s) and get the path of the energy consumption under variable iteration time list

| The speed of the glider | CTS-A*(0.3m/s) | Variable iteration |
|---|---|---|
| Time consumption | 20.5749h | 21.1164 h |
| Energy consumption | $3.1728 \times 10^5$J | $2.5787 \times 10^5$J |
| Corresponding color | green | black |

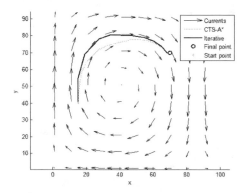

**Fig. 2.** The simulation results of the hyperbolic currents

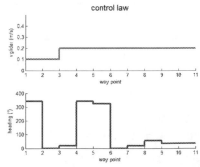

**Fig. 3.** The specific control scheme

Analyzing of simulation results can be found that within a given speed range, using variable iteration scheme, we get a path of energy consumption is only 81.3% of the energy consumption under constant speed, saving nearly 20% of the energy consumption, and time consuming only spent 2.63% more than constant speed situation. We can conclude that using the proposed iterative scheme can obtain a better energy consumption path, and from the point of view of time, it is still acceptable.

# 5     Conclusion and Future Works

This paper presents a novel CTS-A* based iterative path planning algorithm to address underwater gliders with variable speeds, which avoids the deficiency of constant speed path planning. It can maximize the usage of the favorable currents, while overcome the adverse currents as well. Once a path obtained under the maximum speed, the proposed scheme can be implemented. A accurate energy consumption model and a novel heuristic function are included in the path planner, which make the path planning problem far more close to the real situation. From the simulation of hyperbolic currents field, it can be easily found that our novel variable speed scheme can get a path with lesser energy cost compared with constant speed scheme, while the increasing of time is not obvious, and the control law was demonstrated as well, which illustrates the scheme can be actualized not so hard.

In future work we will generalize the scheme into three-dimensional current field and time-varying current field.

**Acknowledgments.** The authors would graciously acknowledge the support for this work provided by the glider research teams of Underwater Research Center, Shenyang Institute of Automation.

# References

1. Blidberg, D.R.: The Development of Autonomous Underwater Vehicles (AUV): A brief summary. In: IEEE International Conference on Robotics and Automation (2001)
2. Eriksen, C.C., et al.: Seaglider: a long-range autonomous underwater vehicle for oceanographic research. IEEE Journal of Oceanic Engineering 26, 424–436 (2001)
3. Sherman, J., Davis, R.E., Owens, W.B., Valdes, J.: The autonomous underwater glider 'Spray'. IEEE Journal of Oceanic Engineering 26, 437–446 (2001)
4. Carroll, K.P., Mc Claran, S.R., Nelson, E.L., Barnett, D.M., Friesen, D.K., William, G.N.: AUV path planning: an A* approach to path planning with consideration of variable vehicle speeds and multiple, overlapping, time-dependent exclusion zones. In: Proceedings of the 1992 Symposium on Autonomous Underwater Vehicle Technology, pp. 79–84 (1992)
5. Tan, C.S., Sutton, R.: J, Chudley. Anincremental stochastic motion planning technique for autonomous underwater vehicles. In: Proceedings of IFAC Control Applications in Marine Systems Conference, pp. 483–488 (2004)

6. Petres, C., Pailhas, Y., Petillot, Y., Lane, D.: Underwater path planning using fast marching algorithms. In: Oceans 2005-Europe, vol. 2, pp. 814–819 (2005)
7. Garau, B., ´Alvarez, A., Oliver, G.: Path Planning of Autonomous UnderwaterVehicles in Current Fields with Complex Spatial Variability: an A* Approach. In: Proc. 2005 IEEE International Conference on Robotics and Automation, pp. 194–198 (2005)
8. Rao, D., Williams, S.B.: Large-scale path planning forUnderwater Gliders in ocean currents. In: Australasian Conference on Robotics and Automation (ACRA), Sydney, December 2-4 (2009)
9. Fernandez-Perdomo, E., Cabrera-Gamez, J., Hernandez-Sosa, D., Isern-Gonzalez, J., Dominguez-Brito, A.C., Redondo, A., et al.: Path planning for gliders using Regional OceanModels: Application of Pinzón path planner with the ESEOAT model and the RU27 trans-Atlantic flight data. In: OCEANS 2010 IEEE - Sydney, pp. 1–10 (2010)
10. Soulignac, M., Taillibert, P., Rueher, M.: Adapting the wavefront expansion in presence of strong currents. In: Proceedings of the 2008 IEEE International Conference on Robotics and Automation, pp. 1352–1358 (2008)
11. Soulignac, M.: Feasible and Optimal Path Planning in Strong Current Fields. IEEE Trans. Robot. 27(1), 89–98 (2011)
12. Soulignac, M., Taillibert, P., Rueher, M.: Time-minimal Path Planning in Dynamic Current Fields. In: Proceedings of the 2009 IEEE International Conference on Robotics and Automation (2009)
13. Isern-Gonzalez, J., Hernandez-Sosa, D., Fernandez-Perdomo, E., Cabrera-Gamez, J., Dominguez-Brito, A.C., Prieto-Maranon, V.: Path planning for underwater gliders using iterative optimization. In: 2011 IEEE International Conference on Robotics and Automation (ICRA), May 9-13, pp. 1538–1543 (2011)
14. Leonard, N.E., Graver, J.G.: Model-based feedback control of autonomous underwater gliders. IEEE J. Ocean. Eng. 26(4), 633–645 (2001)
15. Graver, J.G.: Underwater glider s: Dynamics, control, and design.Ph.D. dissertation, Dept. Mech. Aerosp. Eng. Princeton Univ., Princeton, NJ (2005)
16. Mahmoudian, N., Woolsey, C.: Underwater glider motion control. In: Proc. 47th IEEE Conf. Decision Control, Cancun, Mexico, December 9-11, pp. 552–557 (2008)
17. Yu, J., Zhang, F., Zhang, A., Jin, W., Tian, Y.: Motion parameter optimization and sensor scheduling for the sea-wing underwater glider. IEEE Journal of Oceanic Engineering, 243–254 (2013)
18. Zhu, X., Yu, J., Wang, X.: Sampling path planning of underwater glider for energy consumption. Robot 33(3), 360–365 (2011)

# Visual Features Extraction and Types Classification of Seabed Sediments

Yan Li[1], Chunlei Xia[2], Yan Huang[1,3], Liya Ge[1,3], and Yu Tian[1]

[1] State Key Laboratory of Robotics, Shenyang Institute of Automation,
CAS, Shenyang, China
[2] The Research Center of Coastal Environmental Engineering and Technology
of Shandong Province, Yantai Institute of Coastal Zone Research,
CAS, Yantai, China
[3] University of Chinese Academy of Sciences, Beijing, China
liyan1@sia.cn

**Abstract.** The purpose of this research is to define and extract the visual features of the seabed sediments to improve the autonomous ability of a underwater vehicle while implementing exploring missions. A scheme of seabed image classification is proposed to identify three types of seabed sediments. The texture features of images are stable and robust visual features in underwater environment comparing with general visual features, and which are described by using gray-level co-occurrence matrix and fractal dimension. Subsequently, for purpose of evaluation, a supervised non-parametric statistical learning technique, support vector machines (SVMs), is applied to verify the availability of extracted texture features on seabed sediments classification. The presented results of seabed type recognition justify the proposed features extracted method valid to seabed type recognition.

**Keywords:** Seabed sediments, Underwater vehicle, Visual features, Fractal dimension, Gray-level co-occurrence matrix, SVMs.

## 1    Introduction

Over 70 percent of the earth surface is covered by oceans and extremely large amount of mineral resources is deposited in the oceans. According to the literatures and field survey, approximate 3 trillion tons of polymetallic nodules of the global reserves have been investigated in the ocean areas and more than 1.7 trillion tons of the polymetallic nodules are located in the Pacific ocean area. The polymetallic nodules deposited in the ocean areas contains higher amount of metal elements, such as Mn, Ni, Co and Cu, comparing with the polymetallic nodules in land [1]. China as a developing country has a maritime territory of 3 million square kilometers. Thus, the exploration of ocean and development the ocean resources are rather important and meaningful for China's growth.

Autonomous Underwater Vehicles (AUVs) as a tool to explore the ocean have already been widely applied to oceanic surveys [2], and the autonomous perception to

X. Zhang et al. (Eds.): ICIRA 2014, Part I, LNAI 8917, pp. 153–160, 2014.
© Springer International Publishing Switzerland 2014

surrounding environment of AUVs has attracted more and more attentions from researchers. More than 90% information delivered to human brain is obtained from the visual signals, and similarly, visual techniques provide abundant information to the operations AUVs. The visual system with ability of autonomous perception is one of the most important factors of an intelligent robot system [3]. As the development of robotic vision technology, the robot vision system has become a common exploring device installed at underwater vehicles such as AUVs and ROVs (Remote Operated Vehicle) to complete the various missions for engineering and science studies, e.g., object detection and tracking [4], ecological study [5-6], localization and navigation [7-8], seabed mapping [9] and etc. In the last decade, underwater visual system was also integrated with acoustic sonar system and has demonstrated excellent performance to measure the ocean data.

Identification of seabed sediments is a key issues tor improving the capability of AUVs to perform the autonomous survey of seabed. Accurate and fast identification of sediments is the preliminary study to achieve the ability of autonomous perception for AUVs. And the automatic identification method could also be applied to the other survey tasks by the AUVs.

The seabed sediments have the character with diversity of morphologies and types and another major obstacle to underwater image is arisen by the light absorption and scattering by the marine environment. Therefore, the classification and identification methods in the air based on the features as edge or color are not efficient to the underwater environments. However, the texture feature of images describes the distribution and permutation of pixels and is a measure to express the co-relationship of pixels [10]. Thus, the texture feature can show excellent performance on classification and identification in the underwater environments than other image features. Therefore, in this study a classification and recognition method of seabed sediments based on Support Vector Machines (SVMs) is gained to evaluate the availability of the visual texture features which are extracted based on gray-level co-occurrence matrix (GLCM) and fractal theory.

The sections of this paper are organized as follows: In Sec. 2, we introduce our method of texture features extractions based on GLCM and fractal theory, and a brief introduction of SVMs used for evaluating texture features and recognizing the seabed types. The experiments and results are described in Sec.3, while conclusion is reported in Sec. 4.

## 2      Paper Preparation

### 2.1    Types of Seabed Sediments

In deep sea, the seabed is covered by several types of sediments which mainly consist of polymetallic nodule, sand and benthic community (Fig. 1). The sediments images often lack sharp pronounced features which are useful for feature extraction.

**Fig. 1.** Three types of seabed sediments: polymetallic nodule, sand and benthic community

In this study, these three typical kinds of sediment types are selected to verify the feasible of the method proposed. The images of seabed sediments were captured with a manned submersible named Jiaolong in its scientific expedition to the South China Sea. And the images are normalized to the size 320x240.

## 2.2    Texture Features Extraction from GLCM

GLCM is an estimate of a joint probability density function of gray level pairs in an image. GLCM can be expressed as follow:

$$P_\mu(i, j), (i, j = 0, 1, 2, ..., L-1) \tag{1}$$

where $i,j$ indicate the gray level of two pixels of image, $L$ is the gray level and $\mu$ is the position relation of the two pixels.

Haralick proposed 14 statistical features extracted from GLCM. Considering the calculational complex, for seabed sediments classification, four GLCM features are employed in this study including energy (also called angular second moment), correlation, entropy and contrast. The definitions of the four features are illustrated as following equations.

$$\text{Energy:} \qquad f_1 = \sum_{i=0}^{L-1} \sum_{j=0}^{L-1} P^2(i, j) \tag{2}$$

$$\text{Correlation :} \qquad f_2 = \frac{\sum_{i=0}^{L-1} \sum_{j=0}^{L-1} (ij) \cdot P(i, j) - \mu_x \mu_y}{\sigma_x \sigma_y}, \text{ where}$$

$$\left\{ \begin{array}{l} \mu_x = \sum_{i=0}^{L-1} \sum_{j=0}^{L-1} i \cdot P(i, j), \ \mu_y = \sum_{i=0}^{L-1} \sum_{j=0}^{L-1} j \cdot P(i, j) \\ \sigma_x = \sum_{i=0}^{L-1} \sum_{j=0}^{L-1} (i - \mu_x)^2 \cdot P(i, j), \ \sigma_y = \sum_{i=0}^{L-1} \sum_{j=0}^{L-1} (j - \mu_y)^2 \cdot P(i, j) \end{array} \right\} \tag{3}$$

$$\text{Entropy :} \qquad f_3 = -\sum_{i=0}^{L-1} \sum_{j=0}^{L-1} P(i, j) \log\left( P(i, j) \right) \tag{4}$$

$$\text{Contrast:} \qquad f_4 = \sum_{n=0}^{L-1} n^2 \left\{ \sum_{i=0}^{L-1} \sum_{j=0}^{L-1} P(i, j) \mid |i - j| = n \right\} \tag{5}$$

Directionally invariant texture measures are obtained by calculating the mean of the texture results in four directions ( 0˚, 45˚, 90˚, 135˚ ), and the differences of texture measures of three seabed sediments in four directions are shown as Fig. 2.

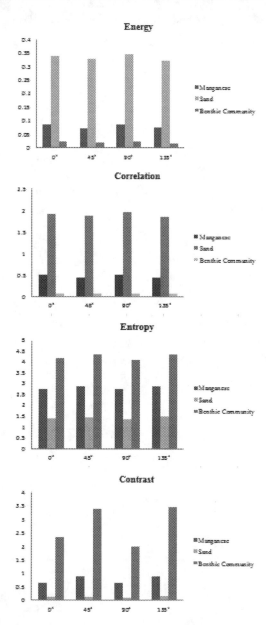

**Fig. 2.** Differences of textural features measured by GLCM

## 2.3    Box-counting Approach to Estimate Fractal Dimension

Self-similar objects and phenomena can be described by a non-integer dimension called the fractal dimension to show the irregular structure of objects [11]. The fractal dimension measures the degree of irregularity and complexity of an object to describe the self-similarity of the object [12]. Therefore, the concept of fractal dimension can be useful in the analysis the texture of images.

The box-counting approach was defined in [13], which is the widely used and common way to estimate fractal dimension [14]. The fractal dimension $FD$ of an image can be derived from the relation [12].

$$FD = \frac{\log(N_r)}{\log(\frac{1}{r})} \tag{6}$$

where $r$ is the size ratio and $N_r$ is the number of boxes with  size ratio  $r$ inside which at least one point from the attractor lies.

## 2.4    Support Vector Machines

SVMs are state-of-the-art large margin classifiers, which have been gained popularity on the fields of visual pattern recognitions [15-16]. A brief review of SVMs will be addressed follow, detailed description and process of the SVMs can be referred to literature [17].

The basic idea behind SVMs is to build a classifier that maximizes the margin between two-class data (the process is also called seeking the optimal separating hyperplane). For an observation  $x \in X$ and a kernel function  $K$ , an SVM, $f(x)$ can be indicated as

$$f(x) = \sum_{i=1}^{N} \lambda_i \xi_i K(x, x_i) + b = \sum_{i=1}^{N} \lambda_i \xi_i \phi(x) \cdot \phi(x_i) + b \tag{7}$$

Here, the  $x_i$,  $\xi_i$ and  $\lambda_i$ are obtained through a training process. The  $x_i$ are called support vectors and the  $\xi_i$ are the target class values: positive and negative examples.

In our case, four textural features: energy, correlation, entropy and contrast were extracted based on GLCM from images of seabed sediments, and their mean and standard deviation of four directions as inputs are fed to SVMs for training. Furthermore, the fractal dimension is regarded as another input to SVMs for training.

# 3    Experiments and Results

Texture features extracted from 81 images of sediments were selected randomly as sample data to train our SVMs framework. Subsequently, the texture features extracted from other 27 images of sediments as testing sample data were fed to the SVMs to evaluate the accuracy of defined features and the performance of recognition. The constitution of training sample data and testing sample data is shown as Tab. 1.

**Table 1.** Constitution of training sample data and testing sample data

|  | Type of seabed sediments | Number |
|---|---|---|
| Training sample data | benthic community | 19 |
|  | polymetallic nodule | 18 |
|  | sands | 44 |
| Testing sample data | benthic community | 12 |
|  | polymetallic nodule | 6 |
|  | sands | 9 |

In our experiments, an efficient open source library toolbox LIBSVM was adopted for classification problem [18]. As recognition results shown in Fig.3, 100% seabed sediment type recognition rate was obtained on the 27 testing sample data. In future work, more testing sample data will be implemented to estimate the performance of our approach.

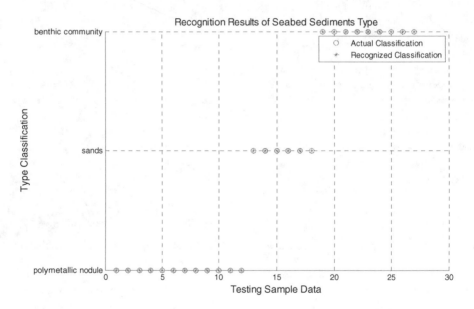

**Fig. 3.** Recognition results of seabed sediments types by implementing SVMs

## 4    Conclusions

A novel classification scheme is proposed based on the texture descriptors. The visual texture features extracted from images by using GLCM and fractal theory contributed excellent performance to the classification of the three types of seabed sediments. The feasibility of our method of feature definition was proved through experiments by SVMs. Consequently, the ability of autonomous sensing of AUVs to seabed environments can be implemented by the proposed method. For the future work, combining visual sensing with acoustics sensing will be considered to enhance the accuracy and stability of the underwater survey.

**Acknowledgment.** The work of this article is in part supported by National Natural Science foundation under grants 61233013 and 41106085, and a self-sponsored project of the State Key Laboratory of Robotics at Shenyang Institute of Automation under grant 2013-Z13.

# References

1. Niu, J.K.: Research and development of oceanic multi-metal nodule. China's Manganese Industry 20(2), 20–26 (2002)
2. Feng, X.S., Li, Y.P., Xu, H.L., et al.: The next generation of unmanned marine vehicles-dedicated to the 50 anniversary of the human world record diving 10912 m. Robot 33(1), 113–118 (2011)
3. Tang, X.D., Zhu, W., Pang, Y.J., et al.: Target recognition system based on optical vision for AUV. Robot 31(2), 171–178 (2009)
4. Kia, C., Arshad, M.R.: Robotics Vision-based Heuristic Reasoning for Underwater Target Tracking and Navigation. International Journal of Advanced Robotic Systems 2(3) (2005)
5. Armstrong, R.A., Singh, H., Torres, J., et al.: Characterizing the deep insular shelf coral reef habitat of the Hind Bank marine conservation district (US Virgin Islands) using the Seabed autonomous underwater vehicle. Continental Shelf Research 26(2), 194–205 (2006)
6. Singh, H., Armstrong, R., Gilbes, F., et al.: Imaging coral I: imaging coral habitats with the SeaBED AUV. Subsurface Sensing Technologies and Applications 5(1), 25–42 (2004)
7. Hao, Y.M., Wu, Q.X., Zhou, C., et al.: Technique and implementation of underwater vehicle station keeping based on monocular vision. Robot 28(6), 656–661 (2006)
8. Gracias, N.R., Van Der Zwaan, S., Bernardino, A., et al.: Mosaic-based navigation for autonomous underwater vehicles. IEEE Journal of Oceanic Engineering 28(4), 609–624 (2003)
9. Rzhanov, Y., Linnett, L.M., Forbes, R.: Underwater video mosaicing for seabed mapping. IEEE International Conference on Image Processing 1, 224–227 (2000)
10. Gao, C.C., Hui, X.W.: GLCM-based texture extraction. Computer system & applications 19(6) (2010)
11. Mandelbrot, B.B.: The Fractal Geometry of Nature. Wh Freeman, New York (1983)
12. Tricot, C.: Curves and fractal dimension. Springer, Heidelberg (1995)
13. Russel, et al.: Dimension of strange attractors. Physical Review Letters 45(14), 1175–1178 (1980)
14. Sarkar, N., Chaudhuri, B.B.: An efficient differential box-counting approach to compute fractal dimension of image. IEEE Trans. Syst. Man. Cybern. A 24(1), 115–120 (1994)
15. Wallraven, C., Caputo, B., Graf, A.: Recognition with local features: the kernel recipe. In: Proc. ICCV, pp. 257–264 (2003)

16. Wolf, L., Shashua, A.: Kernel principal angles for classification machines with applications to image sequence interpretation. In: Proc. CVPR, pp. 635–640 (2003)
17. Hearst, M.A., Dumais, S.T., et al.: Support vector machines. Intelligent Systems and their Applications 13(4), 18–28 (1998)
18. LIBSVM, http://www.csie.ntu.edu.tw/~cjlin/libsvm/

# Cymbal Piezoelectric Micro Displacement Actuator Characteristics Analysis

Hang Xing, Tiemin Zhang*, Sheng Wen, Xiuli Yang, Zhilin Xu, and Fei Cao

College of Engineering
South China Agricultural University, Guangzhou, China

**Abstract.** This paper analyzes the structure and displacement amplification principle of the cymbal piezoelectric micro displacement actuator, researches its static and dynamic voltage-displacement characteristics through experiment, points out its voltage-displacement characteristics is hysteresis nonlinearity, analyzes the influences of the driving voltage amplitude, driving frequency and the number of driving voltage cycles on hysteresis nonlinearity characteristics, provides the basic research for improving its hysteresis nonlinearity and improving the positioning precision.

**Keywords:** Cymbal, piezoelectric, actuator, hysteresis nonlinearity.

## 1 Introduction

Piezoelectric micro displacement actuator has high displacement precision, high resolution, fast response, low consumption, no noise, can be miniaturization and other advantages compared with the previous mechanical, hydraulic, pneumatic, and electromagnetic actuator , then is widely used in the field of precision electronics, precision machining, precision optics and other fields that requires precision positioning. But because the output displacement of piezoelectric ceramic is very small, so in practice, often need to add some structure to improve the output displacement of piezoelectric actuator, such as using Cymbal structure to form the cymbal piezoelectric vibrator, then to increase the output displacement of piezoelectric micro displacement actuator [1].

Cymbal structure can add radial direction displacement caused by piezoelectric ceramic axial displacement to axial displacement, so to produce a larger axial displacement, and if using multiple Cymbal piezoelectric vibrator stacked structure, then can get more larger output displacement. The cymbal piezoelectric system now researched is mainly used in fields of transducer [2] [3] , driver[4] [5]and sensor [6] [7], etc.

When cymbal piezoelectric system applied as a driver, its voltage-displacement characteristics is one of the important performance, Through the theoretical analysis and experimental verification, Li Denghua[8], Gao Quanqin[9], Ma Yongcheng[10] et al. established the cymbal piezoelectric ceramic voltage-displacement relation

---

* Corresponding author.

X. Zhang et al. (Eds.): ICIRA 2014, Part I, LNAI 8917, pp. 161–170, 2014.
© Springer International Publishing Switzerland 2014

model, and gave the influence rules of the key structure parameters to cymbal
lectric ceramic equivalent piezoelectric constant, namely to cymbal piezo
ceramic voltage-displacement characteristics. A lot of researches have done
the factors that affect system performance, such as Sun[11] and Dogan[12]
metal cap size of the cymbal structure, Fernandez[13] and Ochoa[14] stud
adhesive type between metal cap and piezoelectric ceramic, Dogan[15] and
studied piezoelectric ceramic material, and the conclusion is given that the d
type of   piezoelectric ceramic material and size, metal cap material and si
adhesive type has a great influence on the cymbal piezoelectric micro displa
actuator performance. But if these factors are fixed unchanged, the larger in
factor is the input signal, namely, driving voltage. Based on the analysis of
piezoelectric micro displacement actuator static and dynamic characteristic
paper points out the influence of the driving voltage parameters on the syste
tage-displacement characteristics, provide basic research for its application in tl
of precision drive.

## 2     Cymbal Piezoelectric Micro Displacement Actuator Stru

### 2.1   Cymbal Piezoelectric Vibrator Structure

Cymbal piezoelectric micro displacement actuator is composed of multiple
piezoelectric vibrator. Cymbal piezoelectric vibrator is made up of two metal
one piezoelectric ceramic which is thickness direction polarized, as shown in Fi

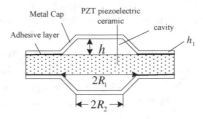

**Fig. 1.** Cymbal piezoelectric vibrator

Pan Zhongming[17] assumes that when piezoelectric vibrator radial
metal cap top radius $R_2$ is basically unchanged, then due to the action of ad
only the cap diameter $R_1$ and the cavity height $h$ changed. Assumes that befo
after deformation, the metal cap surface area is approximately equal, so there are

$$(R_1 - R_2)^2 + (h - h_1)^2 = [(R_1 - R_2) + \Delta R]^2 + [(h - h_1) + \Delta h]^2$$

Where $h_1$ is the thickness of the metal cap( $h_1 < h$ ), $\Delta R$ is the radial displace
the piezoelectric ceramics, $\Delta h$ is axial displacement variation. Simplify and
the high order factor, we can get the metal cap axial displacement variation cau
the radial displacement of piezoelectric ceramics

$$\Delta h = -\frac{R_1 - R_2}{h - h_1} \Delta R \qquad (2)$$

negative sign said when piezoelectric vibrator radial elongate, corresponding metal cap axial compress, and vice versa. So the cymbal piezoelectric vibrator axial total displacement is

$$\Delta H = \Delta z + 2\Delta h \qquad (3)$$

Where $\Delta z$ is axial displacement caused by piezoelectric ceramic $d_{33}$.

It is clear that after adding cymbal structure, as long as to reasonablely design metal cap size, make $R_1 - R_2 \gg h - h_1$, than can greatly improve the axial displacement of the whole system.

This paper selects the thickness direction polarized PZT-5A piezoelectric ceramic to produce cymbal piezoelectric vibrator. Metal cap adopts high elastic beryllium bronze (QBe1.9) [18], Metal cap shape is shown in Fig. 2. a,The underneath gap is used for solder wire before piezoelectric ceramic and metal cap bonding, the metal cap is received from mold cutting after stamping. Before fabrication, Piezoelectric ceramic and metal cap should have a good cleaning to facilitate good adhesion. Solder joints at positive and negative two electrodes surface should welding on the same diameter. Using epoxy resin AB glue 2:1 adhesive evenly coated on the edge bonding surface of the metal cap, Then quickly bond the good welding piezoelectric ceramic and metal cap together. To ensure that the two metal cap and piezoelectric ceramic strictlyconcentric, then put the uniform pressure on it, and after curing 24 hours it can be used, and as far as possible to reduce the thickness of adhesive layer when bonding. So get the cymbal piezoelectric vibrator as shown in Fig. 2. b.

(a)                                (b)

**Fig. 2.** Single Cymbal piezoelectric vibrator.(a)Metal Cap Structure (b)Cymbal piezoelectric vibrator

## 2.2    Cymbal Piezoelectric Micro Displacement Actuator Structure

The single cymbal piezoelectric vibrator's displacement is small, Though it can be improved by increasing driving voltage, but under the high voltage, the piezoelectric ceramics will be breakdown, So in actual application, In order to lower the driving voltage for larger displacement, Usually adopt multiple cymbal piezoelectric vibrator bonding in series, then become stacked structure of piezoelectric vibrator, as shown in Fig. 3. a. Be attention that all the cymbal piezoelectric vibrator should be concentric paste, and the Adhesives should be thin, Every two adjacent cymbal piezoelectric vibrator should be series connection in structure, and parallel connection in electrical. When no load and working in the same electric field, this structure will get the total displacement of multiple piezoelectric vibrator.

This paper adopts the cymbal piezoelectric micro displacement actuator as shown in Fig. 3, using nine cymbal piezoelectric vibrator stacked in series to form composite piezoelectric vibrator, using nut and prepressing spring to adjust the piezoelectric stack preloading, adding driving voltage to piezoelectric ceramic which connected in parallel in electrical, then producing the output displacement through the plunger, so to realize the conversion of electrical energy into mechanical energy. Real photo as shown in Fig. 3 .b.

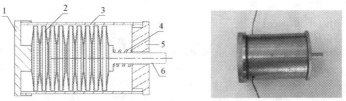

1.base 2. cymbal piezoelectric stack 3.sleeve 4. prepressing spring 5. adjusting nut 6. mandrill

(a)                                              (b)

**Fig. 3.** Cymbal piezoelectric micro displacement actuator. (a) structure chart; (b) real photo.

## 3    Cymbal Piezoelectric Micro Displacement Actuator Test System

Because the cymbal piezoelectric vibrator is built by piezoelectric ceramics and the metal cap bonding together, and cymbal piezoelectric micro displacement actuator is composed of multiple cymbal piezoelectric vibrator bonding together, therefore, cymbal piezoelectric micro displacement actuator mechanical and electrical characteristics are more complex than piezoelectric ceramics. At the same time, every cymbal piezoelectric micro displacement actuator voltage-displacement relationship is different due to the structure size, production process, adhesive type and adhesive thickness maybe is different, so before application, it is necessary to test the cymbal piezoelectric micro displacement actuator characteristics to guide the subsequent application.

The cymbal piezoelectric micro displacement actuator studied in this paper is used in 10 $\mu m$ small travel part which is a part of large travel and high precision positioning system.The driving voltage is 0-120V, Adopt German Polytec company high performance OFV-5000 (OFV-505) single point He-Ne laser vibration meter, frequency range is DC-1.5MHz, working distance is 0.53~100m, displacement resolution is up to 2nm, and it has the advantages of high accuracy, fast dynamic response, wide measuring range and can realize non-contact measurement, the main drawback is that the price is higher and the measuring environment is more strict, and the laser alignment debugging is time-consuming.

Cymbal piezoelectric micro displacement actuator voltage-displacement characteristics test system diagram is shown in Fig. 4. a, the experimental device is shown in Fig. 4. b. Controller DSP28335 is as a signal generator to generate voltage signal, through D/A converter, then put into power amplifier, output the amplified signal to drive the piezoelectric micro displacement actuator, laser vibrameter is to detect the micro displacement signal and transfer it to the oscilloscope, through analysis to get voltage -displacement curve.

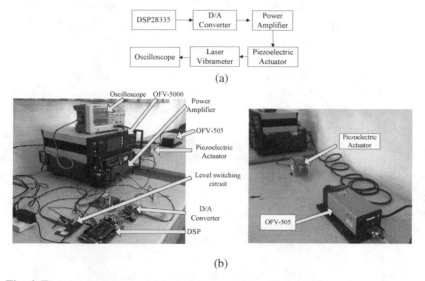

(a)

(b)

**Fig. 4.** Test system structure.(a) test system structure; (b)physical test system diagram

# 4   Cymbal Piezoelectric Micro Displacement Actuator Static Characteristic Test and Analysis

Cymbal piezoelectric micro displacement actuator static characteristic is that the actuator shows the displacement characteristics when the input voltage is static or quasi-static, this characteristics reflects the positioning performance of the system.

In normal temperature conditions and with a good vibration isolation, give cymbal piezoelectric micro displacement actuator 0.01 Hz square-wave driving voltage every 4V up from 0V to 120V, and voltage interval is more than 2 minutes each time, then give every 4V down from 120V to 0V, then get the curve as shown in Fig. 5.

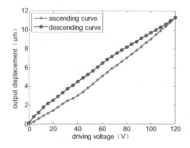

**Fig. 5.** Static voltage-displacement characteristic curve

As can be seen from the Fig.5, cymbal piezoelectric micro displacement actuator output displacement is up to $11.29\,\mu m$ within 0-120V, When the driving voltage up from 0V to 120V, the piezoelectric actuator output displacement curve is nonlinear, the nonlinearity is 3.99%. When driving voltage increases from 0V to 120V, and then decreases from 120V to 0V, the output displacement is not coincidence under the same driving voltage, that is obvious hysteresis, hysteresis error is 13%. Under the same conditions, if the same number of piezoelectric ceramic forming stack actuator, its hysterisis error is 10.38%, The main reason for the increasing hysteresis is the change of the structure, especially the adhesive layer and the cymbal metal cap's mechanical loss.

Now to analyze the influence of static voltage amplitude changes on the displacement characteristics of the system.

In order to inhabit frequency influence, adopt 0.05Hz decay triangle wave to drive cymbal piezoelectric micro displacement actuator, get a voltage-displacement curve shown in Fig. 6. The Fig. 6 shows that the amplitude of the driving voltage is greater, the cymbal piezoelectric micro displacement actuator hysteresis and nonlinearity are greater, as shown in Tab. 1. The principal causes of this phenomenon is when the driving voltage amplitude increases, the internal field strength of the core component of cymbal piezoelectric micro displacement actuator, piezoelectric ceramic, will increase, internal field strength increases can cause piezoelectric hysteresis nonlinearity enhanced, namely system hysteresis and nonlinear degrees increase[21]. So though increase the amplitude of the driving voltage signal can produce a larger output displacement, but can caused hysteresis nonlinearity of the system become more serious, so in the actual application should weigh the selection.

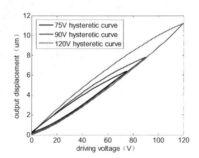

**Fig. 6.** Voltage-displacement curve of different driving voltage amplitude

**Table 1.** Hysterisis error and nonlinearity of different driving voltage amplitude

| driving voltage amplitude(V) | hysterisis error (%) | nonlinearity (%) |
| --- | --- | --- |
| 75 | 15.25 | 5.44 |
| 90 | 16.97 | 5.76 |
| 120 | 17.84 | 6.37 |

## 5    Cymbal Piezoelectric Micro Displacement Actuator Dynamic Characteristics Test and Analysis

Cymbal piezoelectric micro displacement actuator dynamic characteristics is point that when the driving voltage is a dynamic quantity, system shows the displacement characteristics, this reflects the displacement tracking performance of the system. In the positioning error compensation stage, it often needs to real-time adjust the driving voltage, this can be thought of the driving voltage's frequency is changing, and in the system that need to track the cycle track, also needs to study the dynamic characteristics of system.

To input the 1Hz, 10Hz, 20Hz sine voltage into the cymbal piezoelectric micro displacement actuator, get the response curve of the fourth cycle to draw dynamic voltage-displacement curve, and then shown in Fig. 7, the corresponding hysteresis and nonlinearity analysis are shown in Tab. 2.

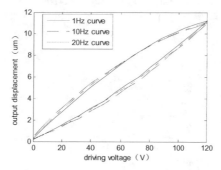

**Fig. 7.** Voltage-displacement curve of different driving voltage frequency

**Table 2.** Hysterisis error and nonlinearity of different driving voltage frequency

| driving voltage frequency (Hz) | hysterisis error (%) | nonlinearity (%) |
|---|---|---|
| 1 | 23.90 | 7.70 |
| 10 | 25.91 | 8.29 |
| 20 | 28.69 | 9.02 |

Fig. 7 and Tab. 2 show that when the frequency of driving voltage increases, the hysterisis error and nonlinearity of actuator all increase, it shows that the cymbal piezoelectric micro displacement actuator voltage-displacement characteristics is frequency-dependent, the main cause of frequency-dependent is that the piezoelectric ceramic dielectric constant is frequency-dependent [19], so in high precision positioning system, it must be considered that the output displacement affected by driving voltage frequency, namely to consider the frequency-dependent of the output displacement to achieve high precision output displacement.

From Fig. 7, it can be seen that the cymbal piezoelectric micro displacement actuator output displacement slightly decreases when the driving voltage frequency increases, this is mainly because piezoelectric micro displacement actuator will slightly fever when driving frequency increasing, so piezoelectric ceramic internal residual polarization decreases, in the external that the output displacement decrease slightly.

From Fig. 5 and Fig. 7, we can see that since each experimental interval keeps more than 2 minutes in the static characteristic shown in Fig. 5, the output displacement of piezoelectric micro displacement actuator is approximate to zero while voltage back to zero, but when give a certain frequency driving voltage, if the voltage reverse back to zero, the output displacement is not back to zero, this is mainly due to the system's creep, creep occurs mainly in a few seconds after the voltage change, about 2 minutes later, creep gradually stop. When frequency is higher, the creep reaction time is shorter, then the effect is relatively larger [20]. How to restrain the influence of creep is need to study to improve displacement tracking precision.

When input 1Hz sine voltage to the cymbal piezoelectric micro displacement actuator, we take out the 1, 20,100,200,500,800 output displacement curves, and analyze the voltage-displacement curve as shown in Fig. 8, the number of driving voltage cycles is equal to the pressure times, the hysterisis error and nonlinearity analysis are shown in Tab. 3.

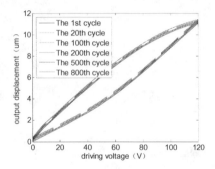

**Fig. 8.** Hysterisis curve of different driving voltage cycles

**Table 3.** Hysterisis error and nonlinearity of driving voltage cycles

| driving voltage cycles | hysterisis error (%) | nonlinearity (%) |
|:---:|:---:|:---:|
| 1 | 24.06 | 7.69 |
| 20 | 23.92 | 7.64 |
| 100 | 23.79 | 7.60 |
| 200 | 23.64 | 7.55 |
| 500 | 23.5 | 7.5 |
| 800 | 23.35 | 7.46 |

Obviously, when the piezoelectric micro displacement actuator driving voltage cycle increases, system hysteresis and nonlinear degrees have weak decrease, this is mainly due to that while the driving voltage cycle increasing, the system will be fever, lead to internal polarization is abate, then cause hysteresis and nonlinearity decrease. At the same time we can see from the Fig. 8, while the driving voltage cycle increasing, hysteresis curve will overall upward drift, namely that the output displacement slight increase, this is because while the piezoelectric micro displacement actuator pressure times increasing, the actuator will gradually fever, though weakened polarization will bring displacement decrease slightly, but the fever phenomenon will cause the coefficient of thermal expansion increases, and this increase is more significant, then cause the output displacement increases slightly, when the pressure times reach a certain value, the displacement decrease caused by polarization decrease and the displacement increase caused by thermal expansion coefficient increase are close to balance, therefore, the hysteresis curve will no longer drift upwards, achieve stability [21].

# 6 Conclusion

From above analysis we can see that the voltage-displacement characteristics of cymbal piezoelectric micro displacement actuator is a kind of typical, serious hysteresis nonlinearity, the reason of this characteristics is very complex, in the cymbal piezoelectric micro displacement actuator, the differences of piezoelectric ceramic material and size, metal cap material and size and adhesives type all will result in different hysteresis nonlinearity. This paper aimed at processed cymbal piezoelectric micro displacement actuator, analyzed its static and dynamic characteristics, and found that the driving voltage parameteanrs (voltage frequency, amplitude, voltage cycles) have a certain influence on the hysteresis nonlinearity, in a high precision displacement system, considering the influence of these factors can better improve the hysteretic nonlinear, and only to better improve the hysteresis nonlinearity, the piezoelectric micro displacement actuator can be better used in precision positioning system.

The work is supported by the National Natural Science Foundation of China (Grant No. 51177053), supported by the Specialized Research Fund for the Doctoral Program of Higher Education of China (Grant No. 2012CXZD0016) and Supported by the Key Project of Department of Education of Guangdong Province (Grant No. 20124404110003).

# References

1. Tong, G.: The research on In-pipe Micro-robot technology actuated by piezoelectric cymbals. Dissertation. Zhejiang University (2005)
2. Kim, H.W., Batra, A., Priya, S., et al.: Energy harvesting using a piezoelectric 'cymbal' transducer in dynamic environment. Japanese Journal of Applied Physics 43, 6178–6183 (2004)
3. Kim, H.W., Priya, S., Uchino, K., et al.: Piezoelectric energy harvesting under high prestressed cyclic vibrations. Journal of Electroceramics 15, 27–34 (2005)
4. Uchino, K.: Advances in Ceramic Actuator Materials. Materials Letters 22, 1–4 (1995)
5. Dogan, A., Uzgur, E.: Piezoelectric Actuator Designs. In: Piezoelectric and Acoustic Materials for Transducer Applications, pp. 341–369. Springer, New York (2008)
6. Denghua, L., Daining, F.: Cymbal Piezocomposites for Vibration Accelerometer Applications. Integrated Ferroelectrics. 78, 165–171 (2006)
7. Sun, C.-L., Lam, K.H.: High sensitivity cymbal-based accelerometer. Review of Scientific Instruments 77, 1–3 (2006)

8.  Denghua, L., Liangying, Z., Xi, Y.: Reseach on Higher Displacement Performance of Pie-zocomposite Actuator with Metal Endcaps"Cymbal". Chinese Journal of Sensors and Ac-tuators 3, 214–217 (1999)
9.  Quanqin, G.: Modeling and Analyzing the Voltage-Displacement Performance of Cymbal Piezoelectric Transducers. Chinese Journal of Sensors and Actuators 25(4), 492–495 (2012)
10. Yongcheng, M., Denghua, L., Lina, W.: Analysis of cymbal piezocomposite transducer's effective piezoelectric coefficients based on ANSYS. Chinese Journal of Scientific Instru-ment 27(6), 1313–1315 (2006)
11. Su, C.Y., Wang, Q., Chen, X., et al.: Adaptive variable structure control of a class of non-linear systems with unknown Prandtl-Ishlinskii hysteresis. IEEE Trans Automatic Con-trol 50(12), 2069–2074 (2005)
12. Dogan, A., Fernandez, J.F., Uchino, K., et al.: The "cymbal" Electromechanical Actuators. In: Proc. IEEE 10th Int. Symp. Applic. Femoelectric, pp. 1213–1216 (1996)
13. Fernandez, J.F., Dogan, A., Fielding, J.T.: Tailoring the performance of ceramic-metal piezocomposite actuators, "cymbals". Sensors and Actuators A-Physical 65(23), 228–237 (1998)
14. Ochoa, P., Pons, J.L., Villegas, M., et al.: Effect of bonding layer on the electromechanical response of the cymbal metal-ceramic piezocomposite. Journal of the European Ceramic Society 27(2-3), 1143–1149 (2007)
15. Dogan, A.: Flextensional Moonie and Cymbal Actuator. Dissertation. Pennsylvania State University (1994)
16. Lam, K.H., Wang, X.X., Chan, H.L.W.: Lead-free piezoceramic cymbal actuator. Sensors and Actuators. A, Physical. 124(2), 395–397 (2006)
17. Zhongming, P., Bo, L.: Mathematical Model of Cymbal-type Piezoelectric Transducers. J. China. Mechanical Engineering 17(3), 283–286 (2006)
18. Sheng, W., Tiemin, Z., Li, L., et al.: Vibration Analysis on Cymbal Transducer Stack. Journal of Vibration, Measurement & Diagnosis 31(3), 295–299 (2011)
19. Mingyuan, Z., Chongyang, X., Changan, W., et al.: Relations of (Ba1-xSrx)TiO3 dielectric constant with temperature and frequency. Journal of Huazhong University of Science and Technology (Natural Science Edition) 31(3), 72–74 (2003)
20. Shuai, T., Hongbai, B., Jianshe, H., et al.: Analysis of the Output Displacement Characte-ristics of Piezoelectric Actuator. Piezoelectrics & Acoustooptics. 32(5), 807–810 (2010)
21. Yuguo, C., Baoyuan, S., Weijie, D., et al.: Causes for hysteresis and nonlinearity of pie-zoelectric ceramic actuators. Optics and Precision Engineering 11(3), 270–275 (2003)

# FEM Analysis and Parameter Optimization of a Linear Piezoelectric Motor Macro Driven

Zhang Tiemin, Cao Fei, Li Shenghua, Liang Li, and Wen Sheng

College of Engineering,
South China Agricultural University, Guangzhou, China, 510642
tm-zhang@163.com

**Abstract.** As the chip packaging industry developed rapidly, high requirements are put forward for positioning accuracy, speed and acceleration of the packaging equipment at the same time. However, the increase of positioning accuracy and the speed are in conflict. This article puts forward to the structure of a new cylindrical piezoelectric linear motor consisted of macro and micro actuate with the character of piezoelectric actuator, on the basis of the organic combination of piezoelectric motor and piezoelectric micro actuator. In addition, it analyzes the working principle of piezoelectric motor, builds finite element model of piezoelectric linear motor macroscopic actuate by using the finite element analysis software, does sensitivity analysis of the structural parameters and identifies the vibration modal automatically. Finally, it takes the symmetric and anti-symmetric modal frequency consistency of the compound vibrator as optimization objective function, redesigns the structure based on the optimization process, in order to reducing the difference frequency of symmetric and anti-symmetric vibration obviously, which can lay the foundation for designing piezoelectric linear motor consisted of macro and micro actuate.

**Keywords:** piezoelectric linear motor, macro-micro drive, finite element, optimization analysis.

## 1 Introduction

With the rapid progress of IC manufacturing process and the urgent demand for tiny chip, chip integration is rising, I/O density is higher and higher, and size of chip, chip lead spacing and bonding pad diameter is decreasing, high requirements are put forward for positioning accuracy, speed and acceleration of the packaging equipment at the same time [1]. However, the increases of positioning accuracy and the speed are in conflict, and the increases of movement speed and acceleration improve inertia force of the mechanism, and also urge inertial force to change quickly [2]. Then the system will be easy to produce elastic deformation and vibration phenomenon, which generates that both mechanism motion precision is collapsed as well as increasing the settling time of the mechanical stability, and the component's fatigue strength is affected as well as increasing motion pair of wear and tear. High precision positioning expects mechanism to move steadily, and high productivity wants that the system can

X. Zhang et al. (Eds.): ICIRA 2014, Part I, LNAI 8917, pp. 171–178, 2014.

move back and forth with high speed as well as starting or stopping at the moment, which causes conflicts between large travel and high precision. The accuracy of the big trip driver and transmission (precision ball screw drive, linear motors, voice coil motor, etc.) generally is limited in micron scale; position precision of piezoelectric ceramic micro drive represented by the micro actuator can reach nanoscale, but travel only tens of microns. How to solve these contradictions, and realize large stroke, high speed precision positioning of mechanical motion systems has become a current problem to be solved a chip packaging industry [3]. Therefore, researching new type of drive and mechanism become inevitable.

Most existing high-resolution linear movements are performed by electromagnetic motor, generally with the help of the rotating motor rotation, and ball screw linear motion, coupled with the micro units, as shown in Fig. 1.

Due to existing intermediate link, a series of adverse consequences [4-8] increase the volume and weight of the whole system, the efficiency and the transmission accuracy lower, structure is complicated ,and the system is difficult to integrate, therefore, this article puts forward to the structure of a new cylindrical piezoelectric linear motor consisted of macro and micro actuate with the character of piezoelectric actuator, on the basis of the organic combination of piezoelectric motor and piezoelectric micro actuator, in order to realize the ultrasonic macro drive and micro drive of static deformation.

In this paper, numerical simulation and parameter optimization of the piezoelectric motor driven vibrator macro drive are carried on, and the influence of the parameters changing on resonance frequency, vibration mode, micro drive displacement range of the motor driven vibrator and the general rule of vibration mode degeneracy are discussed and summarized, in order to satisfy the requirement of the macro drive frequency consistency.

## 2    Working Principle of the Piezoelectric Motor

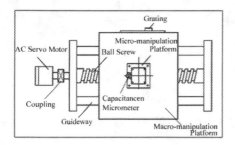
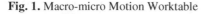
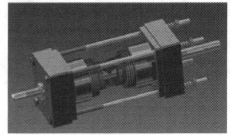

**Fig. 1.** Macro-micro Motion Worktable

**Fig. 2.** Diagram of Macro-micro Motor Structure

As the structure of new piezoelectric linear motor shown in Fig. 2, the working principle of the macro driver is that under the action of high frequency AC voltage, two groups of piezoelectric ceramic plates motivate composite vibrator to produce longitudinal vibration, which will form elliptical movement trajectory on the surface of the composite vibrator, and the double wedge of stator drive mover move drastically along the axial line under the action of friction by applying the pre-tightening force between the mover and the double wedge.

The micro drive of piezoelectric motor means that under the effect of DC high voltage two groups of piezoelectric ceramics in composite vibrator, motivate the mass point on the surface of the double wedge of the vibrator to produce static deformation along the axial line, then the double wedge grips the mover to generate high-resolution linear micro motor.

# 3    Finite Element Analysis of the Composite Vibrator

## 3.1    Finite Element Modelling

Entity model the compound vibrator of the macro and micro driver in piezoelectric linear motor as shown in figure 2, is established by the Ansys finite element software and unit type is selected to match with unit material types, that is to say that unit SOLID5 in the element library is selected for the piezoelectric materials, unit SOLID45 for other materials. At the same time, all dimensions are parameterized to model, and meshing is given priority to with mapping division, the rest with free meshing.

## 3.2    Modal Analysis

To simplify the problem, it is assumed that the stator of the macro and micro driven in cylindrical piezoelectric linear motor situates free boundary, and the initial vibration frequency ranges from 20 kHz to 50 kHz. After calculating, results show that symmetry modal vibration frequency is 31658 Hz, modal of antisymmetric vibration frequency is 31996 Hz, as shown in figure 3 and 4, which means the difference value is 438 Hz. The material point on the surface of the middle of composite vibrator can synthesize elliptical movement trajectory under the appropriate phase in the two modes.

    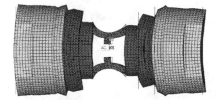

**Fig. 3.** The finite element model of Composite Oscillator Symmetry vibration modal

**Fig. 4.** The finite element model of composite oscillator anti-symmetry vibration modal

# 4    The Sensitivity Analysis of Compound Vibrator and Modal Identification

## 4.1    The Sensitivity Analysis of Compound Vibrator

The structure parameters which have a significant impact on the working frequency, is selected by sensitivity analysis. The composite oscillator has altogether 16 dimensions, including the size of the piezoelectric ceramic fixed, and the other parameters are carried on sensitivity analysis, and the corresponding expressions for the sensitivity as follows.

$$\begin{cases} S_{sj} = \dfrac{\partial f_s}{\partial p_j} = \dfrac{f_{sv} - f_{s0}}{\Delta p_j} \\ S_{aj} = \dfrac{\partial f_a}{\partial p_j} = \dfrac{f_{av} - f_{a0}}{\Delta p_j} \end{cases} \tag{1}$$

In formula (1), $f_s$ is the symmetric vibration modal frequency of the stator; $f_a$ is the anti-symmetric vibration modal frequency of the stator; $f_{s0}$ and $f_{sv}$ are the symmetrical vibration frequency of the initial stator structure and the symmetry vibration frequency of the modified stator structure respectively; $f_{a0}$ and $f_{av}$ are respectively the anti-symmetric vibration frequency of the initial stator structure and the antisymmetric vibration frequency of the modified stator structure; $\Delta p_j$ is the variation of the stator structure parameters $p_i$. Takeing it as 0.001 m, by the formula (1), and sensitivity analysis of structural parameters is carried on by using the finite element model of the stator, as shown in Fig. 5 and Fig. 6.

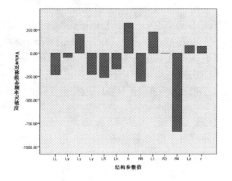
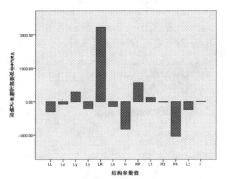

**Fig. 5.** The sensitivity analysis of macro river symmetric vibration modal

**Fig. 6.** The sensitivity analysis of macro river anti-symmetry vibration modal

According to the results of sensitivity analysis, some structure parameters of the middle of elastomer such as arc radius LR, length h, thickness RR and maximum radius R4 show higher sensitivity for the symmetric and anti-symmetric vibration modal frequency, and changing the size of these parameters can easily change the modal frequencies, so these parameters are set as the design variables in optimization design.

## 4.2    Automatic Identification of Modal Vibration Mode

In modal analysis of the compound vibrator, in addition to calculating the symmetric and anti-symmetric vibration modal, there are many other vibration modal, and the two work mode is required to vary with the structure parameters of the stator, as well as the order and degree. Therefore, the symmetric and anti-symmetric modal can be identified by the MAC method (Modal Assurance Criteria) , and MAC value is as follow.

$$(MAC)_{ij} = \frac{(\Phi_{ai}^T \Phi_{bj})^2}{(\Phi_{ai}^T \Phi_{ai})(\Phi_{bj}^T \Phi_{bj})} \tag{2}$$

In type (2), $0 \le MAC \le 1$, $\Phi_{ai} (i=1,2,\cdots,30)$ is the stator vibration mode which is waited recognition calculated by finite element method; $\Phi_{bj}$ ( $j=1,2$ ) is the symmetric and antisymmetric vibration modes selected under the initial stator structure which mean the reference mode. MAC value reflects level of similarity between the vibration mode waited identified and reference mode. The higher the value is, the higher the level will be, in view of the above, the needed symmetric and antisymmetric vibration modal can be identified.

# 5    The Optimization of Composite Vibrator

## 5.1    The Objective Function and the Optimization Process

Taking the symmetric and anti-symmetric modal frequency consistency of the motor stator as the objective function of optimization design, namely

$$\min_x | f_s - f_a | \tag{3}$$

From the analysis of the sensitivity, arc radius, length, thickness and maximum radius of the piezoelectric motor composite vibrator have a greater influence on the vibration frequency. Under the premise that the basic structure isn't affect, the range of the size parameters can be confirmed. Then the optimization design can be realized by the ansys parametric language APDL programming, as the process shown in Fig. 7.

## 5.2    The Results and Analysis of the Structural Parameters Optimization

After several iterative calculation, the results show that the symmetric modal vibration frequency is 35512 Hz, anti-symmetric vibration modal frequency is 35547 Hz, and the frequency difference between the latter and the former is 35 Hz.

In order to verify the above optimization results, based on the vibration frequency calculated by modal analysis, assuming that the voltage amplitude is 30 V, and the frequency ranges from 35 kHz to 37 kHz, harmonic response and transient analysis are carried on for the composite vibrator, as shown in figure 8 and 9.

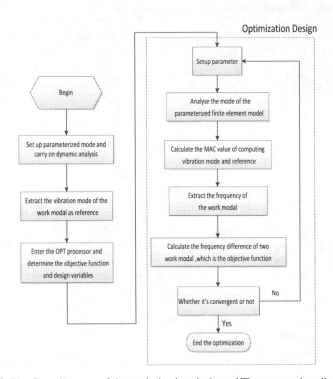

**Fig. 7.** The flow diagram of the optimization design of The composite vibrator

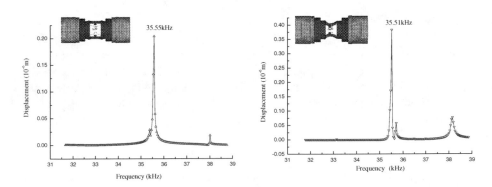

**Fig. 8.** The relationship between displacement and frequency of the symmetric modal

**Fig. 9.** The relationship between displacement and frequency of the anti-symmetric modal

From figure 8 and 9, when the stator keeps vibrating symmetrically, the displacement along the Y direction (vertical) of the node on the end face reachs the maximum for 0.21 mm at 35.51 Hz; Similarly, when the stator keeps vibrating antisymmetrically, the displacement along the Z direction (horizontal) of the node reachs the maximum for 0.38

mm at 35.55 Hz, This suggests that the symmetric and antisymmetric resonance frequency of the motor stator are 35.51 kHz and 35.55 kHz respectively, which fit the results of modal analysis.

# 6    Conclusion

By means of finite element analysis and optimization design for macro drive of macro-micro drive piezoelectric linear motor, it can be seen that the geometry of oscillator, material properties and boundary conditions affect the vibration modal of the compound stator in the piezoelectric linear motor. In order to inspire symmetric and anti-symmetric composite vibration modal at the same frequency and the same time, it is needed to analyze the influence on the two frequencies when material parameters and structure parameters change, so that the structure of the ultrasonic motor stator can be designed reasonably.

(1) It can be seen that reducing the overall size can obviously increase the symmetric and anti-symmetric frequency by the finite element calculation;

(2) Through the optimization design for the structure parameters of the stator, the symmetric and anti-symmetric modal frequency can be as consistent as possible ,without other vibration modal frequency nearby;

(3) The sensitivity analysis of structure parameters and the automatic identification of vibration modal of the new piezoelectric linear motor are carried on. Taking the symmetric and anti-symmetric modal frequency consistency of the compound vibrator as optimization objective function, design the optimization design process and the macro-micro drive of the piezoelectric linear motor on the basis of the process. Results showed that after optimization the frequency difference between symmetric and anti-symmetric vibration of the motor reduces to 35Hz, at the same time the other frequency difference between the adjacent modal vibrations reduces to around 2 kHz.

(4) The optimization design of macro drive is mainly modal analysis, and the optimal design of the micro drive mainly statics analysis. How to let the structure size and material characteristic parameters meet the requirements of the frequency consistency for macro drive and the biggest drive displacement for micro drive at the same time demands further research.

**Acknowledgements.** The work is supported by the National Natural Science Foundation of China (Grant No. 51177053), supported by the Specialized Research Fund for the Doctoral Program of Higher Education of China (Grant No. 2012CXZD0016) and Supported by the Key Project of Department of Education of Guangdong Province (Grant No. 20124404110003).

# References

1. Gwon, H.R., Lee, H.J., Kim, J.M.: Dynamic behavior of capillary-driven encapsulation flow characteristics for different injection types in flip chip packaging. Journal of Mechanical Science and Technology 28(1), 167–173 (2014)

2. Ji, J., Yan, S., Zhao, W.: Minimization of cogging force in a novel linear permanent-magnet motor for artificial hearts. IEEE Transactions on Magnetics 49(7), 3901–3904 (2013)
3. Yao, W., Basaran, C.: Electric pulse induced impedance and material degradation in IC chip packaging. Electronic Materials Letters 9(5), 565–568 (2013)
4. Jiantao, Z., Tiemin, Z.: The advance of ultrasonic motor servo control technology. Electric Machines and Control 13, 879–885 (2009)
5. Hai, X., Chunsheng, Z.: The development and application of linear ultrasonic motor. Chinese Journal of Mechanical Engineering 14, 714–717 (2003)
6. Fung, R.-F., Hsu, Y.-L., Huang, M.-S.: System identification of a dual-stage XY precision positioning table. Precision Engineering 33, 71–80 (2009)
7. Chuan, Y., Qiang, Z., Hairong, W., Zhi, Z.: Study on intelligent control system of two-dimensional platform based on ultra-precision positioning and large range. Precision Engineering 34, 627–633 (2010)
8. Liu, C.-H., Jywe, W.-Y., Jeng, Y.-R., Hsu, T.-H., Li, Y.-T.: Design and control of a long-traveling nano-positioning stage. Precision Engineering 34, 497–506 (2010)

# Finite Element Study on the Cylindrical Linear Piezoelectric Motor Micro Driven

Zhang Tiemin, Cao Fei, Li Shenghua, Wen Sheng, and Liang Li

College of Engineering,
South China Agricultural University, Guangzhou, China, 510642
tm-zhang@163.com

**Abstract.** As the chip packaging industry developed rapidly, high requirements are put forward for positioning accuracy, speed and acceleration of the packaging equipment at the same time. However, the increase of positioning accuracy and the speed are in conflict. This article puts forward to the structure of a new cylindrical piezoelectric linear motor consisted of macro and micro actuate with the character of piezoelectric actuator, on the basis of the organic combination of piezoelectric motor and piezoelectric micro actuator. In addition, it analyzes the working principle of piezoelectric motor, builds finite element model of piezoelectric linear motor microscopic actuate by using the finite element analysis software, carries on the static analysis and the sensitivity analysis of the structural parameters. Finally, it takes the largest static displacement of the micro drive of the composite vibrator as optimization objective function, redesigns the structure based on the optimization process, in order to letting the micro displacement of the linear motor get larger, which can lay the foundation for designing piezoelectric linear motor consisted of macro and micro actuate.

**Keywords:** piezoelectric linear motor, micro drive, finite element analysis Introduction.

## 1    Introduction

With the rapid progress of IC manufacturing process and the urgent demand for tiny chip, chip integration is rising, I/O density is higher and higher, and size of chip, chip lead spacing and bonding pad diameter is decreasing, high requirements are put forward for positioning accuracy, speed and acceleration of the packaging equipment at the same time. However, the increases of positioning accuracy and the speed are in conflict, and the increases of movement speed and acceleration improve inertia force of the mechanism, and also urge inertial force to change quickly. Then the system will be easy to produce elastic deformation and vibration phenomenon, which generates that both mechanism motion precision is collapsed as well as increasing the settling time of the mechanical stability, and the component's fatigue strength is affected as well as increasing motion pair of wear and tear. High precision positioning expects mechanism to move steadily, and high productivity wants that the system can move back and forth with high speed as well as starting or stopping at the moment, which causes

X. Zhang et al. (Eds.): ICIRA 2014, Part I, LNAI 8917, pp. 179–186, 2014.
© Springer International Publishing Switzerland 2014

conflicts between large travel and high precision. The accuracy of the big trip driver and transmission (precision ball screw drive, linear motors, voice coil motor, etc.) generally is limited in micron scale; position precision of piezoelectric ceramic micro drive represented by the micro actuator can reach nanoscale, but travel only tens of microns [1]. How to solve these contradictions, and realize large stroke, high speed precision positioning of mechanical motion systems has become a current problem to be solved a chip packaging industry. Therefore, researching new type of drive and mechanism become inevitable.

Most existing high-resolution linear movements are performed by electromagnetic motor, generally with the help of the rotating motor rotation, and ball screw linear motion, coupled with the micro units, as shown in figure 1.

Due to existing intermediate link, a series of adverse consequences [2-6] increase the volume and weight of the whole system, the efficiency and the transmission accuracy lower, structure is complicated ,and the system is difficult to integrate, therefore, this article puts forward to the structure of a new cylindrical piezoelectric linear motor consisted of macro and micro actuate with the character of piezoelectric actuator, on the basis of the organic combination of piezoelectric motor and piezoelectric micro actuator, in order to realize the ultrasonic macro drive and micro drive of static deformation.

In this paper, numerical simulation and parameter optimization of the piezoelectric motor micro driven are carried on, and the influence of the parameters changing on micro drive displacement range of the motor driven vibrator is discussed and summarized.

## 2     Working Principle of the Piezoelectric Motor

As the structure of new piezoelectric linear motor shown in figure 2, the working principle of its macro driver is that under the action of high frequency AC voltage, two groups of piezoelectric ceramic plates motivate composite vibrator to produce longitudinal vibration, which will form elliptical movement trajectory on the surface of the composite vibrator, and the double wedge of stator drive mover move drastically along the axial line under the action of friction by applying the pre-tightening force between the mover and the double wedge.

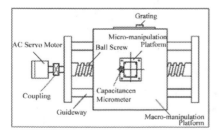

**Fig. 1.** Macro-micro Motion Worktable

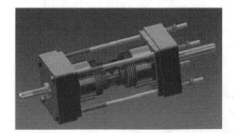

**Fig. 2.** Diagram of Macro-micro Motor Structure

The micro drive of piezoelectric motor means that under the effect of DC high voltage two groups of piezoelectric ceramics in composite vibrator, motivate the mass point on the surface of the double wedge of the vibrator to produce static deformation along the axial line, then the double wedge grips the mover to generate high-resolution linear micro motor.

## 3     The Static Finite Element Analysis of the Composite Vibrator

### 3.1     The Static Finite Element Model

Entity model the compound vibrator of the macro and micro driver in piezoelectric linear motor as shown in figure 2, is established by Ansys finite element software and unit type is selected to match with unit material types, that is to say that unit SOLID5 in the element library is selected for the piezoelectric materials, unit SOLID45 for other materials. At the same time, all dimensions are parameterized to model, and meshing is given priority to with mapping division, the rest with free meshing. The finite element entity model is as shown in figure 3.

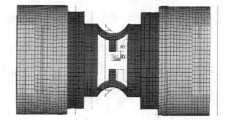

**Fig. 3.** The Finite element modal of the composite oscillator

**Fig. 4.** the static analysis of the composite oscillator

### 3.2     The Static Finite Element Analysis

The displacement of micro drive of the piezoelectric linear motor is related to the static deformation of the piezoelectric ceramics directly, at the same time the static deformation is proportional to the DC voltage applied on both ends of the piezoelectric ceramics. Therefore, the higher the DC voltage is, the larger the deformation of the piezoelectric ceramics gets, and the larger the displacement of micro drive gets. Because of this, the static finite element analysis and calculation are carried on for the micro drive of the composite oscillator. The results show that the shape of the elastomer with the arc tooth structure, the size parameters and the material properties parameters have a great influence on the displacement of micro drive. When the voltage on both ends of the piezoelectric ceramics increases to 500 V, the micro displacement of the double wedge is as shown in figure 4.

# 4    The Optimization Design for Structure Parameters of the Composite Oscillator Micro Drive

## 4.1    The Relation between Drive Range and Voltage

When the DC voltage applies on both ends of the piezoelectric ceramics, it will deform along it's thickness direction, which can be transformed to the micro displacement of the double wedge along the moving direction of the mover by the elastomer with the arc tooth structure, as shown in figure 4, and the value of displacement is the displacement range of micro drive. Therefore, the relationship between the micro displacement of the double wedge and the voltage on both ends of the piezoelectric ceramics needs to analyze by the Ansys finite element software.

**Table 1.** The compare between both ends of the elastomer material

| Displacement (um) | 200v | 300v | 400v | 500v | 600v |
|---|---|---|---|---|---|
| Steel | 0.3 | 0.43 | 0.6 | 0.73 | 0.85 |
| Stainless steel | 0.2 | 0.40 | 0.51 | 0.64 | 0.78 |
| Dural | 0.24 | 0.42 | 0.5 | 0.63 | 0.82 |
| Steel spring | 0.22 | 0.34 | 0.45 | 0.55 | 0.70 |

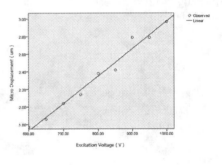

**Fig. 5.** The relationship between the excitation voltage and the micro displacement

To simplify the problem, beryllium bronze is selected as the material of the elastomer, four kinds of materials, like 45 steel, stainless steel, and steel spring respectively as both ends of the elastomer for the finite element analysis and calculation of the micro drive. When different DC voltage is applied on both ends of the piezoelectric ceramic, the micro displacement of the double wedge calculated by Ansys software is as shown in table 1. After carrying on linear fitting for the results of table 1, as shown in figure 5, it can be seen that, the micro displacement of the composite oscillator double wedge increases linearly as the DC voltage applied on both ends of the piezoelectric ceramic, and the double wedge of the composite vibrator grips the mover, which drives the mover to move tidily. Therefore, the micro drive displacement of the mover is also linearly related to the voltage applied on both ends of the piezoelectric ceramic.

## 4.2    The Sensitivity Analysis of the Micro Drive

Taking the displacement of the compound oscillator with the existing optimizing sizes as reference and carrying on the sensitivity analysis, the compound oscillator is applied at the voltage of 800.

$$S_{m,j} = \frac{\partial f_m}{\partial p_j} = \frac{f_{mv} - f_{m0}}{\Delta p_j}$$

(1)

In type (1), $f_{mv}$ is the micro displacement range after the sizes modified; $f_{m0}$ is the displacement value of the original sizes; $\Delta p_j$ is the variation value of the stator structure parameters $p_i$. Taking the variation value $\Delta p_j$ as 0.001 m and the sensitivity analysis of the structure parameter of the compound oscillator finite element modal for the micro displacement is carried on by the formula (1). To simplify the problem, three kinds of materials, like beryllium bronze, stainless steel and dural are selected respectively as the material of the elastomer. When different DC voltage is applied on both ends of the piezoelectric ceramic, the micro displacement of the double wedge calculated by Ansys software is as shown in Tab.2. After carrying on linear fitting for the results of Tab.2, as shown in figure 6, it can be seen that, the structure parameters h, RR, R3, and Ly1 of the compound oscillator have a great influence on the micro displacement range.

**Table 2.** The compare between different the elastomer material

| Displacement (um) | 200v | 300v | 400v | 500v | 600 |
|---|---|---|---|---|---|
| Bronze | 0.26 | 0.37 | 1.51 | 0.67 | 0.85 |
| Stainless Steel | 0.16 | 0.23 | 0.38 | 0.42 | 0.48 |
| Aluminium | 0.15 | 0.31 | 1.55 | 0.62 | 0.93 |

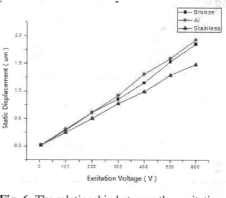

**Fig. 6.** The relationship between the excitation voltage and the micro displacement

## 4.3 The Structure Parameter Optimization of the Micro Drive

According to the sensitivity analysis, the sizes of h, RR, R3, Ly1 have a great influence on the micro displacement range. Because the optimization design for the macro-micro drive piezoelectric linear motor is given priority to the macro drive, the micro drive is designed on the basis of the optimized design of the macro drive.

Therefore, the parameters R3 and Ly1, which have a great influence on the micro drive but little on the macro drive, are selected to carry on the optimization design.

After analyzing and calculating the size parameters R3 and Ly1 by Ansys software, it can be seen that, when the size parameter R3 is increased , the static displacement linearly decreases, otherwise, the static displacement linearly increase; When the size parameter Ly1 is increased , the static displacement linearly decreases, otherwise, the static displacement linearly increase. Therefore, under the premise that the overall structure sizes aren't affected and the consistency of symmetric and anti-symmetric modal vibration frequency is ensured, the size of R3 and Ly1 should be decreased as far as possible, in order to increase the micro displacement. More specifically, R3 changes from 17.2 mm to 17 mm, and Ly1 changes from 12 mm to 11 mm.

# 5    The Material Analysis of the Composite Vibrator

## 5.1    The Material Analysis on Both Ends of the Elastomer Material

Four kinds of materials, like 45 steel, stainless steel, dural and steel spring are selected respectively as the material on both ends of the elastomer, on the basis that the material of the intermediate elastomer is beryllium bronze, and the curves that their static displacement changes as the voltage applyied on both ends of the piezoelectric ceramic is as shown in figure 7. The difference of the static displacement between the four kinds of materials is small: The maximum static displacement is 45 steel, which is easy to rust; the price of Dural is a little high, the corrosion resistance is poor, and high intensity requires high technical. In conclusion, it's appropriate that the cost-effective stainless steel is selected as the material on both ends of the elastomer.

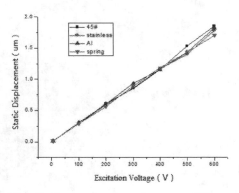
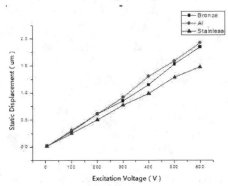

**Fig. 7.** The comparison between different materials on both ends of elastomer

**Fig. 8.** The material selection of difference elastomer

## 5.2    The Material Analysis of the Micro Displacement Amplification Elastomer

Three kinds of materials, like beryllium bronze, stainless steel, and dural are selected respectively as the material on both ends of the elastomer to compare, on the basis that the material on the ends of elastomer is beryllium bronze. Their corresponding static displacement is shown in figure 8, and the material properties of the intermediate elastomer has a greater influence on the static displacement than the material properties on both ends of the elastomer. From figure 8, it can be seen that when the material of the intermediate elastomer selects stainless steel, the micro displacement of the double wedge is small; when the material of the intermediate elastomer selects beryllium bronze and dural, the micro displacement is larger. Compared to beryllium bronze and dural, beryllium bronze has characteristics of high strength, wear resistance, fatigue resistance and corrosion resistance, so beryllium bronze is selected as the material of intermediate elastomer.

## 6     Test System and the Result Analysis

The micro drive test system of the macro-micro drive piezoelectric linear motor consists of the signal generator, the power amplifier, the laser vibration meter, the experimental prototype and the PC computer, as shown in figure 9.

In view that the displacement of the micro drive is small, and measuring accuracy are greatly influenced by the environment, under the existing laboratory environment, the measurement instruments and the experimental prototype are equipped with rubber vibration isolation mat at the bottom, at the same time, the experimental prototype is measured in a quiet environment. When the macro-micro drive piezoelectric linear motor is applied at difference DC voltage, its micro drive displacement changes as the DC voltage increases linearly, which is in line with the change rule of the finite element calcution, and the micro displacement by measured and calculated are at the same order of magnitude

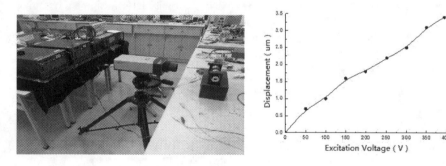

**Fig. 9.** The equipment and the experimental curve

## 7     Conclusion

Through the fininte element analysis and the optimization design of the macro-micro drive piezoelectric linear motor, it can be seen, the geometric structure of the compound vibrator and material characteristics parameters affect the static deformation of compound vibrator of the piezoelectric linear motor. In order to get the large static displacement, it is needed to analyze the influence on the static displacement when the material parameters and structure parameters change, so that the structure of the compound vibrator with the macro drive and the micro drive can be designed reasonably.

(1) It can be seen that the micro displacement of the micro drive has hundreds of nanometer range, when enough DC voltage is applied on both ends of the piezoelectric ceramic, by the finite element analysis of the micro drive of the macro-micro drive piezoelectric linear motor.

(2) The structure parameters h, RR, R3, and Ly1 of the compound vibrator have a great influence on the range of the micro displacement, and the optimization design for the micro drive bases on the optimized design of the macro design. Therefore, the parameters R3 and Ly1 which has a great influence on the micro drive but a little influence on the macro drive are selected to design.

(3) The displacement resolution of the macro drive reaches micron level and the micro drive reaches nanoscale. When the macro drive closes to the end of the big trip, the macro drive with high speed turns into the micro drive with high resolution. In order to achieve the high-resolution positioning of the macro and micro drive and the effective transformation, the displacement range of the micro drive should be as large as possible.

(4) The optimization design process of the micro drive is the static analysis process of the composite vibrator by the finite element software actually. How to adjust the structure size and material parameters, make the conversion accuracy of the macro-micro drive consistent, and satisfy the consistency of the macro drive vibration frequency at the same time needs further research.

**Acknowledgements.** The work is supported by the National Natural Science Foundation of China (Grant No. 51177053), supported by the Specialized Research Fund for the Doctoral Program of Higher Education of China (Grant No. 2012CXZD0016) and Supported by the Key Project of Department of Education of Guangdong Province (Grant No. 20124404110003).

# References

1. Yao, W., Basaran, C.: Electric pulse induced impedance and material degradation in IC chip packaging. Electronic Materials Letters 9(5), 565–568 (2013)
2. Jiantao, Z., Tiemin, Z.: The advance of ultrasonic motor servo control technology. Electric Machines and Control 13, 879–885 (2009)
3. Hai, X., Chunsheng, Z.: The development and application of linear ultrasonic motor. Chinese Journal of Mechanical Engineering 14, 714–717 (2003)
4. Fung, R.-F., Hsu, Y.-L., Huang, M.-S.: System identification of a dual-stage XY precision positioning table. Precision Engineering 33, 71–80 (2009)
5. Chuan, Y., Qiang, Z., Hairong, W., Zhi, Z.: Study on intelligent control system of two-dimensional platform based on ultra-precision positioning and large range. Precision Engineering 34, 627–633 (2010)
6. Liu, C.-H., Jywe, W.-Y., Jeng, Y.-R., Hsu, T.-H., Li, Y.-T.: Design and control of a long-traveling nano-positioning stage. Precision Engineering 34, 497–506 (2010)

# Friction Experiment Study on the Standing Wave Linear Piezoelectric Motor Macro Driven

Li Liang, Tiemin Zhang[*], Penghuan Huang, and Debing Kong

College of Engineering,
South China Agricultural University, GuangZhou, China, 510642
ll-scau@163.com, tm-zhang@163.com

**Abstract.** The rapid development of chip packaging industry has made higher demands for the positioning accuracy, speed of movement and acceleration of packaging equipment at the same time, but the positioning accuracy is contradicted against the improvement of speed of movement. The macro and micro drive standing wave linear piezoelectric motor uses the friction of the contact interface for energy conversion and power transmission, it will convert the vibration of the stator to the macro linear motion of the rotor or convert the static deformation of the stator to the micro linear motion of the rotor, then the motor will drive the load to work, which solves the contradiction of the positioning accuracy and the speed of movement. In order to improve the macro and micro linear piezoelectric motor driving force, enhance the motor operation stability and extend the motor working life, this paper studies the influence rule of the friction pair that consists of different friction materials on the motor performance by applying different friction materials on the contact interface, thus proposes the method of the improve of motor output power and the improve of the stator and the rotor contact state, and provides useful experience to design a kind of macro and micro drive linear piezoelectric motor with a good performance.

**Keywords:** linear piezoelectric motor, macro and micro drive, friction materials, experimental study.

## 1 Introduction

The vast majority of existing high-resolution linear propulsion movement is completed by the electromagnetic motor generally by means of the rotary motion of the rotation motor, with the linear motion of the ball screw, together with the micro-positioner, as shown in Figure 1. Because of the existence of the intermediate transformation that caused a series of adverse consequences, which increases the whole system volume and weight, reduces efficiency, reduces transmission accuracy and the structure is complex that is not easy to be integrated [1]. Therefore, this paper organically combines the piezoelectric motor and the piezoelectric ceramic actuator according to the characteristics

---

[*] Corresponding author.

X. Zhang et al. (Eds.): ICIRA 2014, Part I, LNAI 8917, pp. 187–196, 2014.

of piezoelectric actuator, puts forward a kind of linear piezoelectric motor based on the piezoelectric conversion with macro micro drive in one set, in order to realize ultrasonic macro drive and micro drive with static deformation. This motor uses the friction of the contact interface for energy conversion and power transmission, it will convert the vibration of the stator to the macro linear motion of the rotor or convert the static deformation of the stator to the micro linear motion of the rotor, then the motor will drive the load to work, which solves the contradiction of the positioning accuracy and the speed of movement [2-3]. The friction between the stator and rotor of the motor is dry friction, which not only produces noise, but also accelerates the aging of the friction pair, the wear of the friction interface is more serious. The commonly used friction material of piezoelectric motor is Composite Material, it consists of matrix, reinforcing filler and friction modifier, there are rubber-based friction materials, plastic-based friction materials, and powder metallurgy sintered metal with oil friction material and surface with ceramic coating friction material [4-6]. The performance parameters of the friction material has a great influence on the piezoelectric motor, and the elastic modulus affect no-load speed, output torque, output power and start and stop characteristics of the piezoelectric motor; Increasing the coefficient of friction can improve the output power of the piezoelectric motor, however, too much friction coefficient cannot significantly increase the output power, on the contrary it will increase friction and wear of the motor, increase the motor friction noise and reduce the motor's life. It will affect the motor speed, output force and friction noise [7-9].

In order to improve the linear ultrasonic motor driving force, enhance the running stability of the motor and extend the working life of the motor, this research choose two kinds of friction materials, with the elastomer material of the motor combined into three groups of friction pair, researches the friction operating rules of linear piezoelectric motor by the experiment.

## 2    Piezoelectric Motor Friction Drive Model

When the macro and micro drive standing wave linear piezoelectric motor macro drives, the contact interface is under the conditions of ultrasonic vibration, there exists the phenomenon of ultrasound antifriction, so the friction coefficient of the friction pair constituted by the convex teeth and the mover which are on the composite oscillator of the piezoelectric motor is far less than the conventional static and dynamic friction coefficient. But in the any time of the composite oscillator vibration, the effect of the convex teeth and the friction material layer is same with the friction effect of the conventional case.

According to contact theory, the force of the composite oscillator and the mover at the contact interface can be divided into the normal contact force and the tangential friction component, in the elastic contact range, the normal component of the contact force is proportional to the compression deformation of the compression deformation, the tangential friction force meets the Coulomb friction law.

$$\begin{cases} f_n = -k_n z \\ f_\tau = -\mu_d f_n \end{cases} \qquad (1)$$

In the formula (1), $f_n$ and $f_\tau$ are the distributed normal contact force and tangential friction; $k_n$ is the equivalent distributed spring stiffness coefficient of friction layer; $z$ is the normal contact deformation of the friction layer; $\mu_d$ is the static and dynamic friction coefficient of the stator and the mover contact interface.

## 3    The Selection Principle of the Piezoelectric Motor Friction Material

For the macro and micro drive standing wave linear piezoelectric motor, based on the requirement of the realization of the design of macro micro drive motor, usually the friction materials that have bigger modulus of elasticity and hardness than the composite oscillator elastic body or the rotor material are applied in the convex teeth of the composite oscillator, in the most cases the friction materials are stick to the convex teeth of the composite oscillator, so at the contact interface of the composite oscillator and the rotor, the contact deformation amplitude of the rotor surface is relatively larger. The friction materials used should satisfy the following requirements: (1) The friction coefficient should be as high as possible, so that the vibration can be efficiently transformed into kinetic energy; (2) Good wear resistance, and the wear of the counterpart are very slight; (3) Friction will not change with the change of time, can work steadily for a long time; (4) No friction noise, cannot cause additional vibration of the stator and the rotor; (5) Easy precision machining , good thermal and chemical stability.

The selection of the friction material in addition to considering the motor conversion efficiency, but also the wear, the heat, the noise and so on of the friction surface. Selecting the friction system must consider the abrasion resistance of the both components. Ultrasonic motor friction material requires both wear resistance, and that friction coefficient is too low. Research shows that, the sum of the wear volume of the film itself and the components friction is the minimum, the motor is running smoothly, with large and stable output torque, small heat.

For the convenience of theory analysis, assuming at the contact interface of macro and micro drive piezoelectric motor, the wear resistant materials which is adhesive to the convex teeth of the composite oscillator and the rotor surface constitute a layer of friction material layer. This friction layer is equivalent to a series of parallel line spring, when the piezoelectric motor is running, the composite oscillator and the rotor itself don't make contact deformation, all contact deformation occurs on the friction material layer.

## 4    The Friction Pair Paired Test Scheme

The selection of the friction material in addition to considering the motor conversion efficiency, but also the wear, the heat, the noise and so on of the friction surface.

Selecting the friction system must consider the abrasion resistance of the both components. If only one part has good abrasion resistance, but the other part has bigger wear, this is not a desirable friction material. According to the friction material selection principle, Selection two friction materials, TiN and TiAlN, the two kinds of friction material are applied to the stator driving head by means of the coating method. In order to improve the binding force of the coating and the film-substrate convex teeth, firstly using 320# sandpaper to coarse grind the convex tooth surface, the large convex will be polished, then using 500# sandpaper for semi finishing grinding, finally using 1200# sandpaper for fine grinding, using the microstructure microscope to observe that the convex tooth surface has reached a mirror state. Adopting the vacuum PVD coating machine composed of the vacuum chamber, cathode arc source and magnetic-control ion sputtering source, and using the physical evaporation deposition technology of the vacuum magnetic-control cathode arc and magnetic-control ion sputtering, below 250 degrees Celsius, the convex teeth are placed in a vacuum furnace, and depositing 3~6 hours for coating the convex tooth, the coating film thickness of two kinds of friction materials selected reaches 4um, the coating film hardness of TiN reaches 2500HV, the coefficient of friction is 0.65, the coating film hardness of TiAlN reaches 80HRC, the coefficient of friction is 0.3.

In order to shorten the running in period of the friction pair, at the room temperature, firstly doing reciprocating sliding friction in no less than 300 times on the friction interface as the pre running, through the observation of the metallographic microscope, the surface spike of the convex teeth and the rotor have been worn off, the friction pair enters the better contact state, then assembling the piezoelectric motor and doing the experimental research.

According to the selecting principle of the piezoelectric motor for the friction material, the rotor contact interface materials all are bearing steel GCr15, the convex tooth contact interface materials were selected as: 1# friction pair with uncoated friction material, 2# and 3# were coated friction material on the surface of the stator drive head, material pairing scheme of the friction pair as shown in Table 1.

**Table 1.** Material pairing scheme of the friction pair of the linear piezoelectric motor

| The friction pair number | The convex tooth contact interface materials | the rotor contact interface materials |
|---|---|---|
| 1# | 45 steel | GCr15 |
| 2# | TiN | GCr15 |
| 3# | TiAlN | GCr15 |

## 5    Characteristic Test of Friction Operation

On different friction pairs, conducting the operation characteristics of friction test and test analysis in the following: (1) Surface morphology of the friction pair before and after the grinding; (2) The mechanical characteristics of motor of different friction pairs; (3) The maximum output power; (4) Starting characteristic of motor of different friction pairs.

## 5.1    Surface Morphology

Making the linear piezoelectric motor run in stable condition,  the rotor of motor continuous reciprocating movement, according to the number of friction, the grinding of each friction pair can be divided into three stages: the early, the middle, the late, Observing and recording the grinding of the convex tooth surface in the metallographic microscope under the condition of different friction pairs, and recording the surface morphology of the friction contact interface through the computer image acquisition system.

Figure 1 shows the convex tooth contact surface morphology of the friction pair constituted by 45 steel and GCr15.The rotor of the friction pair is bearing steel(GCr15), its surface hardness is larger, wear occurs mainly on the convex tooth contact surface made by 45 steel.

In the beginning of the friction pair grinding, although there have been pre running, impact of the machinery processing, the contact surface of the convex teeth is still uneven obviously, contacting surface microstructure of the convex teeth shown in figure 1a, the contact of the convex teeth and the rotor surface is a series of point contact, the actual contact area of the friction pair is smaller. With the increase of the preload, drive head engages the surface of the mover, the contact area increases, but the wear is severe, the output force is low, the motor is running extremely unstable.

After a period of interface running, the friction pair enters the interim of grinding, the surface of the convex teeth is polished, the wear of the friction pair is getting smaller, and motor is running stability with greater output power, contacting surface microstructure of the convex teeth shown in figure 1b.

Continue working for some time, the friction pair enters the later period of grinding, the wear of the teeth surface increases further, motor is running instability and output power is reduced, contacting surface microstructure of the convex teeth shown in figure 1c.

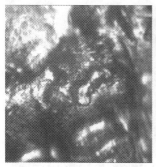  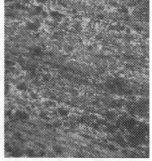

a. The surface morphology of friction pair before grinding

b. The surface morphology of friction pair grinding 500 times

c. The surface morphology of friction pair grinding 1000 times

**Fig. 1.** The surface morphology of friction pair of steel and GCr15 grinding before and after

Figure 2 shows the drive head surface morphology of the friction pair constituted by TiN and GCr15. The rotor of the friction pair is bearing steel (GCr15), its surface hardness is larger, the surface of the convex teeth made by 45 steel materials is plating TiN.

In the beginning of the friction pair grinding, although there have been pre running, with the increase in the number of friction, the TiN material of the convex front is gradually worn away, but most of the area's membrane still exists, the TiN particles worn out draws obvious grooves to the drive head on the contact interface, contacting surface microstructure of the convex teeth shown in figure 2a.

After a period of contact interface running, the friction pair enters the interim of grinding, motor is running stability, the amount of wear is reducing, noise becomes smaller, as shown in Figure 2b, compared with the Friction pair # 1, the motor output power is far greater.

The motor continue works for some time, the friction pair enters the later period of grinding, the coating on the convex teeth surface has been completely worn away, exposing the coated substrate 45 steel, because the presence of TiN hard particles, motor comes into the severe wear stage, as shown in Figure 2c, motor is running instability and, noise increases and output power is reduced.

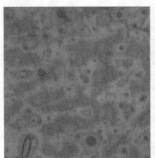  

a. The surface morphology of friction pair before grinding

b. The surface morphology of friction pair grinding 300 times

c. The surface morphology of friction pair grinding 600 times

**Fig. 2.** The surface morphology of friction pair of TiN and GCr15grinding before and after

Figure 3 shows the drive head surface morphology of the friction pair constituted by TiN and bearing steel. The rotor of the friction pair is bearing steel(GCr15), its surface hardness is larger, the surface of the convex teeth made by 45 steel materials is plating TiN.

Figure 3 shows the drive head surface morphology of the friction pair constituted by TiN and GCr15. The rotor of the friction pair is bearing steel(GCr15), its surface hardness is larger, the surface of the convex teeth made by 45 steel materials is plating TiN.

In the beginning of the friction pair grinding, although there have been pre running, with the increase in the number of friction, the film-substrate binding force of TiAlN is larger than the former, under the same conditions, the wear resistance increases, contacting surface microstructure of the convex teeth shown in figure 3a.

After a period of interface running, the friction pair enters the interim of grinding, there are still a large area of coating on the surface of the drive head, the shed TiAlN particles fill the original gully of the coated surface, because the coefficient of static friction of such materials is greater than 1 # friction pair, and the hard film has been deposited on the contact interface, which makes the motor output power greater than the first two Friction pair. As shown in figure 3b.

Continue working for some time, the friction pair enters the later period of grinding, most of TiAlN coating of the convex teeth surface are worn out, the TiAlN particles worn out draws deeper grooves to the surface of the drive head, as shown in figure 3c, motor is start to run instability, the abrasion increases, the output force gradually decreases.

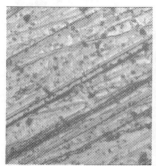  

a. The surface morphology of friction pair before grinding

b. The surface morphology of friction pair grinding 500 times

c. The surface morphology of friction pair grinding 1000 times

**Fig. 3.** The surface morphology of friction pair of TiAlN and GCr15grinding before and after

Through observing and comparing microstructure experiment of grinding before and after of the friction interface morphology of above three groups of friction pairs, it shows that, different coating materials can significantly change the friction coefficient of friction contact interface, TiAlN significantly increases the friction coefficient of Friction pair, so that the output force of the motor increases greatly, transfer efficiency of friction interface increases, Comparing the effects of three friction pairs, 3 # friction pair is the best, 2 # friction pair followed, 1 # friction pair the worst. But there is not fully realizing the purpose of wearable for the convex teeth plated friction materials, the main reasons as follows:

(1) A large friction coefficient of friction materials causes badly worn.

(2) Coating thickness is very small, only 4um, the power transmission of piezoelectric motor totally depends on the friction drive, the friction pair has not yet entered a stable period of wear, the coating has been worn away.

(3) A positive pressure applied between the oscillator and the rotor is 550N, the interface between severely wear.

(4) The coating-off particles have not exhausted out of the contact interface, under the driving force, doing the secondary damage on the contact interface, which accelerates the wear of the interface.

## 5.2    Mechanical Properties

Test for the mechanical properties of piezoelectric linear motors under the effects of three different friction pairs, when test each time, the applying voltage across the motor, two-phase input signal phase difference and frequency are unchanged, load increases 5N each time. Test for motor mechanical properties of each pair of friction under the pre-pressure of 206N, 289N, 371N. As shown in figure 4,5 and 6.

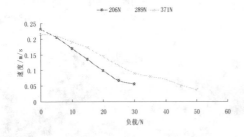

**Fig. 4.** Motor mechanical properties of friction pair of 45 steel and GCr15

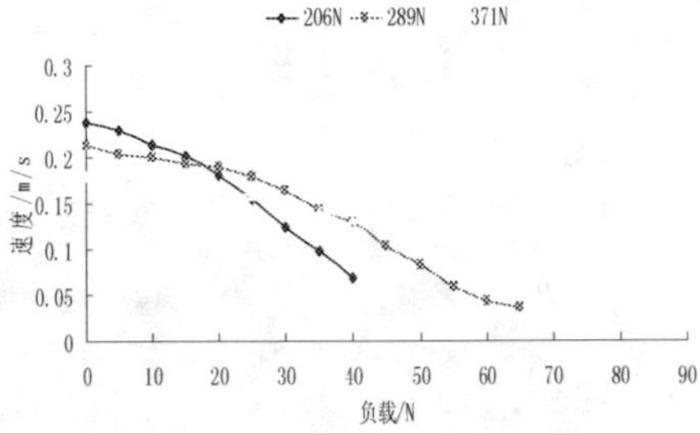

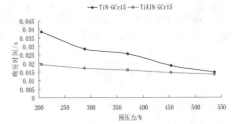

**Fig. 5.** Motor mechanical properties of friction pair of TiN and GCr15

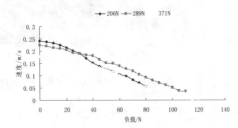

**Fig. 6.** Motor mechanical properties of friction pair of TiAlN and GCr15

**Fig. 7.** Pre-stress effect on the starting characteristics of piezoelectric linear motors

Test results from Figure 4 to Figure 7, we can draw the following conclusions:

(1) The same kind of Friction pair, with the increasing pre-pressure, the motor mechanical properties becomes hardened;

(2) The same kind of Friction pair, with the increasing pre-pressure, the motor output power increases, the running stability deteriorates;

(3) Before and after comparison of friction materials plated the convex teeth, after coating, motors speed is basically the same, the output power substantially increase, compared to the three friction pair, 3 # friction pair is superior to 2 # friction pair, 2 # friction pair is superior to 1 # friction pair.

## 5.3    Characteristics of Transient Response

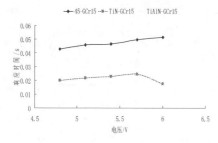

**Fig. 8.** Voltage effects on linear piezoelectric motor starting characteristics

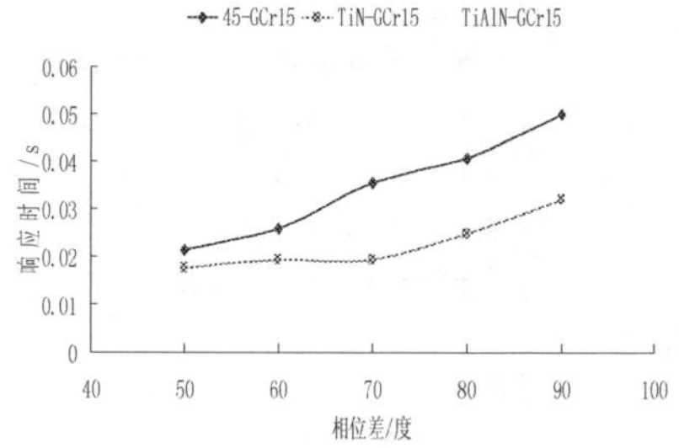

**Fig. 9.** Phase impact on the linear piezoelectric motor starting characteristics

Piezoelectric motors structure is unchanged, driving frequency is unchanged, When the no-load operation, respectively changing the pre-pressure, voltage, and phase, test for the startup characteristics of piezoelectric linear motors under the effects of three different friction pairs, As shown in figure 7-9, it can be found that:

(1) The start-up time of the Linear piezoelectric motors is reducing as the increasing of the preload;

(2) The start-up time of the Linear piezoelectric motors is increasing as the increasing of the voltage;

(3) The start-up time of the Linear piezoelectric motors is increasing as the phase is increasing from 50 to 90 degrees;

(4) In the three kinds of Friction pairs, friction pair of TiAlN and GCr15 has the shortest start-up time, friction pair of TiN and GCr15 has the second long, friction pair of 45 steel and GCr15 has the longest start-up time.

Results of the test show that:

(1) Pressure increases, the driving voltage is reduced, and the phase difference becomes smaller, motors speed is reducing;

(2) The starting response time is related to the steady-state operating speed of the motor, the larger steady speed, the longer response time to start, conversely shorter.

(3) Friction drive characteristics of ultrasonic motor is changed by the change of the friction material of friction pair, among them friction pair of TiAlN and GCr15 has the best wear resistance, with the maximum output power, the shortest start time, and runs most smoothly, thus, in these three groups friction pair, the 3 # friction pair is the best friction pair.

## 6    Conclusion

Through the test of friction pair surface morphology of the early ultrasonic motor assembly and the stable operation, concluded that when design the piezoelectric linear motors, it need to preprocess the friction pair, making the stator and the rotor basely achieve a good contact state;

Using two kinds of friction materials, two kinds of coating wear-resistant materials, 45 # steel and bearing steel to combine into three pairs of friction pair, and doing experiment testing and analysis of the system, the results show, coating friction material with high hardness and the larger static friction coefficient to the convex teeth, which can increase the output power of the motor, reduce the steady-state operating speed of the motor, reduce the start-up response time of the motor. In the three kinds of friction pair of this article, friction pair of TiAlN and GCr15 has the best friction Performance, the maximum output power of the motor, the shortest start-up response time, relatively small wear rate and the most stable operation.

**Acknowledgment.** The work is supported by the National Natural Science Foundation of China (Grant No. 51177053), supported by the Specialized Research Fund for the Doctoral Program of Higher Education of China (Grant No. 2012CXZD0016) and Supported by the Key Project of Department of Education of Guangdong Province (Grant No. 20124404110003).

# References

1. Yamaguchi, D., Kanda, T., Suzumori, K.: Ultrasonic motor using two sector-shaped piezoelectric transducers for sample spinning in high magnetic field. Journal of Robotics and Mechatronics 25(2), 384–391 (2013)
2. Mazeika, D., Vasiljev, P.: Linear inertial piezoelectric motor with bimorph disc. Mechanical Systems and Signal Processing 36(1), 110–117 (2013)
3. He, L.G., Zhang, Q., Pan, C.L.: Piezoelectric motor based on synchronized switching control. Sensors and Actuators A: Physical 197, 53–61 (2013)
4. Guo, M., Pan, S., Hu, J., Zhao, C.S.: A small linear ultrasonic motor utilizing longitudinal and bending modes of a piezoelectric tube. IEEE Transactions on Ultrasonics, Ferroelectrics and Frequency Control 61(4), 705–709 (2014)
5. Luo, Y., Qu, J., Xu, X.: Research on Friction Properties of Ultrasonic Drive. Lubrication Engineering 3, 29–30 (2005)
6. Lu, X., Hu, J., Yang, L.: A novel in-plane mode rotary ultrasonic motor. Chinese Journal of Aeronautics 27(2), 420–424 (2014)
7. Jing, J., Liu, B.: Optimum efficiency control of traveling-wave ultrasonic motor system. IEEE Transactions on Industrial Electronics 58(10), 4822–4829 (2011)
8. Liu, Y., Chen, W., Liu, J.: A cylindrical standing wave ultrasonic motor using bending vibration transducer. Ultrasonics 51(5), 527–531 (2011)
9. Shi, Y., Zhao, C.: A new standing-wave-type linear ultrasonic motor based on in-plane modes. Ultrasonics 51(4), 397–404 (2011)

# Research on Intelligent Mobile Platform
# Base on Monocular Vision and Ultrasonic Sensor

Yuda Mo, Xiangjun Zou[*], Jiaqi Hou, Guichao Lin, Yuhui Long, Nian Liu, Bo Li,
and Zhilin Jiang

Key Laboratory of Key Technology on Agricultural Machine and Equipment
College of Engineering
South China Agricultural University, Guangzhou, China
vettdetmore@live.com, xjzou1@163.com

**Abstract.** In allusion to the automatic obstacle avoidance and path planning problem of the intelligent mobile robot, a kind of intelligent mobile platform based on the cooperation between the monocular vision and ultrasonic sensor was developed. This mobile platform includes the special steering gear, visual navigation control and ultrasonic obstacle avoidance control. In vision navigation module, the color space of HSV and RGB was selected as the processing object , and then split image by using morphological opening-and-closing operation and edge point filtration, detect the road information by Hough transformation. The navigation controller of mobile platform is an adaptive fuzzy PID controller. Utilize the multiple ultrasonic sensors to detect the obstacles around the mobile platform, and achieve the effect of bypassing obstacles through fuzzy controller. The simulation and experiment show that the intelligent mobile robot can turn smoothly and identify road and obstacles well and navigate stably.

**Keywords:** Steering Gear, Vision Guidance, HSV and RGB, Ultrasonic Obstacle Avoidance.

## 1    Introduction

Research of mobile robot began in the late 1960s with the purpose of studying AIT (artificial intelligence technology) and independent reasoning and planning ability of robot system in complex environment [1]. As an earlier developed robot , there were many researches about the structure of the mobile platform, navigation and obstacle avoidance. National Pingtung University of Education, CY Chen et al. proposed a behavioral strategy designed for a humanoid intelligent robot for the purpose of obstacle avoidance with ultrasonic sensors [2]. Gansu University of Technology, Zhiwen Wang et al. proposed that wheeled robot have many advantage such as light weight, large bearing, simple structure, convenient control and fast compared with other forms [3]. Lihua Jiang et al. in Okayama University proposed a method to drive

---

[*] Corresponding author.

X. Zhang et al. (Eds.): ICIRA 2014, Part I, LNAI 8917, pp. 197–203, 2014.

the mobile robot to the target avoiding the obstacles under the uncertain observation information base on the potential field method [4]. The kinematic modeling method for a wheeled mobile robot is studied by Yong Chang et al. and, WCM (Wheel-center modeling) using in that a wheeled mobile robot that moves on uneven terrain is proposed [5]. Jianning Hua et al. introduced a wheel control method based on motion description language [6]. Jun Zhou et al. in NJAU proposed a lateral predictive fuzzy logic control algorithm for wheeled mobile robot navigated by machine vision [7]. A simple fast indoor navigation algorithm for vision guide mobile robot was presented by Mengyin Fu, which used skirting lines as the reference objects to locate the mobile robot [8]. Jingdong Yang proposed the theory of multiple objective optimizations which is introduced and an approach of path planning based on dynamic field weighting [9]. Lei Liu et al. had designed software to implement autonomous navigation and obstacle avoidance mobile robot simulation [10]. From what has been papered above, there are researches on the mobile platform, but the researches on simple and practical automatic obstacle avoidance and visual navigation technology are less.

In this paper, we proposed a special kind of steering gear. To cooperate with this structure, we have studied a road visual identification and steering control algorithm, also proposed a Simple and practical Obstacle avoidance control method based on multiple ultrasonic sensors. The purpose of this study is putting forward a new mobile robot steering structure, a practical obstacle avoidance method and a simple and steady road tracking method base on vision, which make robot's autonomous mobile platform more mature and stable, safety performance better. We have simulated and reality experiments, which had verified the feasibility and practicality of this mobile platform.

## 2    The Structure of Mobile Platform

### 2.1    Position of the Ultrasonic Sensor and Camera

As shown in fig. 1, there are 9 ultrasonic sensors in the mobile platform body, including 5 sensors which the direction of detection is outward and vertical to headstock in front of platform, 2 sensors which the direction of detection is respectively vertical to each flank, the last 2 sensors which the direction is backward and with the automobile body to 45-degree angle. In addition, the camera is installed on top of automobile body and set forward its optical axis with ground to a 45-degree angle.

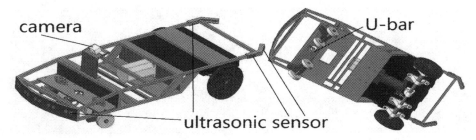

**Fig. 1.** The basic structure of the mobile platform

## 2.2    Structure of Automobile Body

The whole mobile platform is RWD (Rear Wheel Drive). Using gear motor and differential to drive. According to Ackerman steering, designed a trapezoidal steering
configuration which the draw bar is drove by U-bar, and using steering engine can
accurately control the U-bar, and drive veer. Due to the differential and the steering
mechanism with Ackerman steering, center of steering circle between front wheels
and rear wheels are mainly coincide, turning action is smooth.

# 3    Multi-ultrasonic Sensor Detection and Obstacle Avoidance

Obstacle avoidance is very complicated when intelligent mobile platform in indeter-
minacy environment. Because of the obstacle and driving environment is unknown,
it's difficult to build mathematical model to accurately forecast the information of
environment. However, we can realize avoidable system by means of fuzzy control
which depends on practical experience and control planning, and platform will be
more real-time and practicability, it's suitable for obstacle avoidance.

Fuzzy controller's input which detected by 9 sensors is composed of direction and
distance information of obstacle. 9 fuzzy subsets of direction:{incline backward
left(IBL), backward left(BL) left(L), less left(LL), middle(M), less right(LR),
right(R), backward right(BR), incline backward right(IBR)} 4 fuzzy subset of distance:{far(F), middle(M), close(C), Death area(D)}. Output quantity is platform's
steering range, there are 5 fuzzy subsets of steering range:{ turn left(TL), turn left
less(TLL), turn middle(TM), turn right lest(TRL), turn right(TR)}. Fuzzy control is an
application in aspect of "perception – reasoning" action, on the obstacle avoidance, it
according to experiments and experience to accurately reason output result. This
fuzzy controller used 19 rules to make the path mainly stable, sample path as shown
in Fig. 2. a.

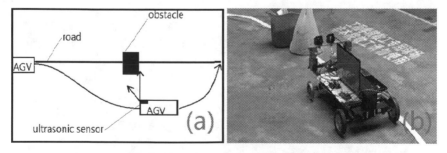

**Fig. 2.** Obstacle avoidance path diagram

According to the basic features of platform described above, work out the mobile
platform used in the experiment (shown in Fig. 2. b). The experimental results show
that stability of obstacle avoidance good, obstacle avoidance path was similar to the
path shown in Fig. 2. a.

# 4    Visual Path Identification and Navigation

## 4.1    Road Identification Method of Digital Image Processing

Because of the long time use, withered gray cement road (shown in Fig. 3, a) will had problems such as surface damage, withered moss, dust pollution, which and unstable light will cause difficulty in road identification or even the misidentification if only use the Grayscale to do image segmentation (shown in fig. 3. b). The road will looks clear (shown in fig. 3. c) by using H channel threshold of HSV color model and RGB threshold limitation and weighted stack. And then, use Canny detector to find the edge point (shown in fig. 3. d).Through filtration by field contrast, field gradient and field standard deviation we can get the edge points of interest(Fig. 3. e). Then image is divided into four areas along the y-axis direction of image coordinate, and using the Hough transform to detect lines randomly appear in each area (Fig. 3. f), and then calculate the coordinates value of two points in each line. Field contrast of edge point " s " can be calculated by

$$s = \frac{\sum_{k=i}^{i+n} I(x_k, y)}{\sum_{k=i}^{i-n} I(x_k, y)} \tag{1}$$

"I (x, y)" is the gray value of edge points. Filed gradient can be calculated by

$$d = \sum_{k=i-n}^{i+n} \frac{f_x'(x_i, y) f_x'(x_k, y) + f_y'(x_i, y) f_y'(x_k, y)}{\sqrt{[f_x'(x_i, y)^2 + f_y'(x_i, y)^2] + [f_x'(x_k, y)^2 + f_y'(x_k, y)^2]}} \tag{2}$$

" $f_x'(x, y)$ " is edge point's partial derivatives of the x direction, " $f_y'(x, y)$ " is the y direction.

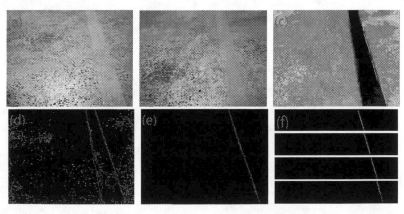

**Fig. 3.** Identification Process of Cement Road

## 4.2    Adaptive Fuzzy PID Controller of Path Tracking

The mobile platform movement mathematical model is difficult to be precise, so fuzzy-PID controller is the best choice. Based on the structure above, a simple and rough motion model of the platform can be found easily. After "Δ t" time, steering Angle of mobile platform is " ":

$$\Delta\alpha = \frac{90}{2\pi} \int_{t}^{t+\Delta t} \frac{v}{L} \sin(\theta(t))d(t) \tag{3}$$

The " L " is the distance of front and rear wheel, " v " is the car's speed at a constant speed (the default), " θ (t) " is the average front wheel steering Angle in the moment "t". Through the camera calibration, can calculate the projection transformation matrix between Mobile platform movement plane and 2D image. Using the matrix can do the coordinate conversion and linear fitting for the point set (as shown in Fig. 3. f) which get in Digital Image Processing. The intersection of line from fitting and The top of the actual coordinate (In our eyes' view, the origin is set at the upper-lift corner, direction of x-axis is Parallel to the right, direction of y-axis is vertical) is "T(Xt,0)". The connect line to the " T " and midperpendicular bottom point "B(w/2,h)" is " X ", the angle between " X " and midperpendicular is " α ° ∈ (- π , π )". Experiments show that " α °" is proportional to "Δ α ".

Control system is established by using these mathematical models and relations above. Fuzzy controller used error " e " and error rate " ec "as the input. Error " e " is distance form " T " to midperpendicular, we can see

$$e = h \times \tan( \alpha \ ^\circ ) \tag{4}$$

Also we can get the relationship between " T " and " α ° "and " e ":

$$T = h \times \tan( \alpha \ ^\circ )+w/2=e+w/2 \tag{5}$$

Through parameter self-tuning and the experiment site debugging, the relationship between the 3 parameter ("$K_p$", "$K_i$", "$K_d$") of PID controller and the input ("e" and "ec") of fuzzy controller can be found. Constantly testing " e " and " ec " during the moving, and change the value of " $K_p$", " $K_i$ " and " $K_d$ " depend on the value of " e " and " ec ". In consideration of the system stability, response speed, overshoot, steady-state error and other control factors, Set 48 rules for the fuzzy PID controller. Also, the initial PID parameters were set by debugging of the live experiment. Control system simulation was run in MATLAB. After the normalization, using unit step signal as input and observe the steady-state and response characteristics (Fig. 4. b).

Adjust the time is about 2 sec, the overshoot amount is 0.02%, The steady state performance good, basically has reached the requirement of practical application. Using sine function input to imitate the changing direction of path detected by vision, observe the tracking ability (see Fig. 4. a). We can see that tracking signal approximately lag of 0.05period compared with the input signal, which can meet the requirement of real-time tracking. Moreover, the practicality experiment using the platform shown in Fig. 2. b which shows that the tracking performance of the platform was stable.

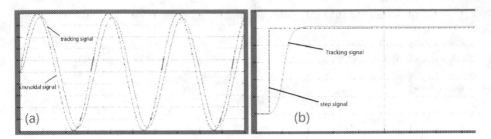

**Fig. 4.** Simulation of Mobile Platform Patch Tracking

## 5    Conclusion

This paper proposed a kind of design of intelligent mobile platform which used visual navigation and ultrasound to accomplish avoidance task. And this design include a kind of steering structure according to Ackerman steering geometry which can be controlled simply and steering smoothly. Steering using multiply ultrasound sensors and fuzzy controller makes the obstacle avoidance process stable and smooth. Moreover, the visual navigation using self-adaption fuzzy PID controller for steering, which road recognition algorithm based on double color space and filterable peripheral points is accurate and fast. Finally, simulation and real experimental results demonstrate that this design have good performance to track and stable effect to bypass obstacle. It follows that the design possess highly industrial prospect.

**Acknowledgements.** This work is supported by Industry-Academia-Research project of Education Department of Guangdong Provincial NO: 2012B091000167.

## References

1. Zhiwen, W., Ge, G.: Present situation and future development of mobile robot navigation technology. J. Robot. 25(5), 470–474 (2003)
2. Chen, C.-Y., Shih, B.-Y., Chou, W.-C., Li, Y.-J., Chen, Y.-H.: Obstacle avoidance design for a humanoid intelligent robot with ultrasonic sensors. Journal of Vibration and Control (2010), doi:10.1177/1077546310381101
3. Jiang, L., Deng, M., Inoue, A.: Obstacle avoidance and motion control of a two wheeled mobile robot using SVR technique. Computing, Information and Control 5(2), 253–262 (2009)
4. Zhu, L., Chen, J.: A Review of Wheeled Mobile Robots. Machinetool and Hydraulics 37(8), 242–247 (2009)
5. Chang, Y., Ma, S., Wang, H., Tan, D., Song, X.: Method of Kinematic Modeling of Wheeled Mobile Robot. Journal of Mechanical Engineering 46(5), 30–36 (2010)
6. Hua, J., Fu, X., Zheng, W., Wang, Y.: Control of Wheeled Mobile Robots Based on Motion Description Languages. Robot 28(3), 316–320 (2006)
7. Jun, Z., Changying, J.: Lateral Predictive Fuzzy Logic Control for Wheeled Mobile Robot Navigated by Machine Vision. Journal of Agricultural Machinery 33(6), 76–79 (2002)

8. Fu, M.-Y., Tan, G.-Y., Wang, M.-L.: An indoor navigation algorithm for mobile robot based on monocular vision. Optical Technique 32(4), 591–593 (2006)
9. Yang, J., Cai, Z., Yang, J.: Efficient Method of Obstacle Avoidance Based on Behavior Fusion for Mobile Robots. Journal of Mechanical Engineering 48(5), 10–14 (2012)
10. Lei, L., Xiaoming, X.: The design of simulation software for autonomous navigation and obstacle avoidance mobile robot. Huazhong Univ. of Sci. & Tech (Natural Science Edition) 39(S2), 196–199 (2011)

# A Novel Robot Leg Designed by Compliant Mechanism

Huai Huang, Yangzhi Chen, and Yueling Lv

School of Mechanical and Automotive Engineering,
South China University of Technology,
Guangzhou, 510640, P.R. China

**Abstract.** A novel flexible robot leg is designed by compliant mechanism, which provides a solution for stable walking. A pseudo-rigid-body model (PRBM) is built to analyze and verify the compliant mechanism. The simulation result indicates that the mechanism can approximately meet the requirements. Furthermore, the robot leg can work in two states by bistable design, namely "open state" and "contraction state". This robot leg is applicable to biped robots or multiped robots in small or micro scale.

**Keywords:** Compliant mechanism, Pseudo-rigid-body model (PRBM), Bistable, Robot leg.

## 1    Introduction

Mobile robots may be classified into wheeled robots, tracked robots, legged robots [1], etc. Legged robots move most commonly by means of bionic legs, such as the one-legged hopping robot, Kenken [2], the planar bipedal compliant legged robot, Spring Flamingo [3], the goat-imitated quadruped legged robot, KOLT [4, 5], the ockroach-inspired hexapod legged robots RHex [6] and iSprawl [7, 8]. Many of them are either multi-actuated or complicatedly designed, which may lead to cumbersomeness, inconvenience for optimization or limitation in size.

Compliant mechanisms are flexible mechanisms that transfer an input force or displacement to another point through elastic body deformation [9]. Compared to traditional rigid-body mechanisms, compliant mechanisms have advantages in simplifying processing and manufacturing, reducing the number of components and assembly time, reducing wear and tear, etc.

This paper presents a novel compliant leg mechanism [12], which is designed and analyzed by means of the pseudo-rigid-body model.

## 2    A Novel Robot Leg Design

In order to design a reliable and efficient leg mechanism for small and micro robot, five objectives are set as follows:

X. Zhang et al. (Eds.): ICIRA 2014, Part I, LNAI 8917, pp. 204–213, 2014.

(1) A single leg applicable to biped robot, quadruped robot or multiped robot;

(2) Driven by only one motor for less weight;

(3) Having a bionic movement locus and pose of foot, which means the back swing of the leg should possibly be stable through the time after touching and before leaving the ground;

(4) Capable of reducing volume, weight, wear and tear by means of simple mechanisms;

(5) Simple but artistic.

## 2.1    Traditional Mechanical Designs of a Single Leg

To contrast with the compliant mechanism presented in this paper, several traditional mechanical designs of a single leg are proposed as shown in Table 1.

**Table 1.** Traditional mechanical designs of a single leg

| Foot shape | Schemes of traditional mechanical designs | | |
|---|---|---|---|
| Planar foot | (a) | (b) | (c) |
| Point foot | (d) | (e) | (f) |

From Table1, the conclusions can be given as follows:

To scheme (a), (b) and (d), the movement of the foot part is barely acceptable;

To scheme (c) and (e), the mechanisms are more valuable because they are conveniently adjustable for optimization. However, there are too many hinges for further machining and assembling in small or micro size.

To scheme (f), the planar four-bar linkage provides the most ideal result of movement locus and speed, but the end of the output link is beyond the whole mechanism. Therefore a parallelogram mechanism is needed for the motion transmission, which will bring the mechanism intricacy as well.

## 2.2     The Design of Compliant Mechanism

To find a mechanism being both simple and capable of meeting the requirements, a mechanism working as scheme (b) in Table 1, is proposed in Fig. 1(b), which is an almost evolution of a normal compliant mechanism in Fig. 1(a) [9]. The mechanism in Fig.1 (a) is used to obtain an approximate linear displacement. It predicts that the mechanism in Fig. 1(b) works in a similar way. In addition, it can imitate the behavior of an animal's leg: swing backward—raise the foot—swing forward—lower the foot.

This mechanism is not only simple, but also convenient for optimization by adjusting the length and the thickness. Additionally, it can be multistable by design, which may offer multifunction.

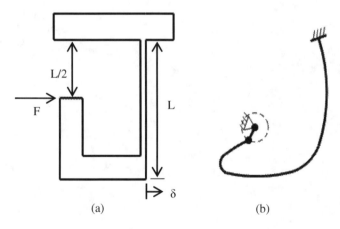

**Fig. 1.** (a) A normal compliant mechanism: the easiest way to obtain an approximate linear displacement from a cantilever beam. (b) A compliant mechanism similar to the mechanism shown in Fig. 1(a)

## 3     Pseudo-Rigid-Body Model (PRBM) of the Robot Leg

In order to design, analyze and optimize the mechanism, a simplified scheme is proposed, as shown in Fig. 2(a). The compliant mechanism is equivalent to three fixed-pinned segments (initially curved cantilever beams) of lengths $l_1$, $l_2$ and $l_3$ respectively, whose joints are added by torsional springs to retain the energy storage properties of compliant mechanisms.

The pseudo-rigid-body model [9-11] is shown in Fig. 2(b).

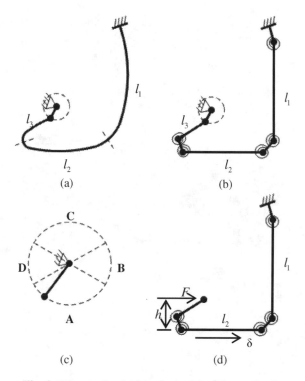

(a)

(b)

(c)

(d)

**Fig. 2.** The pseudo-rigid-body model of the mechanism

The upside end of the PRBM is completely fixed, while the other end is connected to a crank by a hinge, which offers a circular motion. This cyclic circular motion can be divided into four phases according to its different effects, as shown in Fig. 2(c). The compliant leg mechanism swings backward during phase A and swing forward during phase C. Besides, it raises the foot from the ground during phase B and lowers during phase D. To simplify the calculation and analysis, approximately assume that the direction of the input displacement is rightward, as well as the input force. And if the stiffness coefficient of the material is suitable, the support force of the ground can be ignored.

## 4    Calculation and Analysis

As shown in Fig. 2(d), the input force is $F$, which is equivalent to a force $F$ and a moment $M_1$ loading at the free end of segment $l_1$. To segment $l_2$, $F$ is equivalent to a force $F$ and a moment $M_2$ loading at the free end as well. It is shown in Fig. 3(a) and Fig. 3(b).

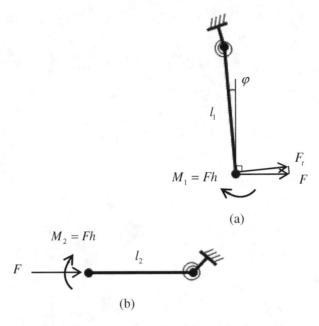

(a)

(b)

**Fig. 3.** Force analysis of the mechanism

(1) In Fig. 3(a), the beam end rotation angle of $l_1$ resulting from $F$ would be obtained by equations [9]:

$$\begin{cases} T = \rho l_1 F_t = \rho l_1 F \cos \varphi \\ T = K_{1F} \theta_{1F} \\ K_{1F} = \rho K_{\theta 1F} \dfrac{E_1 I_1}{l_1} \\ \theta_{1F-end} = c_{\theta F} \theta_{1F} \end{cases} ,$$

where $T$ is the torque on the torsional spring resulting from $F$ at the beam end of $l_1$, $\rho$ is the new characteristic radius factor, $l_1$ is the beam length, $F_t$ is the transverse force of $F$, $\varphi$ is the angle between $F$ and $F_t$, $K_{1F}$ is the torsional spring constant, $\theta_{1F}$ is the pseudo-rigid-body angle, $K_{\theta 1F}$ is the stiffness coefficient, $E_1$ is the modulus of elasticity, $I_1$ is the moment of inertia of the cross-section, $\theta_{1F-end}$ is the beam end angle, the constant $c_{\theta F}$ is termed as the parametric angle coefficient.

$\varphi$ ranges from $-10°$ to $10°$, so $\cos \varphi \approx 1$.

Therefore, the beam end rotation angle of $l_1$ resulting from $F$ is

$$\theta_{1F-end} = \frac{c_{\theta F}}{K_{\theta 1F} E_1 I_1} F l_1^2 \tag{1}$$

(2) In Fig. 3(a), the beam end rotation angle of $l_1$ resulting from $M_1$ would be obtained by equations [9]:

$$\begin{cases} M_1 = Fh \\ M_1 = K_{1M}\theta_{1M} \\ K_{1M} = \gamma_1 K_{\theta 1M}\dfrac{E_1 I_1}{l_1} , \\ \theta_{1M-end} = c_{\theta M}\theta_{1M} \end{cases}$$

where $M_1$ is an equivalent moment by translation of $F$, $h$ is the height of the original $F$ above the ground, $K_{1M}$ is the torsional spring constant, $\theta_{1M}$ is the pseudo-rigid-body angle, $\gamma_1$ is the characteristic radius factor, $K_{\theta 1M}$ is the stiffness coefficient, $\theta_{1M-end}$ is the beam end angle, the constant $c_{\theta M}$ is the parametric angle coefficient.

Therefore, the beam end rotation angle of $l_1$ resulting from $M_1$ is

$$\theta_{1M-end} = \frac{c_{\theta M}}{\gamma_1 K_{\theta 1M} E_1 I_1} Fl_1 h \qquad (2)$$

(3)In Fig. 3(b), $F_t = 0$, and the pseudo-rigid-body angle of $l_2$ resulting from $M_2$ would be obtained by equations [9]:

$$\begin{cases} M_2 = Fh \\ M_2 = K_{2M}\theta_{2M} \\ K_{2M} = \gamma_2 K_{\theta 2M}\dfrac{E_2 I_2}{l_2} \end{cases},$$

where $M_2$ is an equivalent moment by translation of $F$, $K_{2M}$ is the torsional spring constant, $\theta_{2M}$ is the pseudo-rigid-body angle, $\gamma_2$ is the characteristic radius factor, $K_{\theta 2M}$ is the stiffness coefficient.

Therefore, the pseudo-rigid-body angle of $l_2$ resulting from $M_2$ is

$$\theta_{2M} = \frac{1}{\gamma_2 K_{\theta 2M} E_2 I_2} Fl_2 h \qquad (3)$$

(4)If this compliant leg mechanism works stably in walking, the foot part should keep level in a certain range. That means, the rotation angle of the pseudo-rigid-body segment $l_2$ is equal to zero:

$$\theta_{1F-end} - \theta_{1M-end} - \theta_{2M} = 0 \qquad (4)$$

The design equation is given by substituting Eq. (1), Eq. (2) and Eq. (3) into Eq. (4):

$$A \cdot l_1^2 - B \cdot l_1 h - C \cdot l_2 h = 0 \qquad (5)$$

where $A = \dfrac{c_{\theta F}}{K_{\theta 1 F}} \cdot \dfrac{1}{E_1 I_1}$, $B = \dfrac{c_{\theta M}}{\gamma_1 K_{\theta 1 M}} \cdot \dfrac{1}{E_1 I_1}$, $C = \dfrac{1}{\gamma_2 K_{\theta 2 M}} \cdot \dfrac{1}{E_2 I_2}$.

Especially, if the materials and cross-sections of segment $l_1$ and segment $l_2$ are identical, then $E_1 = E_2$ and $I_1 = I_2$.

As a result, the mentioned mechanism that meets the condition of Eq. (5) will have a stable walk on the level.

At the same time, the J-shaped component can be tightly bound around with a caterpillar band, driving by another motor, to move not only on foot but also on the crawler.

## 5     Kinematics and Static Simulation

The input displacement has been assumed to be in the horizontal direction for simplification. When such an input is given, the output is presented by means of finite-element analysis in Fig. 4. The result indicates that the design equation, Eq.(5), approximately meets the requirements.

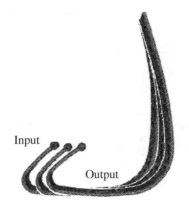

Input

Output

**Fig. 4.** Result of computer simulation

## 6     Bistable Design

Moreover, the robot leg is designed as a bistable mechanism. A bistable compliant mechanism experiences the following changes when working: the potential energy of the system increases from a minimum value at first and then decreases to another [9], as shown in Fig. 5.

**Fig. 5.** Schematic diagram of bistable characteristic [9]

The design shown in Fig. 6(a) is a common bistable compliant mechanism [9], which has two stable states A and B. A schematic mechanism can be generalized from this type of bistable compliant mechanism, as shown in Fig. 6(b). When transferring from state A to state B, because of the limitation of the length of rod $r_1$, the flexible rod end of $r_2$ changes its path from $p_2$ to $p_1$. Therefore rod $r_2$ go through a process in which its potential energy falls after rising. So does the potential energy of the entire system.

According to this principle, a bistable mechanism may be as simple as shown in Fig.6(c), whose energy changing process has been set up.

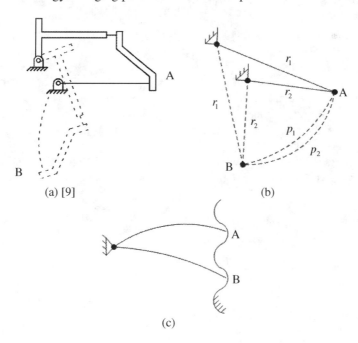

(a) [9]                (b)

(c)

**Fig. 6.** Simplification of a common bistable mechanism

Similarly, the proposed robot leg is designed as a bistable mechanism, with two stable states named "open state" and "contraction state". "Contraction state" makes it much more portable.

As shown in Fig. 7(a), the compliant leg mechanism generally works in the "open state". When the mechanism is forced into the casing, it is affected by the bump shown in Fig. 7(b), which results in a process that the potential energy increases first and then decreases to another state, "contraction state". Moreover, a buckle is set for a firmer state, as shown in Fig. 7(c). It is because a sudden decrease of potential energy at the minimum value point helps to separate the two states.

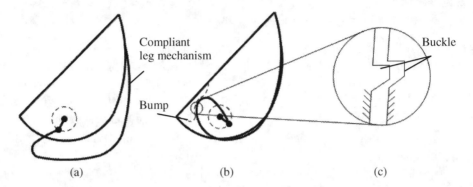

**Fig. 7.** Bistable compliant leg mechanism

## 7    Conclusion

This paper presents a novel robot leg designed by a bistable compliant mechanism being applicable to small or micro robot for stable walk. It is designed by means of the pseudo-rigid-body model and analyzed by computer simulation. The result shows that it approximately meets the requirements, and needs further optimization.

**Acknowledgments.** The work was supported by National Natural Science Foundation of China (No. 51175180). We also really appreciate associate professor Wang N.F.'s help.

## References

1. Zhou, X.D., Bi, S.S.: A survey of bio-inspired compliant legged robot designs. Bioinspiration & Biomimetics 7(4), 1–20 (2012)
2. Hyon, S.H., Emura, T., Mita, T.: Dynamics-based control of a one-legged hopping robot. Proc. Inst. Mech. Eng. I 217, 83–98 (2003)
3. Pratt, J.E., Krupp, B.T.: Series elastic actuators for legged robots. In: Proc. SPIE, vol. 5422, p. 135 (2004)
4. Estremera, J., Waldron, K.J.: Thrust control, stabilization and energetics of a quadruped running robot. Int. J. Robot. Res. 27(10), 1135–1151 (2008)
5. Palmer, L.R., Orin, D.E., Marhefka, D.W., Schmiedeler, J.P., Waldron, K.J.: Intelligent control of an experimental articulated leg for a galloping machine. In: Proc. IEEE Int. Conf. on Robotics and Automation (ICRA 2003), vol. 3, pp. 3821–3827 (2003)
6. Koditschek, D.E., Full, R.J., Buehler, M.: Mechanical aspects of legged locomotion control. Arthropod. Struct. Dev. 33, 251–272 (2004)

7. Cham, J.G., Bailey, S.A., Clark, J.E., Full, R.J., Cutkosky, M.R.: Fast and robust: hexapedal robots via shape deposition manufacturing. Int. J. Robot. Res. 21, 869–882 (2002)
8. Kim, S., Clark, J.E., Cutkosky, M.R.: iSprawl: design and tuning for high-speed autonomous open-loop running. Int. J. Robot. Res. 25, 903–912 (2006)
9. Howell, L.L.: Compliant Mechanisms. Wiley, New York (2001)
10. Howell, L.L., Midha, A.: A method for the design of compliant mechanism with small-length flexural pivots. ASME J. Mech. Des. 116, 280–289 (1994)
11. Howell, L.L., Midha, A.: Evaluation of equivalent spring stiffness for use in a pseudo-rigid-body model of large deflection compliant mechanisms. ASME J. Mech. Des. 118, 126–131 (1996)
12. Huang, H., Chen, Y.Z., Lv, Y.L.: A Robot Leg Device. Chinese Patent, No.201410242401.8, 05 (2014)

# Master-Slave Gesture Learning System Based on Functional Electrical Stimulation

Kaida Chen, Bin Zhang, and Dingguo Zhang*

State Key Laboratory of Mechanical System and Vibration,
Shanghai Jiao Tong University,
Shanghai, China, 200240
dgzhang@sjtu.edu.cn

**Abstract.** Regarding some dexterous skill related to hands, for example, playing musical instruments, it is challenging for common beginners to imitate the motion of the teacher due to unfamiliarity with both the fingering and the music piece. To assist in solving that, we developed a wearable master-slave gesture learning system based on functional electrical stimulation (FES), where a multi-pad FES system with an electrode array (20 pads) was utilized to stimulate the target muscle group accurately. In terms of determining stimulation parameters, we designed a process and used a wearable surface-electromyography (sEMG) acquisition device, SJU-iMYO, to obtain sEMG signal of associated muscles on the master side (teacher); Afterwards we used an algorithm to determine the electrode set to activate and the correspondent current intensity on the slave side (student) automatically. To assess the performance of the device, several experiments were conducted on four subjects with a virtual-reality data glove. All results indicated that the gesture learning system succeeded in imitating the motion of each finger with fairly good accuracy, except for the little finger.

**Keywords:** Master-slave gesture learning, Finger movement control, Functional electrical stimulation (FES), Electrode array, Wearable device.

## 1 Introduction

For a regular beginner in learning dexterous skills of hands, for example, playing musical instrument, it's challenging to imitate the hand movement of the teacher due to unfamiliarity with both the fingering and the music piece. Commonly, beginners have no other choice but keep practicing the music piece over and over again, which is rather time-consuming. Based on a technique termed functional electrical stimulation (FES), we developed a master-slave gesture learning system with wearable devices to help solve this problem. For convenience of attaching electrodes, we introduce the multi-pad electrode array, i.e. array of small electrodes, which has the advanced features of electrode repositioning and dynamic electrode shape and position changing. By activating only an electrode subset

---

* Corresponding author.

X. Zhang et al. (Eds.): ICIRA 2014, Part I, LNAI 8917, pp. 214–223, 2014.

of the array, stimulation can be made highly selective and unwanted muscle activations can be avoided [1].

In 2012, Malesevic et al. [2] developed a multi-pad electrode based FES system, which comprised an electrode composed of small pads that could be activated individually, for restoration of grasp. In the same year, Hoffmann et al. [1] developed a multi-pad FES system (4*4 electrode array), which was capable of determining parameters automatically, for hand opening and closing. Other relevant study also includes Tamakis work of developing PossessedHand that could control motion of 16 joints of the hand [3]. This device encompassed 28 electrode pads to activate the targeted muscle and a forearm belt to inform when and which finger to move. In 2013, Lou et al. [4] first adopted the idea of master-slave control, where the sEMG signal of the master side were used to control the slave side, and applied it in rehabilitation training. However, their work can only realize the motion control of big joints such as wrists, because the large FES electrodes do not have good selectivity and cannot stimulate the small and fine muscles of hands accurately.

In this paper, we developed the master-slave gesture learning system based on the technique of FES, where the multi-pad FES system with a 20-pad electrode array was utilized to stimulate target muscle group accurately. In terms of determining stimulation parameters, we designed a process and used a wearable surface-electromyography (sEMG) acquisition device, SJU-iMYO (SJTU, P.R.C.), to obtain sEMG signal [5,6,7] on the master side (teacher); Afterwards we used an algorithm to determine the electrode set to activate and the correspondent current intensity on the slave side (student) automatically. To assess the performance of the device, several experiments were conducted on four subjects with a virtual-reality data glove. All results indicated that our system succeeded in imitating the movement of each finger with fairly good accuracy, except for the little finger.

## 2    Master-Slave Gesture Learning System

Figure 1 shows the structure of the proposed master-slave gesture learning system, which consists mainly of two parts, i.e. the master unit and slave unit. The master unit was the part that identified the hand movement of the master side by processing the sEMG signal into recognizable movement patterns. The slave unit was the part that implemented finger movement by applying currents to the specific electrode subset on the slave side based on orders from PC.

### 2.1    Master Unit

The master unit was composed of two parts: hardware and algorithm. Its main function included EMG acquisition and pattern recognition (feature extraction and pattern classification, completed with a PC). We achieved the EMG acquisition with a wearable sEMG acquistition device, SJT-iMYO (SJTU, P.R.C.) with eight channels (see Fig. 2). The obtained signals were then analyzed with associated time domain methods [8] in PC.

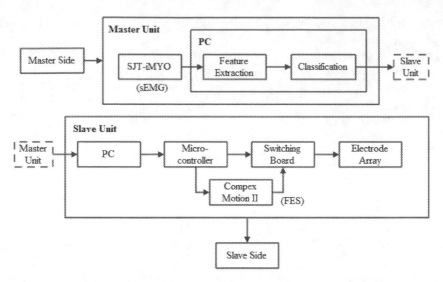

**Fig. 1.** System structure of the proposed master-slave gesture learning system

As for feature extraction, we used mean absolute value (AV), root mean square (RMS) and zero crossing (ZC) as extracted features, which can be calculated as follows:

$$AV = \frac{1}{N}\sum_{i=1}^{N} x_i \tag{1}$$

$$RMS = \sqrt{\frac{1}{N-1}\sum_{i=1}^{N} x_i^2} \tag{2}$$

$$ZC = \sum_{i=1}^{N} \text{sgn}\,(-x_i x_{i-1})\,, \text{sgn}(x) = \begin{cases} 1 & if\, x > 0 \\ 0 & otherwise \end{cases} \tag{3}$$

where $N$ indicates the sample length of the signal; $x_i$ indicates the $i$-th sample.

In terms of classification, we used a machine learning tool, Support Vector Machine (SVM), with a library package developed by Chang et al. [9]. Its basic framework is a binary-class linear classification model, and therefore convenient in implementation.

## 2.2   Slave Unit

The slave unit encompassed a PC terminal, FES device, electrode array and a switching board (see Fig. 3). The PC terminal served as an order controller, whose source could be either manual input or acquired from protocol transmission. A FES device, Compex Motion II (LJO Global Inc. Switzerland) was

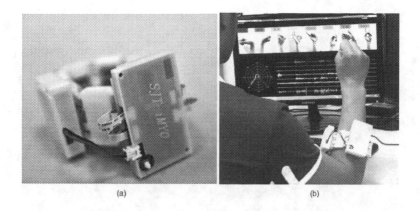

(a)                                                                (b)

**Fig. 2.** SJT-iMYO for sEMG acquisition

utilized to generate low-frequency stimuli to activate muscles. Figure 4 shows
the final design of the electrode array with 20 solid gel pads (12*25 mm) fixed
on two arm belts. Note that, three of them were culled out to compose another
separate belt to activate musculi flexor pollicis longus that has a prominent effect
in the movement of the thumb. Additionally, a solid hydrogel electrode (85*50
mm) was needed as an anode to be attached to the wrist. Detailed information
on electrode placement is shown in Fig. 5. As for the switching board used for
channel switching of the electrode array, it was mainly composed of a microcontroller (Arduino Mega 2560), 20 miniature high power relays (HF115F), a
monolithic high-voltage high-current Darlington transistor array (SN75468) and
a battery.

Besides, for the convenience of determining both the electrode set to activate
and its correspondent current intensity on the slave side, we proposed a new
algorithm termed automatic addressing in the study. In this approach, we had the
master subject's forearm stimulated with each electrode activated respectively
in two different current intensity (7 mA and 8 mA), after which we obtained the
flexion of each finger in each case with the help of a virtual-reality data glove,
5DT Data Glove (5DT, USA). Figure 6 shows the measuring position of flexion
of each finger. Afterwards, data in each case (with identical electrode subset and
identical current intensity) could be reconstructed in the following means:

$$FLEX = [flex1, flex2, flex3, flex4, flex5] \rightarrow CASE = [eleNo, I] \quad (4)$$

where $flex1$ to $flex5$ indicate the flexion of the thumb to the little finger
(varying between 0 and 1) obtained with 5DT Data Glove, respectively; $eleNo$
indicates the activated electrode number; $I$ indicates the current intensity.

For better efficacy of addressing, we defined flexion within interval [0.85,1] as
obvious bending, otherwise no bending. Accordingly, flexion of obvious bending was replaced with a Boolean value, 1, otherwise with a Boolean value, 0.

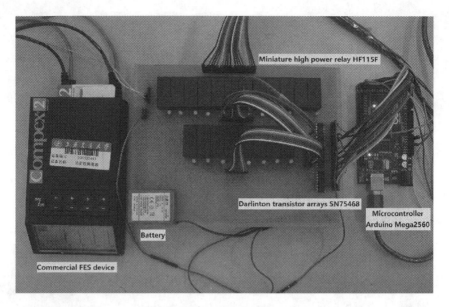

**Fig. 3.** Devices of the slave unit

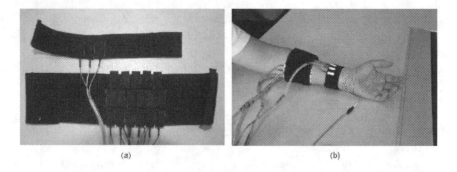

(a)                                                    (b)

**Fig. 4.** Electrode array: (a) Final design; (b) A prototype of the electrode array

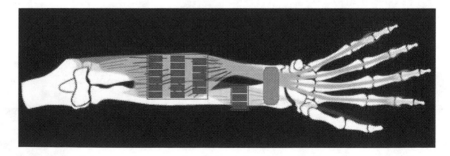

**Fig. 5.** Electrode placement on the forearm
(purple: anode electrode; green: cathode)

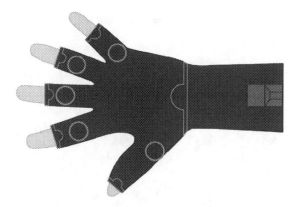

**Fig. 6.** The measuring position of flexion of each finger (marked by blue circles)

Therefore, a 40*5 Boolean matrix mapping to another 40*2 matrix of stimulation information was then obtained. Then in a real master-slave control case, whatever known finger movement, it could be achieved with these two matrices. For example, in case that the slave subject was needed to bend the middle finger and the little finger at the same time, we searched for the vector [0,0,1,0,1] in the 40*5 boolean matrix and its correspondent mapping vector with information of intensity of current and which electrode to activate indicated the way we achieved this movement pattern.

Note that it's important to check out whether the mapping matrix encompassed some basic vectors therefore able to avoid succedent addressing failure. Whenever such unfavorable circumstance occurred, re-adjusting the placement of the electrode array belts might probably help out.

## 2.3    Communicating Protocol

To realize teleoperation, we established a network communication between the master side and the slave side with User Datagram Protocol (UDP). UDP is one of the wireless communication protocols of the reference model of the Open System Interconnection (OSI) and has the advantage of low resource consumption and high processing rate. We implemented this protocol with CSocket class in the C++ environment.

## 3    Experiments

This study was approved by the local Ethics Committee of Shanghai Jiao Tong University. Four healthy subjects (mean age 21±1.44) participated in this experiment. All subjects underwent two independent experiments. The intensity of electric stimulation that we used in the study was either 7 mA or 8 mA (pulse width: 300 $\mu$A; frequency: 40 Hz), determined by the comfort of the subject.

## 3.1   Finger Movement Recognition on the Slave Side

In the first experiment of assessing the completeness of finger movement control, we attached the two electrode array belts to the subject's forearm and activated each channel (each anode electrode of the electrode array) respectively, as mentioned in the addressing algorithm. It could be found that six different finger movement patterns appeared in all 4 subjects' test (see Fig. 7). Table 1 represents statistical results of flexion and numbness of each finger (both varying from 0 to 1) of each subject in the test. The result of Fig. 7 indicated that we were able to achieve independent finger movement control of the thumb, the index finger, the middle finger and the ring finger, but except for the little finger whose movement was always accompanied with the other fingers (the middle finger or the ring finger). Besides, an unfavorable circumstance of numbness without finger movement was possible to occur to the thumb and the index finger.

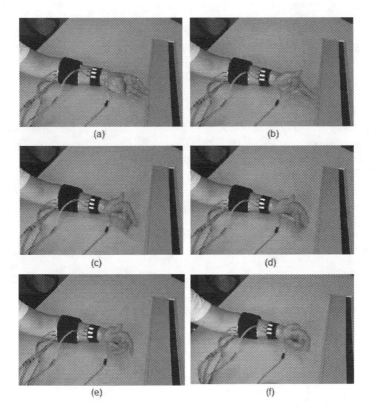

**Fig. 7.** Finger movement patterns appeared in the test
(a)-(d) independent finger movement of the thumb, the index finger, the middle finger and the ring finger respectively; (e) concurrent bending of the ring finger and the little finger; (f) concurrent bending of the middle finger and the little finger.

**Table 1.** Flexion and Numbness of slave subjects

| Subject | | Thumb | Index Finger | Middle Finger | Ring Finger | Little Finger |
|---|---|---|---|---|---|---|
| S1 | Flexion | 0.25 | 0.25 | 1.0 | 1.0 | 0.7 |
| | Numbness | 1.0 | 0.5 | 0.0 | 0.0 | 0.0 |
| S2 | Flexion | 0.0 | 0.25 | 1.0 | 1.0 | 0.5 |
| | Numbness | 1.0 | 0.5 | 0.0 | 0.0 | 0.0 |
| S3 | Flexion | 0.25 | 0.0 | 1.0 | 1.0 | 0.5 |
| | Numbness | 1.0 | 0.25 | 0.0 | 0.0 | 0.0 |
| S4 | Flexion | 0.25 | 0.25 | 1.0 | 1.0 | 0.7 |
| | Numbness | 1.0 | 0.75 | 0.0 | 0.0 | 0.0 |

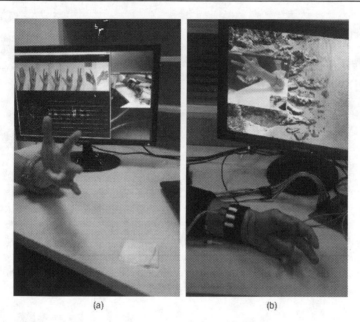

(a)                              (b)

**Fig. 8.** Master-slave control with the gesture learning system
(a) The master side; (b) The slave side

## 3.2   Accuracy of Master-Slave Control

To assess the accuracy of the master-slave control of the proposed system, we
conducted another experiment with a fixed subject (age 22) as the slave side
and four other subjects as the master side. Ahead of the master-slave control,
we obtained all optimal electrode subsets correspondent to finger movement
patterns of the slave subject and trained the master subject with 2 sessions
learning of 6 selected finger movement patterns: {1}, {3}, {4}, {3,4}, {1,4,5},
{} (1-5 represents the thumb to the little finger, respectively; '{}' represents a
pattern of idle condition with no bending at all fingers). Later in real master-
slave control of each subject, the master subject were required to repeat each

of these 6 finger movement patterns 4 times (each lasted for 5 s and separated by 10 s to avoid fatigue). Table 2 represents the accuracy of the test. The result showed that the average accuracy of each selected finger movement pattern was higher than 90%, which indicated a latent capacity for realistic application.

**Table 2.** Accuracy of master-slave control (%)

| Pattern | S1 | S2 | S3 | S4 | Average |
|---------|-----|-----|-----|-----|---------|
| {1} | 95 | 90 | 100 | 85 | 92.50 |
| {3} | 90 | 95 | 9 | 90 | 91.25 |
| {4} | 95 | 90 | 90 | 90 | 91.25 |
| {3, 4} | 100 | 90 | 95 | 90 | 93.75 |
| {1, 4, 5} | 95 | 85 | 95 | 85 | 90.00 |
| { }(idle) | 95 | 95 | 85 | 90 | 91.25 |

## 4   Conclusion and Future Work

In this paper, we developed the prototype of master-slave gesture learning teaching system based on FES, which is capable of selecting optimal electrode subset automatically and achieving master-slave control in finger movement imitation. To assess the performance of the device, several experiments were conducted on four subjects. All results indicated that our system succeeded in imitating the movement of each finger with fairly good accuracy, except for the little finger.

Further work might include improvement of the multi-pad FES system by reducing the area of each electrode pad and increasing the number of electrodes appropriately, or by replacing the solid-gel electric pads with silver/silver-chloride electrode pads to enable independent movement control of the little finger. Moreover, we might also adopt more hand movement pattern for extensive application, for example, master-slave gesture learning in musical instruments, sign language and stroke rehabilitation.

**Acknowledgments.** This work is supported by National Natural Science Foundation of China (Grant No. 51475292), Natural Science Foundation of Shanghai (Grant No. 14ZR1421300) and State Key Laboratory of Robotics and System (Grant No. SKLRS-2012-ZD-04).

## References

1. Hoffmann, U., Deinhofer, M., Keller, T.: Automatic determination of parameters for multipad functional electrical stimulation: application to hand opening and closing. In: 34th Annual International Conference of the IEEE EMB, San Diego, California, August 28-September 1 (2012)
2. Malesevic, N.M., Popovic, L.Z., Ilic, V., Jorgovanovic, N., Bijelic, G., Keller, T., Popovic, D.B.: A multi-pad electrode based functional electrical stimulation system for restoration of grasp. Journal of Neuro Engineering and Rehabilitation 9, 66 (2012)

3. Tamaki, E., Miyaki, T., Rekimoto, J.: PossessedHand: Techniques for controlling human hands using electrical muscles stimuli. In: Proceedings of the SIGCHI Conference on Human Factors in Computing Systems. ACM (2011)
4. Lou, Z., Yao, P., Zhang, D.: Wireless master-slave FES rehabilitation system using sEMG control. In: Su, C.-Y., Rakheja, S., Liu, H. (eds.) ICIRA 2012, Part II. LNCS, vol. 7507, pp. 1–10. Springer, Heidelberg (2012)
5. Kamen, G.: Electromyographic Kinesiology. In: Robertson, D.G.E., et al. (eds.) Research Methods in Biomechanics, Human Kinetics Publ., Champaign (2004)
6. Zhang, D.G., Guan, T.H., Widjaja, F., Ang, W.T.: Functional electrical stimulation in rehabilitation engineering: a survey. In: Proceedings of the 1st International Convention on Rehabilitation Engineering and Assistive Technology, i-CREATe 2007, pp. 221–226 (2007)
7. Graupe, D., Kohn, K.H.: A critical review on EMG-controlled electrical stimulation in paraplegics. CRC Critical Review in Biomedical Engineering 15(3), 187–210 (1987)
8. Chen, X.P., Zhu, X.Y., Zhang, D.G.: Adiscriminant bispectrum feature for surface EMG signal classification. Medical Engineering and Physics 32(2), 126–135 (2010)
9. Chang, C.C., Lin, C.J.: Libsvm: a library for support vector machines (2001), http://www.csie.ntu.edu.tw/~cjlin/libsvm

# A Study of EMG-Based Neuromuscular Interface for Elbow Joint

Ran Tao, Sheng Quan Xie, and James W.L. Pau

Mechanical Engineering, The University of Auckland
20 Symonds Street, Auckland 1010, New Zealand
rtao034@aucklanduni.ac.nz, s.xie@auckland.ac.nz

**Abstract.** With the increase number of limb disable patients, caused by stroke or paralysis, rehabilitation robots and their human-robot interface earned widespread respect. Based on our previous study about Neuromuscular Interface (NI), this research aims at recording and processing EMG signals from test subjects to accurately represent the movement created by these muscle groups on the elbow joint, and through a designed NI control system to control an single-degree-of-freedom (SDOF) exoskeleton arm for flexion and extension. Also, experiments are used to verify the feasibility of the whole interface system. Improvements have been made to achieve accuracy control, real-time processing and wireless transferring.

**Keywords:** Rehabilitation, Interface, EMG, Elbow joint, Neuromuscular, Microcontrollers.

## 1    Introduction

Rehabilitation robotics is a field of research that is vastly expanding with the hope to aid users with robotic rehabilitation methods. One of their major goals is to give an assistive force to the users who may have lost some form of control over their skeletal muscles. These muscles are caused to contract from small electrical signals which are commonly known as Electromyography (EMG) signals. EMG signals reflect the activation level of muscles and contain useful information from the subject's moving intention [1]. They are controlled by the nervous system and remain variable between subjects as it depends on their physical characteristics [2]. Since, advancements made over the years in EMG signals has given rise to the potential of developing exoskeletons that employ EMG control signals generated from muscle contraction, engineers and medical practitioners starts to draw their interest for developing robotics to create assistive rehabilitation devices[1,3, 4].

Compared with other methods such as exteroceptive sensors which can implement torque sensors, encoders or potentiometers, the EMG-based interface does not require additional force by the user, which means, it could suit the patient without fully control of their limbs. There are also some other controller to use a motion capture system or gyroscopes and accelerometers. The problem with the motion capture system

X. Zhang et al. (Eds.): ICIRA 2014, Part I, LNAI 8917, pp. 224–233, 2014.

is that it is not portable at all and the gyroscope and accelerometer positions slowly get worse and worse as they keep adding their position in space together and could have rounding errors[5,6].

In order to reduce the drawbacks from other types of interface, which mainly comes from the electrode placement and variability between subjects, and achieve the goals of accuracy control, real-time processing and wireless transferring, this research develops an EMG-based NI interface and its hardware/software control system and briefly introduces the system methodology in Section 2. Then thought a designed experiment in Section 3, we evaluate the performance of this interface by comparison test result (in Section 4), and finally, draw the conclusions and discussions in Section 4 and 5.

## 2     Methodology

With previous developments of the NI matlab model (Fig.1) [1] and a SDOF joint robot arm (Fig.2) by our research group, this paper mainly introduce the control system which uses the raw EMG signal as the only input to predict elbow joint movement and control SDOF robot arm.

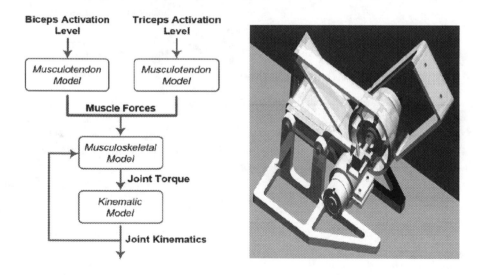

**Fig. 1.** Flowchart of NI matlab model          **Fig. 2.** SDOF joint robot arm

### 2.1     The Layout of the System

The system can be summarized into five main components (A simple flow chart is shown in Fig.3), these components include the individuals arm where EMG signals and gyroscope values are extracted from, the MSP430 where the EMG signal and gyroscope values are read in and then sent out wirelessly, the AVR32 where all signals are received and calculations made, the SDOF arm model which shows the predicted position of the arm and finally the option of having a computer which logs all the data. This entire system however is able to work independently of any computers as the microcontrollers do all the computing work. The only time computers are necessary is when all the data needs to be logged.

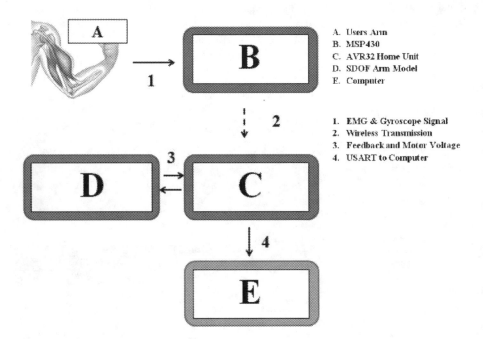

**Fig. 3.** Overview of the whole system. The five letters shows the five main sections which were briefly described above and the numbering system shows signals going to each section and how they interact.

**Sections:**

- **A.** The Users Arm: The users arm has all the electrodes placed on the biceps, triceps and the olecranon. Included on the arm also is the gyroscope which is only used as an on/off signal to note if there is rotation about the elbow joint or not.
- **B.** MSP430: Receiving signals from the electrode leads and the gyroscope leads the MSP430 is used as a central collection point for the arm. It takes all the signals and any future signals from the body and sends it wirelessly away.
- **C.** AVR32: Receiving wireless signals from the MSP430 and with the capability of receiving more wireless signals the AVR32 is the base for all calculations and data handling. It also has a feedback signal from the SDOF arm model which is used for a proportional controller.

- **D.** SDOF Arm model: The arm model is only used as a visual solution to show the user the predicted angle of the arm. It uses a motor driver from Maxon and has a potentiometer to give feed back to the proportional controller on the AVR32.
- **E.** Computer: The computer to log the data is optional and is also not needed for the system to run. It again is placed as a visual representation of the data for the user and is a method of logging all the data that comes from the AVR32.

**Signals:**

- **1.** EMG and Gyroscope signals: The EMG signals from the electrodes are transferred through Lead wires from the arm to the MSP430, also included alongside of the EMG signals is the gyroscope signals.
- **2.** Wireless Transmission: This is where the system on the body/arm is wireless from the main computing unit (AVR32), all the data from the MSP430 is sent out wirelessly.
- **3.** Feedback and motor Voltage: The DAC on the AVR32 shield sends out a voltage as a signal to the motor driver to adjust the motor to a certain position. To know this position there is a feedback loop from the potentiometer that updates the AVR32 to adjust the voltage being sent out.
- **4.** USART to Computer: The AVR32 has been setup to also output all the data it calculates serially to the computer via a USB cable. This is non-compulsory for the system to run.

## 2.2   Hardwires

**Microcontrollers**

Microcontrollers are chosen as the core of this interface system mainly because of its small size. Since the capability of making the system wireless required a secondary microcontroller, a simple, reliable and easy to interface microcontroller was to be chosen. The AVR32 UC3B0256 is a microcontroller developed by Atmel which can operate at a maximum Central Processing Unit (CPU) frequency of 60MHz and has a flash program memory of 256Kb (Fig. 4). As well as having the capability of running all the calculations Atmel provides its customers with free software to program their range of microcontrollers, AVR studio.

However for the wireless transmission the MSP430 Launch Pad from National Instruments was chosen as it is a low range microcontroller for entry level users (Fig.5). The MSP430 is a 16bit microprocessor with a CPU speed of 16MHz. It is able to run serial communication, timers and a 10bit ADC making it perfect to be the transmitter of all filtered EMG signals from the arm. It is also available in surface mount so further miniaturization was possible.

**Fig. 4.** AVR32 UC3B0256 [7]                    **Fig. 5.** MSP430 Texas Instruments [8]

**Attachments**

There are also many attachments for both microcontrollers ranging from LCD screens to wireless modules, a table description is given below.

**Table 1.** Hardware Attachments

| Name | Style |
|------|-------|
| LCD Screen | PCD8544 |
| Wireless Modules | Nordic nrf24L01 transceiver |
| DAC | low power dual 10-bit surface mount DAC |
| Line | USART Line from FTDI |
| Gyroscope | IDG500 |
| Accelerometer | ADXL335 |
| PCB | |

**2.3     Software Development**

AVR studio 5, software provided by Atmel for Atmel's microcontrollers, is the main code for microcontrollers. It is able to run the CPU clock of the microcontroller at 60MHz, run one USART line, various amounts of buttons/LED's and run three SPI interfaces which included two DAC's (one SPI interface for two voltage outputs), a LCD display and a wireless transceiver.   The coding however worked in a very specific arrangement; it flowed smoothly enough to run at a steady 50Hz as the filtered EMG signals have a maximum frequency of approximately 4Hz and according to the Nyquist frequency we are required to sample this signal at 8Hz. It has also accounted for enough time to receive data wirelessly from the MSP430 at a rate greater than 8Hz.

The only other device run by the AVR32 was the SDOF arm model which is run using a proportional controller. A proportional integral controller was trialed but deemed unfit for the purpose of this application. The equation below states how we implemented the digital proportional controller.

$$\text{Output}_x = \text{Output}_{x-1} + K_P \times (\text{Error}_x - \text{Error}_{x-1}) \tag{1}$$

This proportional controller will take the error from the potentiometer on the SDOF arm model and the proportional gain $K_P$ will be set accordingly to achieve the best position control. The output is a voltage sent out to the motor driver to try get the SDOF arm model to the desired position.

### MSP430

The MSP430 uses a custom interface for programming developed by Texas Instruments[8]. It is a plug in and play interface where no intermediate step like the JTAG used in AVR32 is needed, which has made it a very easy to integrate in this research.

The purpose for the MSP430 is to gather the EMG signals which are put through a filtering system and then to transmit them to the AVR32 with these signals there also includes the gyroscope signals which are only used to detect movement in a yes or no state. The code is simple and continuously polls through the ADC and transmitting data meaning it is working at maximum speed.

### SPI

SPI was used with both microcontrollers and all attachments. SPI works serially which reduces the need of multiple connections and is simple enough to understand. SPI uses a clock source which synchronizes both devices and according to the clock source data is sent out or received one byte (8 bits) at a time. This makes it possible for the receiving register to store the byte and possibly be used as an address to store further data into the corresponding register creating the ability to set individual bits in many registers. This is done on the LCD screen as each liquid crystal is a bit in a register and there was no way to have a General Purpose Input/output (GPIO) pin for every liquid crystal, making SPI brilliant for updating registers at high speeds.

## 3    Experiment System Setup

### 3.1    Subjects

Three subjects without arm disability were chosen for this test, and ethics approval was obtained from the University of Auckland Human Ethics Committee.

### 3.2    Electrode Placement

The electrode placement was noted from literature to be over the middle to lower bicep group and in the centre of the triceps group (Fig.6). The ground however was chosen to be positioned over the olecranon which is located just lower of the elbow joint. This location is chosen as there is very little muscle located near this area causing less chance for disturbance such as cross over noise from other muscle groups or unknown artifacts.

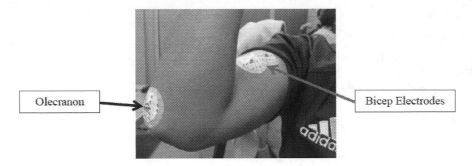

**Fig. 6.** Electrode Placement Example

## 3.3     Settings

### Signal Gains

To tune the filtering system we have a total of four potentiometers that adjust the gain of the EMG signals entering the circuit and the amplitude of the filtered EMG signal that leaves the circuit. The aim of adjusting the gains is to produce the biggest difference in EMG signals from rest to activation. The larger we get the better the resolution as the ADC on the MSP430 is a 10bit ADC that is only able to take in values from 0-3.3V before it is saturated therefore to maximize the signal range between 0 and 3.3V was crucial. The subjects were asked to reach their MVC or maximum muscle contraction, so their EMG signal was at 3.3V. MVC is the EMG signal that relates to the maximum contraction made by the user without putting themselves in harm [1].

### AVR

Once the gain has been adjusted, there is an offset voltage as the two non-inverting amplifiers cause a small zero-offset. This offset is digitally accounted for but as it variable from person to person. Therefore, during set up, the AVR32 will ask the user to input their offset as well as their maximum voltage, if greater than 3.3V it will be set at 3.3V and the AVR32 will normalize all the values between the operators zero-offset and their maximum voltage. There is a separate offset voltage for both the triceps and biceps.

### Motion Capture System

The setup of the spherical marker rig requires two people as placing the rotational part over the users elbow is difficult to position correctly. The setup has to be very precise, or it won't register the users movements on the NDI motion capture system. Figure 7 displays how the rig is to be placed over the arm.

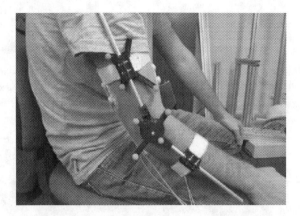

**Fig. 7.** Spherical Arm Rig Positioned Correctly

## 3.4    Testing

The testing consisted of two movements each consisting of three ten second periods
with three test subjects. The two actions included general movement of the lower arm
in a relaxed state and slow movements of the lower arm in a relaxed state. These
states were chosen because if ethics were obtained for others that were not able to
provide enough force into the system then their results would not be comparable to
test subjects who were able to complete all tasks.

## 4    Results and Discussion

The results of all three subjects were gathered showing slightly different results between each individual which was expected. The representative ones are shown in
Fig.8 -9 below.

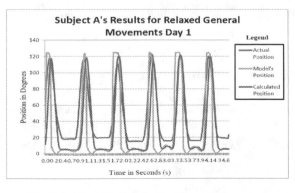

**Fig. 8.** Subject A's Results for General Movement

The result shown above is of subject A in a relaxed position making general movements. From the results we can observe that the actual position does not have as large of a range as the calculated range, this was due to how the testing was carried out as full extensions of the users arms might not have been made causing the angle to reach no lower than 20 degrees. Another note here is that the calculated position generally has a delay in relevance to the actual position though sometimes it is able to accurately follow the movement. However the SDOF arm model is seen to be reacting reasonably but it also seems to reach its peak at an earlier point in time then the actual position of the user's arm this also seems to occur on the way back down.

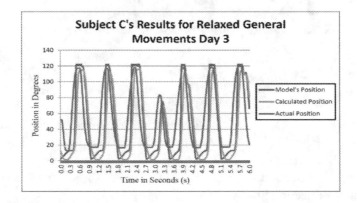

**Fig. 9.** Subject C's Results Day 3 General Movements

Fig.9 displays the general shape of the graph but still to be very delayed in comparison to the actual position of the arm. Another trend starting to be noticed is that the motor is very reactive to changes in the calculated position. This can be seen as the calculated position might change and the motor will propel itself full speed to the position needed. Although they might not be perfectly timed calculated position in Fig.9 seems to have a faster response time on the way down to 0 degrees in comparison to other tests resurfacing the likely hood that fine tuning might actually entice far better results.

## 5    Conclusion and Further Discussion

This research show that: extracted EMG signals have the ability to mimic movements of the user, it is not only controllable, but also leads to some real-time and quite accuracy results; the surface Mount PCB can make the system miniaturized; and wireless transmission of data from the subjects arm to the AVR32 has proven to work at a fast enough rate that no lag is noticed in the model.

However, more work could be done, such as: to develop a very precise EMG filtering system so it is able to detect minute differences in EMG signals such as the users arm resting and being held at 90 degrees; to modularize the AVR32 code even further to ensure it is able to run at capacity. The code of the MSP430 is continuously polling opposed to using interrupts, this would normally not be a problem but as the MSP430 runs on battery, any power saved could mean a longer lasting system.

# References

1. Pau, J.W.L., Saini, H., Xie, S.S.Q., Pullan, A.J., Mallinson, G.: An EMG-Driven Neuromuscular Interface for Human Elbow Joint. In: IEEE RAS and EMBS International Conference on Biomedical Robots and Biomechatronics (BioRob), pp. 145–161 (2010)
2. Nielsen, J.L.G., Holmgaard, S., Jiang, N., Englehart, K., Farina, D., Parker, P.: Enhanced EMG Signal Processing for Simulations and Proportional Myoelectric Control. In: 31st Annual International Conference of the IEEE EMBS, pp. 4335–4338 (2009)
3. Marchal-Crespo, L., Reinkensmeyer, D.J.: Review of Control Strategies For Robotic Movement Training After Neurologic Injury, pp. 1–15 (June 2009)
4. Tao, R., Xie, S., Zhang, Y., Pau, J.W.L.: Review of EMG-based neuromuscular modeling for the use of upper limb control. In: 2012 19th International Conference on Mechatronics and Machine Vision in Practice (M2VIP), November 28-30, pp. 375–380 (2012)
5. Kiguchi, K., Tanaka, T., Fukuda, T.: Neuro-Fuzzy Control of a Robotic Exoskeleton With EMG Signals. IEEE Transactions on Fuzzy Systems 12, 481–489 (2004)
6. Jeon, P.W., Jung, S.: Teleopertated Control of Mobile Robot Using Exoskeleton Type Motion Capturing Device Through Wireless Communication. In: IEEE/ASME International Conference on Advanced Intelligent Mechatronics, pp. 1107–1112 (2003)
7. devtools., http://www.dev-tools.co.kr/ (retrieved September 12, 2012)
8. Texas Instruments, http://www.ti.com/ (retrieved September 14, 2012)

# Towards Enhancing Motor Imagery Based Brain-Computer Interface Performance by Integrating Speed of Imagined Movement

Tao Xie, Lin Yao, Xinjun Sheng*, Dingguo Zhang, and Xiangyang Zhu

State Key Laboratory of Mechanical System and Vibration, Shanghai Jiao Tong
University, 800 Dongchuan Road, Minhang District, Shanghai, China
xjsheng@sjtu.edu.cn
http://bbl.sjtu.edu.cn/

**Abstract.** Left and right motor imagery tasks have commonly been
utilized to construct a two-class Brain-computer Interface system, whilst
the speed property of imagined movement has received less attention.
In this study, we are trying to integrate the types and speed property
of both imagined movement and real movement to further improve the
performance of the two-class BCI system. Thus, real movement session
and imagined movement session were carried out on the separated days.
In real movement session, it has shown that 8 healthy volunteers have
achieved an average accuracy of 67.62% with the same actual left and
right hand clenching speed, and 78.62% with diverse speeds, which was a
significant improvement (p=0.0176). Besides, only three subjects could
pass the 70% accuracy threshold with same actual clenching speed, while
six of them achieved to pass it with diverse speeds. In imagined movement
session, all the subjects with diverse imagined clenching speed achieved
a better control compared with same imagined speed. The proposed idea
of integration of speed information has shown a promising benefit in
two-class BCI construction in this preliminary study.

**Keywords:** Brain-computer interface, motor imagery, clenching speed,
ERD/ERS.

## 1 Introduction

A brain-computer interface (BCI) provides a new non-muscular channel, enabling the brain to directly communicate with to the external world [1]. Motor
imagery based BCI has received intensive interests, and the patients with amyotrophic lateral sclerosis (ALS) syndrome are expected to benefit much from the
advances of this sensorimotor rhythm based BCI system.

* This work is supported by the National Basic Research Program (973 Program) of
China (Grant No.2011CB013305), and the Science and Technology Commission of
Shanghai Municipality (Grant No. 13430721600), and the National Natural Science
Foundation of China (Grant No.51375296).

X. Zhang et al. (Eds.): ICIRA 2014, Part I, LNAI 8917, pp. 234–241, 2014.

Motor imagery (MI) is the most commonly chosen cognitive task in independent BCI application [2][3]. Imagining the movement of left/right hand [4] would lead to predictable changes in the alpha (8-12 Hz) and beta (13-28 Hz) frequency bands, commonly known as Event related (de)synchronization (ERD/ERS) [5][6]. Based on the distinctive topographic distribution of ERD and ERS, different imagined motor types (e.g. left & right hand, left & right foot and tongue) could be discriminated, such as contralateral ERD and ipsilateral ERS could be used for the classification of left and right hand motor imagery.

The topographic distributions of the ERD/ERS are correlated with the different motor imagery types, moreover, the strength of these ERD/ERS has been shown to be correlated with the kinematic property of the imagined movement [7][8]. EEG activity in the alpha and beta frequency bands was found to be linearly correlated with the speed of imagery hand clenching, and similar parametric modulation was found as well during the actual hand movement [9][10].

It has been commonly known that not everyone could efficiently use the MI based BCI system, and there exists the "illiteracy" problem in which BCI control does not work for a portion of users [11][12]. Thus new BCI modalities and hybrid system have been proposed to make the BCI more applicable [13][14]. As EEG contains the kinematic information of MI [7]-[10], we could hypothesize the integration of the types and speed property of the imagined movement might form a more discriminative brain patterns which would be more separable for enhancing a better BCI performance.

This work proposed a motor imagery BCI paradigm that integrated with speed parameter, aiming to increase classification accuracy of BCI. Subjects were asked to imagine left and right hand clenching movement with different speed (e.g. imagining right hand clenching movement with faster speed compared with left hand, or vice versa). As speed of actual movement is more controllable than imagine movement, and both of them has a similar topographies in EEG rhythm[15][16][17], thus we performed two sessions of experiments covering both the actual movement and imagined movement, and preliminary results on MI will be given.

## 2    Materials and Methods

### 2.1    Data acquisition

A SynAmps2 system (Neuroscan, U.S.A) was used for the recording of the 62 channels EEG signals. the electrodes were placed according to the extended 10/20 system. The ground electrode was located on forehead, and the reference electrode was located on the vertex. An analog bandwidth filter of 0.5Hz to 70Hz and a notch filter of 50Hz were applied to the raw signal. EEG signals were digitally sampled at 250Hz.

### 2.2    Subjects

Eight healthy right-handed volunteers participated in these experiments (age range between 22 and 26 years old, one female and seven males). All of them were informed about the whole experiment process. The study protocol was approved by the Ethics Committee of Shanghai Jiao Tong University. informed consent was obtained from all subjects prior to the study.

## 2.3  Experimental Design

During EEG recording, each subject sat in a comfortable armchair in a dim-lighted and electrically shielded chamber room, with forearms and hands resting on the armrest. Eye blink was limited and any facial or arm movements were avoided during the task process.

Real movement session and imagined movement session were arranged in sep-arate days, each session contained two blocks. Subjects were instructed to clench or imagine clenching the left and right hand firstly at the same speed in block one, then at two different speeds in block two. During the imagined movement session, subjects were instructed to mentally simulate kinesthetic movements of their own left or right hands depending on the cues without actually moving them.

For one experiment session, 5 to 10 minutes resting time was permitted be-tween two blocks. Every blocks contained three runs (40 trials for each run), both right hand and left hand task movement (Actual or Imagined) were counterbal-anced. At the beginning of each trial, a fixation cross appeared in the screen center (Fig.1), attracting subject's attention for the subsequent task. Then at the 3rd second, a red cue bar pointing either left or right was presented, which superimposed on the fixation cross and lasted for 1.5s. Subjects should execute given task after appearance of the red cue bar. The given task continued until the disappearing of the fixation cross at the 8th second. Then there was a relax-ation time period lasting for 1s, and a random time period of about 0 to 2s was inserted after the relaxation period to further avoid subject's adaptation. After that the next trial began. During the first run of the two blocks, there was no feedback after the termination of the given task in each trial. For the subsequent two runs of each block, there would be visual feedback according to the on-line classification algorithm implemented within the experiment system. If the task type was decoded right, then the "Right Number" on the screen will plus one, otherwise, the "Right Number" kept constant.

## 2.4  Signal Analysis

Decoding algorithms for both actual and imagined hand clenching was mainly based on Common Spatial pattern (CSP)[18][19]. The raw EEG signal is rep-resented as $X_k$ with dimensions $ch \times len$, where $ch$ is the number of recording electrodes, and $len$ is the number of sample points. The normalized spatial co-variance of the EEG can be obtained from

$$C_k = \frac{X_k X_k^T}{trace(X_k X_k^T)} \tag{1}$$

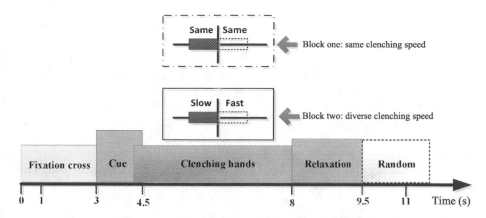

**Fig. 1.** Trial procedure of the experiment. Subjects executed the actual or imagined hand clenching movement according to the red cue randomly presented in the screen. The experiment was carried out on two blocks (same speed in block one, and diverse speeds in block two), every block contained three runs (40 trials for each run). Real movement session and imagined movement session were arranged in separate days.

where $X_k^T$ denotes the transpose of the matrix $X_k$, and $trace(X_k X_k^T)$ is the sum of the diagonal elements of the matrix $X_k X_k^T$. Let

$$C_l = \sum_{k \in S_l} C_k \qquad C_r = \sum_{k \in S_r} C_k \qquad (2)$$

where $S_l$ and $S_r$ are the two index sets of the separate classes.

The projection matrix $W$ could be gained from augmented generalized decomposition problem, $(C_l + C_r)W = \lambda C_r W$. The rows of $W$ are called spatial filters, and the columns of $W^{-1}$ are called spatial patterns. To the $k$-th trial, the filtered signal $Z_k = W X_k$ is uncorrelated. In this work, the log variance of the first three rows and last three rows of $Z_k$ corresponding to largest three eigenvalues and smallest three eigenvalues are chosen as feature vectors, and linear discriminative analysis (LDA) was used as the classifier.

## 3    Results

The time interval for on-line and off-line analysis (actual and imagined clenching) was chosen from the 4th second to the 7th second at the beginning of the trial. The frequency band was chosen to cover the alpha and beta band from 8Hz to 26Hz, applying 4th-order butterworth filter. Table 1 outlines the on-line classification accuracy of the 2nd and 3rd run (the 1st run was used as calibration data set). A 10×10 fold cross validation was adopted to evaluate the classification accuracy between left and right hand, and its detail is described as following: first trials were randomly permutated (60 trails for left hand and 60 trails for right hand), then equally divided into ten partitions. Each partition

was used as an unknown test set which was classified by the classifier trained
with the remaining nine partitions, and a classification accuracy for each partition was achieved. This process was repeated ten times, and 100 classification
accuracy indexes were generated. Fig.2 shows the discrimination accuracy of
seven subjects respecting to actual left and right hand clenching with same and
diverse speeds. One-way ANOVA shows a significant group effect (p=0.0176).
The average accuracy is significantly higher when the left and right hand are
clenching at diverse speeds. For the same speed, only three subjects could pass
the 70% accuracy threshold, and six of the eight subjects achieve to pass the
70% accuracy threshold with diverse speeds. Moreover, 3 subjects with diverse
speed reach a better control of above 70% as compared with a chance level of
50% with same speed.

Fig.3 shows the discrimination accuracy of five subjects respecting to imagined
hand clenching with same and diverse speeds. One-way ANOVA shows a significant group effect (p=0.0014). This preliminary results shows a consistent results
with actual movement. All the subjects with diverse imagined speed achieve a
better control results compared with same imagined speed.

**Table 1.** On-line Classification Accuracy of Actual Clenching Movement Across the
Two Sessions

|     | Block One | | Block Two | |
| --- | --- | --- | --- | --- |
|     | 2nd Run | 3rd Run | 2nd Run | 3rd Run |
| S1  | 0.450 | 0.475 | 0.700 | 0.725 |
| S2  | 0.525 | 0.575 | 0.625 | 0.750 |
| S3  | 0.875 | 0.925 | 0.900 | 0.950 |
| S4  | 0.500 | 0.400 | 0.500 | 0.275 |
| S5  | 0.700 | 0.775 | 0.975 | 1.000 |
| S6  | 0.850 | 0.775 | 0.900 | 0.900 |
| S7  | 0.450 | 0.600 | 0.625 | 0.625 |
| S8  | 0.500 | 0.600 | 0.550 | 0.650 |

## 4    Discussion

The discrimination results with same actual clenching speed in block one and
those with diverse clenching speed in block two, shows that subjects could achieve
a better control with diverse speeds. The discrimination accuracy with diverse
speeds can be increased significantly, compared with same speed. Only 37.5% of
the subjects in our study have an accurate (more than 70%) classification (left
and right) when speed is identical, whereas 75% of the subjects have an accurate (more than 70%) classification (left and right) when the speed is different
for left and right hand, i.e. the integration of speed property could form a more
discriminative brain patterns that would be more separable for a better BCI performance. For imagined clenching movements, the results shows that the speed is
a crucial factor for the discrimination accuracy, and all the subjects with diverse

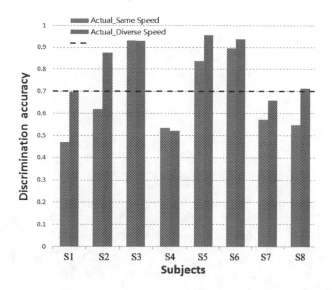

**Fig. 2.** Comparison between actual hand clenching with the same speed and diverse speeds. The blue bar indicates discrimination accuracy of left and right hand with the same clenching speed. The red bar indicates the discrimination accuracy of left and right hand with the diverse clenching speed. The black dash-dotted line indicates 70% accuracy.

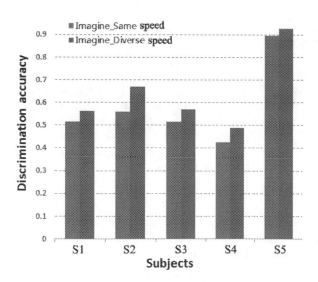

**Fig. 3.** Comparison between imagined hand clenching with the same speed and diverse speeds. The blue bar and red bar indicate the discrimination accuracy of left and right hand with the same and diverse clenching speed correspondingly.

imagined speed achieve a better control compared with same imagined speed. Study in [9] has showed topologies of average EEG changes in the alpha and beta frequency band using time and frequency windows, and a larger decrease of EEG activity was observed as speed was increased. A consistent result is found in our research, as shown in Fig.4, enhanced rhythm in upper alpha band (10 to 13 Hz) and beta band (13 to 26 Hz) can be achieved with fast hand clenching speed. This phenomenon, in some case, explains the reason of a more separable BCI performance with the integration of speed factor.

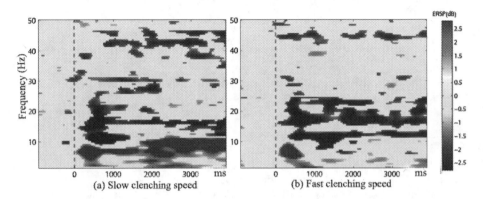

**Fig. 4.** Powerspectrum across the time and frequency at channel C4 corresponds to left actual hand clenching. (a) Actual left hand clenching with slow speed. (b) Actual left hand clenching with fast speed.

## 5 Conclusion

This work proposed a motor imagery BCI paradigm that integrated with speed parameter, results indicated that this paradigm could significantly increase classification accuracy of BCI. With diverse clenching speed of left and right hand, the discrimination accuracy was increased. For imagined clenching, result showed that the speed was a crucial factor for the discrimination accuracy, and further study will be focused on the imagined clenching speed. The proposed idea of integration of speed information has shown a promising benefit in two-class BCI construction in our preliminary study.

**Acknowledgments.** The authors thank Guangquan Liu and Gan Huang for their preliminary work in the setting up of the BCI system, and all the volunteers for their participation in the study.

## References

1. Wolpaw, J.R., Birbaumer, N., McFarland, D.J., Pfurtscheller, G., Vaughan, T.M., et al.: Brain-computer interfaces for communication and control. Clinical neurophysiology 113(6), 767–791 (2002)

2. Millan, J.R., Renkens, F., Mourino, J., Gerstner, W.: Noninvasive brain-actuated control of a mobile robot by human eeg. IEEE Transactions on Biomedical Engineering 51(6), 1026–1033 (2004)
3. Pfurtscheller, G., Müller, G.R., Pfurtscheller, J., Gerner, H.J., Rupp, R.: Thought – control of functional electrical stimulation to restore hand grasp in a patient with tetraplegia. Neuroscience Letters 351(1), 33–36 (2003)
4. Pfurtscheller, G., Neuper, C., Flotzinger, D., Pregenzer, M.: Eeg-based discrimination between imagination of right and left hand movement. Electroencephalography and Clinical Neurophysiology 103(6), 642–651 (1997)
5. Pfurtscheller, G., Neuper, C., Brunner, C., Lopes da Silva, F.: et al. Beta rebound after different types of motor imagery in man. Neuroscience Letters 378(3), 156 (2005)
6. Pfurtscheller, G., Lopes da Silva, F.H.: Event-related eeg/meg synchronization and desynchronization: basic principles. Clin. Neurophysiol. 110, 1842–1857 (1999)
7. Jerbi, K., Vidal, J.R., Mattout, J., Lecaignard, F., Ossandon, T., Hamame, C.M., Dalal, S.S., Bouet, R., Lachaux, J.-P.: Inferring hand movement kinematics frommeg, eeg and intracranial eeg: From brainmachine interfaces to motor rehabilitation. IRBM 32, 8–18 (2011)
8. Robinson, N., Vinod, A.P., Kai, K.A., Keng, P.T., Guan, C.T.: Eeg-based classification of fast and slow hand movements using wavelet-csp algorithm. IEEE Transactions on Biomedical Engineering 60(8), 2123–2132 (2013)
9. Yuan, H., Perdoni, C., He, B.: Relationship between speed and eeg activity during imagined and executed hand movements. J. Neural. Eng. 7, 26001 (2010)
10. Yuan, H., Perdoni, C., He, B.: Decoding speed of imagined hand movement from eeg. In: 2010 Annual International Conference of the IEEE Engineering in Medicine and Biology Society (EMBC), pp. 142–145. IEEE (2010)
11. Guger, C., Daban, S., Sellers, E., Holzner, C., Krausz, G., Carabalona, R., Gramatica, F., Edlinger, G.: How many people are able to control a p300-based brain? computer interface (bci)? Neuroscience Letters 462(1), 94–98 (2009)
12. Allison, B.Z., Brunner, C., Kaiser, V., Müller-Putz, G.R., Neuper, C., Pfurtscheller, G.: Toward a hybrid brain–computer interface based on imagined movement and visual attention. Journal of Neural Engineering 7(2), 026007 (2010)
13. Yao, L., Meng, J., Zhang, D., Sheng, X., Zhu, X.: Selective sensation based brain-computer interface via mechanical vibrotactile stimulation. PLoS ONE 8(6), e64784 (2013)
14. Yao, L., Meng, J., Zhang, D., Sheng, X., Zhu, X.: Combining motor imagery with selective sensation towards a hybrid-modality bci. IEEE Transactions on Biomedical Engineering PP(99), 1 (2013)
15. Schupp, H.T., Lutzenberger, W., Birbaumer, N., Miltner, W., Braun, C.: Neurophysiological differences between perception and imagery. Cogn. Brain Res. 2, 77–86 (1994)
16. Pfurtscheller, G., Neuper, C.C.: Motor imagery activates primary sensorimotor area in humans. Neurosci. Lett. 236, 65–68 (1997)
17. McFarland, D., Miner, L., Vaughan, T., Wolpaw, J.: Mu and beta rhythm topographies during motor imagery and actual movements. Brain Topography 12(3), 177–186 (2000)
18. Fukunaga, K.: Introduction to statistical pattern recognition, 2nd edn., pp. 1–2. Academic Press (1990)
19. Ramoser, H., Muller-Gerking, J., Pfurtscheller, G.: Optimal spatial filtering of single trial eeg during imagined hand movement. IEEE Transactions on Rehabilitation Engineering 8(4), 441–446 (2000)

# An Exoskeleton System for Hand Rehabilitation Based on Master-Slave Control

Zhangjie Chen[1], Shengqi Fan[2], and Dingguo Zhang[1,*]

[1] School of Mechanical Engineering, Shanghai Jiaotong University,
Dongchuan Road 800, 200240 Shanghai, China
[2] University of Michigan – Shanghai Jiaotong University Joint Institute,
Shanghai Jiaotong University, Dongchuan Road 800, 200240 Shanghai, China
dgzhang@sjtu.edu.cn

**Abstract.** Most of patients with hand injury lose only one side of the function, a master-slave rehabilitation system can help them to recover by training the disabled hand using the healthy one. In this paper, such a system was presented with several components. A data glove based on bending sensor and a reliable hand recognition algorithm used to capture real-time locomotion data and send data to exoskeleton system through wireless Bluetooth. The drive and control circuit with PWM control strategy kept the shape-memory-alloy (SMA) actuator working well and ensured reliability and safety of the system. An adaptive dorsal metacarpal base with 15 degree of freedom was designed to be attached to patients' palm tightly. Finally, the exoskeleton was fabricated with 3D printing technology, and the performance of the whole system was tested and analyzed.

**Keywords:** Master-Slave control, PWM, hand rehabilitation, shape memory alloy, exoskeleton.

## 1 Introduction

Hands are the main part of human body for people to deal with different kinds of work. However, hands are vulnerable. Patients with hand injury need long-term, high strength and high repeatability of rehabilitation treatment in order to recover [1], [2]. One of the main methods is to use medical rehabilitation facilities, which enables patients to do the treatment by their own. With robotic technology introduced to medical field, many exoskeletal rehabilitation robots were put into use [3]. And, since most of the patients only lose one side of their hand function, self-rehabilitation with mirror control of the healthy hand under the aid of exoskeleton system is a feasible way.

The most commonly adopted robot assisted rehabilitation at present is Continuous Passive Motion (CPM) [4], [5]. Its main theory is that in order to accelerate recovery, affected limbs should prevent stiffness and promote blood circulation by doing pas-

---

* Corresponding author.

X. Zhang et al. (Eds.): ICIRA 2014, Part I, LNAI 8917, pp. 242–253, 2014.
© Springer International Publishing Switzerland 2014

sive exercise for long time and at all ranges with aid of rehabilitation facilities. For now, Passive Range of Motion (ROM) [6] has been a common part in clinical treatment of stroke patients. Other robot assisted rehabilitation including Robotic Active-assistance Exercise (RAAE), Robotic Progressive-assistance Exercise (RAPE) [7], and Virtual Reality Task-oriented Training (VRTOT) [8].

Hand Mentor system developed by Deaconess [9] was the first active rehabilitation system put to use in clinical treatment. Patients can actively bend or stretch their wrists and fingers based on the instruction shown on the screen. Hand Mentor only assists the exercise when it is necessary. It was an exoskeleton system of one degree of freedom, so the joints of hand could not be trained separately. Hand motion assistant robot developed by Kawasaki H. et all [10] could aid the hand to do exercise of 18 degrees of freedom. The system also adopted master-slave control, which controlled the affected hand by exoskeleton based on the motion data of the healthy hand sent by the data glove. The deficiency of the system was poor portability. Wege A. et all designed a hand exoskeleton [11], which could control every joint separately. It adopted closed-loop control using surface electromyography signal. But it had a very high requirement for wearing accuracy. The hand exoskeleton designed by Zhang Q.C. [12] controlled the system using motors to pull wires. It had good portability and high flexibility. Liu Z.W. also developed a rehabilitation robot [13]. Its highlight was using air muscle connected with motor and sensors to realize closed-loop control. But, it also had poor portability.

In our design, we used shape memory alloy (SMA) instead of traditional motor as the driving component. SMA is a kind of alloy that shaps under low temperature, and when it is heated to some critical temperature, it can return to its original shape. It has the so-called shape memory effect. This effect is mainly caused by the transition of the two different phases in the material: martensite and austenite. Compared with traditional motor actuator, SMA actuator has several advantages: high ratio of power and mass, simple structure, no pollution and no noise. [14]. We already had a primary design of a hand exoskeleton rehabilitation system based on SMA spring actuator [15]. It had high portability but still had some deficiencies such as lack of master-slave control mode and low in control accuracy. In this paper, we mainly developed a master-slave control system for the rehabilitation facilities, built the system, and did some tests.

## 2    Methodology

This work focused on the master-slave rehabilitation paradigm. For hemiplegic patients, the intact hand could be the master side to guide the hand motion of paralyzed (salve) side. For tetraplegic patients, the therapist's hand could be the master side to guide the hand motion of patients (slave side). The designs of master side and slave side were introduced separately as follows.

### 2.1    Design of Master Side

Data glove is the most common tool in virtual simulation. It can transmit the real-time status of the hand to virtual environment through bending sensors, and improve the interactivity and immersivity. The data gloves now in the market are pretty mature and can provide abundant data of hand. However, they are too costly to be a part of arehabilitation system. Since most patients with hand injury cannot complete complicated tasks, the requirement for data collection is not very high. A relatively simple data glove was designed as the master part for the system. By sewing the bending sensors on the outer surface of the glove, the real-time bending data of five fingers could be sent to the upper-computer or the slave side through Bluetooth.

**FLEX Bending Sensor.** Flex 4.5 inches sensor, which was a kind of one directional sensor, was chosen to measure the locomotion of the fingers. As the flex sensor was bent in one direction, the resistance of it gradually increased. Range of resistance of the sensor was between 10K and 40K depending upon the degree of the flex. Five sensors were used to measure the data of five fingers. Since the resistance of the sensor varied when it is bent, a bleeder circuit was used to transfer the resistance change to voltage signal.

**Data Collection Circuit.** We used a minimum system based on STC15F2k60S2 SCM as the control circuit. The AD module was used to transfer the analog signal to digital signal. A 1602LCD screen could show the real-time data. By using a HC-05 Bluetooth module, the data could be sent to the exoskeleton system as well as the upper-computer. The circuit diagram is shown in Fig. 1.

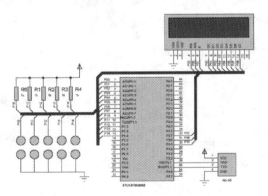

**Fig. 1.** Master data glove circuit diagram

The photo of the master data glove is shown in Fig. 2, where "1" denotes bending sensors, and "2" denotes the 1602LCD. "3" represents the Bluetooth module, and "4" represents the STC15 SCM powered by a portable power source.

### Data Glove Software Design

Fig. 3 (a) shows the program flow chart. The timer was set to call AD function every 50ms. The five groups of data collected were sent to the exoskeleton through serial ports. To ensure the accuracy and safety of the data, a 0x64 value was sent first as the standard data. The program was written in Keil.

**Fig. 2.** Actual photo of data glove in master side

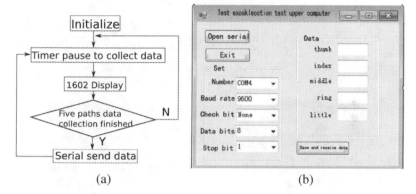

(a)                                          (b)

**Fig. 3.** (a) Program flow chart of master part. (b) Program interface of upper computer.

To record and analyze the data as well as adjust the system, an upper computer program was written in visual C# with serial Port Control. The interface is shown in Figure 3 (b).

## 2.2. Design of Slave Side

The mechanism of hand exoskeleton was driven by SMA. This section introduced the design of mechanical system and control system.

**Optimization of Exkoskeleton Structure.** A 3D hand model was established in UG software to help for design and analysis the structure of exoskeleton system. The structure of the hand exoskeleton is shown in Fig. 4. There were two four bar linkages in each finger. Actuator drived R1 and through the gears all the linkages were driven.

The transmission efficiency of the optimized exoskeleton was high and the system was comfortable and easy to wear. Since the size and appearance of human hands vary from person to person, a 15-degrees-of-freedom adaptive dorsal metacarpal base can be tightly attached to the back of patients' palm, assuring the efficiency of the hand exoskeleton rehabilitation system.

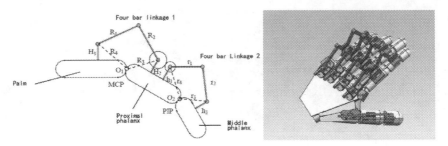

**Fig. 4.** Structure of hand exoskeleton

**Design and Analysis of SMA Actuator.** Fig. 5 shows the running process of the SMA actuator. When heating SMA1, it contracts and drags the joint of exoskeleton as well as SMA2. When SMA1 cools down, ohmic heat SMA2 spring and the joint will be dragged down, thus completing a circulation.

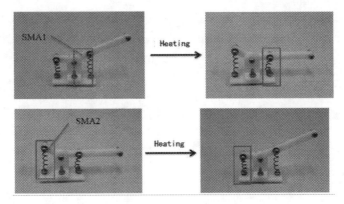

**Fig. 5.** Running process of SMA actuator

**SMA Position Feedback.** Controlling the SMA actuators only by controlling the time of heating can cause overheating or power shortage. Feedback of SMA position is demanded. Since our system adopted a drive mode of single degree of freedom coupling, only the rotation angle of the drive rod should be measured. We chose the SV01A103AEA01R00 potentiometer to be the feedback component. As shown in Table 1, the potentiometer is light in mass, small in size, easy to install, and low in resistance. So it is suitable for our system.

**Table 1.** Parameters of SV01A103AEA01R00 potentiometer

| Total resistance | Effective rotational angle | Voltage rating | Torque | Mass |
|---|---|---|---|---|
| 10K | 330 degree | 5V | 2mN/m | 0.36g |

When connected as in Figure 6, the inner ring of the potentiometer could be rotated together with the drive rod. And then, the position signal could be obtained.

**Fig. 6.** Assembly of potentiometer

**Design of Drive Circuit.** Since SMA actuators generate restoring force by changing material characteristic when heated, the heating current is very large and usually beyond the capability of IO port. It is necessary to design a drive circuit to heat the SMA actuators. In order to power the actuators separately, we used IRF3205 field effect transistor to build the on-off circuit because of its high reliability and short delay. 18650 lithium batteries were chosen to be the power source. Assume that the exoskeleton runs in a rehabilitation mode of circulation from thumb to little finger, every circulation last for 30s and requires 20.83mAh. A battery has 2600mAh, so it can support the exoskeleton for 5 hours, meeting the power requirement for a rehabilitation system. LEDs were used in indication circuit to show if the SMA was being heated.

**Design of Controller Circuit.** The hand exoskeleton system used the same SCM as the data glove (see Fig. 7). The P4 port and P3.2, P3.3 pin generated10-way control signals to control the SMA's heating current on and off. The hand exoskeleton circuit had a 1602LCD as well, which could show the motion status of both master and slave parts.

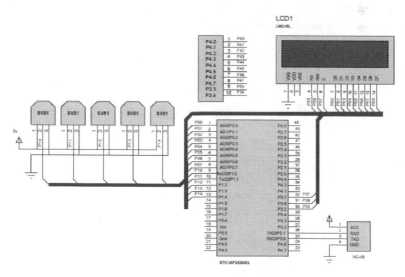

**Fig. 7.** Control circuit diagram for exoskeleton.

**PMW Control Strategy.** When SMA spring is heated to its phase transition temperature, the shape memory effect will be activated and the SMA spring will be compressed automatically. However, the SMA spring should not be heated above the transition finishing temperature or the shape memory effect may be lost or changed. Therefore, the temperature of the SMA needs to be strictly controlled. A PWM based strategy was introduced to control the SMA's heating current. Since the power source of the exoskeleton system was Lithium battery, which was a kind of constant voltage source, the best method to control the SMA heating current was pulse-width modulating (PWM).

We used the following way to control the heating of the SMA spring: (1) Heating the SMA with full current (100% PWM) to raise temperature. (2) Lowering the current to make the temperature of SMA constant. (3) Waiting the position feedback signal to stop heating. By using this method, the SMA actuator had quick response, stable force output and less electricity consumption.

Through the differential equation of SMA heating process, we get the heating equation and cooling equation of spring respectively:

$$T = T_0 + \frac{i^2 R}{hA}(1 - e^{-t\frac{hA}{mc_p}})  \tag{1}$$

$$T = T_0 + (T_s - T_0)e^{-t\frac{hA}{mc_p}}  \tag{2}$$

where $T_0$ is initial temperature of SMA, $i$ is the heating current, $h$ is convection constant and $C_p$ is the specific heat of SMA. Considering the voltage of Lithium battery was 3.7V and the maximum heating current was 3A, we took 1.5 second as heat time to use maximum current to make SMA completely shift to austenite, which could maximize the stiffness of SMA spring. This stiffness made driving force meet the requirement.

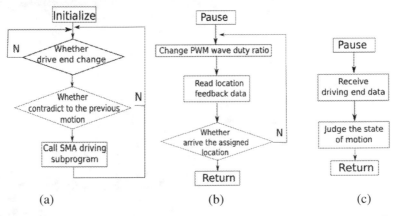

**Fig. 8.** Programs in slave side. (a) Main program. (b) Timer pause program. (c) Serial port pause program.

**Design of Software**. In the program, timer0 produced a 50Hz PWM signal and timer1 worked as the baud rate generator. The data of position feedback was measured every 100ms. The status of master was distinguished first and then the judge function was run. The judge function could ensure every SMA has at least six seconds to cool down. Then, it would run the drive function to heat the SMAs. According to the position feedback signal, the program would stop heating the SMAs and came back to distinguish the change of master status.The flow charts are shown in Fig. 8.

## 3    Test and Analysis

### 3.1    Motion Recognition in Master Side

Because of the coupling of finger motion, in our design we only needed to recognize if the finger was bending by setting a threshold value. If the data from sensors were greater than the threshold value, it should be recognized as bending. We used thumb, index, middle, ring and little to represent the value for each finger. Fig. 9 shows the data collected during single-finger-motion circulation, from which we can know that the threshold value of the bending finger must be greater than the maximum value of other fingers in order to avoid identification error. Consequently, we set thumb>138, index>142, middle>141, ring>135, little>139.

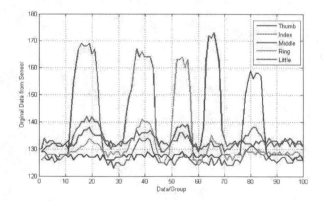

**Fig. 9.** Data of single-finger-motion circulation

Fig. 10 shows the situation that two fingers are moving together. Since the coupling of thumb and other fingers was negligible, it was not concerned. We only considered the most common situations, which were index and middle, middle and ring, ring and little moving together. When two fingers bended together, they must be recognized as moving at the same time. So we chose the relationship of the two threshold values as the average value of the difference in two fingers, as shown in Fig. 11, that was, index-middle=2.18, middle-ring=2.31, ring-little=12.18.

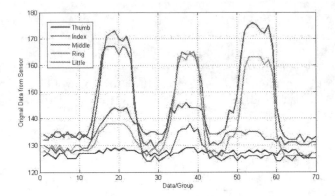

**Fig. 10.** Data of two fingers moving together

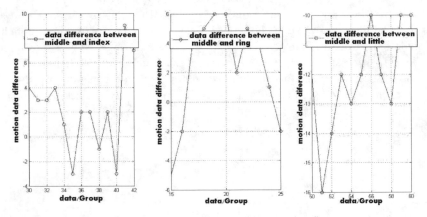

**Fig. 11.** Difference in data of two moving fingers

Then, we did the same thing when three fingers and four fingers were moving together. We got other two sets of correlation. Since we got the data in form of integer, the final correlation was set at the average value that index-middle=1, middle-ring=3, ring-little=-10. So the threshold values are thumb=145, index=150, middle=149, ring=146, little=156.

By using this set of threshold values, the master side data glove can efficiently recognize the movement of each finger, as well as multiple fingers moving together, and effectively decrease the possibility of recognition error.

## 3.2    Test on Prototye

The main structure of the hand exoskeleton was produced by Object-3D printer (Stratasy Company, Israel). Several experiments were designed to test the performance of hand exoskeleton system. Here we take the bending of index single finger as an example. We used the 1602LCD screen to monitor the real time data. The 6 sets of data shown in Fig. 12 were (1) motion-flag, (2) judge-flag, (3) slave-status, (4) motion-status, (5) master-status and (6) control-status, respectively. Detailed explanations are in Table 2.

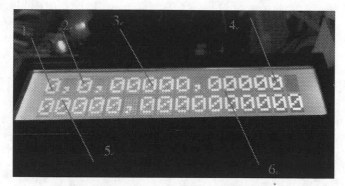

**Fig. 12.** 1602LCD screen indication

**Table 2.** Details of the data on 1602LCD screen

|  | 0 | 1 | Digit Explanation |
|---|---|---|---|
| Motion-flag | No motion | Motion recognized | / |
| Judge-flag | No error | Movement error | / |
| Slave-status | Straight | Bending | thumb-index-middle-ring-little |
| Motion-status | Static | Actuating | thumb-index-middle-ring-little |
| Master-status | Straight | Bending | thumb-index-middle-ring-little |
| Control-status | Static | Heating | (SMA1)thumb-index-middle-ring-little-(SMA2)thumb-index-middle-ring-little |

Fig. 13 shows the initial state of the test. The index finger of master side stayed straight, and the exoskeleton system also stayed static. All of the indications on the screen were zero.

(a)                                               (b)

**Fig. 13.** Initial state of index bending test (a) Master side. (b) Slave side.

Then we started the test by bending the index finger at the master side. When the data collected by the sensor was beyond the threshold value of the index, the system recognized the motion of index finger and went into driving state. The indication data were 1, 0, 00000, 01000,01000, 0100000000, which meant the system completed the recognition and was actuating the slave side. When the slave side received the signal through Bluetooth and the index of the exoskeleton was moved to the designed spot, the actuating was completed. And, the screen showed 0, 0, 01000, 00000,01000, 0000000000, which implied that the slave side had completed the actuating and was ready for the next instruction.

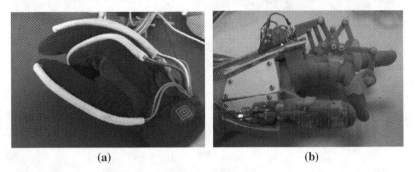

(a)                                                  (b)

**Fig. 14.** (a) Actuating state (b) Actuate completed state of index bending test.

There were also tests about bending of each finger and multiple fingers. In all the tests, the movement recognition ratio was about 100%, and all the actuating was completed within the designated upper limit time, which was 5 seconds. The results of the tests showed that the system identification performance was excellent, the Bluetooth transmission was stable and driving response was fast, which all met the design requirements. The only deficiency was that when there was shift between different gestures, the motion frequency was quite low, since the cooling speed of the SMA actuators cannot be controlled.

## 4     Conclusion

A rehabilitation system based on master-slave control was presented in the paper. Design of a data glove based on bending sensor and tests of relevant circuits were completed. The master data gloves can capture the real-time locomotion data of the fingers and send the data to the exoskeleton system through wireless Bluetooth. Also a stable and feasible hand motion recognition algorithm was designed to fulfill the need of controlling.

The drive and control circuit was also designed. PWM control strategy was introduced to keep the SMA actuator working well. A 15-degrees-of-freedom adaptive dorsal metacarpal base, which can be tightly attached to the back of patients' palm was also designed. Finally, the 3D printing technology was used to fabricate the exoskeleton structure. The performance of the master-slave hand exoskeleton system's was tested and analyzed through a series of experiments.

There were still some problems in the system. The future work will focus on the following aspects. The SMA actuators presented have an effective way to control its heating speed, while its cooling speed cannot be controlled. Therefore, the motion frequency is quite low, which is the key limitation of SMA actuators. The temperature is the main factor to influence the performance of SMAs, so temperature is the best way to know the status of SMAs. The circuit with a higher voltage power may lead to a quick respond speed and less weight, because only a single battery will be needed and a much higher current is available, meaning the SMA can be heated much quickly.

**Acknowledgement.** This work is supported by the National Natural Science Foundation of China (51475292), the Natural Science Foundation of Shanghai (14ZR1421300) and the State Key Laboratory of Robotics and System (SKLRS-2012-ZD-04).

# References

1. Taub, E., Miller, N., Novack, T., Cook, E., Fleming, W., Nepomuceno, C., Connell, J., Crago, J.: Technique to improve chronic motor deficit after stroke. Archives of Physical Medicine and Rehabilitation. 74(4), 347–354 (1993)
2. Mark, V.W., Taub, E.: Constraint-induced movement therapy for chronic stroke hemiparesis and other disabilities. Restorative Neurology and Neuroscience 22(3-5), 317–336 (2004)
3. Patton, J.L., Mussa-Ivaldi, F.A.: Robot-assisted adaptive training: custom force fields for teaching movement patterns. IEEE Transactions on Biomedical Engineering. 51(4), 636–646 (2004)
4. Salter, R.B., Field, P.: The Effects of Continuous Compression on Living Articular Cartilage. The Journal of Bone and Joint Surgery. 42(1), 31–90 (1960)
5. O'Driscoll, S.W., Giori, N.J.: Continuous Passive Motion (CPM): Theory and Principles of Clinical Application. Rehab. Res. Dev. 37(2), 179–188 (2000)
6. Flowers, K.R., LaStayo, P.C.: Effect of Total End Range Time on Improving Passive Range of Motion. Journal of Hand Therapy 25(1), 48–55 (2012)
7. Hogan, N., Krebs, H.I., Rohrer, B.: Motions or Muscles? Some behavioral factors underlying robotic assistance of motor recovery. Rehab. Res. Dev. 43(5), 605–618 (2006)
8. Oblak, J., Cikajlo, I., Matjacic, Z.: Universal Haptic Drive: Robot for Arm and Wrist Rehabilitation. IEEE Trans. Neural Syst. Rehab. Eng. 18(3), 293–302 (2010)
9. Hand Physical Therapy with The Hand Mentor,
   http://www.kineticmuscles.com/hand-physical-therapy-hand-mentor.html
10. Kawasaki, H., Ito, S.: Development of a Hand Motion Assist Robot forRehabilitation Therapy by Patient Self-Motion Control. In: Proc. IEEE 10th Int. Conf., pp. 234–240. Rehabil. Robotics, Noordwij (2007)
11. Wege, A., Kondak, K., Hommel, G.: Development and Control of a Hand Exoskeleton for Rehabilitation of Hand Injuries. In: IEEE/RSJ Int. Conf. on Intelligent Robots and Systems, pp. 3046–3051 (2005)
12. Zhang, Q.C.: Design and Research of the Mechanical System for a Hand Rehabilitation Robot (in Chinese) Master Degree Dissertaion, Harbin Institute of Technology (2011)
13. Liu, Z.W.: Design of Structure and Control System of Hand Rehabilitation Robot (in Chinese), Master Dissertation, Huazhong University of Science and Technology (2008)
14. De Laurentis, K.J., Mavroidis, C., Pfeiffer, C.: Development of a shape memory alloy actuated robotic hand. In: Proceeding of the ACTUATOR 2000 Conference, Bremen, Germany, pp. 19–21 (2004)
15. Tang, T., Zhang, D.G., Xie, T., Zhu, X.Y.: An Exoskeleton System for Hand Rehabilitation Driven by Shape Memory Alloy. In: The 2013 IEEE International Conference on Robotics and Biomimetics (ROBIO), Shenzhen, China, pp. 756–761 (2013)

# Anthropometric and Anthropomorphic Features Applied to a Mechanical Finger

Alejandro Prudencio, Eduardo Morales, Mario A. García, and Alejandro Lozano

Centro de Investigación en Ciencia Aplicada y Tecnología Avanzada,
Unidad Querétaro Cerro Blanco #144, Santiago de Querétaro, México

**Abstract.** This work presents the dimensional synthesis for a mechanical finger, based in four bar mechanisms, where the anthropometric and kinematic constrictions of the human finger are satisfied; the basic criteria used for the dimensional synthesis is the maximum rotation angle for each phalanx, so a human like motion may be achieved. The synthesis of each finger is obtained via Freudenstein's methodology, assuming that the links lengths are constant, and fixed to a main link, the pivot of the system is the knuckle. The only variable which may modify the displacement behavior of the system is the coupler links for each four-bar mechanism; therefore its length is directly related to the position of the finger, such a mechanism has to be able to withstand the average loads and impacts that may happen in a unstructured environment.

**Keywords:** four bar mechanism, kinematics, under actuated finger, Freudenstein Method.

## 1 Introduction

The human hand itself is extremely difficult to model; when attempting to duplicate the high dexterity of its actions, its lightness, and hight rigidity of the structure, are of equal concern. The internal bone structure of the human hand and finger are constructed in a segmented pattern as shown in Figure 1 the proximal, intermediate and distal phalanges of the finger, and the metacarpal bones within the palm, have a system of tendons that connect to the base of each of the phalanges which serve to curl the fingers to form a hand gesture. The *digitorum superficialis* and *profundis tendons*, seen in Figure 2, are connected from the base of the phalanges to the flexor digitorum muscles in the upper forearm. When in contraction this muscle will pull on the tendons, which in turn will curl the fingers.

From an engineering viewpoint, the full range of dexterity available in the human hand is amazing; each of the phalange joints acts as an independent revolute joint as each one is controlled by its own flexor digitorum muscle in the forearm., The joint between the metacarpal bones and the proximal phalanges, the knuckle, is similar to a limited range ball and socket joint as the lumbrical muscles, Figure 1, in between each of the fingers provide limited adduction and abduction of the fingers.

X. Zhang et al. (Eds.): ICIRA 2014, Part I, LNAI 8917, pp. 254–265, 2014.
© Springer International Publishing Switzerland 2014

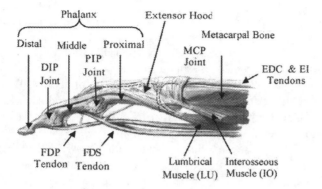

**Fig. 1.** The finger anatomy (18)

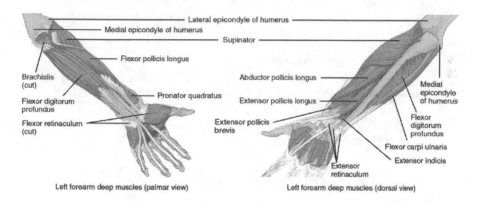

**Fig. 2.** Muscle structure of the forearm (1)

For a relatively simplistic robotic hand, in order accurately mimic just a few of the possible motions of the human one. It is excessive and beyond requirements a design of a highly complex mechanical hand, which sole propose are grasping tasks, because it does not have to mimic all the possible gestures of the human hand.

However articulations in the hand involve in grasp, the robotic hand must be designed to allow these, to appear as natural as possible, while robustness is required for industrial and possible prosthetic application. The parts designed, would also be easily manufactured.

This approach is quite beneficial in the reproduction of human like extremities, especially regarding the structure and degrees of freedom involved in the grip and manipulation functions of the hand. Some relevant hand designs which relay in the after mention classic concept include the UTHA/MIT hand (2)and the Stanford/JPL hand (3), however their designs are bulky and complex due to the high number of actuators and sensors, deriving in a complex structure. Therefore a new approach has been taken, leading to the implementation of simpler mechanisms (4), while maintaining a natural movement and appearance of the finger (5) with a smaller number of actuators.

There are two main strategies for implementing such mechanisms, the use of tendons and bar linkages. Tendon transmission is limited to small grasping forces where as the bar linkage transmission is far more suitable for applications in which human like forces may be required. In both cases active actuators are replaced by passive elements such as mechanical breaks, physical limiters and different kinds of springs. When applied to mechanical fingers these kinds of mechanism leads to auto adaptability.

When those characteristics are applied to mechanical fingers these kinds of mechanism leads to auto adaptability in the grasp of an object

The present work proposes the dimensional synthesis by Freudenstein's equations of a mechanical finger capable of generate a good approach to the human finger trajectory, while maintaining a compact, humanoid and efficient design. Analysis of its trajectories by geometric analysis of the mechanism, finally achieving a single trajectory, similar to a logarithmic curve, which due to anatomical analysis as show in (6) are the ones that mostly resemble the one in human fingers during a grasp task

## 2    Materials and Methods

In order to obtain the parameters for a human like mechanical finger, it is necessary to know the anthropomorphic and anthropometric parameters of it, also its mobility range for each finger, which can be found in the anatomy and mechanics of the human hand as presented by Scharz (6).

### 2.1    Review of the Human Finger Physiology

A human finger consists of finger bones called phalanges which are three in any finger as stated by Tubiana (7). They are named from farthest to the closest to the palm distal, medial and proximal phalanx respectively. Keeping the same order, the distal and middle phalanx are join by the Distal Interphalangeal joint (DIP), the middle and proximal phalanx are connected thru the Proximal Interphalangeal Joint (PIP) while the proximal phalange and the palm are keep together by the Metacarpal phalange joint (MCP) the position of these phalanges and joints are show in Figure 1 .

### 2.2    Finger Morphology

In order to know the morphology of the finger, the range of movement, dimensions and nomenclature is needed, it is assumed in the mechanical synthesis of each four-bar mechanism to be the proximal, middle and distal phalanx for the mechanical model.

The MCP joint is the first one to be analyzed, this joint can rotate up to an approximate angle of 90° clockwise from the metacarpal bone which is taken as the relative origin, and the complete extension of the finger measure from this joint can be as much as 45° counterclockwise from the origin of the joint, follow by the PIP which can't realize extension movements, but presents a flexion of about 100°, the flexion of the fingers in the DIP Joint can form an angle of about 90° and the finger extension of the finger gives a maximum of 10° in the opposite direction. The movement range of these joints is summarized in the Table 1 (7).

**Table 1.** Joint range in Degrees

| Joint | Movement Range |
|-------|----------------|
| MCP | Hyperextension 0° a 45° |
|  | Flexion     0° a 90° |
| PIP | Extension    0° |
|  | Flexion     0° a 100° |
| DIP | Extension    0° a 10° |
|  | Flexion     0° a 90° |

The anthropometric data for the dimension of each finger, is summarized in Table 2, base on a Latin American average

**Table 2.** Anthropometric dimensions in mm of the human fingers

| Finger | Proximal Phalanx | Medial Phalanx | Distal Phalanx |
|--------|------------------|----------------|----------------|
| **Thumb** | 17.1 | - | 12.1 |
| **Index** | 21.8 | 14.1 | 8.6 |
| **Middle** | 24.5 | 15.8 | 9.8 |
| **Ring** | 22.2 | 15.3 | 9.7 |
| **Little** | 17.2 | 10.8 | 8.6 |

This information is taken as the input to develop a mechanism capable of emulating the displacement of a human finger to do so an underactuated model is proposed by Gosselin (8), which is capable of mimic its trajectory while decreasing the number of actuators and maintaining a human like movement.

## 2.3    The Human Finger Trajectory during Grasp

Grasping is defined as the property of a system to prevent motion of an object while manipulating is defined as the capability of a system to control the position and orientation of an object.

The starting point for the present work is therefore the trajectory of a single finger during a grasping task, and a simplified approach to it. However, the fingertip trajectories with respect to the hand during the reach-to-grasp tasks have been well described by (9) which study attempted to characterize the motion of the digits during grasping tasks in the human hand. A direct approach to provide a mechanical finger capable of generate a good approach to the after mention trajectory, while maintaining a compact, humanoid and efficient design is the focus of this paper.

## 2.4    Comparison between Some Mechanical Fingers

There are several finger models based on four-bar and pulley mechanism's, although there are similar in the sense of mobility and connectivity, as discussed in (10) they are not the same, from the structural an dimensional point of view each one is  a proposal of its own, for example the Manus-hand which  main concerns is weight , size and grasping ability is not concert with dextrousity or force from a human point of view , therefore the  proposed mechanism is based on pulleys rather than links, to ease the control and overall design of the hand (11); while a four-bar mechanism approach is taken by (12) in its multi-segmented finger design for a experimental prosthetic hand, where each phalanx is a four-bar mechanism, increasing force transmission and reducing mobile parts, being required an actuator per finger versus the three required by the Manus hand. On the industrial side, other mechanical designs approach, is the Naist Hand (13) and the one by LiCheng (14), which are based in links and  differential mechanisms, providing force control and dexterity, but increasing weight an space, therefore a link base mechanism is propose due to simplicity and ease of configuration to emulate  a pre determinate trajectory, previously discussed, via  type and dimensional synthesis of four link mechanisms.

## 2.5    Mechanism Description

The mechanical finger is inspired by the structure presented in (12). Figure 5 shows the schematic of the design; which is divided in three four-link mechanisms. First a slider crank mechanism coupled to a linear actuator thought link 4 provides the linear displacement, converted to angular displacement at link 1. The second mechanism, a rocker crank type, will take the exit angle of link 5 as its input, generating an angular displacement on link 2, the last mechanism, takes its input from links 6 and 2 generating an angular displacement in link 3 which describes the final trajectory of the mechanism.

# 3    Dimensional Synthesis of the Mechanical Finger

For the proposed model the type and maximum size parameters are defined, because of this a dimensional synthesis is restricted by the anthropometrical dimensions previously stated.  To define the most suitable size for the coupler links while maintaining the desire trajectory in the finger Freudenstein equations are used. While the proposed architecture is restricted to a unique trajectory, which should be an approximation of a human like one, the grasping task should be possible according to section 2.3.

## 3.1    Synthesis of the Four Bar Mechanism for the Entry Displacement of the Finger

Each mechanism shall be analyzed individually to obtain the lengths for the coupling link which comply the desire trajectory while maintaining the anthropometry of the finger, at the end of the dimensional synthesis the proposed model will provide a trajectory similar to the one stated in (9), the proposed finger is shown in Figure 3.

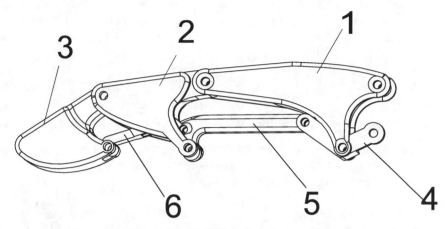

**Fig. 3.** Side view of the proposed finger

### 3.2    Synthesis of the Four Bar Mechanism for the Entry Displacement of the Finger

To obtain the values for the first mechanism, the coupling link length $a_3$ been the lengths of $a_2$ and $a_4$ are limited by the anthropometric conditions established in Table 2.

In the mechanism green lines show the ground of the system, the arrow represents the linear actuator and its direction, the red link is the crank and finally the blue line is the coupler which fully describes the first mechanism.

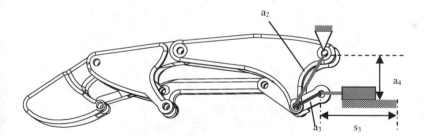

**Fig. 4.** Slider Crank Mechanism

To find the missing value for this mechanism the Freudenstein equation is employ as follows, first in its complex form and then in its trigonometric components.

$$a_2 e^{i\theta_2} + a_3 e^{i\theta_3} = s - ia_4$$

$$a_2(cos\theta_2 + isin\theta_2) + a_3(cos\theta_3 + isin\theta_3) = s - ia_4 \tag{1}$$

Now we separate the real and complex part, getting two equations and three variables, where $s$ is the position of the slider. Solving the system for a know value of $s$ the value of $a_3$ is:

$$a_3{}^2 = s^2 + a_4{}^2 + a_2{}^2 + 2a_2{}^2(a_4 sin\theta_2 - s\ cos\theta_2) \tag{2}$$

### 3.3    Synthesis of the Four Bar Mechanism for the proximal phalanx

The next mechanism is located in the proximal phalanx and is a four bar, rocker/crank type, which as described, takes its input from the link $b_4$, and starts the rotation of the finger with the knuckle as a fixed pivot, link $b_1$ in

Figure 5 represents the ground and link $b_5$ is the couple link which length is unknown. First the close loop equation has to be found, and then it is separated in its real and complex components. From the vector sum the equations are:

$$b_4(\cos\theta_4 + sin\theta_4) + b_3(\cos\theta_5 + sin\theta_5) = b_1 + b_2(\cos\theta_6 + sin\theta_6) \tag{3}$$

$$b_4(\cos\theta_4) + b_3(\cos\theta_5) = b_1 + b_2(\cos\theta_6) \tag{4}$$

$$b_4(\sin\theta_4) + b_3(\sin\theta_5) = b_2(\sin\theta_6) \tag{5}$$

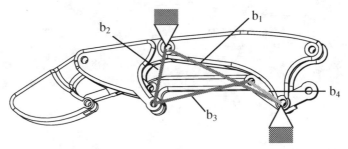

**Fig. 5.** Four link mechanism for the proximal phalanx

As in the last mechanism the only unknown mechanism is the coupler, so we solve the equation system for $b_3$ which in this particular case represents the coupler, obtaining the equation:

$$b_3{}^2 = b_1{}^2 + b_2{}^2 + b_4{}^2 + 2b_1 b_2 cos\theta_6 - 2b_1 b_2 cos\theta_4 - 2b_2 b_4(cos\theta_6 cos\theta_4 + sin\theta_6 sin\theta_4) \tag{6}$$

The mechanism in the distal phalanx, shown in Figure 6 is analyzed as the one in the proximal phalanx.

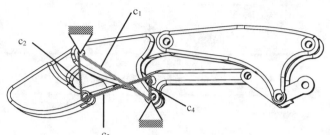

**Fig. 6.** Four link Medial phalanx mechanism

The displacement of the proximal phalanx generates a proportional displacement in the medial and distal ones, which may be expressed as a constant relationship between the medial and distal positions, therefore generating a single trajectory. The after mention trajectory is an approximation to that of the human finger, based in the dimension of the links presented in Tables 3 to 5, and the trajectory of the obtain mechanism is shown in Figure 7. Such an approach is based in the use of the geometrical method to describe a planar mechanism

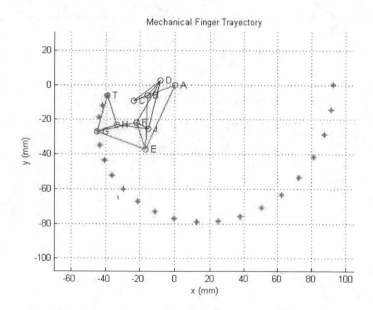

**Fig. 7.** Trajectory of the finger tip with the proposed dimension synthesis

The position in the Cartesian plane for each (X,Y) coordinates is calculated by a program, obtaining the trajectory shown in Figure 7, which can be seen is in accordance of that stated by (9) therefore obtaining a suitable trajectory for grasp tasks on a unstructured environment.

**Table 3.** Dimensions of each of the links in the first mechanism

| Link | Length [mm] |
|------|-------------|
| $a_2$ | 17 |
| $a_3$ | 8.77 |
| S | 20 |

**Table 4.** Dimensions of each of the links in the second mechanism

| Link | Length [ mm] |
|------|--------------|
| $b_1$ | 36.3 |
| $b_2$ | 10 |
| $b_3$ | 28 |
| $b_4$ | 12.5 |

**Table 5.** Dimensions of each of the links in the third mechanism

| Link | Length [ mm ] |
|------|---------------|
| $c_1$ | 23.5 |
| $c_2$ | 18.7 |
| $c_3$ | 21.5 |
| $c_4$ | 14.12 |

Each of the stated dimensions in tables 3 thru 5 are in accordance with the average size of a male adult hand in latinamerica , therefore length of each link may vary for a different region, gender and ethnicity. Once the appropriate length of the couplers links is calculated, the displacement of each phalanx in relation to the knuckle, the system reference, it follows a logarithmic curve therefore proving, the proposed hypothesis while the results are satisfactory the kinematic calculations obtained in the program are based on a geometrical model not an analytical one, and they are faster than the later approaches, where the input angles for each mechanism had to be calculated as in (15) or an iterative method as in (12) while the after mention methodology is valid it can be improved with the use of a single equation which describes the whole model, this haven been done in similar models based on the geometric analysis of the model the duction and abduction movements can be simulated by the use of a rotational matrix for the z axis of the model, generating a surface swept of the points in space that the finger is able to reach.

As the conditions to design a proper model for a mechanical finger are set, thru the mechanism type and dimensional synthesis, the mechanical part of the proposal may take place, fist a proposition according with the measures of the found links and anthropometric constrains is sketch in Solidworks™ which then have to take into consideration the fabrication process where tolerances, ease of manufacture and mechanical resistance is take into account, been the main design concern the deformation of the system due to the forces exerted in the system.

**Table 6.** Reaction Forces

| Group Selection | Units | Sum X | Sum Y | Sum Z | Result |
|---|---|---|---|---|---|
| Entire Body | N | 0.0153701 | 85.7848 | -51.975 | 100.302 |

**Table 7.** Normal stress , Displacement

| Name | Tipo | Min | Max |
|---|---|---|---|
| Normal Stress | Von Mises Stress | 0.172N/m^2 <br><br> Node: 25718 | 1.22201e+008 N/m^2 <br><br> Node: 26441 |
| Displacement | URES: | 0 mm <br> Node: 4516 | 0.104016 mm <br> Node: 451 |

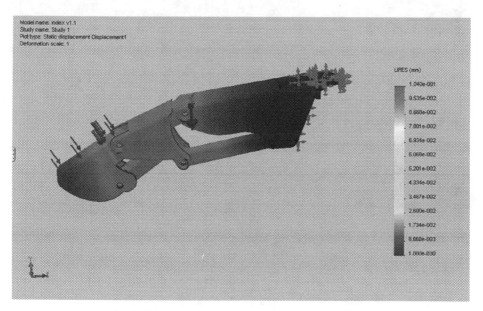

**Fig. 8.** Displacement in the model and its distribution

For a full stress of any given body it is needed to evaluate six variables for each of the differential particles e.g. three normal and three shearing strains, those variables are evaluated through an iterative process where changes in the model lead to a new behavior of the body subject to different loads, improving the resistance of the model in different load conditions

To analytically determine the strain at a specific point in a body with complex shape, boundary and properties, the FEM (Finite Element Method) is a common and appropriate approach. Base on such methodology a simulation of different loads, of 10, 100 and 200 Newtons on the model, therefore showing the possible deformations

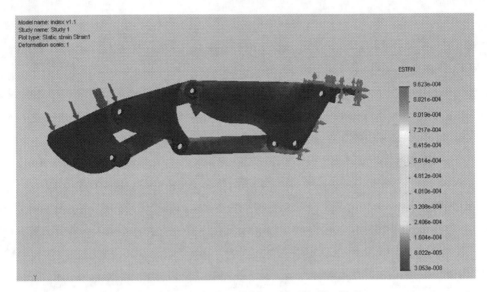

**Fig. 9.** Strain in the model and its distribution

in the model, the designed model is shown if all the proposed material is standard aluminum 1060, with no additional treatment, the material is choose  due  to its  a viability and ease of manufacture.

The conditions for the static analysis are as follows: the fixed point for the analysis is the knuckle, the  finger is  supposed to be in its fully extended position, therefore the link connected to the  actuator is supposed to be locked, where the finger is supposed to be connected to the palm which will support the entire assembly.

## 4    Conclusions

The anthropometry of the design  is  base  in the length of the main links 1, 2 and 3, therefore  the only  admissible  variables  should be  links 4, 5  and 6  which lengths must  be  obtain in compliance  with the desire trajectory for  grasp tasks, and  within the  boundaries in section 2.3, and  the  length of the overall finger. Other  variable  is the  position of the  junctures where  such links may be  assembled, as  shown by the simulation obtained and  shown in Figure 7, the proposed  synthesis  accomplish all the previous, After which a validation of the  model via FEM is  done.

As can be seen thru the realized simulations, de proposed design, does not present a meaningful deformation, even with a load 10 times the ones it may receipt in service. The maximum deformation  due  to  the  previously stated  load  is  of 0.104016mm , as  shown in  figure 9,  the load  is  100N uniformly distributed  on the upper side  of  the  finger  tip, this  decision is taken due  to the probability of  a collision by this particular  part in an under structure environment,  therefore  the rigidity of the model  is  adequate  for the previously considered applications, as a robotic manipulator or a prosthetic structure.

# References

1. OpenStax College. Muscles of the Pectoral Girdle and Upper Limbs(June 28, 2014), `http://cnx.org/content/m46495/latest/content_info#cnx_attrib ution_header` (Cited: July 14, 2014)
2. Jacobsen, S.C., Iversen, E.K., Knutti, D.F., Johnson, R.T., Biggers, K.B.: Design of the UTHA/M.I.T. dextrous hand. In: International Conference on Robotics and Automation, pp. 1520–1532. IEEE Xplore Press (1986)
3. Salisbury, J.J.: Articulated Hands: Force control and Kinematics issues. Int. J. Robot. Res. 4, 17 (1982)
4. Christopher, L., Xu, Y.: Actuability of UNderactuated Manipulators. Carnegie Mellon University, Pittsburgh (1994)
5. Caffaz, A., Cannata, G.: The design and development of the DIST-hand Dextrous Gripper. In: International Conference on Robotics and Automation, pp. 2075–2080. IEEE (1998)
6. Taylor, C.L., Schwarz, R.J.: The Anatomy and Mechanics of the Humand Hand. Artificial Limbs: A Review of Current Developments, pp. 2:22–2:35 (May 1955)
7. Tubiana, R.: Physiologie des Mouvements et Prehension, Paris. Massons (1980)
8. Gosselin, Laliberté, Clement: Simulation and design of Underactuated Mechanicla Hands. Mechanism and Machine Theory 33(1), 39–57 (1998)
9. Kamper, D.G., Cruz, E.G., Siegel, M.P.: Stereotypical Fingertip Trajectories During Grasp. J. Neurophysiol. 90, 3702–3710 (2003), doi:10.1152/jn.00546.2003
10. Mason, M.T., Salisbury, J.K.: Robot Hands and the Mechanis of manipulation. MIT (1985)
11. Pons, J.L., Rocon, E., Ceres, R.: The MANUS HAND dextrous robotics upper limb prosthesis: Mechanical and manipulation aspects. Autonomous Robots 16, 143–163 (2004)
12. Dechev, N., Cleghorn, W.L., Numman, S.: Multiple Finger Pasive Adaptative Grasp Prosthetic Hand. Mechanism and Machien Theory (1999)
13. Jun, U., Masahiro, K., Tsukasa, O.: The Multifingeres Naist hand system for robot in-hand manipulation. Mechanism and Machine Theory, 224–238 (2012)
14. Wu, L.C., Guiseppe, C., Marco, C.: Designing an underactuated mechanism for a 1 active DOF finger operation. Mechanism and Machine Theory, 336–348 (2009)
15. Velázquez-Sánchez, A.T., et al.: Síntesis de un mecanismo subactuado a partir de la función decriptiva del dedo índice, vol. 13(2), pp. 1665–0654. Instituto Politécnico Nacional, Mexico (2009)
16. Otto Bock HealthCare GmbH. Fascinated. With Michelangelo® – Perfect use of precision technology. Duderstadt/Germany: Otto Bock HealthCare GmbH (2011)
17. Gosselin, L., Birglen, C.: Grasp stability of 2-phalanx underactuated fingers. Journal of Robotic Research (2004)
18. The Hand Doctor Is. climbers finger. wordpress.org, `http://balourdas.com/wp/` (Cited: July 14, 2014)

# Design and Development of a Rotary Serial Elastic Actuator for Humanoid Arms

Xijian Huo, Yanfeng Xia, Yiwei Liu[*], Li Jiang, and Hong Liu

State Key Laboratory of Robotics and System, Harbin Institute of Technology,
150001 Harbin, China
lyw@hit.edu.cn, huoxijian@163.com

**Abstract.** The paper discusses the design of a novel serial elastic actuator (SEA) which is used as the revolute joint in humanoid arms. First, characteristics and requirements of SEA are illustrated. And the elastic element is further analyzed using the design of experiment (DOE) method. Then, a discoid elastic element with larger flexibility is developed. Based on this, an integrated SEA containing the position sensors and the torque sensor is developed. And also, a 7-DOF humanoid arm with joint flexibilities is designed. Finally, experiments of SEA demonstrate the approximation of the stiffness between the desired SEA and the prototype.

**Keywords:** Humanoid arm, Flexible joint, Serial elastic actuator, Joint stiffness, Design of experiment.

## 1 Introduction

Recently, robots are increasingly used in human environments to assist or co-work with people. Different from traditional industrial ones, the safety and compliance are the major aspect of these types of robot. By introducing compliant actuators, service robots can run into humans without generating forces high enough to cause damage. And exoskeleton robots for rehabilitation can assist human movements through restoring and releasing energies of patients. Thus, compliant actuators are important to guarantee better performance of robots in unstructured environments.

Many implementations of compliant actuators have been developed. And series elastic actuator (SEA) is one typical form. Various SEAs have been developed to meet specified requirements. A prismatic SEA with linear springs is developed in the walking robot [1]. Also, Knox B. T. *et al.* designed the rotary SEA containing a spiral torsion spring which is used in the biped robot [2]. Similarly, Lagonda C. *et al.* designed an electric SEA with a double spiral spring for a robot that would assist in gait rehabilitation training [3]. And a velocity sourced SEA is designed for human-robot interaction. The series elastic element containing springs is treated as a modular component that might be used with a range of motor systems [4]. In addition, Veneman J. F. *et al.*

---

[*] Corresponding author.

X. Zhang et al. (Eds.): ICIRA 2014, Part I, LNAI 8917, pp. 266–277, 2014.
© Springer International Publishing Switzerland 2014

constructed a torque actuator for exoskeleton robots by combining the standard SEA with a Bowden-cable-transmission [5]. Joonbum B. *et al.* proposed a rotary SEA for assisting human motions. This type of SEA consists of a DC motor, a torsional spring and two encoders [6]. And Oblak J. *et al.* developed a system for the arm and wrist rehabilitation using an SEA [7]. Besides, different rotary SEAs also developed for legged robot [8-9]. Paine N. *et al.* developed the University of Texas Series Elastic Actuator (UT-SEA) with the compact structure and light weight to enable high speed locomotion in actuated legged systems [10].

This paper is motivated by the compliant operations of the humanoid arm in human environments. And a 7-DOF humanoid arm with rotary SEAs is developed. The paper is organized as follows. In Section 2, the requirements of the SEA are proposed. And the analysis and modeling of the elastic element are performed in Section 3 and Section 4 respectively. Based on this, an integrated SEA is designed in Section 5. Experimental validations are demonstrated in Section 6. And Section 7 presents the conclusions.

# 2 Characteristic and Requirements of SEA

SEAs are actuators with an elastic element in series between the actuator and the actuated link. And the elastic element serves as the compliant part. The deflection of the elastic element corresponds directly with the force applied from the actuated link to the actuating link. Measuring the deflection gives the component of the force or torque acting along the joint. The concept of SEA is illustrated in Fig. 1.

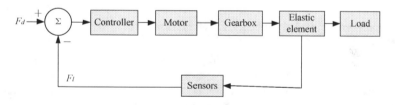

**Fig. 1.** Series elastic actuator topology

## 2.1 Characteristics of SEAs

Different from rigid actuators, SEAs provide better performances in special situations because of the elastic element. And the advantages are given as follows [11-12]:

(1) SEAs have lower passive impedance and forgiving shock tolerance when collisions occur.
(2) SEAs exhibit lower back-drivability. The dynamic effects of the motor inertia and gear train friction are nearly invisible at the output.
(3) The force transmission fidelity of the gear reduction or piston is no longer critical, allowing inexpensive gear reduction to be used. SEAs greatly increasing the fidelity and stability of force control.

(4) The motor's required force fidelity is drastically reduced, allowing inexpensive motors to be used. As a result, motors with large torque ripple can be used.

(5) Energy can be stored and released in the elastic element, potentially improving efficiency in harmonic applications.

## 2.2     Design Requirements

An integrated SEA prototype is developed for the 7-DOF humanoid arm. According to the human arm capability, a rotary SEA with multiple sensors is developed which is treated as the elbow joint. The desired specifications of the SEA are illustrated in Table 1.

**Table 1.** Specifications of the integrated SEA

| Parameter | Value |
|---|---|
| Nominal Output Torque | 70 Nm |
| Max angular velocity | 0.5 rad/s |
| Max angular acceleration | 1 rad/s$^2$ |
| Out Meter | < 120 mm |
| Length | < 150 mm |
| Mass (Kg) | < 2 Kg |
| Positional accuracy | ± 0.01° |
| Torque accuracy | < 2% |
| Stiffness | ± 10°(with the nominal torque) |

# 3     Design and Analysis of Elastic Elements

The stiffness of the integrated SEA is affected by the harmonic drive (gearbox), the torque sensor and the elastic element (torsional spring). With respect to the former two parts, the stiffness of the elastic element is lower. And the flexibility of the integrated SEA can be equivalent to that of the elastic part.

## 3.1     Parameters and Material

According to specifications in Table 1, the deflection of SEA is $10°$ corresponding to the nominal output torque. Hence, the desired stiffness of the elastic element is given by

$$K_{elastic} \approx K_{SEA} = \frac{\Delta T_N}{\Delta \theta} = 401.07 \left( Nm/rad \right) \tag{1}$$

The property of the elastic element of SEA is similar to that of torsional springs. Thus, the spring steel 50CrVA is the optimal selection for the elastic element. This material provides good mechanical properties and high hardenability. By adding the vanadium, the steel grain is refined and the heat sensitivity is reduced. Hence,

50CrVA has better strength and higher yield strength. The specifications of 50CrVA are shown in Table 2.

**Table 2.** Specifications of 50CrVA

| Parameters | Value |
|---|---|
| Density | 7850 Kg/m$^3$ |
| Poisson ratio | 0.33 |
| Young's modulus | $2 \times 10^5$ Mpa |
| Yield strength | 1320 Mpa |

## 3.2    Structure Design and Optimization

The elastic element of SEAs has different forms and structures, and the typical parts include the discoid element and the cylindrical element. In order to achieve design optimizations of the structure and parameters, design of experiments (DOE) in Workbench is adopted. Based on the Monte Carlo sampling techniques, DOE calculates the response of the parameterized input parameters. And then, the response surface corresponding to the design variable can be constructed using quadratic interpolation function. As an optimization method, DOE performs the hypothetical research that many input parameters varies within a certain range. And the evaluations of each group of parameters should be performed and considered. With respect to the relationship between responses and variables, optimal parameters can be determined directly.

## 3.3    Finite Element Analysis of the Elastic Element

Rotary discoid elastic elements usually have the larger diameter/length ratio. The element is located between the torque sensor and the output, as shown in Fig. 2.

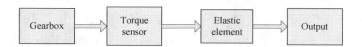

**Fig. 2.** Assemble relationship of the discoid elastic element

This type of elastic element has simple structure. It consists of the input and output connections, and several elastic areas, as shown in Fig. 3. The flexibility of the element originates mainly from the elastic area around the circumference. When the load exceeds the nominal output, collisions between the elastic area and the output connection occur to keep from being damaged.

In order to determine optimal structure of the elastic part, finite element analysis (FEA) is performed. The structural parameters contain four variables including the thickness $t$ of the element, the width $b$ of the elastic area, the angular interval $\alpha$ and the angular interval $\beta$, as shown in Fig. 3. Ranges of the variables are defined as $t = 9 \sim 12 \text{(mm)}$, $b = 1.9 \sim 2.1 \text{(mm)}$, $\alpha = 305° \sim 315°$, $\beta = 80° \sim 90°$.

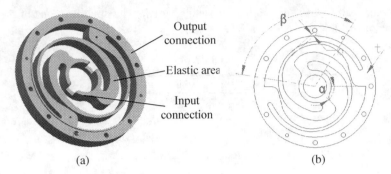

(a)                                                    (b)

**Fig. 3.** Type I elastic element including (a) CAD model of the part, (b) structural parameters

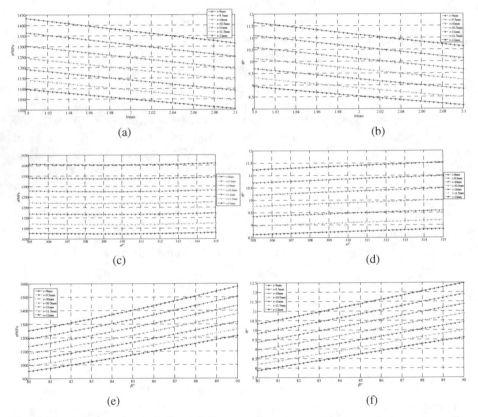

(a)                                                    (b)

(c)                                                    (d)

(e)                                                    (f)

**Fig. 4.** Simulation results, including (a) changes of stress caused by $t$ and $b$ corresponding to $\alpha = 310°$, $\beta = 85°$; (b) changes of the angular defection caused by $t$ and $b$ corresponding to $\alpha = 310°$, $\beta = 85°$; (c) changes of stress caused by $t$ and $\alpha$ corresponding to $b = 1.95(\text{mm})$, $\beta = 85°$; (d) changes of the angular defection caused by $t$ and $\alpha$ corresponding to $b = 1.95(\text{mm})$, $\beta = 85°$; (e) changes of stress caused by $t$ and $\beta$ corresponding to $b = 1.95(\text{mm})$, $\alpha = 310°$; (f) changes of the angular defection caused by $t$ and $\beta$ corresponding to $b = 1.95(\text{mm})$, $\alpha = 310°$.

Based on this, static simulations using Workbench are performed corresponding to the nominal output load and structural variables. Using the DOE method, the relationship between the stress and structural parameters is expressed in Fig. 4. The figure also illustrates changes of the angular deflection simultaneously.

With respect to Fig. 4, it is impossible to achieve the desired stiffness within the Yield strength of the material. And summarization can be expressed as follows.

(1) Corresponding to a specified $\alpha$, $\beta$, $b$, the maximal stress $\sigma$ and the angular deflection $\theta$ decrease with increasing the variable $t$;

(2) Corresponding to a specified $\alpha$, $\beta$, $t$, the maximal stress $\sigma$ and the angular deflection $\theta$ decrease with increasing the variable $b$;

(3) Corresponding to a specified $b$, $\beta$, $t$, the maximal stress $\sigma$ increases slightly with increasing the variable $\alpha$, and the angular deflection $\theta$ increases with increasing the variable $\alpha$;

(4) Corresponding to a specified $\alpha$, $t$, $b$, the maximal stress $\sigma$ and the angular deflection $\theta$ increase significantly with increasing the variable $\beta$.

Based on above mentioned analysis, the structural parameters of the elastic element are defined as $t = 9.5(\text{mm})$, $b = 1.95(\text{mm})$, $\alpha = 315°$ and $\beta = 82°$. The simulation results in Workbench are performed in Fig. 5. It can be found that the maximal angular deflection is about 10° while the stress is about 1180Mpa corresponding to the nominal output load. Hence, this type of discoid elastic element can meet the design requirements of SEAs.

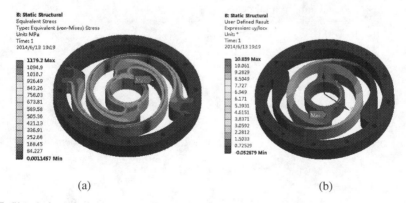

(a)                                                    (b)

**Fig. 5.** Simulation results corresponding to the nominal output load, including (a) the stress $\sigma$ and (b) the angular defection $\theta$

# 4    Stiffness of the Elastic Element

In order to determine the property of the stiffness, the angular deflection $\theta$ is simulated in Workbench corresponding to different output torque within the nominal load. The output torque ranges from -70Nm to 70Nm. And the corresponding deflections are illustrated in Table 3.

**Table 3.** Angular deflection corresponding to different output torque

| Output torque / Nm | Angular deflection $\theta / °$ |
|---|---|
| 70 | 10.839 |
| 60 | 9.182 |
| 50 | 7.559 |
| 40 | 5.976 |
| 30 | 4.432 |
| 20 | 2.920 |
| 10 | 1.443 |
| -70 | -9.215 |
| -60 | -7.987 |
| -50 | -6.733 |
| -40 | -5.448 |
| -30 | -4.133 |
| -20 | -2.787 |
| -10 | -1.410 |

## 4.1    Analysis of Constant Stiffness

According to Table 3, the relationship between the output torque $T$ and the deflection $\theta$ is determined. And the explicit formulation about the output torque can be derived using the least square method.

$$T = 399.84|\theta| - 2.31(\mathrm{Nm}) \tag{2}$$

And the linearity is given by

$$\rho = \frac{\max|\Delta T|}{T} \times 100\% = \frac{3.381}{70} \times 100\% = 4.83\% \tag{3}$$

Since the stiffness of the SEA approximate that of the elastic element, thus

$$K_{SEA} \approx K_{elastic} = 399.84(\mathrm{Nm/rad}) \tag{4}$$

With respect to (1) and (4), the stiffness of the SEA closely approximates that of the elastic element. And the designed element can meet the stiffness requirement of the SEA.

## 4.2    Analysis of Variable Stiffness

In order to achieve better approximation of the stiffness, the model of variable stiffness is established. According to Table 3, relationship between the output torque $T$ and the corresponding stiffness is determined. And the explicit formulation about the output torque can be derived using the least square method.

$$K_{SEA} = -0.466|T| + 402.105 (\text{Nm/rad}) \tag{5}$$

And the linearity is given by

$$\rho = \frac{\max|\Delta K|}{K} \times 100\% = \frac{0.533}{435.237} \times 100\% = 0.12\% \tag{6}$$

Hence, the model with variable stiffness has better linearity. The stiffness of the elastic element is proportionate to output torque. And the stiffness of the SEA approximate that of the elastic element, thus

$$K_{SEA} \approx K_{elastic} = -0.466|T| + 402.105 (\text{Nm/rad}) \tag{7}$$

With respect to (1) and (7), the stiffness of the SEA approximates that of the elastic element. Based on a degree of oversensitivity, the designed element can meet the stiffness requirement of the SEA.

## 5    Development of Integrated SEA

In order to achieve better performance during works, the SEA is a highly integrated with mechanics, electronics and multisensory systems, such as torque sensor, position sensors, etc., as shown in Fig. 6. The integrated SEA can be divided into two subsystems: the input and the output. The stator of motor is fixed to the input, while the rotator is connected to the output via the gearbox. Sensors integrated into SEAs gather information about positions and output torque. The fail-safe brake is fixed to the input and used to stop the output revolution against the input while power on. The electronics and controls, including the drive board and the control board, are also integrated into the SEA. The internal connection between adjacent elements is shown in Fig. 7

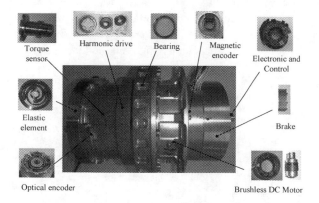

**Fig. 6.** The critical component selected in the integrated SEA

Based on the integrated SEA, a 7-degree-of-freedom humanoid arm is designed. The humanoid arm is built to handle 10 Kg payload with full arm extension in 1-g condition. And total weight is less than 18.5 Kg. The configuration of the humanoid arm is shown in Fig. 8.

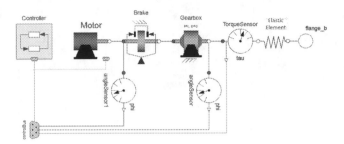

**Fig. 7.** The internal connection of the integrated SEA

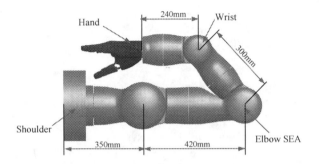

**Fig. 8.** CAD model of the humanoid arm

## 6    Experimental Validation

Experiments are carried out to validate the performance of the integrated SEA. The testing platform is shown in the Fig. 9. The platform mainly contains five parts which are the CPU, power system, data collection and disposing system, the integrated SEA, the oscillograph and the PCI card. The input of the SEA is fixed to the platform and the motor is locked while the output connecting with the load.

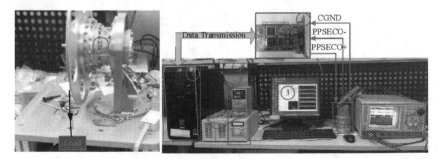

**Fig. 9.** Testing platform of the integrated SEA

The load and the corresponding change of the output position are observed in experiments. Based on this, the stiffness of the SEA can be demonstrated.

If the stiffness of the SEA is considered as a constant, the linear relationship between the output torque and the angular deflection can be illustrated in Fig. 10. And the explicit formulation about the two variables can be derived using the least square method.

$$T = K_s |\theta| + d = 342.122|\theta| - 0.950 \, (\text{Nm}) \tag{8}$$

In indicates that the real stiffness $K_s$ of the SEA is $342.122 (\text{Nm/rad})$. With respect to the desired stiffness of the SEA in (1), the relative error is given by

$$\varepsilon = \left| \frac{K_s - K_{SEA}}{K_{SEA}} \right| \times 100\% = 14.69\% \tag{9}$$

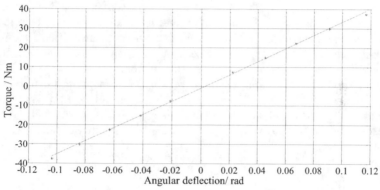

**Fig. 10.** Relationship between the load and the corresponding angular deflection

If the stiffness of the SEA is considered as a variable, the linear relationship between the output torque and the stiffness can be illustrated in Fig.11. And the explicit formulation about the two variables can be derived using the least square method.

$$K_s = -0.505|T| + 348.877 \, (Nm / rad) \tag{10}$$

And the linearity of the stiffness is given by

$$\rho = \frac{\max |\Delta K_s|}{K_s} \times 100\% = \frac{3.760}{365.589} \times 100\% = 1.03\% \tag{11}$$

According to (1) and (11), it is clear that the linear relationship between the stiffness and the output torque performs better. With respect to the desired stiffness of the SEA, the relative error is given by

$$\varepsilon_{\max} = \max \left| \frac{K_s - K_{SEA}}{K_{SEA}} \right| \times 100\% = 21.83\% \tag{12}$$

Whether the stiffness of the SEA is considered as a constant or variable, it is obvious that the real stiffness of the SEA is significantly lower than the desired analysis.

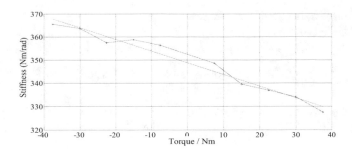

**Fig. 11.** Relationship between the load and the stiffness

This discrepancy is probably due to the actual properties of the material being different from the nominal used in the simulation and the imperfections in the model and mesh used in the analysis. And the mismachining tolerance also affects the stiffness of the elastic element. Besides, the stiffness of the SEA contains the effect of the torque sensor, the harmonic drive and the elastic element.

Another experiment for output position tracking corresponding to different loads with the constant velocity control is performed, as shown in Fig. 12. It is obvious that the tracking performance with no load is better than that of output load. This is because the chatter caused by the SEA flexibility goes against the stability of the SEA control.

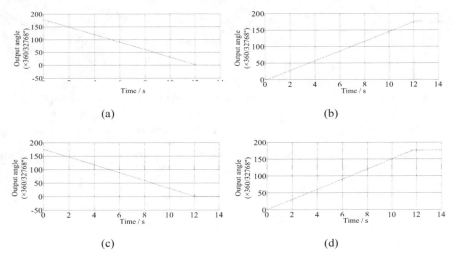

**Fig. 12.** Position tracking of the SEA with deferent loads, including: (a) output position without loads in the positive direction; (b) output position without loads in the opposite direction; (c) output position with a load of 15Nm in the positive direction; (d) output position with a load of 15Nm in the opposite direction.

## 7    Conclusions

In this paper, an integrated SEA with electronics and multisensory systems is designed and developed for the humanoid arm. According to the requirements of the humanoid arm operations, the discoid rotary elastic element is designed. Based on the DOE method, optimal structural parameters of the elastic element are determined. And models with the constant stiffness and the variables stiffness are completed respectively. And then, the integrated SEA and the 7-DOF humanoid arm are designed. Experiments demonstrate the stiffness of the elastic element approximates that of the SEA prototype. This SEA prototype is adequate for applications of the humanoid arms.

**Acknowledgments.** This work was supported in part by National Natural Science Foundation of China (51175106) and National 973 Program (2011CB013306).

# References

1. Robinson, D.W., Pratt, J.E., Paluska, D.J., et al.: Series Elastic Actuator Development for a Biomimetic Walking Robot. In: International Conference on Advanced Intelligent Mechatronics, pp. 561–568. IEEE Press, Atlanta (1999)
2. Knox, B.T., Schmiedeler, J.P.: A Unidirectional Series-Elastic Actuator Design Using a Spiral Torsion Spring. Journal of Mechanical Design 131, 125001 (2009)
3. Lagonda, C., Schouten, A.C., Stienen, A.H.A., et al.: Design of an Electric Series Elastic Actuated Joint for Robotic Gait Rehabilitation Training. In: International Conference on Biomedical Robotics and Biomechatronics, pp. 21–26. IEEE Press, Tokyo (2010)
4. Wyeth, G.: Control Issues for Velocity Sourced Series Elastic Actuators. In: Proceedings of the Australasian Conference on Robotics and Automation (ACRA). Australian Robotics and Automation Association Inc. (2006)
5. Veneman, J.F., Ekkelenkamp, R., Kruidhof, R., et al.: A Series Elastic- and Bowden-Cable-Based Actuation System for Use as Torque Actuator in Exoskeleton-Type Robots. International Journal of Robotics Research 25(3), 261–281 (2006)
6. Joonbum, B., Kyoungchul, K., Masayoshi, T.: Gait Phase-Based Smoothed Sliding Mode Control for a Rotary Series Elastic Actuator Installed on the Knee Joint. In: 2010 American Control Conference (ACC), pp. 630–6035. IEEE Press, Baltimore (2010)
7. Oblak, J., Cikajlo, I., Matjacic, Z.: Universal Haptic Drive: A Robot for Arm and Wrist Rehabilitation. IEEE Transactions on Neural Systems and Rehabilitation Engineering 18, 293–302 (2010)
8. Simon, C., Brian, T.K., James, P.S., et al.: Design of Series-Elastic Actuators for Dynamic Robots with Articulated Legs. Journal of Mechanical Design 1, 011006 (2009)
9. Yin, P., Li, M., Guo, W., et al.: Design of a Unidirectional Joint with Adjustable Stiffness for Energy Efficient Hopping Leg. In: Proceeding of the IEEE International Conference on Robotics and Biomimetics (ROBIO), pp. 2587–2592. IEEE Press, Shenzhen (2013)
10. Painc, N., Sentis, L.: A New Prismatic Series Elastic Actuator with Compact Size and High Performance. In: Proceeding of the IEEE International Conference on Robotics and Biomimetics (ROBIO), pp. 1759–1766. IEEE Press, Guangzhou (2012)
11. Arumugom, S., Muthuraman, S., Ponselvan, V.: Modeling and Application of Series Elastic Actuators for Force Control Multi Legged Robots. Journal of Computing 1, 26–33 (2009)
12. Paluska, D., Herr, H.: Series Elasticity and Actuator Power Output. In: Proceedings of the 2006 IEEE International Conference on Robotics and Automation (ICRA), pp. 1830–1833. IEEE Press, Florida (2006)

# Human Manipulator Shared Online Control Using Electrooculography

Jinhua Zhang, Baozeng Wang, Jun Hong[*], Ting Li, and Feng Guo

State Key Laboratory for Manufacturing Systems Engineering
Xi'an Jiaotong University
Xi'an, 710049, China
jhong@mail.xjtu.edu.cn

**Abstract.** This paper presents a shared online control method of 7 Degrees-of-Freedom (DOF) articulated manipulator based on electrooculography (EOG). Firstly, based on the previous signal offline analysis research in "Linear Decoding of Eye Gazing Target Continuous Motion Information via Electrooculography", the signal online processing methods are proposed here including online calibration, subsection processing and incremental output. Then, the interactive interface is designed and the control strategy is made to realize the manipulator is controlled by smooth pursuit eye movement and blink. Finally, the experiments of the manipulator end motion path control are carried out to verify the control scheme. The simulation and experimental results showed a good fit with the ideal path and demonstrated the effectiveness of control human manipulator. The new methods are expected to be widely used in control human manipulator with EOG to help disabled patients in practical clinical application to improve the quality of life of handicapped people.

**Keywords:** Human-machine interaction, Electrooculography (EOG), Articulated Manipulator, Shared control.

## 1    Introduction

There has been an increase in the development of human-computer-interaction (HCI) based on human body bioelectric signals which included electrooculography (EOG) , electromyography (EMG), electroencephalography (EEG) and so on.[1]. As the simplest bioelectric signal, EOG allows us to detect eye movements by measuring the voltage difference between the cornea and the retina. Owing to the ease of detection and high signal-to-noise ratio (SNR), it has great study potential and good prospects for wide application, so EOG signals are used to control external devices in Brain–Computer Interfaces (BCI) [2].

Mainly, the research using EOG-based HCI focuses on two aspects. The first is the pattern recognition [3] [4] of EOG signals produced by different eye movements especially for blinks and saccades. But for the less information, the controls are li-

---

[*] Corresponding author.

X. Zhang et al. (Eds.): ICIRA 2014, Part I, LNAI 8917, pp. 278–287, 2014.

mited and complex. The second is the relationship of the learning models between eye movements and EOG signals. For example, a linear model of eye saccade angles and EOG amplitudes using linear fitting was built by He et al. [5]. These researches are all applied to the control of robots [6], [7], wheelchairs [8], [9] and so on. However, most researches ignore the smooth pursuit eye movement and cannot map the EOG voltages on the eye gazing target positions or eye movement information directly.

A shared control manipulation system combines some of the autonomy of supervised systems with the presence found in direct master-slave bilateral systems which has a wide range of application in BCI. To help operators control robotic equipment, it is usually used with multi-agent systems on cooperative tasks [10], [11]. Shared control provides a framework for extending the capabilities of an immersive manipulation system [12].Shared control can increase safety and decrease between the handicapped people and assistive robot, and it has also has increased the level of user acceptance, due to the personal dynamics of the control policy.

In this work, the blink detection algorithm was introduced and the eye movement decoding model was proposed to reconstruct the human eye gazing positions in 2D space from smooth pursuit EOG signals [1]. On the basis of them, the online processing of the smooth pursuit EOG signals was focused on, and a share control of an articulated manipulator was presented by this paper. This could eventually allow disabled people control robots to help them in their daily activities.

The outline of the paper is presented as follows: The second part describes the EOG signal processing and online output realization. The third part shows the human manipulator shared control strategies and the control experiment. The fourth part shows the experiment results and discussion and the fifth part draws the final conclusions.

## 2    System Architecture

The control platform system could be seen in Figure 1. To measure eye signals, two pairs of electrodes are typically placed either above and below the eye or to the left and right of the eye. The EOG signals acquired from the human eye movement are recorded by amplifier and processed by computer. Then the extracted signal features

**Fig. 1.** Control platform system

are transformed into the control commands according to the control strategy, and the control commands applied to control the manipulator (RBT-6T/S01S, SuZhouBoShi Robotics Technology Co., Ltd., China). The manipulator motion responses are captured by the camera (DLC300-L, Digital lab Inc., China) and the image frames are transferred back to the computer screen. The screen frames is fed back to user's eyes through visual sense to come into being a closed-loop control system. So the control platform system was built complete and Users can continue rotate of their eye to execute next control via the control platform system.

## 3    Signal Acquisition and Processing

### 3.1    Signal Acquisition, Pre-processing, and Decoding

In the research of "Linear Decoding of Eye Gazing Target Continuous Motion Information via Electrooculography", the authors have introduced the acquisition, blink detection algorithm, pre-processing, and decoding model of the EOG signals.

By placing two pairs of skin electrodes on the opposite sides of the eye, the EOG signals are recoded through a NeuroscanNuAmps Express System (Compumedics Ltd., VIC, Australia). The horizontal EOG signal (HEOG) and the vertical EOG signal (VEOG) could be obtained by channel difference:

$$\begin{cases} U_H(t) = U_{HR}(t) - U_{HL}(t) \\ U_V(t) = U_{VU}(t) - U_{VL}(t) \end{cases} \tag{1}$$

where: $U_H$ is the horizontal EOG, $U_V$ is the vertical EOG, $U_{HR}$, $U_{HL}$, $U_{VU}$, and $U_{VL}$ are the signal voltages collected by electrodes HR, HL, VU, and VL as shown in Figure 2.

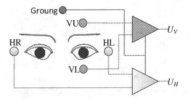

**Fig. 2.** EOG signal acquisition and electrodes layout

Based on the wavelet transform and threshold processing, the blink detection algorithm could separate the voluntary single-blink and double-blink information from raw VEOG, and obtain the blink (including voluntary and involuntary blink) removed VEOG signals by linear interpolation (Figure 3).

After blink detection, both HEOG and VEOG signals are de-noised by median filter with 150ms window size, then they would be used to decode. Then subsection processing and increment control were carried out to control the human manipulator. Most EOG researches ignore the smooth, gradual EOG signals. Hence, we tried to use a linear model to extract movement information of the observable from EOG signals. The model is shown below:

$$\begin{cases} h(t) = a_0 + \sum_{k=1}^{n-1} a_k U_H(t-k) + \varepsilon_h(t) \\[2em] v(t) = b_0 + \sum_{k=1}^{n-1} b_k U_V(t-k) + \varepsilon_v(t) \end{cases} \tag{2}$$

where $h(t)$ and $v(t)$ are the decoded space positions of the observable at time $t$ in the horizontal and vertical directions; $n$ is the number of lags, or the order of the model, here set to 10; $UH(t-k)$ and $UV(t-k)$ are the preprocessed and blink-removed EOG signals in both directions at time $(t-k)$; $a$, $b$ are the parameters; and $h$, $v$ are the residual errors in both directions.

The adaptive Recursive Least Square algorithm was used to calibrate the model parameters, and the cross-validation procedure was used to assess the model feasibility. The decoding accuracy was within 2 cm. It could not be applied to accuracy control, but preferable retained the smooth pursuit eye movement trend and could be used to external control with the aid of the interactive interface.

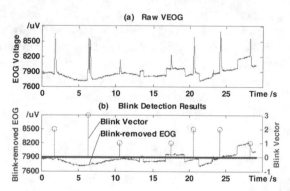

**Fig. 3.** Blink detection. In blink vector, a value of 1 means an involuntary blink, 2 is a voluntary single-blink, and 3 is a voluntary double-blink

## 3.2    Online Processing

The acquisition system provided a network-service/client-based interface to help with reading the recorded data in real-time. Self-developed processing and control algorithms were added to produce online control by the control system. The sample rate was 500 Hz, and one sample block contained 20 points for each channel.

The users need to face the computer screen, and their eyes should gaze at the screen center horizontally without strabismus. To perform online control, several problems should be resolved. First, the decoding and control effects were not very stable because of individual variation. Second, only when the number of blink activities is counted accurately, the detection algorithm works well, especially for double-blinks. Third, the baseline drifting is far-reaching in online handling and cannot avoid in decoded positions. To solve these problems, online calibration, subsection processing, and incremental output are proposed.

### 3.3     Online Calibration

Before control starting, a calibration dialog is shown to begin the calibration procedure as shown in Figure 4. Users initially face the screen center without head movement. After 5 seconds of stillness, a circular cursor which initialized in the screen center begins moving slowly along the predefined trajectory, and the participant should follow it smoothly with their eye gazing the circular cursor. The EOG signals are saved into a text file simultaneously. When the cursor reaching the end of the trajectory, after another 5 seconds of stillness, the calibration dialog is closed automatically and the saving of EOG signals is terminated. Then, using the adaptive Recursive Least Square algorithm, the model parameters were solved and saved. The duration time of the calibration is about 16 s.

### 3.4     Subsection Processing

In online processing, a blink may be across two data segments, especially for a double-blink where one blink is in the first segment, and another in the second. This cannot be correctly detected. So, the subsection processing is introduced. The system takes 30 blocks data as one frame, and the frame shift is one-third frame length, i.e. 10 blocks data. As shown in Figure 5, one frame is divided into three sections on average. The whole frame length data are processed by blink detection and decoding, but only the results of the middle section are output to control human manipulator. After 10 new sample blocks are read, i.e. the frame shift length, a frame of the newest 30 blocks data is processed as well.

The output information is continuous without any loss of information. To be convenient for output control and to reduce the control frequency, the average of the decoded positions of the middle section is taken and used as the output for this time point. The time interval between two outputs is about 0.4 s, and the time delay of the control is about 0.8 s.

### 3.5     Incremental Output

The decoded positions from the blink removal and de-noising singles can be seen as in absolute coordinates with baseline shifting and errors accumulating unavoidably. To correct for these errors, the incremental control which adopts relative coordinates is adopt. Initially, users have to face the screen center without strabismus as mentioned before. The current output position should subtract the previous one to get an increment or relative coordinate. The first position decoding is not output. The incremental control can eliminate the drift efficiently and improve the control flexibility. Moreover, the observable resetting to the origination, the accumulated errors could be eliminated by resetting triggered through a voluntary single-blink and meanwhile eye gazing coming back to the screen center.

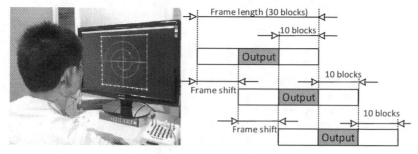

**Fig. 4.** Online calibration          **Fig. 5.** Segments processing

# 4    Manipulator Control

## 4.1    Control Interface and Strategy

The EOG signals which were processed are used to control an articulated manipulator. To facilitate the interactive control, the interface and control strategy are designed to improve the inactive performance. Expecting the images transferred by the camera in the interface, the assisting elements are added. Three modes are designed to accurate control with their labels in right of the interface: "Reset", "End Control", and "Re-calibration". The virtual scene modeling methods in terms of virtual scene modeling method based on Visual C++ and virtual scene modeling method based on MATLAB [13]. The chosen mode label is in red and green color. Before the control starts, the online calibration is conduct first to obtain the decoding parameters as has been said in the third section. Then the articulated manipulator control interface will be shown in the Figure 6.

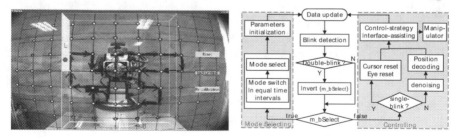

**Fig. 6.** Interface under the "End Control" mode          **Fig. 7.** Control strategy

The control strategy can be seen in Figure 7. The control mode selecting state and the controlling state can be switched by double-blink.

If the system is in mode selecting state, the interface will be shown "Mode Select" to hint user, and the three control mode will be loop switched in equal time intervals of 1 s. When the mode the user needed is in selected, user should be double-blink to switch into controlling state, i.e., user will come into the selected control mode and the interface will be refreshed into the corresponding style.

When the system state is controlling, if the mode is "Reset", the articulated manipulator will be reset to the initialized pose. If the mode is "Re-calibration", the online calibration will obtained again the new decoding parameters. After the procedure of the two mode finished, the system will be come back to the mode selecting state.

If the mode is "End Control", as shown in Figure 6, a red circle cursor is initialized in the screen center and controlled by incremental output positions decoded from user's smooth pursuit eye movements. The interface will be surrounded into 7 areas by the closed boundary expect for the center area. These areas are tagged by "L/R/U/D/F/B/Paw" which means 7 different commands for articulated manipulator control. When the cursor comes into these areas, the corresponding part will become yellow. Anytime when the cursor enters into the center area, the articulated manipulator will be stop. The area of "Paw" means the open/close of the paw in the end of the articulated manipulator. If the cursor come into the area of "L/R/U/D/F/B" and stay in the area for 0.5 s, the cursor will be locked in this area, and the articulated manipulator end will be triggered to motion towards the direction of "left/right/up/down/ front/ back" in the given speed until the user conduct a voluntary single-blink. Then the cursor will be reset to the center of the screen, i.e. the center area, the articulated manipulator will be stopped. Meanwhile user eyes should be gazing back at the cursor on the screen center, and do next eye movement control. This can correct and reduce the error accumulation.

However, for most unavoidably effectors, for example, the articulated manipulator has 7 degrees of freedom (DOF), a more sophisticated control strategy is required to accomplish the control articulated manipulator at more complex level. The strategy is enough to do a basic inverse kinematics control of the articulated manipulator [01].

### 4.2    Control Experiments

To verify the feasibility of every scheme, the articulated manipulator end motion path control experiment was conducted. An assisting transparent panel was placed between the articulated manipulator and camera, with a red path drawn on its surface (Figure 1). The path can be seen in the screen interface clearly (Figure 6).

Three subjects were required to control the cursor on the interface into different areas with their eye movements to trigger the articulated manipulator end motioning along the path on the interface. Each subject conducts the task once. They have 10 min to be familiar with the control scheme.

## 5    Results and Discussion

### 5.1    Control Results

To get better control articulated manipulator, we checked the control system by carrying cursor control experiments. One control result of the three randomly for each participant was extracted and the cursor movement trajectories were shown. Control based on EOG saccade signals can also be used with similar tasks.

Different areas of the interface which were orthogonal targets (top, bottom, left, and right) and corner targets (left top, left bottom, right top, and right bottom) were designed as shown in Figure 6. Users control the cursor quick access between distinct areas. Five participants controlled the cursor by slowly moving their eyes to hit the predefined target areas according to user's intention. Meanwhile, the cursor movement trajectories were recorded with the acquisition system. The control trials were run three times for each participant and the statistical results can be seen in Table 1. The hit rate was 100% for the four orthogonal targets, and 96.7% for the four corner targets, which means that the targets in the orthogonal directions are more likely to be hit. The overall hit rate was 98.3%, showing a good control effect.

**Table 1.** Hit rates of the cursor control

|  | Orthogonal targets | Corner targets | Total |
|---|---|---|---|
| Hit times/total | 60/60 | 58/60 | 118/120 |
| Hit rate | 100% | 96.7% | 98.3% |

The articulated manipulator initialization pose and the ideal path reconstructed by the software of Matlab can be seen in Figure 8 (a). And the control results of the three subjects are shown in Figure 8 (b). All control paths had a good fit with the ideal path, which indicate the feasibility of the scheme.

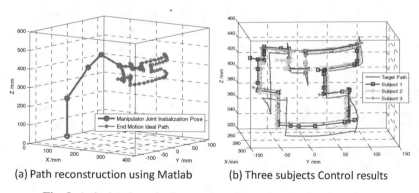

(a) Path reconstruction using Matlab      (b) Three subjects Control results

**Fig. 8.** Articulated manipulator end motion path control and result

One participant finished the path control cost about 9 min at first time, but only 5 min at the second time, which showed the system's easy to use and master. Furthermore, compared to saccade eye movement which needs the eye muscle violent contraction and tends fatigue, the smooth pursuit are relaxed for human eyes and can be used for long time operation.

## 5.2    Control Robustness

For the interactive system, the control stability is affected by the normalization of the user's eye movement and the acquisition of the signal processing. The eye movements should be standard to tend to be easy detected. For the users, the bioelectricity signals are acquired from every user at the surface of the skin, and eye movement are closely related to the mental activity. Affected by physiological and psychological conditions, the signals themselves have certain instability.

For the external environment, environmental variation or external interference may cause attention diversion or head moving, which will influence the control effects.

When the control is not ideal, the user should suspend it and adjust their psychology and mentality to reduce the impact of the subjective factors. On the other hand, some improvement in optimizing the control strategy, improving the signal processing algorithm and practicing more can effectively reduce unstable factor, the control robustness will be further enhanced.

## 6    Results and Discussion

The online calibration, subsection processing, and incremental output to realize the signal online processing were proposed based on the previous research. Then the articulated manipulator was control by the eye movement to do an end path motion to verify the scheme with the aid of the interface.

This technique could eventually allow severely disabled people to control robots only with the movements of their eyes, which can help them in activities of daily life.

**Acknowledgment.** The research work described in this paper is supported by a Grant from the National Nature Science Foundation of China (No.50905136) and Academy of Fundamental and Interdisciplinary Sciences, Xi'an Jiaotong University.

## References

1. Jinhua, Z., Feng, G., Jun, H., et al.: Linear Decoding of Eye Gazing Target Continuous Motion Information via Electrooculography. Journal of Xi'an Jiaotong University 47(12), 123–129 (2013)
2. Deng, L.Y., Hsu, C.L., Lin, T.C., et al.: EOG-based Human–Computer Interface system development. Expert Systems with Applications 37(4), 3337–3343 (2010)
3. Bulling, A., Ward, J.A., Gellersen, H., et al.: Eye movement analysis for activity recognition using electroocu-lography. IEEE Transactions on Pattern Analysis and Machine Intelligence 33(4), 741–753 (2011)
4. Lv, Z., Wu, X., Li, M., et al.: A novel eye movement detection algorithm for EOG driven human computer interface. Pattern Recognition Letters 31(9), 1041–1047 (2010)
5. He, Q., Zhang, J., Li, T., et al.: Position Feedback of Bionic Manipulator Following Electrooculography Signals. Journal of Xi'an Jiaotong University 1(011) (2012)
6. Iáñeza, E., Azorína, J.M., Fernándezb, E., Úbeda, A.: Interface based on electrooculography for velocity control of a robot arm. Applied Bionics and Biomechanics 7(3), 199–207 (2010)
7. Ubeda, A., Ianez, E., Azorın, J.M.: Wireless and Portable EOG-Based Interface for Assisting Disabled People. IEEE/ASME Transactions on Mechatronics 16(5), 870–873 (2011)

8. Barea, R., Boquete, L., Mazo, M., et al.: EOG guidance of a wheelchair using neural networks, pp. 668–671. IEEE (2000)
9. Barea, R., Boquete, L., Mazo, M., López, E.: Wheelchair Guidance Strategies Using EOG. Journal of Intelligent and Robotic Systems 34(3), 279–299 (2002)
10. Iáñez, E., Úbeda, A., et al.: Assistive robot application based on an RFID control architecture and a wireless EOG interface. Robotics and Autonomous Systems 60, 1069–1077 (2012)
11. Yongsheng, G., Jie, Z., Jihong, Y., Yanhe, Z., Hegao, C.: Multi-robot teleoperation system based on multi-agent structure. In: Proceedings of the IEEE International Conference on Mechatronics and Automation, vol. 4, pp. 1726–1730 (2005)
12. Thomas, B.: Sheridan: Space teleoperation through time delay: review and prognoisis. IEEE Transaction on Robotics and Automation 9(5), 592–606 (1993)
13. Xiao, X., Li, Y., Tang, H.: Kinematics and Interactive Simulation System Modeling for Katana 450 Robot. In: IEEE International Conference on Information and Automation (ICIA), Yinchuan, Ningxia, China, August 26-28 (2013)
14. Wang, J., Li, Y., Zhao, X.: Inverse Kinematics and Control of a 7-DOF Redundant Manipulator Based on the Closed-Loop Algorithm. International Journal of Advanced Robotic Systems 7(4), 1–9 (2010)

# ET Arm: Highly Compliant Elephant-Trunk Continuum Manipulator

Yunfang Yang[1] and Wenzeng Zhang[1,2]

[1] Dept. of Mechanical Engineering, Tsinghua University, Beijing 100084, China
[2] State Key Laboratory of Tribology, Tsinghua University, Beijing 100084, China
yangyunfang12@gmail.com, wenzeng@tsinghua.edu.cn

**Abstract.** This paper proposes a novel elephant-trunk continuum manipulator, called ET Arm. ET Arm is composed of 2 muscle segments. Each of the muscle segments is composed of elastic skeletons, an artificial skin and 3 artificial muscles. In one muscle, a motor mounted to a base is connected to a flexible rod, and the end of the rod is connected to an end disk by screw driving. When one motor rotates, the distance from the base to the end disk will change, the arm bends. The experimental results show that ET Arm is valid and it can bend to any point in its workspace, which is useful for many applications.

**Keywords:** Continuum manipulators, biologically inspired robots, artificial muscles, elephant trunk.

## 1    Introduction

The design aim of robots is to replace human to accomplish tasks, which usually needs a pair of upper-limbs. A human arm, which belongs to serial cantilever structure, has 7 degrees of freedom and highly-developed muscles, and is able to generate great strength and move at high speeds. These features make developing the robot arm very difficult. Over the last twenty years, inspired by elephant trunk and octopus tentacle, researchers have developed soft manipulators that provide new capabilities compared to traditional rigid manipulators.

Cieslak et al [1] presented an elephant trunk type elastic manipulator mainly for the transportation of bulk and liquid materials. The manipulator is able to move flexibly, and carry large loads through its hydraulic system.

Gravagne et al [2] discussed the Clemson Tentacle Manipulator (CTM), which uses highly elastic rod as its backbone and antagonistic cable pairs to deform its shape. This kind of structure is light in weight and can adopt numerous poses for a given actuator displacement, which can be very useful in complicated environments.

Hannan et al [3] designed an "elephant trunk" manipulator. It consists of 16 2-DOF joints for the backbone and can be actuated by a hybrid cable and spring servo system. This kind of manipulator has a wide range of configurations, and can create grasping through bending.

Tsukagoshi et al [4] designed a soft robot arm called Active Hose (AH). Unlike the traditional vertical use of tube, this structure connects short units in series and uses each

X. Zhang et al. (Eds.): ICIRA 2014, Part I, LNAI 8917, pp. 288–299, 2014.

of the 2 rotational DOFs independently. Experimental results show that it can also bend smoothly.

Bukinghan et al [5] from OC Robotics UK, designed and manufactured snake-arm robots, which are specifically designed to perform remote handling operations in confined and hazardous spaces.

OC Robotics Company [6] developed well-known snake-like robots by tendon remote drive for nuclear, aerospace and security.

Simaan et al [7-8] designed one kind of continuum robot, which contains one primary backbone, three identical secondary backbones, some spacer disks, one base disk and one end disk. This device is quite small and dexterous. More amazing research has been conducted to use the device for surgical operation.

McMahan et al [9-10] presented an Air-Octor (AO), which is a pneumatically pressurized central chamber sealed by end caps. The cable guides outside the central tube are used to actuate the manipulator. The Air-Octor has a large amount of space in the center of the manipulator, and it also has great advantages in bending and load capacity.

Walker et al [11-16] designed a soft robotic manipulator, called OCTARM. It adopts McKibben air muscles as actuators and has a total of 12 DOFs. His team also discussed motion planning and new approaches to autonomous grasping. OCTARM can grasp things by wrapping tightly around it. Therefore, this kind of robotic manipulator has accurate grasping configurations and can move flexibly along the ideal path. It has great advantages in avoiding obstacles and grasping heavy load.

Degani et al [17] designed a snakelike robot called HARP (Highly Articulated Robotic Probe), and it is mainly used in minimally invasive surgery. HARP uses four cables to actuate the probe and has been tested successfully for surgery in the lab.

Webster III et al [18] designed a miniature snakelike surgical robot, called Active Cannulas (AC). It consists of several telescoping pre-curved super-elastic tubes and use elastic energy stored in the backbone to create bending. Unlike the traditional structure, this robot can be very small and may be quite useful in the near future.

FESTO Company [19] developed a Bionic Handling Assistant (BHA). This manipulator is made out of polyamide and pneumatically actuated. It consists of three segments. The BHA can move stably and bear heavy loads. FESTO also presents a model for the simulation and experiment with BHA.

Deng et al [20] designed a cable-driven soft robot for cardiac ablation. By using 8 cables to control two segments of soft material, a relative movement can be realized. This surgery robot has a small volume and can avoid obstacles flexibly.

The bio-inspired continuum manipulators mentioned above adopted different principles, structures and actuation methods. The cable-drive systems have been used widespread for a long time and their physical designs are crucial. These systems use motors to actuate the cables, which makes the joints rotate. Cable-drive systems are re-mote control system, which are small in volume, light in weight and cheap in cost, so cable-drive systems are suitable for some special environments such as minimal surgery. However, it also has the weaknesses of low precision and low stiffness. There is another type of continuum manipulator driven by pneumatic actuators. However, air compressors, which are necessities in these systems, tend to take up too much space.

In addition, these equipment are complex in structure, high cost in manufacturing and maintenance as well as difficulty with sealing and high energy consumption.

The multiple tasks in industrial and medical fields require more demands in function and quality for manipulators such as high compliance, high flexibility, high admissible loads, low cost and simple control. There is still a lot of research to be done in this area. This paper proposes a novel design of a continuum manipulator, elephant-trunk arm, which is called ET Arm.

## 2    Design of ET Arm

ET Arm proposed in this paper is shown in Fig. 1. ET Arm with one segment, shown in Fig. 1a, includes a base, a reduced motor, three flexible rods, multiple middle disks, multiple springs, multiple couplings, three screws, one end disk with threaded holes, a partition and connectors.

ET Arm with two segments is composed of 2 muscle segments, each of muscle segments is made of multiple elastic skeletons, an artificial skin and 3 artificial muscles. When actuated by a motor mounted in a base, the flexible rod in each artificial muscle will rotate around its center axis, and will transmit a torque to a screw in the other side of the flexible rod, which will lead to the bending or stretching of ET Arm. The elastic skeletons will help hold the arm straight. The mechanism and work principle of ET Arm are described below in detail.

(1) The artificial muscle of ET Arm

The novel artificial muscle designed includes a base, an actuator, couplings, a flexible rod and an end disk. The actuator is connected to the flexible rod by the coupling, and may actuate the flexible rod to rotate. The other side of the flexible rod is connected to a screw, the screw is connected to the end disk with threaded holes.

When the screw rotates, the distance from the end disk to the basement along the flexible rod will change, if the distance decreases, the muscle shortens and the arm bends; if the distance increases, the muscle stretches and the arm bends towards the opposite direction.

The flexible rod is supposed to be bendable but difficult to stretch or shrink. Experimental results show that the rods with great elasticity are not suitable for the flexible rod. Through a lot of experiments, polytetrafluoroethylene (PTFE) is adopted to be the material of the flexible rod.

(2) The muscle segment of ET Arm

The muscle segment includes 3 artificial muscles that are evenly distributed in a circle, several middle disks, springs and wire-embedded-bellows. Each of the flexible rods in the 3 artificial muscles goes through one hole of all middle disks. By doing this, the muscle can be restrained by skeletons (middle disks and springs). Each middle disk is separated by 3 springs that are also evenly distributed along the circle. The muscle segment is wrapped by artificial skin made by wire-embedded-bellows.

The spring can support the arm and have limited influence to the compliant bending of the muscles. To lighten the weight, 3D printing technology is used to manufacture the middle disk in our prototype.

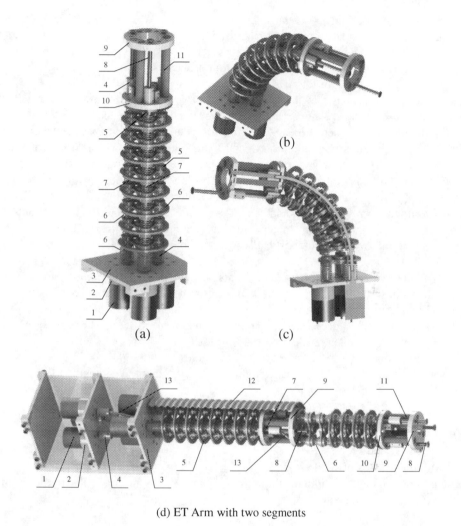

(d) ET Arm with two segments

**Fig. 1.** Design of ET Arm. (a) 1 segment, (b) ET Arm bends, (c) Section view, (d) 2 segments. 1-motor; 2-reducer; 3-base; 4-coupling; 5-middle disk; 6-spring; 7-1$^{st}$ rod; 8-screw; 9-end disk, 10-partition, 11-connector, 12-artificial skin, 13-2$^{nd}$ rod.

When shortening of one muscle happens, due to the constraint of the middle disks, the other 2 muscles will correspondingly bend certain angle towards the direction of the shortened muscle. This leads to bending towards certain direction of the whole arm, and the bend principle towards other directions is the same. One may also actuate 2 or 3 muscles at one time as well.

(3) The work principle of artificial muscles

Originally, the muscle segment is straight up. When actuating the motor in the first artificial muscle, the coupling connects between the motor and the rod will drive the rod to rotate, which will transmit the torsion force to the screw. Screw drive will change the distance from the end disk to the basement along the flexible rod, therefore this leads to the shortening or extension of the first artificial muscle. However, the second muscle and the third muscle do not change their length when the first muscle shorten, so the whole muscle segment will bend towards the direction of the first artificial muscle.

(4) The artificial skin of ET Arm

Experimental results show that ET Arm without artificial skin, namely constructed with only muscles and skeletons, cannot finish the bend behavior due to the flexible rod will twist badly under heavy load. After a series of analysis and experiments, our solution to this problem is to add an artificial skin. The springs will help improve the carrying capacity. The artificial skin, which is made of wire-embedded-bellows, is bound from point to point to the middle disks. This kind of material cannot twist easily, which can help counterbalance the twisting force by restricting the position of middle disks while the arm is actuated, and can also provide protection to the inner structure from accidental damage. Also when the load is light and the number of middle disk is rather small, ET Arm will not twist badly even without the artificial skin.

(5) ET Arm

ET Arm is composed of 2 muscle segments, and can be designed for even more than 3 segments in series. Simple connection in series makes a highly compliant continuum manipulator, however, the gravity of the actuators in higher level will increase the load of lower segment and decrease the stability of the whole system. Therefore, a better design is that all actuators put into the bottom of ET Arm and transmission parts (flexible rod or steel wire) through the central holes in the middle disks of the lower muscle segment to deliver torque to higher muscle segment. This method solves the interfering problem between different segments.

## 3   Motion Analysis of ET Arm

(1) Motion analysis of single muscle

The motion principle of shortening a single muscle is shown in Fig. 2. The meanings of the symbols are listed below:

$i$ : a certain moment;

$s_{10}$ , $s_{20}$ , $s_{30}$ : the length of 3 flexible rods at the beginning;

$s_1$ , $s_2$ , $s_3$ : the length of 3 flexible rods at time $i$;

$O$ : the center of the base;

$O_i$ : the center of the arc made by muscle segment at time $i$;

$O_m$ : the center of the arc made by artificial muscle 1 at time $i$;

$O_n$ : the center of the arc made by artificial muscle 2 at time $i$;

$O_q$ : the center of the arc made by artificial muscle 3 at time $i$;

$H$ : the center of end disk at time $i$;

$R$ : $\overline{OO_i}$ ; $R_1$ : $\overline{MO_m}$ ; $R_2$ : $\overline{NO_n}$ ; $R_3$ : $\overline{QO_q}$ ;

$\theta$ : the angle between vertical plane and the normal of end disk;

$r$ : $\overline{MO}$ , called the radius of ET Arm;

{A} : the world frame, its origin is $O$ , $OO_i$ is $x$ axis at the beginning, vertical axis is $z$ .

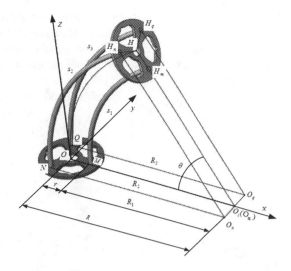

**Fig. 2.** Motion principle of shortening one muscle

Let $\Delta s_1 = s_{10} - s_1$, $\Delta s_2 = s_{20} - s_2$, $\Delta s_3 = s_{30} - s_3$.
At the beginning, the 3 flexible rods are all straight and have the same length:

$$s_{10} = s_{20} = s_{30}. \tag{1}$$

At time $i$, the length of one rod shortened of $\Delta s_1$, while the lengths of other 2 rods remain the original length. Because of the restraint of end disk, the 3 flexible rods will form different arcs shown in Fig. 2. The center of 3 arcs are $O_m$, $O_n$, $O_q$, and 3 points are in the same line. One has

$$s_1 = s_{10} - \Delta s_1 = (R - r)\theta, \tag{2}$$

$$s_2 = s_{20} = (R + \frac{r}{2})\theta. \tag{3}$$

Using the eq. (1), (2), (3), one gets

$$\theta = \frac{2\Delta s_1}{3r}, \quad R = \frac{3s_{10}r}{2\Delta s_1} - \frac{r}{2}.$$

At this moment, the position of $H$ is

$$\begin{cases} x_H = R - R\sin(\frac{\pi}{2} - \theta), \\ y_H = 0, \\ z_H = R\cos(\frac{\pi}{2} - \theta). \end{cases} \tag{4}$$

While decreasing the length of one muscle of $\Delta s_1$, the muscle segment will form an arc whose radius is $R$ and central angle is $\theta$. The relationship between $R$ and $\Delta s_1$ is shown in Fig. 3. When $\Delta s_1$ is less than 5mm, the radius will decrease rapidly with the average increase of $\Delta s_1$. When $\Delta s_1$ is more than 5mm, the speed of decrease of R will slow down. Besides, as shown in Fig. 3, with the increase of the original length of the muscle, the radius will be correspondingly shorter.

The relationship between $r$ and the range of motion of $H$ is shown in Fig. 4. If the radius of ET Arm is smaller, the bending will be more obvious and the range of motion will be larger.

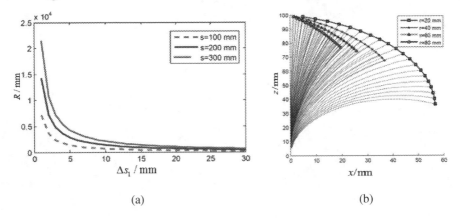

(a)                                                        (b)

**Fig. 3.** Relations. (a) The relation between $R$ and $\Delta s_1$. (b) The relation between $r$ and $H$.

(2) Motion analysis of two muscles

The principle of two muscles is shown in Fig. 4. The symbols are listed below:

$\alpha$ : the angle between $OO_i$ and $x$ axis at time $i$;

$\{B\}$ : the robot frame, its origin is $O$, $OO_i$ is $x_B$ at time $i$, the vertical axis is $z_B$.

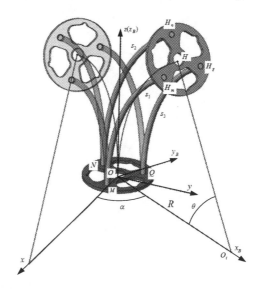

**Fig. 4.** Motion principle of shortening two muscles

In Fig. 4, the left arm shows its position when shortening one muscle, while the right arm shows its position when shortening two muscles, the angle between these two positions is $\alpha$.

The homogeneous transformation matrix of the world frame and the robot frame is:

$$
{}^{B}T_{A} = ({}^{A}T_{B})^{-1} =
\begin{pmatrix}
\cos\alpha & \sin\alpha & 0 & 0 \\
-\sin\alpha & \cos\alpha & 0 & 0 \\
0 & 0 & 1 & 0 \\
0 & 0 & 0 & 1
\end{pmatrix}.
\tag{5}
$$

In frame $\{A\}$, the coordinates of $M, N, Q$ are respectively:

$$
{}^{A}P_{M} = \begin{pmatrix} r & 0 & 0 & 1 \end{pmatrix}^{T}, \quad
{}^{A}P_{N} = \begin{pmatrix} r/2 & -\sqrt{3}r/2 & 0 & 1 \end{pmatrix}^{T}, \quad
{}^{A}P_{Q} = \begin{pmatrix} -r/2 & \sqrt{3}r/2 & 0 & 1 \end{pmatrix}^{T}.
$$

Transform the coordinates of the three points in (8), (9), (10), one has the coordinates of $M, N, Q$ in the frame $\{B\}$:

$$
{}^{B}P_{M} = {}^{B}T_{A} \cdot {}^{A}P_{M} = \begin{pmatrix} r\cos\alpha & -r\sin\alpha & 0 & 1 \end{pmatrix}^{T},
$$

$$
{}^{B}P_{N} = {}^{B}T_{A} \cdot {}^{A}P_{N} = \begin{pmatrix} \dfrac{-r\cos\alpha - \sqrt{3}r\sin\alpha}{2} & \dfrac{r\sin\alpha - \sqrt{3}r\cos\alpha}{2} & 0 & 1 \end{pmatrix}^{T}.
$$

$$
{}^{B}P_{Q} = {}^{B}T_{A} \cdot {}^{A}P_{Q} = \begin{pmatrix} \dfrac{-r\cos\alpha + \sqrt{3}r\sin\alpha}{2} & \dfrac{r\sin\alpha + \sqrt{3}r\cos}{2}\alpha & 0 & 1 \end{pmatrix}^{T}.
$$

Therefore, in the frame $\{B\}$, the coordinates of $M, N, Q$ are:

$$
{}^{B}x_{M} = r\cos\alpha, \quad {}^{B}x_{N} = -r\cos\alpha/2 - \sqrt{3}r\sin\alpha/2, \quad {}^{B}x_{Q} = -r\cos\alpha/2 + \sqrt{3}r\sin\alpha/2.
$$

When bending the muscle segment, the radiuses $R_{1}$, $R_{2}$ and $R_{3}$ of the arcs made by each of three flexible rods are:

$$
R_{1} = R - r\cos\alpha, \quad R_{2} = R + r\cos\alpha/2 + \sqrt{3}r\sin\alpha/2, \quad R_{3} = R + r\cos\alpha/2 - \sqrt{3}r\sin\alpha/2.
$$

The length of three flexible rods $s_{1}$, $s_{2}$ and $s_{3}$ can be shown as:

$$
R_{1}\theta = s_{10} - \Delta s_{1}, \quad R_{2}\theta = s_{20} - \Delta s_{2}, \quad R_{3}\theta = s_{30} - \Delta s_{3},
$$

$$
\frac{R_{1}}{R_{2}} = \frac{S_{1}}{S_{2}}, \quad \frac{R_{1}}{R_{3}} = \frac{S_{1}}{S_{3}}.
$$

In addition, let $K = \dfrac{S_{1}}{S_{2}}$, $J = \dfrac{S_{1}}{S_{3}}$,

$$
K = \frac{R - r\cos\alpha}{R + r\cos\alpha/2 + \sqrt{3}r\sin\alpha/2},
\tag{6}
$$

$$
J = \frac{R - r\cos\alpha}{R + r\cos\alpha/2 - \sqrt{3}r\sin\alpha/2}.
\tag{7}
$$

Therefore,

$$(K-1)R = -\frac{rK\cos\alpha}{2} - \frac{\sqrt{3}rK\sin\alpha}{2} - r\cos\alpha,\tag{8}$$

$$(J-1)R = -\frac{rJ\cos\alpha}{2} + \frac{\sqrt{3}rJ\sin\alpha}{2} - r\cos\alpha.\tag{9}$$

(i) When $J = K = 1$, $s_1 = s_2 = s_3$, $\theta = 0$,

$$^B p_H = \begin{pmatrix} 0 & 0 & s_1 & 1 \end{pmatrix}^T.\tag{10}$$

(ii) When $K = 1$, $J \neq 1$, $s_1 = s_2$, $s_1 \neq s_3$,

$$\begin{cases} \alpha = -\dfrac{\pi}{3}, & (s_3 \geq s_1 = s_2) \\[2mm] \alpha = \dfrac{2\pi}{3}. & (s_3 < s_1 = s_2) \end{cases}\tag{11}$$

$$R = \frac{rJ\cos\alpha - \sqrt{3}rJ\sin\alpha + 2r\cos\alpha}{2(1-J)}.\tag{12}$$

(iii) When $J = 1$, $K \neq 1$, $s_1 \neq s_2$, $s_1 = s_3$,

$$\begin{cases} \alpha = \dfrac{\pi}{3}, & (s_2 \geq s_1 = s_3) \\[2mm] \alpha = -\dfrac{2\pi}{3}. & (s_2 < s_1 = s_3) \end{cases}\tag{13}$$

$$R = \frac{rK\cos\alpha - \sqrt{3}rK\sin\alpha + 2r\cos\alpha}{2(1-K)}.\tag{14}$$

(iv) When $J \neq 1$, $K \neq 1$, $s_1 \neq s_2$, $s_1 \neq s_3$,

$$\alpha = \arctan\frac{\sqrt{3}(J-K)}{J+K-2JK},\tag{15}$$

$$R = \frac{rK\cos\alpha - \sqrt{3}rK\sin\alpha + 2r\cos\alpha}{2(1-K)}.\tag{16}$$

In (ii), (iii), (iv),

$$\theta = s_1 / (R - r\cos\alpha).\tag{17}$$

The position of $H$, the center of the end disk is

$$\begin{cases} x = R(1-\cos\theta)\cos\alpha, \\ y = R(1-\cos\theta)\sin\alpha, \\ z = R\sin\theta. \end{cases}\tag{18}$$

While decreasing the length of the first muscle for $\Delta s_1$ and the length of the second muscle for $\Delta s_2$, the muscle segment will form an arc whose radius is $R$ and central angle is $\theta$. The relationship between $\theta$ and $\Delta s_1$, $\Delta s_2$ is shown in Fig. 5. These results are similar to the situation of shortening a single muscle. The range of motion of $H$ is shown in Fig. 5. While actuating one muscle will cause the movement of $H$ along an arc, actuating two muscles will expand the range of motion to a curved surface.

The analysis mentioned above gives the important relationships between the range of motion of ET Arm and its parameters. These analysis results are rather crucial for the design of a real continuum manipulator.

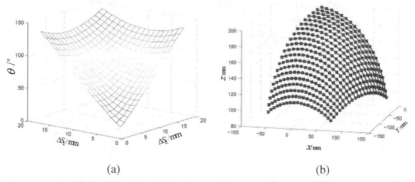

(a)                                                    (b)

**Fig. 5.** Relations. (a) The relation among $\theta$ , $\Delta s_1$ and $\Delta s_2$ . (b) The range of motion of $H$.

## 4    Experiments of ET Arm

A prototype of ET Arm has been developed, shown in Fig. 6. In ET Arm, 6 DC motors are applied for driving 6 flexible rods. The output speed of the reducer is 500 rpm. Polyethylene rod ($\phi$4 mm) is used as the material of flexible rods and the screws are all M6.

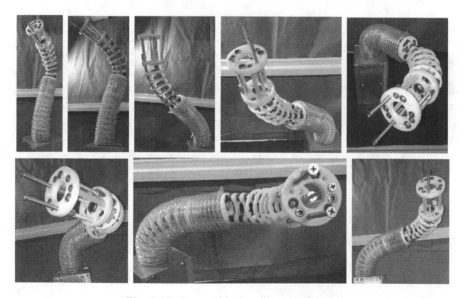

**Fig. 6.** ET Arm and its bending experiments

ET Arm is composed of 2 muscle segments, each of the muscle segments is made of multiple elastic skeletons, one artificial skin and 3 artificial muscles. Each of artificial muscles includes the basement, one reduced DC motor, two couplings, one flexible rod and one end disk. Each of the flexible rods goes through each hole at the same side in each middle disk. Each adjacent middle disk is separated by 3 springs that are also evenly distributed along the circle. The muscle segment is wrapped by artificial skin made by wire-embedded-bellows. The bellow is connected to the middle disks (skeletons) from point to point to counterbalance the twisting forces. The end disk in the first segment is connected with the bottom middle disk in the second segment. The transmission parts (flexible rod) of the second segment go through the central holes in the middle disks of the first muscle segment to deliver torque to higher level. The middle disk is manufactured by a 3D printer.

Multiple sensors are embedded into the ET Arm to detect the positions of multiple muscles. A camera and a gripper are mounted in the end of the manipulator. Arduino controller and motor driver-circuit modules are used for control. ET Arm has a whole length of more than 800 mm, each of the muscle segment is about 400mm. The radius of ET Arm is 24 mm. The maximum shortening length of artificial muscle is around 65 mm. The maximum bending angle of one muscle segment is 90 degree. The inner radius of the artificial skin is about 70 mm. Each segment has 6 to 10 middle disks.

The motion experiments of ET Arm are shown in Fig. 6. Fig. 6 shows the implementation of bending of the first and the second segment. The experimental results show that ET Arm can bend flexibly to a certain point in the space, form complex configuration such as S shape or C shape. ET Arm can be widely used in many kinds of automatic system that needs manipulators.

## 5    Conclusion

A novel elephant trunk continuum manipulator (ET Arm) is proposed. The mechanism and principle of ET Arm are described in detail. Unlike the traditional pneumatic muscle, the artificial muscle in ET Arm adopts low-cost flexible rod and screw drive to create the shortening and extension of the muscle, and it can move stably. In ET Arm with 2 segments, all motors are embedded into the bottom box and flexible rods are used to go through all the central holes in the middle disks of the lower segment to deliver torques to the higher segment. ET Arm is rather flexible, highly compliant and safe for human use. It can be widely used in many kinds of automatic system that needs manipulators.

**Acknowledgements.** This research was supported by National Training Program of Innovation and Entrepreneurship for Undergraduates (No. 201410003006).

## References

1.  Cieslak, R., Morecki, A.: Elephant trunk type elastic manipulator - a tool for bulk and liquid materials transportation. Robotica (01), 11–16 (1999)
2.  Gravagne, I.A., Walker, I.D.: Manipulability, Force, and Compliance Analysis for Planar Continuum Manipulators. IEEE Trans. on Robotics and Automation 18(3), 263–273 (2002)

3. Hannan, M., Walker, I.D.: Analysis and experiments with an elephant's trunk robot. Adv. Robot. 15, 847–858 (2001)
4. Tsukagoshi, H., Kitagawa, A., Segawa, M.: Active Hose: an Artificial Elephant's Nose with Maneuverability for Rescue Operation. In: 2001 IEEE Int. Conf. on Robotics & Automation, Seoul, Korea, pp. 2454–2459 (May 2001)
5. Buckingham, R., Graham, A.: Snaking around in a nuclear jungle. Industrial Robot: An Int. J. 32(2), 120–127 (2001)
6. Buckingham, R., Chitrakaran, V., Conkie, R., et al.: Snake-arm robots: a new approach to aircraft assembly (2007)
7. Simaan, N., Taylor, R., Flint, P.: A dexterous system for laryngeal surgery. In: IEEE Int. Conf. on Robotics and Automation, New Orleans, LA, USA, pp. 351–357 (April 2004)
8. Xu, K., Simaan, N.: An investigation of the intrinsic force sensing capabilities of continuum robots. IEEE Trans. on Robototics 24(3), 576–587 (2008)
9. McMahan, W., Jones, B., Walker, I., et al.: Robotic Manipulators Inspired by Cephalopod Limbs. In: Inaugural CDEN Design Conf., Montreal, Quebec, Canada, pp. 1–10 (July 2004)
10. Jones, B., Walker, I.D.: Kinematics for Multisection Continuum Robots. IEEE Trans. on Robotics 22(1), 43–57 (2006)
11. Walker, I.D., Dawson, D., Flash, T., et al.: Continuum robot arms inspired by cephalopods, SPIE, vol. 5804, pp. 303–314 (2005)
12. McMahan, W., Chitrakaran, V., Csencsits, M., et al.: Field Trials and Testing of the OctArm Continuum Manipulator. In: IEEE Int. Conf. on Robotics and Automation, Orlando, Florida, USA, pp. 2336–2341 (May 2006)
13. Kapadia, A., Walker, I.D.: Self-Motion Analysis of Extensible Continuum Manipulators. In: IEEE Int. Conf. on Robotics and Automation, Karlsruhe, Germany, May 6-10, pp. 1980–1986 (2013)
14. Walker, I.D.: Continuous Backbone "Continuum" Robot Manipulators. SRN Robotics 726506, 19 pages (2013)
15. Neppalli, S., Csencsits, M., Jones, B.: A Geometrical Approach to Inverse Kinematics for Continuum Manipulators. In: IEEE/RSJ Int. Conf. on Intelligent Robots and Systems (2008)
16. Li, J., Teng, Z., Xiao, J., et al.: Autonomous Continuum Grasping. In: IEEE/RSJ Int. Conf. on Intelligent Robots and Systems, Tokyo, Japan, November 3-7, pp. 4569–4576 (2013)
17. Degani, A., Choset, H., Wolf, A., et al.: Highly Articulated Robotic Probe for Minimally Invasive Surgery. In: IEEE Int. Conf. on Robotics and Automation, Orlando, Florida, USA, pp. 4167–4172 (May 2006)
18. Webster III, R., Okamura, A., Cowan, N.: Toward Active Cannulas: Miniature Snake-Like Surgical Robots. In: IEEE/RSJ Int. Conf. on Intelligent Robots and Systems, Beijing, China, pp. 2857–2863 (October 2006)
19. Rolf, M., Steil, J.: Constant curvature continuum kinematics as fast approximate model for the Bionic Handling Assistant. In: IEEE/RSJ Int. Conf. on Intelligent Robots and Systems, Vilamoura, Algarve, Portugal, pp. 3440–3446 (October 2012)
20. Deng, T., Wang, H., Chen, W., et al.: Development of a New Cable-driven Soft Robot for Cardiac Ablation. In: IEEE Int. Conf. on Robotics and Biomimetics, Shenzhen, China, pp. 728–733 (December 2013)

# Design of an Anthropomorphic Prosthetic Hand with EMG Control

Nianfeng Wang, Kunyi Lao, and Xianmin Zhang

Guangdong Province Key Laboratory of Precision Equipment and Manufacturing Technology, South China University of Technology, Guangzhou, Guangdong 510640, P.R. China

**Abstract.** This paper presents an anthropomorphic prosthetic hand with elastic joints. Helical springs are used as elastic joints and the joints of each finger are coupled by tendons. Although each finger has only one degree of freedom, the prosthetic hand is able to complete the basic actions of hand, such as gripping, pinching, and so on. The overall structure design of prosthetic hand and arrangement of driving motors are completed basing on the motion law and actual requirements of hand. Also, the overall control system for the EMG prosthetic hand has been built, which can control the prosthetic hand to carry objects with EMG signal.

**Keywords:** Anthropomorphic prosthetic hand, compliant hinge, myoelectrical control.

## 1 Introduction

The human hand is the most complex and dexterous tool of men, and it has played an important role during evolution. Men use their hands to grab and move objects, moreover, they explore, contact and perceive the external circumstances with their hands to obtain large amount of information, such as the shape, material, temperature, speed and such physical properties of the objects. As contacting the natural, the hand is a perfect tool for human communication. The human hand makes the meaning and emotion of the language richer during communication, it can also replace language at times through fondling or hug for example.

The lack of human hand will hinder man from communicating with nature and society, damage the human ability of reforming the nature society, and make man suffer from great mental traumas. In the last few years, the amount of the disabled has rapidly increased due to traffic accidents, disease, industrial injury, etc. Wearing a prosthesis is one of the important means for the disabled man to have the normal daily living.

The ideal prosthetic hand is supposed to be the same with the human hand in the shape and features. The prosthetic hand has to replace the feeling and movement of human hand and complete a certain amount of human hand operation. Also, the appearance of the prosthetic hand needs to be similar to human hand. There are 27 degree-of -freedoms (DOFs) in the upper limbs of human,

X. Zhang et al. (Eds.): ICIRA 2014, Part I, LNAI 8917, pp. 300–308, 2014.

and among the finger parts there are 21 DOFs [1]. The recent researches have not invented a prosthetic hand which has 21 DOFs yet. The prosthetic hands of the researches in America and Japan have 7 to 11 DOFs [1], But many of the clinic commercial prosthetic hands have 3 DOFs at most, which can only complete 4 of the most common human hand motions including finger stretch, finger flexion, pronation and supination.

Electromyography (EMG) prosthetic hand [2] is an anthropomorphic prosthetic hand controlled by electrical signals produced by muscles. It has better appearance and functions than other prosthetics. The EMG prosthetic hand utilizes the electrical signals collected by the electrodes which are placed above the muscle on the surface of the residual arm to drive the prosthetic hand to move.

This paper describes an EMG controlled anthropomorphic prosthetic hand based on flexure hinges, which has small volume, light weight and most grasp functions of human hand in daily living. The anthropomorphic prosthetic hand can help the disabled users to some extent. The disabled can take care of themselves in daily living with the hand, improving their abilities to live.

## 2   Design Objectives

The important factors in the performance of an anthropomorphic prosthetic hand are: 1) with cosmetic appearance similar to human hand and natural shape which is acceptable to users; 2) light weight, compact structure and can be processed easily; 3) with a good accuracy of identifying signals and grasp gestures; 4) with sufficient grasp and reach gesture models to meet the needs of patients; 5) patients can control the artificial hand by means of certain ways; 6) appropriate grasp speed and force; 7) low cost, low noise; etc [3].

According to the research on the skeleton of human hand, the human hand has excellent adaptability due to the 21 DOFs. Human hand is flexible that it has different grasp postures to grasp all kind of objects. This requires that the prosthetic hand should have the similar appearance and shape to human hand and more DOFs.

In most of the existing prosthetic hand designs, the number of the joints will be more than the number of the independent actuators. Generally, the relevant joints will be coupled into a compound motion which can be driven by just one independent actuator [4]. In this way, the discrepancy between the number of independent actuators and the number of joints is accommodated. Another way is coupling motions through adaptive underactuation methods in which one independent actuator controls several independent DOFs [5–7]. Underactuation, variable compliance couplings and module design are adopted in the multi-DOF hands nowadays which have functions similar to the human hand though having the less DOFs than the human hand.

In the activities of daily living, the main functions of human hand are grasp, pull, push and raise the objects that the thumb plays an important role. The thumb effects about 40% of all the human hand functions [8]. In most of the prosthetic hand design, the main motions of the thumb are stretch, flexion and rotation around the palm. The rotation of the thumb can switch the prosthetic hand grasp posture mode between lateral and strength or accuracy grasp [9].

About the grasp posture modes, many researchers tend to use six basic types of prehension present by Taylor and Schwarz [10], which are cylindrical, hook, tripod, spherical, tip and lateral. The six types of prehension can complete most grasp motions in activities of daily living. This paper defines eight grasp posture modes (Fig. 1) based on the six types of prehension: 1) cylindrical, for grasping cylindrical objects, e.g., for grasping a cylindrical bottle or a glass; 2) hook, for carrying or pull objects through four fingers, e.g., for carrying a briefcase; 3) lateral, for a grasp posture between the thumb and the lateral of the forefinger, e.g., for grasping a key or a card; 4) point, for showing directions or punching buttons; 5) rest, a platform posture, for holding a plate or a book; 6) spherical, for grasping objects by the whole hand, e.g., for grasping a tennis ball; 7) tripod, for grasping small objects between the thumb, forefinger, and the middle finger, e.g., for grasping a bottle cap; 8) tip, for pinching smaller objects like pins. These eight grasp posture modes contain most human hand motions which can meet the needs of the disabled in activities of daily living.

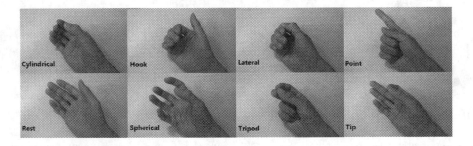

**Fig. 1.** Eight grasp posture modes

Physical design of the prosthetic hand is an important part in the hand design which decides the size and dexterity of the hand. The design objectives include the shape and size of the hand, the motions of the finger and the functions which needs to be considered. The weight of the hand should be within 500g [11, 12]. The shape and the size of the hand is supposed to be similar to the human hand.

## 3 Hand Design

The EMG prosthetic hand (Fig. 2) designed in this paper has 5 fingers, 15 joints and 4 independent actuators which are settled in the palm. Module design is adopted that the forefinger, middle, ring and little fingers have the same structure in order to process conveniently and reduce the cost. Flexure hinges are adopted in the finger joints and the finger motions are coupled by tendons.

The finger of the hand has three phalanxes (Fig. 3), i.e., proximal, middle and distal phalanx. The joint of the finger is a flexure hinge which is connected by

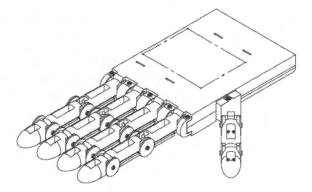

**Fig. 2.** The EMG prosthetic hand

2 equal length extension spring placed side by side. The open and close finger motion is driven by tendons, through pulling and releasing the ropes.

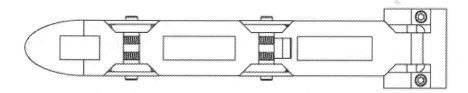

**Fig. 3.** The top view of the finger

The joint of the prosthetic hand is designed with a compliant mechanism. When the finger flexes, the extension spring will bend in a certain range, having a relatively large displacement without any permanent deformation or torsion. In order to limit the transverse motion of the flexure hinge, a certain amount of spring is set abreast. Therefore, the whole joint is limited to make movements only in one plane since it is equal to have rotation hinge in only one DOF.

Compliant mechanism is adopted in the joint design of the EMG prosthetic hand which plays an important role. The joint will need just one positive actuator to ensure the basic motions like grasping of the finger, as additional tendons for finger extension are not necessary. The structure and system of the prosthetic hand is simplified. Further, the flexure hinge is a simple and reliable structure and the hinge could hardly have any abrasion which is easy to setup. Finally, the flexure hinge can give a cushioning effect to the system when it receives impact or overloads, protecting the prosthetic hand and its user by improving the stationarity of the hand.

It is not necessarily better when the anthropomorphic prosthetic hand has got more DOFs. When the hand has more DOFs, it needs more actuators to run the device and hence the complexity, weight and volume of the structure will increase comparatively. Therefore, the DOF of prosthetic hands should be minimized as long as it fits its purposes [13]. In fact, a finger only needs one DOF to compete basic movements like grasping and pinching driven by only one independent actuator. Therefore, tendon driven is able to satisfy this requirement.

Tendon driven comes to be practical by pulling the ropes which are going through each phalanx. In the compliance coupling which the proximal phalanx joint moves following the palm joint, the tendon goes through the whole proximal phalanx and is fasten to both of the proximal and distal phalanx (Fig. 4).

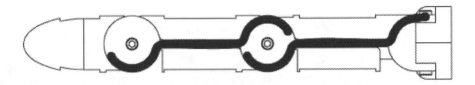

**Fig. 4.** The tendon route of the finger

The EMG prosthetic hand is a highly integrated system with mechanical system, sensors, control and drive system. The drive system of the prosthetic hand is setup inside a palm with 4 independent actuators in order to control 5 fingers with 4 DOFs including the rotation of thumb around palm, the flexion and stretch of the thumb, forefinger and the other 3 coupled fingers. Fig. 5 the shows internal view of the palm. The rotation of thumb, flexion and stretch motion of the forefinger and the other 3 fingers are driven by independent actuators through gear trains while the flexion and stretch motion of thumb is through a gear train and a pulley. Fig. 6 shows the 8 grasp postures of the EMG prosthetic hand.

## 4    The Driving and Control System

A key point of creating the EMG prosthetic hand is building the driving and control system. Therefore, the suitable hardware should be selected in order to satisfy the motions of the fingers of the prosthetic hand. The basic requirements for the driving and control system [14] are: 1) the system will be able to communicate with the EMG signals that have been processed in order to realize the control with EMG signals directly; 2) the structures of hardware should have the necessary ability for real-time computing and process the data immediately; 3) the hardware should be modularized in order to add or change different kind of port and so on. In addition, with the rapid development of computer technology in recent years, the driving and control system based on PC is mainly used in the research of the EMG prosthetic hand. The controller and motor driver from ACS Motion Control company are selected in this paper.

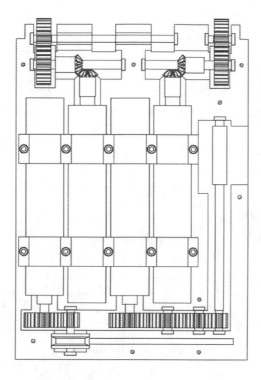

**Fig. 5.** The internal view of the palm

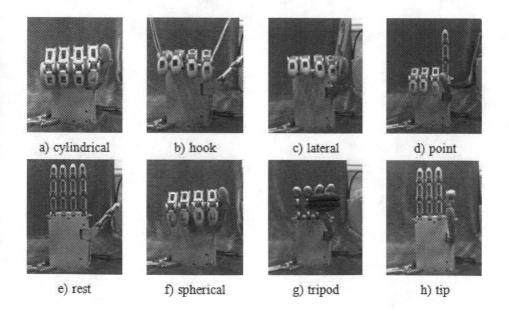

a) cylindrical          b) hook          c) lateral          d) point

e) rest          f) spherical          g) tripod          h) tip

**Fig. 6.** The 8 grasp posture of the EMG prosthetic hand

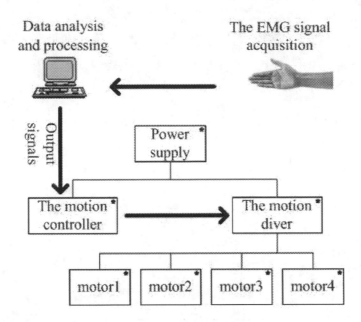

**Fig. 7.** The overall control system of the EMG prosthetic hand

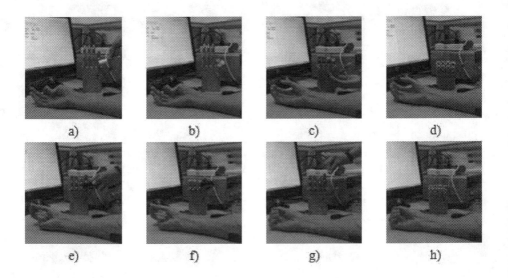

**Fig. 8.** The movement of the EMG prosthetic hand

The structure of the overall control system (Fig. 7) for EMG prosthetic hand mainly includes 2 basic parts: 1) the part for EMG signals acquisition and data analysis including feature extraction and pattern recognition; 2) the part for outputing signals and motion control commands for the fingers of the prosthetic hand. In this paper, the SPiiPlus NTM motion controller from ACS motion control company is adopted to control the prosthetic hand. The SPiiPlus UDMLC motion driver is adopted as the motion driver of the prosthetic hand. The DC graphite brush micro motors from Maxon company are adopted as the motors for the prosthetic hand, and planetary gearboxes are adopted as the reducers for the micro motors. Encoders with 2 channels and line driver from Maxon company are adopted as the encoders for the micro motors. First, the connection between the PC, motion controller, motion diver and the EMG prosthetic hand is built. Then complete the corresponding settings of the motion controller through the software SPiiPlus MMI Application Studio and program for the motions of the motor at last. Fig. 8 shows the movement of the EMG prosthetic hand controlled by the EMG signals in real time.

## 5     Conclusion and Future Work

This paper presents an anthropomorphic prosthetic with elastic linkage and builds the control platform of the EMG prosthetic hand. Based on the analysis of the prosthetic hand design objectives, the flexure hinge is adopted in the finger joint and the motions of the joints are coupled by tendon driven. With the overall control and drive system of the EMG prosthetic hand, the hand can be controlled by the EMG signal and achieve the eight grasp posture modes.

Future work includes the statics and kinetics analysis of the EMG prosthetic hand and the design of a new version of the prosthetic hand which is more similar to the real human hand, especially the part of palm.

## References

1. Zheng, X.J., Zhang, J., Chen, Z.W.: Research status of the emg prosthetic hand. Chinese Journal of Rehabilitation Medicine 18(3), 168–170 (2003)
2. Nishikawa, D., Yu, W., Yokoi, H., Kakazu, Y.: Emg prosthetic hand controller discriminating ten motions using real-time learning method, Kyongju, South Korea, vol. 3, pp. 1592–1597 (1999)
3. Yin, Y., Zhu, L.J., Bao, H.T.: Research status and developing tendency of intelligent prosthetic hand. Chinese Journal of Rehabilitation Medicine 19(1), 52–53 (2004)
4. Pons, J.L., Rocon, E., Ceres, R., Reynaerts, D., Saro, B., Levin, S., Moorleghem, W.V.: The manus-hand dextrous robotics upper limb prosthesis: Mechanical and manipulation aspects. Autonomous Robots 16(2), 143 (2004)
5. Dollar, A.M., Howe, R.D.: The sdm hand as a prosthetic terminal device: A feasibility study. In: 2007 IEEE 10th International Conference on Rehabilitation Robotics, ICORR 2007, Noordwijk, Netherlands, p. 978 (2007)

6. Dollar, A.M., Howe, R.D.: The highly adaptive sdm hand: Design and performance evaluation 29, 585–597 (2010)
7. Dollar, A.M., Howe, R.D.: Joint coupling design of underactuated grippers, vol. 2006, Philadelphia, PA, United States (2006)
8. Ouellette, E.A., McAuliffe, J.A., Carneiro, R.: Partial-hand amputations: Surgical principles. In: Atlas of Limb Prosthetics, Surgical, Prosthetic, and Rehabilitation Principles, 2nd edn., pp. 199–216. St. Louis: Mosby-Year Book (1992)
9. Coert, J., van Dijke, G.H., Hovius, S., Snijders, C., Meek, M.: Quantifying thumb rotation during circumduction utilizing a video technique. Journal of Orthopaedic Research 21(6), 1151–1155 (2003)
10. Schwarz, R.J.: The anatomy and mechanics of the human hand. Artificial Limbs, 22 (1955)
11. Light, C., Chappell, P.: Development of a lightweight and adaptable multiple-axis hand prosthesis. Medical Engineering and Physics 22(10), 679–684 (2000)
12. Vinet, R., Lozac'h, Y., Beaudry, N., Drouin, G.: Design methodology for a multifunctional hand prosthesis. Journal of Rehabilitation Research and Development 32(4), 316–316 (1995)
13. Massa, B., Roccella, S., Carrozza, M., Dario, P.: Design and development of an underactuated prosthetic hand, Washington, DC, United States, vol 4, pp. 3374–3379 (2002)
14. Hao, X.: The Research of EMG Prosthetic Hand Control System. PhD thesis (2002)

# Hexapod Walking Robot CG Analytical Evaluation

X. Yamile Sandoval-Castro[*], Eduardo Castillo-Castaneda,
and Alejandro A. Lozano-Guzman

Instituto Politecnico Nacional, CICATA. Cerro Blanco #141. Colinas del Cimatario,
76090, Queratro, Mexico
`yamile.sandoval.castro@gmail.com`

**Abstract.** A methodology for calculating a hexapod robot CG is presented. The CG coordinates related to a global reference frame are obtained as a function of the angular position of the joints and the thorax orientation. The equations presented were validated using a CAD software. Using the Normalized Energy Stability Margin (SNE) criterion the robot stability was analyzed simulating the robot CG position on an inclined plane.

**Keywords:** stability, center of gravity, hexapod walking robot, recovery strategy.

## 1 Introduction

The use of robots for construction, maintenance, inspection, cleaning, rescue and repair of tall buildings, nuclear plants, repository tanks and bridges is becoming a standard practice. This is due to the economic and safety advantages compared with the onsite human involvement. So in order to carry out the tasks needed, robots must have good ability to adapt to their environment both in gait and the ascent and descent.

Researches on walking robots and climbers (WCR) have been reported for different types of environments, terrain, walls and slopes. That is the case of the so called BigDog that has emerged as a robot with very good dynamic stability [1]. The SCORPION robot introduced by Spenneber, can adapt to different environments [2]. Robots able to adapt to uneven terrain have been developed by Kirchnere, Kimura and Kar [3], [4], [5].

WCR use different strategies to perform their tasks in the presence of several types of disturbances such as changes in terrain level or wind. The TITAN VI robot, [6] can adapt itself to slight changes in ground level using a passive mechanism attached at the end of each limb. Using an array of wind sensors, a robot able to walk against the wind is reported in [7]. In [8], a robot that can adapt itself to different types of terrain, slopes and vertical surfaces is presented. However, in several applications, static and dynamic stabilities remain to be studied.

A walking robot is statically stable if the projection of its center of gravity (CG) is within the support polygon which is formed by the robot support points [9]. The gradient

---

[*] The authors acknowledge the support of CONACYT.

X. Zhang et al. (Eds.): ICIRA 2014, Part I, LNAI 8917, pp. 309–319, 2014.

stability margin evaluates robot CG tilt when robot lean over [10]. Tipover stability margin is similar to gradient stability margin, though in the tip over concept, all external forces including gravity are considered to locate the robot CG [11]; energy stability margin considers CG maximum potential energy. This criterion assesses robot stability by the difference between CG maximum and initial potential energy [12]. The dynamic energy stability margin considers in the stability analysis evaluation, all forces including gravity [13]. The normalized energy stability margin (SNE) is defined as the vertical distance between the CG initial and maximum positions [14]. This approach offers advantages over that reported on [12] because increase in weight does not necessarily lead to an increase in stability. So stability of a robot can be better evaluated in terms of mm instead of Jouls.

In this paper the static stability of a hexapod robot is analyzed. All the stability criteria are based on the knowledge of the CG position at all times, so a methodology for computing the CG is presented, based on the angular position of all joints and the orientation of the robot´s chest. The criterion of stability for SNE hexapod robot for different terrain´s slopes was implemented. This stability criterion might be applicable for robots of more than 4 legs.

Once the CG is calculated, it is possible to develop robot stability recovery strategies, either based on adding special mechanisms, sensor fusion, implementation of intelligent algorithms or using odometry.

## 2    Methodology for Calculating the Robot CG

### 2.1    Robot Architecture

Figure 1 shows the basic robot architecture. The reference frame {b} is located in the center of the chest, {0} indicates the global reference frame. The robot has six identical limbs that are symmetrically distributed around the thorax. Each leg has 3DF and are composed of three links which represent the coxa, femur and tibia. These elements are connected by rotational joints, and at the end of the tibia, a suction cup is placed to secure attachment to vertical surfaces. In this figure, the wire configuration of the ith leg is also shown. The frameworks were placed according to Denavit-Hartenberg convention, as well as the distances among links.

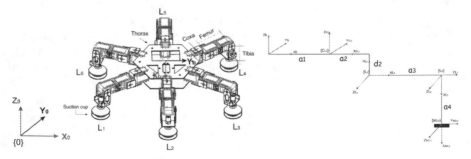

**Fig. 1.** Walking robot morphology

## 2.2    Methodology for Calculating the Robot CG

The CG was calculated for the current position of the robot. The angular positions of the 18 joints and their geometric parameters are used as inputs. Assuming that a body can be decomposed into its components and subsystems [15], if the weight and position of each CG of all components is known, it is possible to determine the entire body CG. For calculating the coordinates $\bar{x}$, $\bar{y}$, $\bar{z}$, and the CG, it is required that the resulting weight is equal to the weight of all the n components, ie. $W_R = \Sigma W$. In order to calculate the CG of a composite body, a reference frame is set and each subsystem is referred to it. A particle mass ($m_j$) is associated with the $j$ element and its CG coordinates ($x_i$, $y_i$, $z_i$) associated to the reference frame are set. CG determination is carried out as if the body were a set of material points. So,

$$X_G = \frac{\Sigma_{j=1}^{n} m_j x_i}{W}$$   (1)

$$Y_G = \frac{\Sigma_{j=1}^{n} m_j y_i}{W}$$   (2)

$$Z_G = \frac{\Sigma_{j=1}^{n} m_j z_i}{W}$$   (3)

**Robot Components and CG Evaluation Referred to a Specific Frame**
For the hexapod robot, it is divided into 5 main components, thorax, servo, coxa, femur and tibia. Components of the ith leg are servo-coxa-servo-femur-servo-tibia. Each component and its reference frame are represented graphically. The CG is localized for each component. Considering its light weight compared with the robot legs, the suction cup mass was not taken into account.

Figure 2 shows the thorax CG position. This element is made of two parallel plates. The reference frame for this component was set in the center of the upper plate.

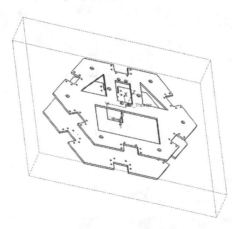

**Fig. 2.** Thorax component

Figure 3 shows the ith leg. In this figure, reference frame for each component can be seen. CG for each component (*servo*, *coxa*, *femur* and *tibia*) are marked with an asterisk.

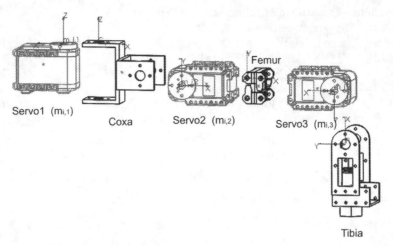

**Fig. 3.** Components of the ith leg

## Robot Components and CG Evaluation

The CG of all components must be calculated with respect to the global reference frame. The $^{0}R_{th}$ matrix (includes, Pitch, Yaw and Roll) describes the shift between the global reference frame and the one associated with the thorax. $^{0}P_{th}$ vector represents the translation between them.

$$
{}^{0}_{}R_{th} = \begin{bmatrix} c\phi c\theta & c\phi s\theta s\psi - s\phi c\psi & c\phi s\theta c\psi + s\phi c\psi \\ s\phi c\theta & s\phi s\theta s\psi + c\phi c\psi & s\phi s\theta c\psi - c\phi s\psi \\ -s\phi & c\theta s\psi & c\theta c\psi \end{bmatrix}, \quad {}^{0}_{}P_{th} = \begin{bmatrix} v_x \\ v_y \\ v_z \end{bmatrix}
$$

Calculations presented below are those corresponding to the ith leg.

The procedure presented here, is applicable to any other leg, just applying a 60° rotation around the z axis between the reference frame $\{th\}$ and the i-th link. Figure 4 (a) shows the CG wire representation of the thorax and figure (b) the servo CG $m_{i,1}$ referred to $\{0\}$. Vector $^{th}P_{cg\text{-}th}$ represents the thorax CG position refereed to $\{th\}$ and $^{0}P_{cg\text{-}th}$ gives this CG position related to $\{0\}$. $^{th}P_{mi,1}$ represents the translation from $\{m_{i,1}\}$ to $\{th\}$. $^{mi,1}P_{cg\text{-}mi,1}$ is the servo CG related to its reference frame and $^{0}P_{cg\text{-}mi,1}$ is the CG position of $m_{i,2}$ related to the global reference frame $\{0\}$.

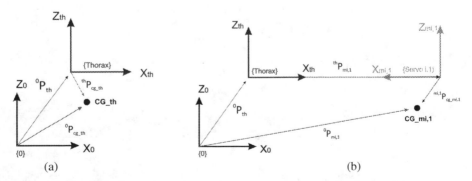

**Fig. 4.** (a) Thorax CG related to $\{0\}$.    (b) $m_{i,1}$ servo CG related to $\{0\}$.

Equations (4) and (5) define the CG servo $m_{i,1}$ and thorax calculation related to {0}. Matrix $^{th}R_{mi,1}$ represents the shift between reference frames {$th$} and {$m_{i,1}$} (180 ° around the $Z_{th}$ axis).

$$P^0_{cg\_th} = R^0_{th}P^0_{th} + P^{th}_{cg\_th}$$ (4)

$$P^0_{cg\_m_{i,1}} = R^0_{th}P^0_{th} + R^{th}_{m_{i,1}}P^{th}_{m_{i,1}} + P^{m_{i,1}}_{cg\_m_{i,1}}$$ (5)

Figure 5 shows the wire representation of the CG position of the coxa referred to {0}. $^{mi,1}P_{ci}$ represents the shift of {$m_{i,1}$} to {$ci$}. $^{ci}P_{cg-ci}$ and $^0P_{ci}$ show the position of coxa CG related to {$ci$} and {0}.

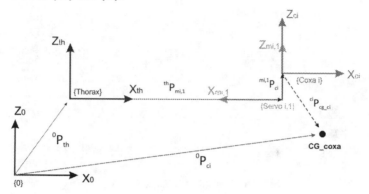

**Fig. 5.** Coxa CG related to {0}

Equation (6) defines the position of the coxa CG position related to {0}, with $^{th}R_{mi,1}$ indicating the orientation change between {$m_{i,1}$} and {$ci$}; given by a $90°$ rotation around $Z_{mi,1}$ axis.

$$P^0_{cg\_ci} = R^0_{th}P^0_{th} + R^{th}_{m_{i,1}}P^{th}_{m_{i,1}} + R^{m_{i,1}}_{ci}P^{m_{i,1}}_{ci} + P^{ci}_{cg\_ci}$$ (6)

Figure 6 shows the servo2 CG ($m_{i,2}$) of the i-th leg. Vectors $^{mi,2}P_{cg-mi,2}$ and $^0P_{cg-mi,2}$ show the servo2 CG related to {$m_{i,2}$} and {0}. $^{ci}P_{mi,2}$ represents shift between {$m_{i,2}$} and {$ci$}.

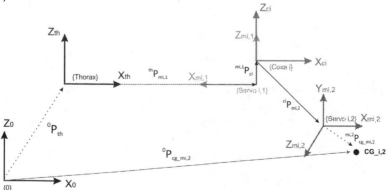

**Fig. 6.** CG of $m_{i,2}$ related to {0}

Equation (7) defines the servo2 CG position related to the thorax, being $^{ci}R_{mi,2}$ the rotation matrix associated with the angular shift ($90°$ around $X_{ci}$ axis) between $\{m_{i,2}\}$ and $\{ci\}$.

$$P^0_{cg\_mi,2} = R^0_{th}P^0_{th} + R^{th}_{m_{i,1}} P^{th}_{m_{i,1}} + R^{m_{i,1}}_{ci} P^{m_{i,1}}_{ci} + R^{ci}_{m_{i,2}} P^{ci}_{m_{i,2}} + P^{m_{i,2}}_{cg\_m_{i,2}} \quad (7)$$

Figure 7 shows the wire representation of the femur CG related to $\{0\}$, with $^{mi,2}P_{fi}$ representing shift between $\{m_{i,2}\}$ and $\{fi\}$; $^{fi}P_{cg\text{-}fi}$ and $^{0}P_{cg\text{-}fi}$ represent the femur CG related to $\{fi\}$ and $\{0\}$. Equation (8) gives the femur CG.

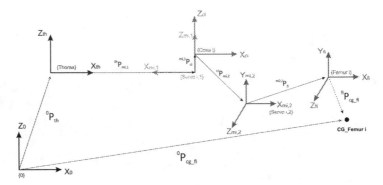

**Fig. 7.** Femur CG related to $\{0\}$

$$P^0_{cg\_fi} = R^0_{th}P^0_{th} + R^{th}_{m_{i,1}} P^{th}_{m_{i,1}} + R^{m_{i,1}}_{ci} P^{m_{i,1}}_{ci} + R^{ci}_{m_{i,2}} P^{ci}_{m_{i,2}} + P^{m_{i,2}}_{fi} + P^{fi}_{cg\_fi} \quad (8)$$

Figure 8 shows the servo3 ($m_{i,3}$) CG referred to $\{0\}$, with $^{fi}P_{mi,3}$ giving the shift between $\{ m_{i,3}\}$ and $\{fi\}$, $^{0}P_{cg\text{-}mi,3}$ and $^{mi,3}P_{cg\_mi,3}$ representing the servo3 CG shift related to $\{0\}$ and $\{m_{i,3}\}$. Equation (9) defines the servo3 ($m_{i,3}$) CG related to $\{0\}$. Matrix $^{fi}R_{mi,3}$ shows the angular shift ($180°$ around the $Z_{fi}$ axis) between $\{ m_{i,3}\}$ and $\{fi\}$.

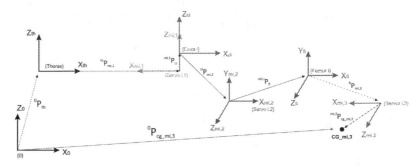

**Fig. 8.** Servo3 ($m_{i,3}$) CG related to $\{0\}$

$$P^0_{cg\_mi,3} = R^0_{th}P^0_{th} + R^{th}_{m_{i,1}} P^{th}_{m_{i,1}} + R^{m_{i,1}}_{ci} P^{m_{i,1}}_{ci} + R^{ci}_{m_{i,2}} P^{ci}_{m_{i,2}} + P^{m_{i,2}}_{fi} + R^{fi}_{m_{i,3}} P^{fi}_{m_{i,3}} + P^{m_{i,3}}_{cg\_mi,3} \quad (9)$$

Figure 9 shows the wire diagram of the tibia CG position related to {0}. Equation (10) defines this CG position. Vector $^{mi,3}P_{ti}$ represents shift along Z axis between { $m_{i,3}$ } and { $ti$ }; $^{ti}P_{cg\text{-}ti}$ and $^{ti}P_{cg\text{-}ti}$ define the tibia CG related to { $ti$ } and {0}.

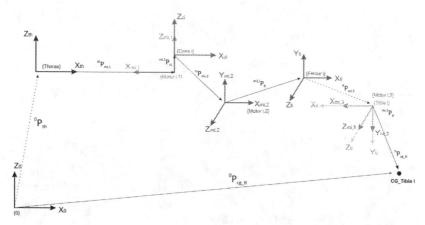

**Fig. 9.** Tibia CG related to {0}

$$P^0_{cg\_ti} = R^0_{th}P^0_{th} + R^{th}_{m_{i,1}}P^{m_{i,1}}_{m_{i,1}} + R^{ci}_{ci}P^{ci}_{ci} + R^{m_{i,2}}_{m_{i,2}}P^{m_{i,2}}_{m_{i,2}} + P^{fi}_{fi} + R^{fi}_{m_{i,3}}P^{fi}_{m_{i,3}} + P^{m_{i,3}}_{ti} + P^{ti}_{cg\_ti} \quad (10)$$

**Robot CG Coordinates Evaluation**

Once you know the CG of each component relative to {0}, it is possible to apply Equations (1), (2) and (3). So, now using equations (11), (12) and (13), it is possible to know $X_{cg}$, $Y_{cg}$ y $Z_{cg}$.

$$X_{CG} = \frac{m_{th}x_{th} + m_s x_{mi,1} + m_c x_{ci} + m_s x_{mi,2} + m_f x_{fi} + m_s x_{mi,3} + m_{ti}x_{ti}}{W} \quad (11)$$

$$Y_{CG} = \frac{m_{th}y_{th} + m_s y_{mi,1} + m_c y_{ci} + m_s y_{mi,2} + m_f y_{fi} + m_s y_{mi,3} + m_{ti}y_{ti}}{W} \quad (12)$$

$$Z_{CG} = \frac{m_{th}z_{th} + m_s z_{mi,1} + m_c z_{ci} + m_s z_{mi,2} + m_f z_{fi} + m_s z_{mi,3} + m_{ti}z_{ti}}{W} \quad (13)$$

In the above equations, $m_{th}$, $m_s$, $m_c$, $m_f$, and $m_{ti}$ are the mass associated to the thorax, servos, coxa, femur and tibia respectively. $x_{th}$, $y_{th}$ and $z_{th}$ are the thorax CG coordinates related to {0}. $x_{mi1}$, $y_{mi,1}$, $z_{mi,1}$, $x_{mi2}$, $y_{mi,2}$, $z_{mi,2}$ and $x_{mi3}$, $y_{mi,3}$, $z_{mi,3}$ are the CG coordinates of servo1, servo2, and servo3 respectively. $x_{ci}$, $y_{ci}$, $z_{ci}$, $x_{fi}$, $y_{fi}$, $z_{fi}$ and $x_{ti}$, $y_{ti}$, $z_{ti}$ are the coxa, femur and tibia CG coordinates.

# 3    Stability Criterion

Once robot CG as a function of the joints angular positions related to {0} is known, it is possible to set a static stability criterion in order to ensure robot balance while walking. In this case the SNE (Normalized Energy Stability Margin) [14] was chosen. This criterion is based on CG height changes, taking into account the minimum and maximum support surface. This is expressed by Equation (14),

$$SNE = \overrightarrow{PG}_{\max}\Big|_{zo} - \overrightarrow{PG}\Big|_{zo} \tag{14}$$

where, $\overrightarrow{PG}\Big|_{zo} = h\cos\theta + \sin\theta_e$ and $\overrightarrow{PG}_{\max}\Big|_{zo} = \sqrt{h^2+d^2}+\cos\theta_s$, $h$ is the height and $\theta$ is the support surface slope angle. SNE criterion is explained in detail in [14].

# 4    Numerical Example

Equations (11), (12) and (13) were programmed, for plotting robot CG position and a drawing of robot´s components using a CAD software for different thorax positions. Table 1 shows some characteristic values for different thorax orientations. These values are used as an example of the numerical evaluation methodology.

**Table 1.**    Example numerical values used for the robot CG localization

| Component | Description | Mass (grs.) | Leg Number (i) | Leg position | | |
|---|---|---|---|---|---|---|
| | | | | $q_{i,1}$ | $q_{i,2}$ | $q_{i,3}$ |
| $th$ | Thorax | $m_{th} = 81.12$ | 1 | 0° | -45° | 90° |
| $m_{i,j}$ | Servo | $m_s = 57.1$ | 2 | 0° | 0° | 90° |
| $c_i$ | Coxa | $m_c = 12.307$ | 3 | 0° | 0° | 90° |
| $f_i$ | Femur | $m_f = 8.33$ | 4 | 0° | -45° | 90° |
| $ti$ | Tibia | $m_{ti} = 25.778$ | 5 | 0° | 0° | 90° |
| | | | 6 | 0° | 0° | 90° |

Example CG coordinates related to each local coordinate system are shown in Table 2.

**Table 2.** Robot components CG coordinates related to the local and global reference frame

| Component | CG related to the local frame | | | CG related to the global frame | | |
|---|---|---|---|---|---|---|
| | $X_{cg}$ | $Y_{cg}$ | $Z_{cg}$ | $X_{cg}$ | $Y_{cg}$ | $Z_{cg}$ |
| Thorax | 0.41 | 0.01 | -22.34 | 0.41 | 0.01 | -22.34 |
| Servo1 | -16.5 | 0.0 | -16.94 | 86.62 | 0.0028 | -14.7 |
| Coxa | 19.0 | 0.0 | -22.5 | 122.13 | 0.0 | -18.76 |
| Servo2 | -16.5 | 0.0 | -16.94 | 157.64 | -4.06 | -18.75 |
| Femur | -5.5 | -8.0 | -19.0 | 182.63 | -0.76 | -18.76 |
| Servo3 | -16.5 | 0.0 | -16.94 | 207.62 | -4.06 | -18.763 |
| Tibia | -34.56 | 0.0 | -16.94 | 224.65 | -0.075 | -53.32 |

Figure 10, (a) and (b) shows the robot´s CG and the example configuration. For this specific case, robot is supported on legs number 2, 3, 5, and 6. According to the stability criterion chosen, in order that a hexapod robot to be stable in a horizontal plane, at least three supporting points have to be active.

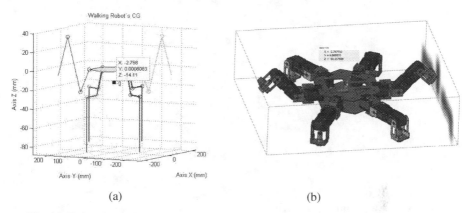

(a)                                      (b)

**Fig. 10.** Robot CG localization. (a) CG using MatLab. (b) CG using a CAD software

Table 3 shows the example position CG localization, applying Matlab and a CAD software.

**Table 3.** CG localization using MatLab and CAD software

| CG lozalization | $X_{cg}$ | $Y_{cg}$ | $Z_{cg}$ |
|---|---|---|---|
| (MatLab) | -2.768 | 0.0008083 | -14.11 |
| CAD software | -2.7677 | 0.00057 | -14.107 |

For this example, the stability criteria (SNE) was applied, taking as initial position that shown in figure 10. Figure 11, shows robot initial position, supported on 4 legs.

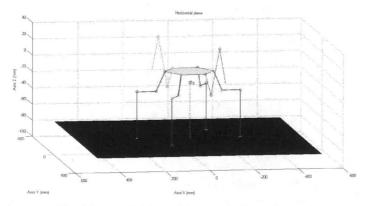

**Fig. 11.** Robot initial position on a horizontal plane

Figure 12 shows the robot on a θ=15°, inclined plane. CG position can be seen in both figures (a), (b), for the specified time.

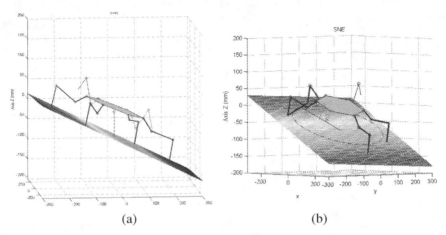

(a)                                        (b)

**Fig. 12.** (a) Robot position for a specified time. θ=15°. (b) SNE contour for the specified time

Table 4 shows the SNE evaluation (CG height and its x, y, z coordinates).

**Table 4.** CG data for the example presented

| Position | Height | CG (mm) | | |
|---|---|---|---|---|
| | | $X_{cg}$ | $Y_{cg}$ | $Z_{cg}$ |
| Initial | 67.8878 | -2.768 | 0.0008083 | -14.11 |
| Final | 60.198 | -2.888 | 0.1056 | -10.7112 |

## 5     Conclusions

The methodology to evaluate the CG position of a hexapod walking robot was presented. This methodology is based on the composite bodies procedure. CG position was calculated applying the equations presented and verified using a CAD software. The advantage of the presented method is that the CG can be calculated for any thorax orientation and it is possible to refer the nth position $\{th\}$ to the reference frame $\{0\}$. .

Simulation approach of the SNE stability for a hexapod robot, was presented. This simulation corresponds to a motion from a horizontal plane to $\theta$=15° inclined plane. For the presented case the hexapod robot stability is maintained because the CG height stays within the SNE stability criteria.

From the results presented, it can be seen that the SNE stability criterion works well in robots with more than four legs.

The results of this paper can be taken as a precedent for designing recovery strategies to ensure hexapod robots static stability for either a horizontal, inclined, vertical surfaces or bumpy ground.

# References

1.  Playter, R., Buehler, M., Raibert, M.: BigDog. In: Proceeding of SPIE, vol. 6230 (2006)
2.  Kirchnere, F., Spenneber, D.: Scorpion: A biometric walking robot. Robotik (2000)
3.  Kimura, H., Fukuoka, Y., Cohen, A.H.: Adaptative dynamic walking of a quadruped robot on natural ground based on biological concepts. International Journal of Robotics Research 26(5), 475–490 (2007)
4.  Kar, D.C.: Design of a Statically Stable Walking Robot: A review. International Journal of Robotics System 20, 671–686 (2003)
5.  Luck, B.L., Collie, A.A., Piefort, V., Virk, G.S.: Robug 3: A tele-operated climbing and walking robot. In: Proceeding of UKACC International Conference of Control, UK (1996)
6.  Fukuda, Y., Hirose, S., Okamoto, T., Mori, J., Hodoshima, R., Doi, T.: Development of TITAN VI: A quadruped walking robot to work on slopes-desing of system and mechanism. In: IEEE International Conference of Robot Systems IROS (2010)
7.  Taniwaki, M., Iida, M., Kang, D.H., Tanaka, M.: Walking Behaviour of a hexapod robot using a wind direction detector. Biosystems Engineering 100, 516–523 (2008)
8.  Loc, V.-G., Roh, S.-G., Koo, I.M., Tran, D.T., Kim, H.M., Moon, H., Choi, H.R.: Sensing and gait planning of quadruped walking and climbing robot for traversing in complex environment. Robotics and Autonomous Systems 58, 666–675 (2010)
9.  McGhee, R.B., Frank, A.A.: On the stability properties of quadruped creeping gaits. Mathematical Bioscience 3, 331–351 (1968)
10. Hirose, Iwasaki, Umetani: Static Stability Criterion for Walking Vehicles. In: 21st SICE Symposium, pp. 253–254 (1978) (in Japanese)
11. Papadopoulos, E.G., Rey, D.A.: A New Measure of Tipover Stability Margin for Mobile Manipulators. In: IEEE International Conference on Robotics and Automation, pp. 3111–3116 (1996)
12. Messuri, D.A., Klein, C.A.: Automatic Body Regulation for Maintaining Stability of a Legged Vehicle During Rough Terrain Locomotion. IEEE Journal on Robotics and Automation RA 1(3) (1985)
13. Ghasempoor, A., Sepehri, N.: A Measure of Machine Stability for Moving Base Manipulators. In: IEEE International Conference on Robotics and Automation, pp. 224–2254 (1995)
14. Hirose, S., Tsukagoshi, H., Yoneda, K.: Normalized Energy Stability Margin and its Contour of Walking Vehicles on Rough Terrain. In: Proceedings of International Conference on Robotics and Automation (2001)
15. Riley, W.F., Sturges, L.D.: Engineering Mechanics, Statics, 2nd edn. John Wiley & Sons, Inc. (2004)

# The Design and Analysis of Pneumatic Rubber Actuator of Soft Robotic Fish

Jinhua Zhang, Jiaqing Tang, Jun Hong[*],Tongqing Lu, and Hao Wang

State Key Laboratory for Manufacturing Systems Engineering
Xi'an Jiaotong University
Xi'an, 710049, China
jhong@mail.xjtu.edu. cn

**Abstract.** This paper describes a pneumatic rubber actuator, achieving forward locomotion in water by swinging like fish tail. First, according to the anatomical structure of fish, the geometry model of the actuator is designed. Second, the finite element method is used to analyze the designed structure and the structure is improved based on the stress distribution. Finally, the modified actuator is fabricated using silicone rubber. When pressurized, the actuator has large bending deformation and realizes swing motion similar to fish tail by alternatively inflating and deflating the channels. The designed structure achieves the expected performance. So the finite element method is an effective approach to the design and analysis of structures made in nonlinear materials.

**Keywords:** Soft robotics, Pneumatic rubber actuator, Finite element method.

## 1    Introduction

The robotics community defines "soft robots" as machines made of soft-often elasto-meric-materials [1]. Soft materials have the characteristic of large strain and low stiffness. The robots made of soft materials can deform continuously so that they have infinite degrees of freedom and can grab objects with different sizes. Due to low stiffness, soft robots can operate on fragile objects with high human-computer interaction security, and are adapted to moving in unstructured environment, such as go through small gap. Therefore, soft robots have broad application prospects in the field of reconnaissance, detection, search and rescue.

There are mainly three actuation modes, including EAP (Electroactive polymer), SMA (Shape memory alloy) and pneumatic actuation [2]. Octopus tentacles made of EAP can have a contraction of 20% at the voltage of 2000V, so the disadvantage of EAP is that it requires high voltage. As for SMA, it has low response speed, low efficiency and is easily aging. The latest actuation mode is pneumatic actuation, which uses the compressed air as power source. The pneumatic actuation has the advantages of high response speed and high power density. The representatives of pneumatic soft robots are

---

[*] Corresponding author.

X. Zhang et al. (Eds.): ICIRA 2014, Part I, LNAI 8917, pp. 320–327, 2014.

as follows. George M. Whitesides in Harvard made a quadruped robot [3], which can crawl at a speed of 0.3cm/s. The length of the robot is 140cm. Cagdas D. Onal, Daniela Rus in MIT made a bionic snake robot [4], supported by DARPA, which can move at a speed of 1.9cm/s. Besides, they also made a soft robotic fish [5]. Although the soft fish can swim fast relatively, it lacks of design and analysis before the actual fabrication. It mainly focuses on the work of the onboard integration and the analysis after fabrication by high-speed camera. As an emerging field, the pneumatic soft robots are mainly made from hyperelastic materials, with the characteristics of large deformation and nonlinearity. So it is difficult to establish an accurate model for the use of soft materials.

In this paper, a pneumatic actuator is designed by comprehensively considering the mechanical characteristics of materials and the geometric structure of channels, using finite element software Abaqus. Analysis shows the designed actuator has large bending deformation, similar to the swing of fish tail. Finally, the actuator is made out using silicone rubber Ecoflex 0030 and PDMS. Experiment shows that the designed actuator can move forward in water by swing.

## 2    Design of the Actuator

### 2.1    The Working Principle of Fish Movement

As the first kind of vertebrate animals which appeared in water after million years of evolution, fish has special moving ability, such as quickness, high efficiency and high mobility. According to the part of body used when fish swims, there are two propulsion modes, namely body/caudal fin (BCF) and median/paired fin (MPF). According to the movement characteristics, there are two kinds of swimming patterns, fluctuation pattern and swing pattern [6]. Swing pattern often refers to the swing of caudal fin. Fish can move forward by periodic swinging. Compared to fluctuation pattern, swing pattern has higher propulsion efficiency and is suitable for long-time and long-distance swimming. So this paper mimics fish, using BCF propulsion mode and adopting swing pattern, to design the pneumatic rubber actuator.

### 2.2    The Working Principle of Actuator

The working principle of pneumatic soft robots: A hollow chamber is formed with two layers of materials. The two layers are in different stiffness, restraint layer in high stiffness and elastic layer in low stiffness. The two layers will have different elongation when the chamber is inflated, resulting in bending deformation. The main influence factors of bending deformation are material properties and chamber geometry [7], as shown in Fig. 1. The chambers in the first row are in the same material but with different layer thickness. The thin layer expands and bends towards the thick layer when the chamber is inflated. The chambers in the second row are in different materials. The direction of the bending deformation mainly depends on properties of the two materials. The layer formed by low stiffness material will expand and bend towards the opposite side. Different deformation can be achieved by controlling the inflation pressure and the material and geometry of chambers.

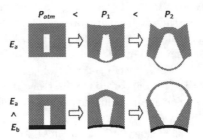

**Fig. 1.** Deformation of a single chamber [7]

In order to achieve large bending deformation of the actuator, the size of a single chamber can be increased or multiple parallel chambers can be adopted. When the size of a single chamber is increased, large bending deformation is achieved, accompanied with large expansion and large stress that even exceeds the tensile strength of the material. The stress and bending deformation of a structure under 30 kPa are shown in Fig.2. The structure is 48*16*8 mm, and the upper layer is Ecoflex 0030, the lower layer is PDMS. Combining single chambers into complex channels, Fig.3 shows the deformation of multiple parallel chambers under pressure of 50 kPa. It can be found that higher pressure is applied to the structure with added walls between chambers, but the stress decreases obviously.

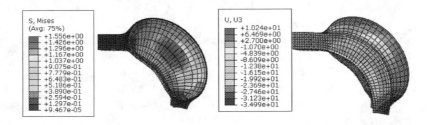

**Fig. 2.** Single chamber formed with Ecoflex and PDMS

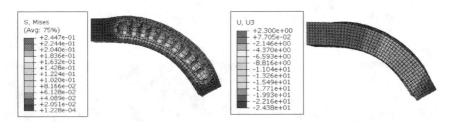

**Fig. 3.** Multiple parallel chambers formed with Ecoflex and PDMS

The actuator mimics the shape of real fish and is designed in cone shape, with multiple parallel chambers in it as shown in Fig.4. The green and gray parts are Ecoflex layers with grooves, the red part in the middle is PDMS layer. The green and red parts form the channels for inflation on one side, and the gray and red parts on the other side. When one side is inflated, the actuator will bend towards the opposite side. The swing motion of the actuator can be achieved by alternately inflating one side and deflating the other side.

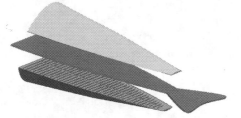

**Fig. 4.** Model of the actuator

# 3    Finite Element Analysis of the Actuator

The soft materials used in this paper are Ecoflex 0030 and PDMS, both of which are silicone rubber. The materials are hyperelastic, with the characteristics of large elastic deformation and nonlinearity. So it is difficult to analyze the deformation of rubber structures coupling with geometric nonlinearity and material nonlinearity. Finite element method is used to solve the problems. Abaqus, the FEM software used in this paper, uses hyperelastic models to describe the complex stress-strain relationship. The properties of the materials could be acquired by the uniaxial tensile test. Then the data is used to analyze the bending deformation of the actuator.

## 3.1    Uniaxial Tensile Test

According to the GB/T528-2009, specimens in dumbbell-shape are used in the uniaxial tensile test to acquire the stress-strain data. Through the experiments, the nominal stress-nominal strain curves are obtained as shown in Fig. 5. Compared to Ecoflex 0030, PDMS is hard and brittle, so it is suitable to be the restraint layer. Ecoflex 0030 is the opposite. According to their different characteristics, different material modes are used, Yeoh mode for Ecoflex 0030, and Neo hooke mode for PDMS.

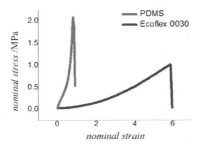

**Fig. 5.** Nominal stress-nominal strain curves of Ecoflex 0030 and PDMS

## 3.2    Deformation Analysis of the Actuator

In order to simplify the analysis, only half of the actuator is analyzed. The half actuator can only achieve one-way swing. The model of the half actuator is symmetric, so
1/4 model is used to analyzed. The geometric model shown in Fig. 6 is 160 mm long,
50 mm wide and 25 mm high in the left end. The thickness of Ecoflex layer and
PDMS layer are both 2 mm, the chambers are 2 mm wide and the walls between the
chambers are 2 mm thick. The material data is input to Abaqus，and employing
Yeoh model for Ecoflex and Neo hooke model for PDMS. For the nearly incompres-
sible mechanical behavior of the materials, hybrid element is adopted in the mesh. Fix
the left end and set symmetry constraint about XZ plane to the face in the up, as
shown in Fig. 7. Applying pressure load 6 kPa, the results of the half actuator are
shown in Fig. 8, including the stress, strain and the displacement in the Z direction.

**Fig. 6.** 1/4 Model of the actuator    **Fig. 7.** Boundary conditions

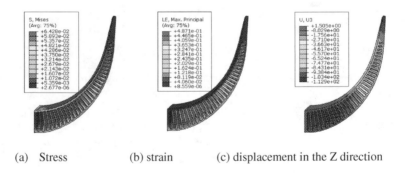

(a)    Stress              (b) strain          (c) displacement in the Z direction

**Fig. 8.** The stress, strain and displacement of actuator under 6 kPa

Fig. 8b shows when pressurized, both Ecoflex layer and PDMS layer will be elon-
gated. For they are of the same thickness, the softer Ecoflex layer extends much longer
than PDMS layer. So the structure bends. Fig. 8c shows the biggest displacement in the
Z direction in the end of the actuator is up to 113 mm. And the displacement is mainly
from the deformation of the front several chambers, as shown in Fig. 8b. The chambers
in the end are relatively small and deform a little. In order to have a more uniform
deformation, the size of the chambers in the front should be decreased. In this paper, the
height of the chambers is reduced as shown in Fig. 9. The upper surface is changed
from cone to plane.

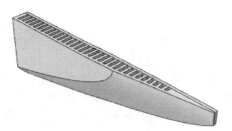

**Fig. 9.** The modified model of actuator

The same material model and boundary conditions are applied to the modified model. When the highest height of the chambers decreases from 21 mm to 10 mm, under pressure of 8 kPa, the results are shown in Fig. 10, including the stress, strain and the displacement in the Z direction.

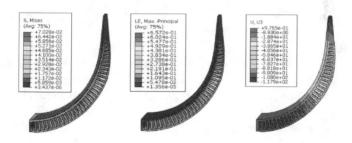

**Fig. 10.** The stress, strain and displacement of modified actuator under 8 kPa

As Fig. 10 shows, compared to conical structure, the modified structure has more uniform stress distribution. For the decrease of the front several chambers, the structure needs lager pressure to get the similar bending deformation. So here, the pressure load increases to 8 kPa.

Comparing the two models, the modified structure can form large bending deformation and has more uniform stress and strain distribution. In addition, the modified structure is material saving and easy to manufacture. Therefore, two of the modified models are combined to form a whole actuator. The whole actuator can achieve two-way swing by alternately inflating one side and deflating the other side. When one side is inflated, the actuator bends to the other side and vice versa. The side which is not inflated will be compressed a little, and it decreases the whole deformation to some degree. When the PDMS is 2 mm thick, applying pressure 10 kPa, the maximum value in U3 is 109 mm. When the thickness of PDMS layer decreases to 1 mm, under the same pressure, the maximum value in U3 is 114 mm, the results are shown in Fig. 11. The lower side is inflated and the upper is not.

**Fig. 11.** The stress, strain and the displacement of whole actuator under 10 kPa

Fig. 11 shows the whole actuator has a reduced displacement for the constraint of the side not inflated. In order to get the similar large deformation, a higher pressure is applied.

# 4    Experiments and Results

The soft pneumatic actuator is made by cast. The elastic layer with grooves on the two sides is cast in Ecoflex 0030, and the flat layer in the middle is cast in PDMS. The molds used to cast are made by 3D printer, using photosensitive resin. The Ecoflex layer and the PDMS layer are got after curing. Then seal them together with a thin PDMS film and make a hole with a needle for inserting inlet pipe, so is the same with the other side. Finally, the actuator is made out. In this paper, the modified structure is fabricated, and the PDMS layer is 1 mm thick.

Air compressor is used as power supply. But the air from the compressor is often too high in pressure and too big in flow for the soft actuator. Valves are used to reduce the pressure and flow. Also, two solenoid valves are used for inflating and deflating to control the bending direction. When one side is pressurized, the actuator bends gradually towards to the opposite side according to the pressure and flow. Adjusting the pressure to 10 kPa, the final deformation of the real actuator is shown in Fig. 12.

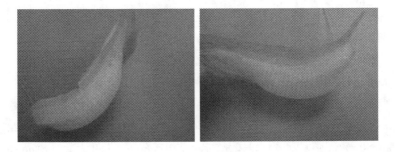

**Fig. 12.** Experimental results under 10 kPa

Fig. 12 shows the experimental results are similar to the simulation. Both are bending continuously, in the similar shape under the same pressure. The front chambers deform uniformly for the reduced in height, and the chambers in the end deform a little. By alternatively inflating and deflating, the actuator achieves a two-way swing and can move forward slowly in water. The small difference between experiments and simulation may be for the difference between the fabricated soft actuator and the model. Another reason may be that the material model does not describe the real mechanical properties accurately for their complex stress-strain relationship. The accuracy of the simulation can be promoted by improving the material model through more test data. For the structures with large deformation and nonlinearity of material and geometry, finite element method is a good approach to design and analysis.

## 5    Conclusion

Comprehensively considering the mechanical characteristics of materials and the geometric structure of channels, an actuator is designed by finite element method according to the fish tail. Different from rigid robots in tradition, the actuator is made of silicone rubber so that it can deform continuously and is more biomimetic. Robot fish with this kind of actuator has a potential to swim more quickly. Besides, the actuator is simple in structure and easy to make. The structure of the actuator can be further modified based on the study of the kinematics and dynamics models of the fish to achieve faster moving. Finally the finite element method is a good approach to analyzing the structure with nonlinearity problems, including geometric nonlinearity and material nonlinearity.

**Acknowledgments.** This work was supported by "the Fundamental Research Funds for the Central Universities".

## References

1. Crespi, A., Badertscher, A., Guignard, A., Ijspeert, A.J.: Robot. Auton. Syst. 50, 163–175 (2005)
2. Cao, Y., Dhang, J., Liang, K., et al.: Review of the Research Status of Soft robot. Journal of Mechanical Engineering 48(3), 25–33 (2012)
3. Shepherd, R.F., Ilievski, F., Choi, W., et al.: Multigait soft robot. Proceedings of the National Academy of Sciences 108(51), 20400–20403 (2011)
4. Onal, C.D., Rus, D.: Autonomous undulatory serpentine locomotion utilizing body dynamics of a fluidic soft robot. Bioinspiration & Biomimetics 8(2), 026003 (2013)
5. Onal, C.D., Rus, D.: Autonomous Soft Robotic Fish Capable of Escape. Soft Robotics 1(1) (2014)
6. Xie, G., He, C.: Independent biomimetic robotic fish, p. 25. Harbin Engineering University Press (2013)
7. Llievski, F., Mazzeo, A.D., Shepherd, R.F., et al.: Soft robotics for chemists. Angewandte Chemie 123(8), 1930–1935 (2011)

# Flexible Flying-Wing UAV Attitude Control Based on Back-Stepping, Adaptive and Terminal-Sliding Mode

Yinan Feng[1], Xiaoping Zhu[2], Zhou Zhou[1], and Yanxiong Wang[1]

[1] Science and Technology on UAV Laboratory, Northwestern Polytechnical University,
710065 Xi'an, China
[2] UAV Research Institute, Northwestern Polytechnical University,
710065 Xi'an, China
anan83030899@tom.com, zhouzhou@nwpu.edu.cn,
wangyanxiongli@163.com,

**Abstract.** Flexible Flying-wing UAV dynamics parameters changing fiercely results from Flying-wing UAV flight envelope and wing-body fusion exhibition string special aerodynamic layout, Aerodynamic nonlinear and nonlinear dynamics is very serious, the coupling effects, rigid/elastic coupling,between the channels and the coupling effect between the rudder surfaces is very serious, which show strong multivariable coupling and nonlinear, Moreover，other unknown disturbance factors lead to uncertainty. In order to solve such problems, the paper on the basis of such control methods as back stepping, the adaptive and non-singular terminal sliding mode，design a creative UAV attitude adaptive nonlinear sliding mode controller, Simulation examples suggest that the control method has strong nonlinear control performance and robust performance.

**Keywords:** nonlinear control, flexible flying-wing UAV, adaptive, terminal sliding mode, backstepping.

## 1    Introduction

Flying-wing UAV with considerable lift-to-drag ratio, which is good for loading task and invisibility has drawn more and more attention in recent years. But due to its special layout,large flight envelope environment and light materials selection Aerodynamic nonlinearity and nonlinear dynamics is very serious, the coupling effects, rigid/elastic coupling between the channels and the coupling effect between the rudder surfaces is very serious, which show strong multivariable coupling and nonlinear, Moreover,other unknown disturbance factors lead to uncertainty, Ever think rigid modal and flexible modal separation, the aircraft was established as the assumption of rigid body is no longer. So, it is necessary for the flexible flying wing UAV design strong robust and strong nonlinear controller of considering elastic. There has been some nonlinear control method applied such as，inverse system method, sliding mode control, dynamic programming, Back stepping control etc. the inverse system method has significant limitations because this method do not consider the system's

X. Zhang et al. (Eds.): ICIRA 2014, Part I, LNAI 8917, pp. 328–339, 2014.

uncertainty[1];Although the sliding mode control is very powerful in solving control problems about the uncertain nonlinear systems, it requires uncertainties satisfy the matching condition[3];Meanwhile,the adaptive inverse design method can well handle non-matching conditions of uncertainty, but it only takes parameter uncertainty into account, requiring controlled system[2]; basic backstepping control theories are not able to suppress external disturbance and modeling uncertainties[4].the paper on the basis of such control methods as back stepping,the adaptive and non-singular terminal sliding mode,design a creative UAV attitude adaptive nonlinear sliding mode controller.

The paper is organized as follows. Firstly, the dynamic model of Flexible flying-wing aircraft  is established. Secondly, the nonlinear controller based on Back-stepping,Adaptive and Terminal-sliding mode is designed. Finally, the flight simulation showed the controller designed at one trim condition could perform excellent in large flight envelope and under . parameter perturbation condition .

## 2    Nonlinear Dynamic Model of Flexible Flying-wing UAV

In order to nonlinear controller design for flexible flying-wing UAV.the nonlinear dynamic model of flying-wing aircraft is provided.[5]firstly. The paper  simplifies the flying wing UAV as a non-uniform elastic beam, According to the Lagrange equation and principle of virtual work,the model is deduced   for elastic body/rigid body coupling dynamics model of UAV control.Nonlinear model of the UAV is shown below.

$$mV\frac{d\beta}{dt}=-T\sin\beta\cos\alpha+C-mV(-p\sin\alpha\sin\beta+r\cos\alpha)+G_{ya}$$

$$+(\dot{q}\sum_{i=1}^{n}\lambda_i\eta_i+2q\sum_{i=1}^{n}\lambda_i\dot{\eta}_i+pr\sum_{i=1}^{n}\lambda_i\eta_i)\sin\beta\cos\alpha$$

$$-(-\dot{p}\sum_{i=1}^{n}\lambda_i\eta_i+2p\sum_{i=1}^{n}\lambda_i\dot{\eta}_i-qr\sum_{i=1}^{n}\lambda_i\eta_i)\cos\beta+(\sum_{i=1}^{n}\lambda_i\ddot{\eta}_i-q^2\sum_{i=1}^{n}\lambda_i\eta_i-p^2\sum_{i=1}^{n}\lambda_i\eta_i)\sin\beta\sin\alpha \tag{1}$$

$$mV\cos\beta\frac{d\alpha}{dt}=-T\sin\alpha-L+mV\left(-p\cos\alpha\sin\beta+q\cos\beta-r\sin\alpha\sin\beta\right)+G_{za}$$

$$+(\dot{q}\sum_{i=1}^{n}\lambda_i\eta_i+2q\sum_{i=1}^{n}\lambda_i\dot{\eta}_i+pr\sum_{i=1}^{n}\lambda_i\eta_i)\sin\alpha-(\sum_{i=1}^{n}\lambda_i\ddot{\eta}_i-q^2\sum_{i=1}^{n}\lambda_i\eta_i-p^2\sum_{i=1}^{n}\lambda_i\eta_i)\cos\alpha \tag{2}$$

$$\dot{\mu}=p\frac{\cos\alpha}{\cos\beta}+r\frac{\sin\alpha}{\cos\beta}-\frac{g\cos\gamma\cos\mu\tan\beta}{V}+\frac{L}{mV}(\sin\mu\tan\gamma+\tan\beta)+\frac{C\cos\mu\tan\gamma}{mV}$$

$$+\frac{T_y}{mV}\tan\gamma\cos\beta\cos\mu+\frac{T_x\sin\alpha-T_z\cos\alpha}{mV}(\sin\mu\tan\gamma+\tan\beta) \tag{3}$$

$$-\frac{T_x\cos\alpha+T_z\sin\alpha}{mV}(\cos\mu\tan\gamma\sin\beta)$$

$$\dot{p}=\left(\frac{1}{a_1}+\frac{a_2}{a_1}\frac{a_6}{a_1\Sigma}\right)\bar{A}+\frac{a_2}{a_1}\frac{a_4}{a_3\Sigma}\bar{B}+\frac{a_2}{a_1}\frac{1}{\Sigma}\bar{C}+\left(\frac{1}{a_1}+\frac{a_2}{a_1}\frac{a_6}{a_1\Sigma}\right)l+\frac{a_2}{a_1}\frac{a_4}{a_3\Sigma}m+\frac{a_2}{a_1}\frac{1}{\Sigma}n \tag{4}$$

$$\dot{q}=\frac{a_4}{a_3}\frac{a_6}{a_1\Sigma}\bar{A}+\left(\frac{1}{a_3}+\frac{a_4}{a_3}\frac{a_4}{a_3\Sigma}\right)\bar{B}+\frac{a_4}{a_3}\frac{1}{\Sigma}\bar{C}+\frac{a_4}{a_3}\frac{a_6}{a_1\Sigma}l+\left(\frac{1}{a_3}+\frac{a_4}{a_3}\frac{a_4}{a_3\Sigma}\right)l+\frac{a_4}{a_3}\frac{1}{\Sigma}n \tag{5}$$

$$\dot{r}=\frac{a_6}{a_1}\frac{1}{\Sigma}\bar{A}+\frac{a_4}{a_3}\frac{1}{\Sigma}\bar{B}+\frac{1}{\Sigma}\bar{C}+\frac{a_6}{a_1}\frac{1}{\Sigma}l+\frac{a_4}{a_3}\frac{1}{\Sigma}m+\frac{1}{\Sigma}n \tag{6}$$

Where, $a_1=\left(I_{yy}+\sum_{i=1}^n \eta_i^2\right)$ , $a_2=\left(I_{xz}+\sum_{i=1}^n \psi_{xi}\eta_i\right)$ , $a_4=\sum_{i=1}^n \psi_{yi}\eta_i$

$a_3=\left(I_{yy}+\sum_{i=1}^n \eta_i^2\right)$ , $a_5=I_{zz}$ , $a_6=\left(I_{xz}+\sum_{i=1}^n \psi_{xi}\eta_i\right)$

$A=L+\bar{A}$ , $B=M+\bar{B}$ , $C=N+\bar{C}$

$$\bar{A}=-\left(I_{zz}-I_{yy}-\sum_{i=1}^n \eta_i^2\right)qr+I_{xz}qp-2p\sum_{i=1}^n \eta_i\dot{\eta}_i$$

$$+\left(\dot{v}+ru\right)\sum_{i=1}^n \lambda_i\eta_i-\left(r^2-q^2\right)\sum_{i=1}^n \psi_{yi}\eta_i-\sum_{i=1}^n \psi_{yi}\ddot{\eta}_i+pq\sum_{i=1}^n \psi_{xi}\eta_i$$

$$\bar{B}=-\left(I_{xx}-I_{zz}\right)pr-\left(p^2-r^2\right)I_{xz}-\left(\dot{u}+qw-vr\right)\sum_{i=1}^n \lambda_i\eta_i$$

$$-2q\sum_{i=1}^n \eta_i\dot{\eta}_i+\sum_{i=1}^n \psi_{xi}\ddot{\eta}_i-pr\sum_{i=1}^n \eta_i^2-\left(p^2-r^2\right)\sum_{i=1}^n \psi_{xi}\eta_i-pq\sum_{i=1}^n \psi_{yi}\eta_i$$

$$\bar{C}=-\left(I_{yy}-I_{xx}\right)pq-I_{xz}qr+pr\sum_{i=1}^n \psi_{yi}\eta_i$$

$$-qr\sum_{i=1}^n \psi_{xi}\eta_i+2p\sum_{i=1}^n \psi_{xi}\dot{\eta}_i+2q\sum_{i=1}^n \psi_{yi}\dot{\eta}_i$$

According to the established flexible UAV model, The radiation nonlinear equation of UAV for nonlinear control is derived by Taylor theorem theorem and the mean value theorem at $[\bar{p},\bar{q},\bar{r}]^T=0$ .With this, air flow angle equation can be transformed to affine nonlinear equation .

## 2.1    Air Flow Angle Affine Nonlinear Equation

Equation(1-3) are rewritten in the standard form for the application of nonlinear   controller considering the unmodeled dynamics and disturbance.

$$\dot{x_1} = F_1(x) + G_1(x)x_2 + \Delta_1 \tag{7}$$

Where $x_1 = [\beta, \alpha, \mu]^T$, $x_2 = [p, q, r]^T$, $x = (\beta, \alpha, \mu, p, q, r, \eta_i, \dot{\eta}_i)^T$, $\Delta_1$ is modeling discrepancy caused by unmodeled dynamics and disturbance. $F_1(x)$ is coupling nonlinear force vector function.which is defined as $F_1(x) = [f_\beta(x), f_\alpha(x), f_\mu(x)]^T$. $G_1(x)$ relates the fast states to the slow states.Reference(1-3),The expressions of $F_1(x) = [f_\beta(x), f_\alpha(x), f_\mu(x)]^T$ and $G_1(x)$ may be given.

$$
\begin{aligned}
f_\beta(x) = & \frac{\overline{q}SC_Y(\alpha, \beta, Ma)}{MV_t} + \frac{g \sin\mu \cos\gamma}{V_t} - \frac{T_z \sin\alpha \sin\beta}{MV_t} + \\
& \frac{(\dot{q}\sum_{i=1}^{n}\lambda_i\eta_i + pr\sum_{i=1}^{n}\lambda_i\eta_i)\sin\beta\cos\alpha}{MV_t} - \frac{(-\dot{p}\sum_{i=1}^{n}\lambda_i\eta_i - qr\sum_{i=1}^{n}\lambda_i\eta_i)\cos\beta}{MV_t} \\
& + \frac{(\sum_{i=1}^{n}\lambda_i\ddot{\eta}_i - q^2\sum_{i=1}^{n}\lambda_i\eta_i - p^2\sum_{i=1}^{n}\lambda_i\eta_i)\sin\beta\sin\alpha}{MV_t}
\end{aligned}
\tag{8}
$$

$$
\begin{aligned}
f_\alpha(x) = & -\frac{\overline{q}SC_L(\alpha, \beta, Ma)}{MV_t\cos\beta} + \frac{g\cos\mu\cos\gamma}{V_t\cos\beta} - \frac{T_x\sin\alpha}{MV_t\cos\beta} + \\
& \frac{(\dot{q}\sum_{i=1}^{n}\lambda_i\eta_i + pr\sum_{i=1}^{n}\lambda_i\eta_i)\sin\alpha}{MV_t\cos\beta} - \frac{(\sum_{i=1}^{n}\lambda_i\ddot{\eta}_i - q^2\sum_{i=1}^{n}\lambda_i\eta_i - p^2\sum_{i=1}^{n}\lambda_i\eta_i)\cos\alpha}{MV_t\cos\beta}
\end{aligned}
\tag{9}
$$

$$
\begin{aligned}
f_\mu(x) = & \frac{\overline{q}SC_L(\alpha, \beta, Ma)}{MV_t}(\sin\mu\tan\gamma + \tan\beta) \\
& + \frac{\overline{q}SC_Y(\alpha, \beta, Ma)\cos\mu\tan\gamma}{MV_t} \\
& - \frac{g\cos\gamma\cos\mu\tan\beta}{V_t} + \frac{T_y}{MV_t}\tan\gamma\cos\beta\cos\mu \\
& + \frac{T_x\sin\alpha - T_z\cos\alpha}{MV_t}(\sin\mu\tan\gamma + \tan\beta) \\
& - \frac{T_x\cos\alpha + T_z\sin\alpha}{MV_t}(\cos\mu\tan\gamma\sin\beta)
\end{aligned}
\tag{10}
$$

$$
G_1(x) = \begin{bmatrix}
-\sin\alpha + 2\cos\beta\sum_{i=1}^{n}\lambda_i\eta_i & 2\sum_{i=1}^{n}\lambda_i\dot{\eta}_i & -\cos\alpha \\
-\tan\beta\cos\alpha & 1 + 2\sum_{i=1}^{n}\lambda_i\dot{\eta}_i & -\tan\beta\sin\alpha \\
\sec\beta\cos\alpha & 0 & \sec\beta\sin\alpha
\end{bmatrix}
\tag{11}
$$

## 2.2    Angle Velocity Affine Nonlinear Equation

Equation(4-6) are rewritten in the standard form for the application of nonlinear   controller considering the unmodeled dynamics and disturbance.

$$\dot{x}_2 = F_2(x) + G_2(x)\delta + \Delta_2 \tag{12}$$

Where, $x_2 = [p, q, r]^T$, $x = (\beta, \alpha, \mu, p, q, r, \eta_i, \dot{\eta}_i)^T$,

$\Delta_2$ is modeling discrepancy caused by unmodeled dynamics and disturbance. $F_2(x)$ is coupling nonlinear force vector function.which is defined as $F_2(x) = [f_p(x), f_q(x), f_r(x)]^T$ . $G_2(x)$ is $3 \times 3$ matrix relates the three control variables.Reference(4-6),The expressions of $F_2(x) = [f_p(x), f_q(x), f_r(x)]^T$ and $G_2(x)$ may be given.

$$f_p(x) = \left(\frac{1}{a_1} + \frac{a_2}{a_1}\frac{a_6}{a_1\Sigma}\right)\bar{A} + \frac{a_2}{a_1}\frac{a_4}{a_3\Sigma}\bar{B} + \frac{a_2}{a_1}\frac{1}{\Sigma}\bar{C} + \left(\frac{1}{a_1} + \frac{a_2}{a_1}\frac{a_6}{a_1\Sigma}\right)\hat{l} + \frac{a_2}{a_1}\frac{a_4}{a_3\Sigma}\hat{m} + \frac{a_2}{a_1}\frac{1}{\Sigma}\hat{n} \tag{13}$$

$$f_q(x) = \frac{a_4}{a_3}\frac{a_6}{a_1\Sigma}\bar{A} + \left(\frac{1}{a_3} + \frac{a_4}{a_3}\frac{a_4}{a_3\Sigma}\right)\bar{B} + \frac{a_4}{a_3}\frac{1}{\Sigma}\bar{C} + \frac{a_4}{a_3}\frac{a_6}{a_1\Sigma}\hat{l} + \left(\frac{1}{a_3} + \frac{a_4}{a_3}\frac{a_4}{a_3\Sigma}\right)\hat{m} + \frac{a_4}{a_3}\frac{1}{\Sigma}\hat{n} \tag{14}$$

$$f_r(x) \frac{a_6}{a_1\Sigma}\bar{A} + \frac{a_4}{a_3\Sigma}\bar{B} + \frac{1}{\Sigma}\bar{C} + \frac{a_6}{a_1\Sigma}\hat{l} + \frac{a_4}{a_3\Sigma}\hat{m} + \frac{1}{\Sigma}\hat{n} \tag{15}$$

$$G_f(x) = \begin{bmatrix} \left(\dfrac{1}{a_1} + \dfrac{a_2}{a_1}\dfrac{a_6}{a_1\Sigma}\right) & \dfrac{a_2}{a_1}\dfrac{a_4}{a_3\Sigma} & \dfrac{a_2}{a_1}\dfrac{1}{\Sigma} \\[2ex] \dfrac{a_4}{a_3}\dfrac{a_6}{a_1\Sigma} & \left(\dfrac{1}{a_3} + \dfrac{a_4}{a_3}\dfrac{a_4}{a_3\Sigma}\right) & \dfrac{a_4}{a_3}\dfrac{1}{\Sigma} \\[2ex] \dfrac{a_6}{a_1\Sigma} & \dfrac{a_4}{a_3\Sigma} & \dfrac{1}{\Sigma} \end{bmatrix} \tag{16}$$

# 3    Nonlinear Controller Design

The flying wing UAV  flight control law design goal is to ensure the air flow angle can asymptotic track the expect order signal $\beta_c, \alpha_c, \mu_c$ .no matter the situation of airflow , the model parameter and the un-modeled dynamic conditions. The basic idea is[5]: Firstly, based on the given reference signal of $\beta_c, \alpha_c, \mu_c$ , then use the adaptive backstepping slide mode control method to design the expect pitch rate $p_c, q_c, r_c$ of the UAV, which can make $\beta, \alpha, \mu$ global convergence to $\beta_c, \alpha_c, \mu_c$ very rapidly. Secondly, consider the $p_c, q_c, r_c$ as the reference input, and design control law $u$ , to make the $p, q, r$ can tracks the $p_c, q_c, r_c$ very quickly.

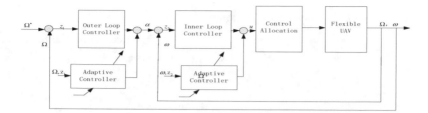

**Fig 1.** Schematic of  UAV control system

Assuming control object satisfies the following assumptions :

Assumption 1: There exists an unknown positive number $p_i$ and Non-negative smooth function $\delta_i(x_i,t)$, thus:

$$\|\Delta_i(x_i,t)\| \le \rho_i\delta_i(x_i,t), i = 1,2 \qquad (17)$$

Assumption 2: The expectations bounded tracking signal $y_d(t)$ is continuous and has the derivative until $n$ order,  and the expectation's $n+1$ dimensions tracking vector is bounded on $y_d(t)=\left[y_d,\dot{y}_d\cdots y_d^n\right]..$

### 3.1    The Virtual Control Law Design

**Step 1** : define the error variable of the UAV's angle of attack $z_1=x_1-y_d$,  then exist:

$$\dot{z}_1=\dot{x}_1-\dot{y}_d=f_1(x_1)+\varphi_1^T(x_1)\theta+g_1(x_1)x_2+\Delta(x_1,t)-\dot{y}_d \qquad (18)$$

Set  $\hat{\theta}_i$ are the estimate value of  $\theta_i$ , so we can get the equations like:  $\tilde{\theta} = \theta - \hat{\theta}$ . We can modify the equation (17) as:

$$\dot{z}_1=\dot{x}_1-\dot{y}_d=f_1(x_1)+\psi_1^T\hat{\theta}+g_1(x_1)x_2+\Delta(x_1,t)-\dot{y}_d+\psi_1^T\tilde{\theta} \qquad (19)$$

Where  $\psi_1(x_1)=\varphi_1(x_1)$

For the sub-system(7), define Lyapunov function

$$V_1 = \frac{1}{2}z_1^T z_1 + \frac{1}{2}\tilde{\theta}^T\Gamma^{-1}\tilde{\theta} \qquad (20)$$

Among them:  $\Gamma=\Gamma^T>0$ is parameter adaptive gain matrix,  derivation for  $V_1$, we can get:

$$\dot{V}_1 = z_1\left(f_1(x_1)+ g_1(x_1)x_2 + \varphi_1^T\hat{\theta} + \Delta_1(x_1,t)- \dot{y}_d\right)$$
$$+\tilde{\theta}^T\Gamma^{-1}\left(\Gamma\varphi_1 z_1 - \dot{\hat{\theta}}\right) \qquad (21)$$

Define $\tau_1 = \Gamma\varphi_1 z_1$ , design virtual control

$$\alpha_1 = \frac{1}{g_1(x_1)}\left(-f_1(x_1) - \varphi_1^T\hat{\theta} - c_1 z_1 - \frac{z_1\rho_1^2}{2\varepsilon} + \dot{y}_d\right) \tag{22}$$

$c_1$ is positive constant, $\dfrac{z_1\rho_1^2}{2\varepsilon}$ is nonlinear damping term.

Define the angular velocity error variables.

$$z_2 = x_2 - \alpha_1 = x_2 + \frac{1}{g_1(x_1)}\left(f_1(x_1) + \varphi_1^T\hat{\theta} + c_1 z_1 + \frac{z_1\rho_1^2}{2\varepsilon} - \dot{y}_d\right) \tag{23}$$

Make(23)substitute(22)

$$\dot{z}_1 = \dot{x}_1 - \dot{y}_d = -c_1 z_1 + g_1(x_1)z_2 + \varphi_1^T\tilde{\theta} + \Delta_1(x_1,t) - \frac{z_1\rho_1^2}{2\varepsilon} \tag{24}$$

Make $\alpha_1$ substitute $\dot{V}_1$ Then we can get:

$$\begin{aligned}
\dot{V}_1 &= -c_1\|z_1\|^2 + z_1\varphi_1^T\tilde{\theta} + \frac{1}{g_1(x_1)}z_1 z_2 + z_1\left(\Delta_1 - \frac{z_1\rho_1^2}{2\varepsilon}\right) + \tilde{\theta}^T\Gamma^{-1}\dot{\hat{\theta}} \\
&= -c_1\|z_1\|^2 + z_1\varphi_1^T\tilde{\theta} + g_1(x_1)z_1 z_2 + z_1\left(\Delta_1 - \frac{z_1\rho_1^2}{2\varepsilon}\right) + \tilde{\theta}^T\Gamma^{-1}\dot{\hat{\theta}}
\end{aligned} \tag{25}$$

### 3.2     The Final Control Law Design

**Step 2** : Design the actual controlling variable $u$, to make angular velocity tracking error variable $z_2$ converge to zero in a finite time. In order to improve the error system's convergence speed and steady-state tracking accuracy, we us the non-singular terminal sliding surface.

**Assumption 1:** The uncertainty conditions are satisfy the matching situation and its derivative is bounded,  thus: $|\dot{\Delta}_i| \leq l_f$ Among  $s = z_2 + \gamma\dot{z}_2^{p/q}$ : $p, q$ are positive odds , $1 < p/q < 2; \gamma > 0$ 。 Assume that at the time $t_r, s(t_r) = 0$  then $z_2$ and $\dot{z}_2$ will converge to zero in finite time,  the converging time point is

$$t_s = t_r + \gamma^{q/p}\left(\frac{p}{p-q}\right)|z_2(t_r)|^{\frac{p-q}{p}},$$ then , the error system enter in 2-order sliding mode situation $z_2 = \dot{z}_2 = 0$.

Now we can design the control law base on sliding mode equivalent control method, so we can get $u = u_{eq} + u_c$. $u_{eq}$ is the equivalent control, which is used to offset the existing non-linear characteristics ; $u_c$ is the sliding mode, which is used to the impact of non-parametric uncertainties.

Then design compensation control law

$$u_{eq} = -g_2^{-1}\left[ f_2\left(x_1, x_2\right) + \varphi_2 \hat{\theta} - \frac{\partial \alpha_1}{\partial \hat{\theta}}\dot{\hat{\theta}}\right] \tag{26}$$

And design sliding mode control law

$$u_c = -g_2^{-1}\int_0^t\left[\frac{1}{\gamma}\frac{p}{q}\dot{z}_2^{2-p/q} + \left(l_f + \eta\right)\mathrm{sgn}\left(s\right)\right]d\tau \tag{27}$$

derivate for $z_2$ , we can get:

$$\dot{z}_2 = \dot{x}_2 - \dot{\alpha}_1 = f_2(x_1, x_2) + \phi_2^T\theta + g_2(x_1)u + \Delta_2(x,t) - \dot{\alpha}_1 \tag{28}$$

Substitute the control law ,we can get:

$$\dot{z}_2 = \dot{x}_2 - \dot{\alpha}_1 = g_2 u_c + \Delta_2 - \theta^T\frac{\partial\alpha}{\partial x_1}\varphi_1 \tag{29}$$

Derivate $\dot{z}_2$ again

$$\ddot{z}_2 = \Delta g_2\dot{u} + \dot{\Delta}_2 - \theta^T\left[\frac{\partial F}{\partial x_1}x_2 + \frac{\partial F}{\partial\hat{\theta}}\dot{\hat{\theta}} + \frac{\partial F}{\partial x_1}\varphi_1^T\theta\right]$$

$$G_1 = \frac{\partial F}{\partial x_1}x_2 + \frac{\partial F}{\partial\theta}\dot{\hat{\theta}}, G_2 = \frac{\partial F}{\partial x_1}\varphi_1^T, F = \frac{\partial\alpha_1}{\partial x_1}\theta \tag{30}$$

Consider Lyapunov function below: $V_2 = 0.5S^2$ derivate for $V_2$ :

$$\dot{V}_2 = s\dot{s} = s\left(\dot{z}_2 + \frac{p}{q}\dot{z}_2^{p/q-1}\ddot{z}_2\right)$$

$$= s\gamma\frac{p}{q}\dot{z}_2^{p/q-1}\left(\ddot{z}_2 + \frac{1}{\gamma}\frac{q}{p}\dot{z}_2^{2-p/q}\right)$$

$$= s\gamma\frac{p}{q}(\dot{x}_2 - \dot{\alpha}_1)^{p/q-1}\left(\ddot{z}_2 + \frac{1}{\gamma}\frac{q}{p}\dot{z}_2^{2-p/q}\right)$$

$$= s\gamma\frac{p}{q}\left(f_2 + \phi_2^T\theta + g_2 u + \Delta_2 - \frac{\partial\alpha}{\partial x_1}x_2 - \frac{\partial\alpha}{\partial\hat{\theta}}\dot{\hat{\theta}} - \theta^T\frac{\partial\alpha}{\partial x_1}\varphi_1\right)^{p/q-1}\left(\ddot{z}_2 + \frac{1}{\gamma}\frac{p}{q}\dot{z}_2^{2-p/q}\right) \tag{31}$$

So:

$$\dot{V}_2 = s\dot{s} = s\gamma\frac{p}{q}\dot{z}_2^{p/q-1}\left[\left(\dot{\Delta}_2 - \theta^T G_1 - \theta^T G_2\theta\right) - \right.$$

$$= \left(l_f + \eta\right)]\mathrm{sgn}\left(s\right)\le -\gamma\frac{p}{q}\dot{z}_2^{p/q-1}\eta|s| \tag{32}$$

$$\le 0$$

# 4     Simulation

Control system's simulation is conducted aiming  at the flexible flying-wing UAV dynamics model  Control goal is to make the UAV to track a given  Instructions and to ensure the elastic modal not to spread.  In order to verify the effectiveness and robust of the control strategy, Firstly,Simulation at different state point of  large flight envelope which is based on the one trim condition controller;sencondly,Simulation under the circumstance that 30% perturbations of all the aerodynamic coefficients are assigned.The simulation initial conditions are:

$$x = y = 0m, H = 9.0km, M = 5300kg, V = 0.5ma, \chi = 0, \gamma = 0, \alpha = 1.14^{o}$$

$$\beta = 0^{o}, \mu = 0^{o}, p = q = r = 0rad / s, \eta_{i} = 0.01, \dot{\eta}_{i} = 0.01, (i = 1, 2 \cdots)$$

## 4.1     Different Control Law Simulation

The controller is tested in the others flight condition points using the same parameters designed in one trim condition point.The attitude command is $\beta_{c} = 0^{o}, \alpha_{c} = 5^{o}, \mu_{c} = 2^{o}$ .the result of numerical simulation is shown below:

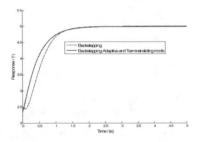 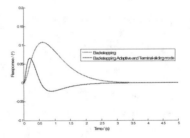

**Fig. 2.** Response of alpha angle          **Fig. 3.** Response of beta angle

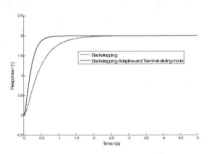 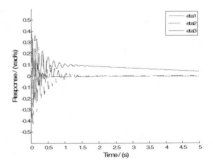

**Fig. 4.**   Response of  mu angle

**Fig. 5.** Response of flexible State under different status

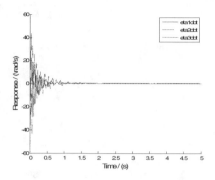

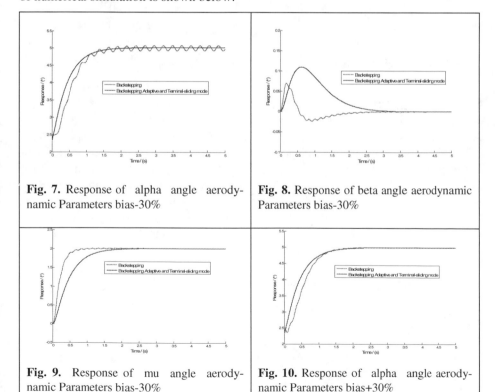

**Fig. 6.** Response of   derivatives of Flexible States   under different status

## 4.2    Robust Simulation

Simulation under the circumstance that 30% perturbations of all the aerodynamic coefficients are assigned.The attitude command   is $\beta_c = 0^o, \alpha_c = 5^o, \mu_c = 2^o$ . the result of numerical simulation is shown below:

**Fig. 7.** Response of   alpha   angle   aerodynamic Parameters bias-30%

**Fig. 8.** Response of beta angle aerodynamic Parameters bias-30%

**Fig. 9.**   Response of   mu   angle   aerodynamic Parameters bias-30%

**Fig. 10.** Response of   alpha   angle aerodynamic Parameters bias+30%

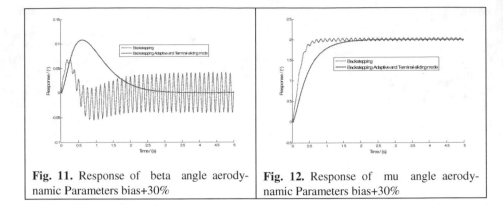

**Fig. 11.** Response of beta angle aerodynamic Parameters bias+30%

**Fig. 12.** Response of mu angle aerodynamic Parameters bias+30%

The simulation results show that,when the controller track the order signal, The tracking performance of adaptive back-stepping nonlinear sliding mode controller is better than back-stepping.it will response the order very quickly,The outputs of the actual system are matched with the order signals, moreover, there is no overshoot during the tracking $\beta_c, \alpha_c, \mu_c$.it Can effectively restrain the aerodynamic parameter perturbation,Shows that the robustness of the controller is good. UAV elastic modal can converge to a steady state in short time, and there is no overshoot phenomenon, shorter adjusting time. the simulation results indicate that the paper proposes adaptive back-stepping nonlinear sliding mode controller which is a good way to control flying wing UAV aircraft motion, and to steady aerodynamic elasticity mode.

## 5    Conclusion

Aiming at the strong coupling nonlinear model of flying wing UAV,The paper Combined with the back-stepping, adaptive and sliding mode control method, a novel flying wing UAV adaptive back-stepping sliding mode controller is design. control system has good dynamic quality,tracking performance and strong robustness. Depending on the robustness of adaptive sliding mode control, control system solves the traditional back-stepping adaptive control problem of the computational expansion without introducing differential or filter for virtual control input, simplified controller designing .and Without known aerodynamic parameter perturbation bounds, overcame the difficult to accurately predict in practical engineering parameter perturbation range of difficulties, have engineering application value.

## References

1. Zhang, J., Jiang, C.S., Fang, W.: Variale structure near space vehicle control characteristics of large flight envelope. Journal of Astronautics 30(2), 2390–2394 (2009)
2. Han, T.T., Ge, S.S., Lee, T.H.: Adaptive neural control for a class of swithed nonlinear systems. System and Control Letter 58(2), 109–118 (2009)

3.  Pai, M.C.: Observer-based adaptive sliding mode control for nonlinear uncertain systems. Automatic 45(8), 1923–1928 (2009)
4.  Zhou, Y., Wu, Y., Hu, Y.: Robust backstepping sliding mode control of a class of uncertain MIMO nonlinear systems. In: 2007 IEEE International Conference on Control and Automatic, Guzhou, China (2007)
5.  Newman, B., Buttrill, C.: Conventional flight control for an aeroelastic, relaxed static stability high-speed trasnsport. In: AIAA Guidance, Navigation, and Control, pp. 717–720 (1995)

# Landing Control System Design for a Flying-Wing Aircraft Based on ADRC

Yanxiong Wang[1], Xiaoping Zhu[2], Zhou Zhou[1], and Zhuang Shao[1]

[1] Science and Technology on UAV Laboratory, Northwestern Polytechnical University,
710065 Xi'an, China
[2] UAV Research Institute, Northwestern Polytechnical University,
710065 Xi'an, China
{wangyanxiongli,shaozhuang233@163.com}, zhouzhou@nwpu.edu.cn

**Abstract.** In this paper, the flight simulation and control of a flying-wing aircraft in landing is presented. Significant dynamic nonlinearity and atmospheric disturbance are main problems for a flying-wing aircraft in landing phase. Therefore, a nonlinear controller with strong robustness is required. Active disturbance rejection control technique is used to design the controller which could estimate and compensate the uncertainties. A landing condition in which the wind shear encounters and ground effect before touchdown is considered, and a nonlinear longitudinal dynamic model of flying-wing aircraft in variable wind field is established. Simulations are conducted in MATLAB and the controller is shown to suppress the wind effects and ground proximity.

**Keywords:** nonlinear control, active disturbance rejection control, control allocation, wind encounters, ground effect, flying-wing aircraft.

## 1 Introduction

The main problems for a flying-wing aircraft landing control system design are from significant dynamic nonlinearity and atmospheric disturbance. In accordance with the nonlinear object, nonlinear control is needed. Nonlinear control theory has been a hot area of research for several years, domestic and overseas scholars developed many nonlinear control theories. Nonlinear dynamic inversion (NDI) [1] based on feedback linearization and back stepping theory [2] and [3] based on Lyapunov stability theory has been used for flight control. Defects of nonlinear dynamic inversion and basic backstepping control theories are accurate model required and not able to suppress external disturbance and modeling uncertainties. In landing phase, as low altitude and airspeed, any disturbance would put aircraft at risk, including wind shear encounters and ground effect. A nonlinear controller with strong robustness is required.

In this paper, active disturbance rejection control (ADRC) technique [4] is proposed to solve these problems. The core part of ADRC is nonlinear extended state observer (ESO), which affords a means of estimating external disturbance and modeling uncertainties. To integrate the ESO with feedback compensator make the controller possess strong robustness.

X. Zhang et al. (Eds.): ICIRA 2014, Part I, LNAI 8917, pp. 340–351, 2014.

The paper is organized as follows. Firstly, the dynamic model of flying-wing air-craft in variable wind field is established. Secondly, the landing control system based on ADRC is designed. Finally, the flight simulation is presented in a landing condition considers wind shear encounters and ground effect before touchdown, the responses and the performance of the landing control system are shown in figures and discussed.

## 2  Nonlinear Longitudinal Dynamic Model of Flying-Wing Aircraft in Variable Wind Field

To set this work in proper context, the nonlinear longitudinal dynamic model of fly-ing-wing aircraft in variable wind field is provided firstly. By ignoring earth rotation and elastic deformation, the longitudinal dynamic model could be described by follow first order differential equations [5] and [6].

$$\dot{V}_A = \frac{1}{M}[-D - Mg\sin\gamma + T\cos\alpha] - \dot{u}_{Wg}\cos\gamma + \dot{w}_{Wg}\sin\gamma \tag{1}$$

$$\dot{\gamma} = \frac{1}{MV_A}[L - Mg\cos\gamma + T\sin\alpha] + \frac{1}{V_A}[\dot{u}_{Wg}\sin\gamma + \dot{w}_{Wg}\cos\gamma] \tag{2}$$

$$\dot{\alpha} = \frac{1}{MV_A}[-L + Mg\cos\gamma - T\sin\alpha] + q - \frac{1}{V_A}[\dot{u}_{Wg}\sin\gamma + \dot{w}_{Wg}\cos\gamma] \tag{3}$$

$$\dot{q} = \frac{1}{I_{yy}}m \tag{4}$$

$$\dot{z}_g = -V_A\sin\gamma + w_{Wg} \tag{5}$$

where $u_{Wg}$ and $w_{Wg}$ are the horizontal wind speed and vertical wind speed, $V_A$ is airspeed, $\gamma$ and $\alpha$ are the flight-path angle and attack angle , $q$ is pitch rate, $z_g$ is the displacement on the principal axis of inertia $z$ direction, $M$ and $g$ are the mass of aircraft and gravitational acceleration, $I_{yy}$ is inertia moment on $y$ direction of body axis, $m$ is the pitch moment.

The pitch moment is composed of body pitch moment $m_b$, damping moment $m_q$ and controlling moment $m_\delta$.

$$m = m_b + m_q + m_\delta \tag{6}$$

By the influence of wind shear, the damping moment applied on flying-wing air-craft in variable wind field is diffident [5].

$$m_q = Qsc_A C_{mq}(q + w_{Wx})\frac{c_A}{2V_A} \tag{7}$$

where $Q$ is dynamic pressure, $s$ is wing surface, $C_A$ is mean aerodynamic chord, $C_{mq}$ is damping moment coefficient, $w_{Wx}$ is wind gradient.

# 3     Active Disturbance Rejection Controller

## 3.1     Schematic of ADRC Controller

The Schematic of landing control system is shown in Fig. 1. The altitude and flight path angle loop are designed using ADRC as a first order cascade controller according to the time scale separation principle. In order to improve response speed, the control law of attack angle and pitch rete is design as a second order back stepping controller combining with ESO.

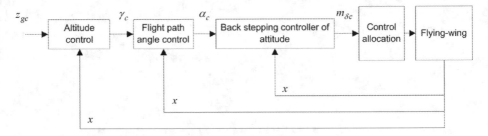

**Fig. 1.** Schematic of ADRC controller

## 3.2     ADRC Controller of Outer Loops Design

Equation (2) and (5) could be expressed as standard form of first-order system.

$$\dot{x}_i = f_i(x) + w_i(x,t) + g_i(x)u_i \tag{8}$$

where $f_i(x)$ is known part, $w_i(x,t)$ is unknown part including modeling uncertainties and wind disturbance, $g_i(x)$ is nonlinear functions, $u_i$ is input, subscript $i$ represents states including $\gamma$ and $z_g$.

The control algorithm of ADRC used in altitude and flight path angle loop is given below.

$$\dot{z}_{1i} = z_{2i} - \beta_{1i}(z_{1i} - x_i) + f_i(x) + g_i(x)u_i \tag{9}$$

$$\dot{z}_{2i} = -\beta_{2i} fal(z_{1i} - x_i, \varepsilon, \Delta) \tag{10}$$

$$u_{0i} = \beta_{0i}(v_{ci} - z_{1i}) \tag{11}$$

$$u_i = g^{-1}(x)(u_{0i} - f_i(x) - z_{2i}) \tag{12}$$

where $v_{ci}$ is the control command signal computed by outer loop, $z_{1i}$ and $z_{2i}$ are the estimated values of state $x_i$ and uncertainties $w_i(x,t)$ computed by ESO, $\beta_{1i}$ and $\beta_{2i}$ are parameters of ESO, $\beta_{0i}$ is gain.

The *fal* function is given below.

$$fal(e,\varepsilon,\Delta) = \begin{cases} \dfrac{e}{\Delta^{1-\varepsilon}} & |e| \le \Delta \\ |e|^{\varepsilon} \, sign(e) & |e| > \Delta \end{cases} \tag{13}$$

To choose appropriate parameters make ESO converge, $z_{1i}$ would be tend to $x_i$, and $z_{2i}$ would be tend to $w_i(x,t)$. Due to the accurate estimated value of uncertainties, feedback compensator can be realized in equation (12), which results in better robustness.

When flight-path angle is small, below relationship exists.

$$\sin \gamma \approx \gamma \tag{14}$$

The $f_i(x)$ and $g_i(x)$ in each loop are designed below.

$$f_\gamma(x) = \frac{1}{MV_A}[QsC_{L0} + L_q - Mg\cos\gamma] \tag{15}$$

$$f_z(x) = 0 \tag{16}$$

$$g_\gamma(x) = \frac{1}{MV_A}[QsC_L^\alpha + \mathrm{T}] \tag{17}$$

$$g_z(x) = -V_A \tag{18}$$

where $C_{L0}$ is the lift ratio while $\alpha = 0$.

The control inputs in each loop are chosen as $\alpha_c$ and $\gamma_c$. Subscript $c$ represents command signal.

According to the time scale separation principle, the $\beta_{0i}$ in each loop are designed as $\beta_{0\gamma} = 1$ and $\beta_{0z_g} = 0.3$.

## 3.3    Attitude Controller Design

The design procedure is show below.

First step, the error of attack angle is defined as

$$l_1 = \alpha_c - \alpha \tag{19}$$

Then

$$\dot{l}_1 = \dot{\alpha}_c - \dot{\alpha} = \dot{\alpha}_c - f_\alpha(x) - w_\alpha(x,t) - q \tag{20}$$

where

$$f_\alpha(x) = \frac{1}{MV_A}[-L + Mg\cos\gamma - T\sin\alpha] \tag{21}$$

The Lyapunov function is defined as

$$V_1 = \frac{1}{2}l_1^2 \tag{22}$$

Then

$$\dot{V}_1 = l_1\dot{l}_1 = l_1(\dot{\alpha}_c - f_\alpha(x) - w_\alpha(x,t) - q) \tag{23}$$

Set

$$q = c_1 l_1 + \dot{\alpha}_c - f_\alpha(x) - w_\alpha(x,t) - l_2 \tag{24}$$

where $l_2$ is virtual control, $c_1$ is constant and $c_1 > 0$.

The $l_2$ is defined as

$$l_2 = c_1 l_1 + \dot{\alpha}_c - f_\alpha(x) - w_\alpha(x,t) - q \tag{25}$$

According to equation (23),

$$\dot{V}_1 = -c_1 l_1^2 + l_1 l_2 \tag{26}$$

Second step, the Lyapunov function is defined as

$$V_2 = V_1 + \frac{1}{2}l_2^2 \tag{27}$$

According to equation (25),

$$\dot{l}_2 = c_1\dot{l}_1 + \ddot{\alpha}_c - \dot{f}_\alpha(x) - \dot{w}_\alpha(x,t) - f_q(x) - w_q(x,t) - g_q(x)m_{\delta c} \tag{28}$$

Then

$$\dot{V}_2 = -c_1 l_1^2 + l_1 l_2 + l_2[c_1\dot{l}_1 - f_q(x) - w_q(x,t) - g_q(x)m_{\delta c} - \dot{f}_\alpha(x) - \dot{w}_\alpha(x,t) + \ddot{\alpha}_c] \tag{29}$$

where

$$f_q(x) = \frac{1}{I_{yy}}[m_b + Qsc_A C_{mq}q\frac{c_A}{2V_A}] \tag{30}$$

$$g_q(x) = \frac{1}{I_{yy}} \tag{31}$$

According to equation (29), the control law is defined as

$$m_{\delta c} = \frac{1}{g_q(x)}[c_1\dot{l}_1 + c_2l_2 + l_1 + \ddot{\alpha}_c - \dot{f}_\alpha(x) - f_q(x) - z_{2q} + \eta sign(l_2)] \tag{32}$$

where $c_2$ is constant and $c_2 > 0$, $\eta$ is constant and $\eta > 0$, $z_{2q}$ is the estimated values of uncertainties $w_q(x,t)$ computed by ESO, the structure of ESO is the same as equation (9) and (10).

According to the control law

$$\dot{V}_2 = -c_1 l_1^2 - c_2 l_2^2 + (z_{2q} - w_q(x,t)) - (\eta sign(l_2) + \dot{w}_\alpha(x,t)) \tag{33}$$

Therefore, when $z_{2q}$ tends to $w_i(x,t)$ and $\eta > |\dot{w}_\alpha(x,t)|$, $\dot{V}_2 \le 0$ exists.

To eliminate vibration, replace $\eta sign(l_2)$ to $\eta\dfrac{l_2}{|l_2| + \tau}$, where $\tau$ is constant and $\tau > 0$.

Parameters of back stepping controller are chosen as $c_1 = 4$, $c_2 = 5$, $\eta = 1$ and $\tau = 0.001$.

## 3.4   Parameter Tuning of ESO

Huang Yi proposed a proof of convergence for nonlinear second order continuous extended state observer based on self-stable region, and a parameter tuning method was given [8].

The estimation errors of ESO should be equal to:

$$e_{1i} = z_{1i} - x_i \tag{34}$$

$$e_{2i} = z_{2i} - w_i(x,t) \tag{35}$$

According to equation (8)-(10):

$$\dot{e}_{1i} = e_{2i} - \beta_{1i}e_{1i} \tag{36}$$

$$\dot{e}_{2i} = -\dot{w}_i(x,t) - \beta_{2i} fal(e_{1i}, \varepsilon, \Delta) \tag{37}$$

Assuming that function $w_i(x,t)$ and its derivate are bounded, $|\dot{w}_i(x,t)| < W$, $W > 0$. If the relationship below between parameters of ESO $\beta_{1i}$ and $\beta_{2i}$ exists, $\beta_{1i}^2 > 4c_2\beta_{2i}|dfal(e_{1i}, \varepsilon, \Delta)/de_{1i}|$, $c_2 > 1$, estimation errors would converge to the boundary of region $G_0$, and stay in the region.

The boundary of $G_0$ should be equal to:

When $\dfrac{k_1 c_2 W}{\beta_{2i}(c_2-1)} \geq \Delta$,

$$e_{1i}^* = \max\{|e_{1i}|\} = \left(\frac{k_1 c_2 W}{\beta_{2i}(c_2-1)}\right)^{\frac{1}{\varepsilon}} \tag{38}$$

And, when $\dfrac{k_1 c_2 W}{\beta_{2i}(c_2-1)} < \Delta$.

$$e_{1i}^* = \max\{|e_{1i}|\} = \Delta^{1-\varepsilon}\left(\frac{k_1 c_2 W}{\beta_{2i}(c_2-1)}\right) \tag{39}$$

$$e_{2i}^* = \max\{|e_{2i}|\} = \beta_{1i} e_{1i}^* - \frac{(k_1-1)c_2 W}{\beta_{1i}(c_2-1)} \tag{40}$$

where $k_1$ is constant and $k_1 > 1$, $0 < \varepsilon < 1$, $\Delta > 0$[8].

According to the converge theory above, the convergence condition of ESO is shown below.

$$\beta_{1i}^2 > 4c_2 \beta_{2i} \left| dfal(e_{1i}, \varepsilon, \Delta)/de_{1i} \right| \tag{41}$$

It is equal to:

$$\beta_{1i}^2 > 4c_2 \beta_{2i} \Delta^{\varepsilon-1} \tag{42}$$

According to equation (38)-(40) and $1/\varepsilon > 1$, to get estimation errors as small as possible, $\beta_{2i}$ should meet:

$$\beta_{2i} > \frac{k_1 c_2 W}{c_2-1} \tag{43}$$

According to parameter tuning rule above, parameters of ADRC are adjusted below by simulation.

**Table 1.** Parameters of second order extended state ob-server

| Canister number | $i$ | $\beta_{1i}$ | $\beta_{2i}$ |
| --- | --- | --- | --- |
| 1 | $q$ | 200 | 300 |
| 2 | $\gamma$ | 100 | 50 |
| 3 | $z_g$ | 150 | 50 |

Parameters of $fal$ function are chosen as $\varepsilon = 0.5$ and $\Delta = 0.001$.

## 3.5   Control Allocation

According to control redundancy of flying-wing aircraft, a control allocator based on generalized inverse [7] is designed to solve this problem. The analysis formula of elevator deflection is shown below.

$$\delta_1 = \frac{C_m^{\delta_1}}{C_m^{\delta_1 2} + C_m^{\delta_2 2}} \cdot \frac{m_{\delta c}}{QsCA} \tag{44}$$

$$\delta_2 = \frac{C_m^{\delta_2}}{C_m^{\delta_1 2} + C_m^{\delta_2 2}} \cdot \frac{m_{\delta c}}{QsCA} \tag{45}$$

where $C_m^{\delta_1}$ and $C_m^{\delta_2}$ are the control effectiveness.

## 4    Simulation

### 4.1    Wind Shear Model

A microburst is a strong localized downdraft that strikes the ground, creating winds that diverge radially from the impact point. An airplane penetrating the core of a microburst typically encounters an increasing wind first, and then this headwind disappears and quickly followed by a downdraft and tailwind [9]. This is the most serious wind shear in landing phase. Domestic and overseas scholars developed many models of this wind shear. According to JAWS data, a two dimensional model for microburst is established, and the schematic is shown in Fig. 2.

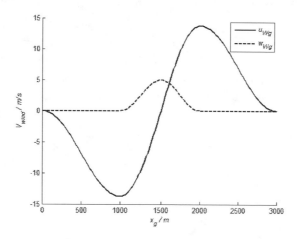

**Fig. 2.** Schematic of microburst model

### 4.2    Ground Effect

During take-off and landing, the distances between the airplane and the ground are relatively small in comparison with the dimensions of the aircraft, and the airplane aerodynamic coefficients are influenced by the ground proximity [10]. For flying-wing aircraft with high lift-drag ratio, the influence of ground effect shows mainly in lift coefficient. In Fig.3 when $\alpha = 2°$, increment caused by ground effect is 25.3% of lift coefficient without the effect. Clearly, lift coefficient changes rapidly in a small altitude range. Due to this effect, especially when it is unmolded, the disturbance caused by ground effect cannot be ignored.

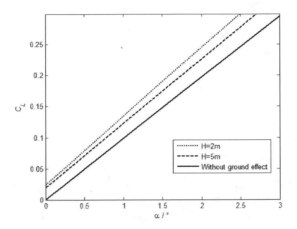

**Fig. 3.** Influence of ground effect

## 4.3    Flight Simulation

The landing control system based on active disturbance rejection control technique is designed. The initial flight conditions in the simulation are: altitude $H = 300m$, mach $Ma = 0.2$, the mass of flying-wing aircraft is $3500kg$, mean aerodynamic chord is $2.9m$, wing surface is $45m^2$. The simulation step size is $1ms$, simulation time is $200s$.

The responses of ADRC and NDI are compared in figures below.

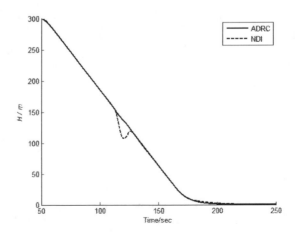

**Fig. 4.** Time history of altitude response

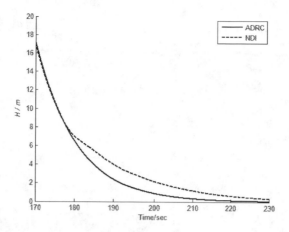

**Fig. 5.** Time history of altitude response in flare phase

Simulation results show that the ADRC controller has better performance in serious disturbance. In Fig.4, altitude loss is less than $2m$ when the aircraft encounters a microburst at the height of $150m$, at the same time, altitude loss of NDI is more than $30m$. It means the ADRC controller could suppress wind effects and keep altitude loss at a relatively low level.

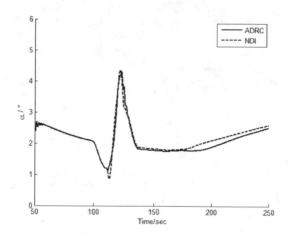

**Fig. 6.** Time history of attack angle response

In Fig.5, when the aircraft approaches the ground, lift coefficient increases rapidly. The ADRC controller estimates and compensates the uncertainties caused by ground effect, the command of reducing attack angle is given, and there is almost no vertical deviation, and at the time vertical deviation is about $2m$. Attack angle response is given in Fig.6, after altitude less than $6m$, attack angle response of ADRC is clearly less than the response of NDI, it means the ADRC controller suppresses the influence of ground effect effectively.

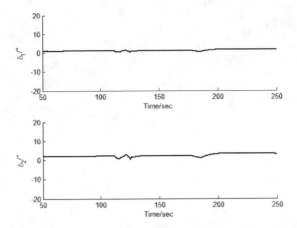

**Fig. 7.** Time history of elevator deflection

Time history of elevator deflection is given in Fig.7. It is stay in normal range and proves the engineering feasibility of the controller.

## 5     Conclusion

Because of the significant dynamic nonlinearity and atmospheric disturbance, a nonlinear control theory with strong robustness is proposed to design the landing control system for a flying-wing aircraft. Simulation results indicate that the performance of ADRC controller is satisfied. With this controller, the flying-wing aircraft could keep altitude effectively under the most serious wind disturbance, and make altitude loss less than $2m$ at the height of $150m$. While aerodynamic coefficients change rapidly caused by ground effect, the controller could assure adequate landing precision.

## References

1. Snell, S.A., Nns, D.F., Arrard, W.L.: Nonlinear Inversion Flight Control for a Supermaneuverable Aircraft. AIAA Journal of Guidance, Control, and Dynamics 15(4), 976–984 (1992)
2. Sonneveldt, L., Chu, Q.P., Mulder, J.A.: Nonlinear Flight Control Design Using Constrained Adaptive Backsteepping. AIAA Journal of Guidance, Control, and Dynamics 30(2), 322–336 (2007)
3. Zhu, T.F., Li, M., Deng, J.H.: Nonlinear Flight Control System Based on Backstepping Theory and Supermaneuver. Acta Aeronautica Et Astronautica Sinica 26(4), 430–433 (2005)
4. Han, J.Q.: Active Disturbance Rejection Control Technique. National Defense Industry Press, Beijing (2009)
5. Xiao, Y.L., Jin, C.J.: Flight Theory of Atmospheric Disturbance. National Defense Industry Press, Beijing (1993)
6. Du, Y.L.: Study of Nonlinear Adaptive Attitude and Trajectory Control for Near Space Vehicles. Nanjing University of Aeronautics and Astronautics (2010)

7. Shi, J.P.: Research on Control Allocation Methods in the Advanced Aircraft. Northwestern Polytechnical University (2009)
8. Huang, Y., Han, J.Q.: Analysis and Design For Second Order Nonlinear Continuous Extended States Observer. Chinese Science Bulletin 45(13), 1373–1379 (2000)
9. Mulgund, S.S., Stengel, R.F.: Optimal Recovery from Microburst Wind Shear. AIAA Journal of Guidance, control, and Dynamics 16(6), 1010–1017 (1993)
10. Boschetti, P.J., Cardenas, E.M., Amerio, A.: Stability and Performance of a Light Unmanned Airplane in Ground Effect. In: 48th AIAA Aerospace Sciences Meeting Including the New Horizons Forum and Aerospace Exposition, AIAA 2010-293, Orlando, Florida (2010)

# Smartphone-Controlled Robot Snake for Urban Search and Rescue

Yifan Luo[1], Jinguo Liu[1,2,*], Yang Gao[2], and Zhenli Lu[3]

[1] State Key Laboratory of Robotics, Shenyang Institute of Automation,
Chinese Academy of Sciences, Shenyang, China
liujinguo@sia.cn
[2] Surrey Space Centre, Faculty of Engineering and Physical Sciences, University of Surrey,
Guildford, GU2 7XH, UK
[3] Institute of Electronics and Telematics Engineering of Aveiro, University of Aveiro, Aveiro
3810-193, Portugal

**Abstract.** Search and rescue robots would benefit from versatile locomotion ability and hence cope with varying environments. Robot snakes, with hyper-redundant body and unique gaits, offer a promising solution to search and rescue applications. This paper presents a portable design of robot snakes that can be controlled from commercial mobile devices like the smartphones. The control results are validated and demonstrated using hardware prototypes.

**Keywords:** Search and rescue, Robot snake, Motion planning, Smartphone based control.

## 1 Introduction

Search and rescue robots are widely studied in recent years. In previous work, many robots have been developed to reduce the damage of disaster[1-13]. Intentions of using snake-like robots in search and rescue have attracted a lot of attention from the robotics community due to the locomotion capabilities that the bio-inspired robots can offer[3-17]. Many researchers investigated the structural design and motion planing, and the control system design for collecting information and dealing with the feedback [15-20].

The aim of developing bio-inspired robot is to carry out bionics study and learn the features and functions of the animals, and through the mechanical and electronic devices to mimic and reproduce these functions. It combines biological function, mechanical principles, control technology and computational intelligence. Snakes have the characteristics of slim and legless body, low center of gravity, multiple redundancies, sealing structure. With different gaits, such as serpentine locomotion, concertina locomotion, side winding locomotion, snakes can adopt itself in various environments[1,5,14,16].

A snake-like robot has a hyper-redundant body which makes the design of its control system more challenging. To meet task requirements in a search and rescue scenario, a portable robot snake solution is proposed in this work that uses

X. Zhang et al. (Eds.): ICIRA 2014, Part I, LNAI 8917, pp. 352–363, 2014.

commercially available, compact mechanical, control and communication system designs. The proposed system has the controller implemented in a mobile APP to reduce cost and size. This allows the system to be portable, prevalent and used by non-professional operators in search and rescue operation.

The remainder of the paper is organized as follows: Section 2 presents design requirements of the robot snake within this work. Section 3 describes the mechanical and control system of the robot. Section 4 explains motion planning of the robot snake. Section 5 presents and discusses experimental results on a hardware prototype of the proposed system. Finally, Section 6 concludes the paper and gives perspectives of future work.

## 2    Development Goal

The goals of the portable snake-like robot designed for urban search and rescue are provided as follows:

➢    Modularization. The robot should be composed of identical modules which makes it be easily assembled and disassembled.

➢    Remote control.   Snake-like robot can get rid of the shackles of cable which make it easy to hide in search task and reach a deeper position in rescue task.

➢    Prompt feedback. The operator can use the collected information to adjust of the movements of the robot more easily and gather the required information.

➢    Easy to operate. Non-professionals can quickly operate the snake-like robot so as to enlarge the range of applications.

➢    Larger field of view. To gather more comprehensive information, the robot needs a larger vision to accomplish the task.

## 3    System Introduction

The snake-like robot in this paper is connected by 10 joints, and adopts portable control system platform, which includes WIFI module, camera, controller, battery, smartphone. The system structure is shown in Figure 1.

### 3.1    Hardware Development

Hardware of the snake-like robot mainly includes wireless module and robot module.

Wireless module is responsible for the communication between the client-side and snake-like robot. The client-side sent instructions to snake-like robot through wireless module. The snake-like robot sent video, audio and information of itself back to client-side through wireless module.

WIFI wireless network includes two kinds of topological form, Infra and Ad-hoc network. Infra network is  based on the AP (wireless access point) and founded by AP and many STA (station, wireless network terminal). The characteristics of this type of network is that AP is the center of the whole network, all the communications in the network are fulfilled through the AP. The ad-hoc network is composed of two or more STA. There is no AP in the network, this type of network is a loose structure, thus all the STAs in the network can communicate with each other directly. The Infra network form, 802.11g/n protocol is adopted in this work. In this WIFI module, the snake-like robot is taken as AP and the portable equipment such as mobile phones is used as STA.

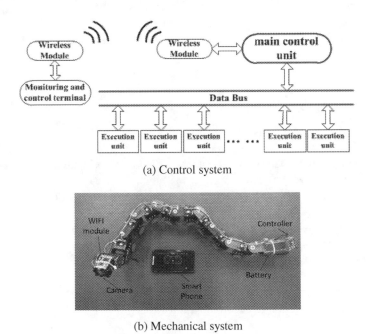

(a) Control system

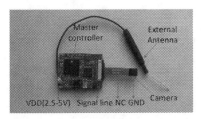

(b) Mechanical system

**Fig. 1.** Outline of the snake-like robot

The WIFI module and camera module are shown in Figure 2, and the parameters are shown in table 1. This module is responsible for the communication between the client side and robot side.It can also send the control commands from the client side to the main controller through the signal line.

**Fig. 2.** WIFI module and camera module

The snake-like robot adopts modular design, is linked together by same type of joints. Each joint is equipped with the sensors for the position feedback, temperature, load and input voltage sensor to collect the current state of the joint in real time. The robot is also equipped with an alarm system, which can warn the user so as to automatically handle the problem when the internal parameters (e.g. internal voltage, torque and voltage, etc.) deviate from the work value. The internal structure of the single joint is shown in Fig.3, which includes the 1:254 reducer, driver and detection plate of the motor. The connecting devices between the joint structures are shown in Figure 4. The plane and orthogonal connections are being used to reduce the frictional resistance during movement. The scroll wheel is installed below the snake-like robot.

**Table 1.** Main parameters of WIFI module

| Specifications | Details |
| --- | --- |
| Protocols and standards | 802.11g/n、802.11b |
| Power Consumption | 1.5W |
| Pixels of Camera | 30W |
| Output sample rate of Module | 30W Frame |
| Operating mode | AP mode |
| Transmission rate | Maximum 150MHz |
| Sensitivity @PER | <-65dbm |
| RF Power | -2482 |
| Transmit Power | 14DBM, ≥30 meters |
| Operating Voltage | 3~5V |
| Operating Temperature | -10℃~+70℃ |

(a) 1:254 gear reducer          (b) Driver and detection plate of motor

**Fig. 3.** Internal structure of a single joint

(a) Plane connection                    (b) Vertical connection

**Fig. 4.** Joint connection

**Table 2.** Specification of mechanical system

| Specifications | Details |
|---|---|
| Number of joints | 10 |
| Size of link[mm] | 30×48×48 |
| Weight of link[kg] | 0.03035 |
| Motion range of yaw angle[deg] | [-115, +115] |
| Actuators | RC servo motor |
| Sensor | Temperature, Speed , Position sensor |

The parameters of mechanical system of the robot are shown in Table II. Three cables for the signal of power, ground and data are developed   between the joint motor and the controller. The internal connection of the hardware system of each joint is shown in Figure 5, where the yellow line means   ground ginal , the red line means 12V power signal,and   the blue line means the control data signal.

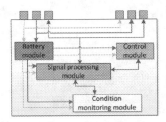

**Fig. 5.** Internal structure of the hardware system on each joint

As mentioned previously, distributed control method is developed in order to save the resources of the controllers among motors. A plurality of parameters can be set and fed to each motor in a command packet.

Distributed feedback compensation control is also used as the control method. In the control process, the input value is values for the position p(t) and velocity v(t). The feedbacks of position pn(t) and the speed vn(t) are read in real-time. Here the compensations for position and velocity are ep(t) and ev(t). The control system diagram is shown in Figure 6.

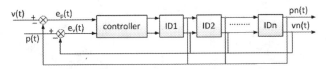

**Fig. 6.** Control system diagram

### 3.2    Software System

High level control system is developed in the client to interface with the controller of snake-like robot. The client instructions are shown in Figure 7. The GUI(Graphic User Interface) is suitable for operators to grasp how to control the robot quickly for the search of rescue task . The software of the client is integrated on a smartphone.

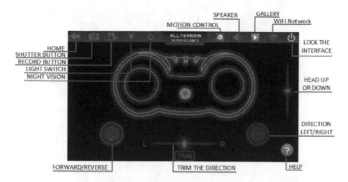

**Fig. 7.** APP Interface

WIFI module is responsible for transmitting information among the robot and the client. The video captured by the camera is transmitted to the client through the WIFI module in real time which can be displayed in GUI. The  pictures or video can be recorded  via GUI. The GUI can also be used by the gravity sensing devices on portable devices. The control system flowchart of low level layer on robot is shown in Figure 8.

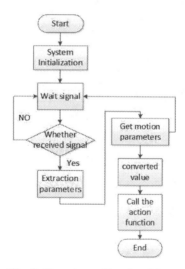

**Fig. 8.** Flowchart of low level layer

The operating principle are shown as following: when the WIFI module receives the information from the client, it will send them to the low level controller, and the controller will extract corresponding parameters from the data package and calculates the motion data of each joint, then sent to each joint through the serial data bus to complete the corresponding action.

# 4      Motion Planning of Snake-Like Robot

## 4.1      Motion Planning in a Line Direction

Snake can use a variety of gaits(serpentine locomotion, concertina locomotion, rectilinear locomotion and sidewinding locomotion) so as to make it adopted in different environments. Serpentine locomotion with highest efficiency is the most typical of gait of snake[5,14,16,17,20]. According to serpentine crawling, a serpentine curve can be described in Figure 9, which proposed by S.Hirose[14].

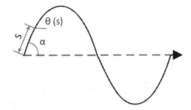

**Fig. 9.** Serpentine curves and trajectories

The curvature of the serpenoid curve is given by:

$$\rho = -\alpha b \sin(bs) \tag{1}$$

Where

α—Amplitude angle (rad) ;

b—Constant of proportionality (rad/m) ;

s—length of Serpentine curve (m)

By integrating (1), tangential angle with respect to s is derived and given by

$$\theta_{(s)} = \alpha \cos(bs) \tag{2}$$

Its relative value is thus obtained from (2) as

$$\varphi = \theta_{(s+1)} - \theta_{(s-1)} = -2\alpha \sin(bl) \sin(bs) \tag{3}$$

Where

l—Half the length of the snake-like robot unit.

By rewriting the relative angle as function of time, we have

$$\varphi_i(t) = A \sin(\omega t + (i-1)\beta) \tag{4}$$

Where $A = -2\alpha \sin(bl)$ ;

   $\omega t = bs$ ;

$\beta{=}2bl$ ;

$i =1,\ldots,$ n is the nth joint ;

$t$—time ;

$n$—number of robot joints

In order to make the initial movement of the snake-like robot gradually approach the serpentine curves and remove the jitter, the input function of motion equation is:

$$\varphi_i(t) = A(1-e^{-\lambda t})\sin(\omega t + (i-1)\beta) \tag{5}$$

where $\lambda$ is the attenuation factor.

In general, when, snake moving, its head must always direct to the forward direction in order to get more information about front condition. The head unit relative angle for driving head to the forward direction is thus given by

$$\varphi_h = -\theta(s) = -\alpha\cos(bs) \tag{6}$$

By rewriting (5) as function of time, we have:

$$\phi h(t) = -\alpha(1-e^{-\lambda t})\cos(\omega t + (n-1)\beta + b1) = \alpha(1-e^{-\lambda t})\cos(\omega t + (n-0.5)\beta) \tag{7}$$

The simulation of serpentine curve is shown in Figure 10.

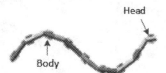

**Fig. 10.** Serpentine curves of forward head

Equation (6) ensures the snake-like robot head parallel to the symmetrical line of the body. Thus, the screen receiving the picture from the camera fixed on the head is the front direction. The operator can identify environments well.

## 4.2    Turning Mode of Snake-Like Robot

Due to its flexible architecture, it's hard for the snake-like robot to make a turning movement like other robot. To ensure the snake-like robot can achieve high efficiency in turning movement.C.Ye proposed several methods for the turning motion of snake-like robot[21]. In this paper, the initial movement can make the snake-like robot gradually approach the serpentine curves and remove the jitter, as shown in the follow.

**Symmetrical Line Modulation Method**

$$\varphi_i(t) = A(1-e^{-\lambda t})\sin(\omega t + (i-1)\beta) + \gamma \tag{8}$$

where y is the scalar of modulation.

This method, by adding a constant to the angle function, the snake-like robot makes its shape unsymmetrical and moves along a curve. In fact, this angle function is obtained by modifying the (1) as

$$\rho = -\alpha b \sin(bs) + c \qquad (9)$$

Where c is a constant for turning motion.(rad/m), Disadvantages of this method are as follows: 1) the input angle is not continuous when robot starts to turn from the forward movement, 2) efficiency is reduced for the track is not serpenoid curve, and 3) the angle of turn is obscure. Advantages of this method are as follows: 1) the radius of turn is small, and 2) the control is simple.

## Amplitude Modulation Method

$$\varphi_i(t) = A(1 - e^{-\lambda t})(1 + \Delta A sign(\sin(\omega t + (i-1)\beta))) \times \sin(\omega t + (i-1)\beta) \qquad (10)$$

Where A is the scalar of modulation.

Relating the amplitude to the phase of the serpenoid curve enables the snake to move along a new track. The key to this method is to find a location on the serpenoid curve where the acceleration is at zero to guarantee its continuity while amplitude changes. Equation(10) expresses that the amplitude is increased or decreased when the input angle value changed from positive to negative. This method can realize turning motion.

**Fig. 11.** Circle locomotion by Amplitude modulation

Advantages of this method are as follows: 1) input angle value is continuous when it moves from the forward movement to turning movement, 2) The control is simple, and 3) the angle of the turn can be exactly realized. While in this method, the turn angle is confined to the amplitude of serpenoid curve, and the radius of turn is big. A simulator performing circle locomotion by using amplitude modulation method is shown in Figure 11.

## Phase Modulation Method

movement, the phase modulation method is presented from morphology. By which, the snake can complete the direction regulation exactly and at the same time keep serpenoid curve shape,as shown in (11). The idea is that the snake moves forward σ = 0, when start to turn, let the snake's head direction parallel to the turn direction, the CT changes to new serpenoid curve. Where σ is a scalar for turning motion.

$$\varphi_i(t) = A(1 - e^{-\lambda t})\sin(\omega t + (i-1)\beta + \sigma) \tag{11}$$

Disadvantage of this method is that the input angle value is not continuous. Especially in some worse condition, the gap is so big that the snake slides much. At this time, it is necessary to modulate the amplitude A and frequency to get better performance. In application, for the phase modulation method of regulating the angle as discontinuous and the limitation of the amplitude and the frequency, the snake-like robot input angle value is interpolated for the continuous movement. With these modifications, the performance of movement is better than the other two. Advantages of this method are as follows: 1) the movement of turning can be realized at once, 2) the angle of turn is exactly, and 3) the serpenoid curve shape is guaranteed.

## 5    Experiment

The operability and maneuverability of the proposed solutions for the snake-like robot are validated through experiment. In the experiment, different motion tasks to move the snake-like robot forward, turn around and head up are carried out so as to avoidance obstacle in search and rescue tasks. The operator can observe the information in front of the robot through the client interface and use the GUI shown in Figure 6 to control the robot.

In the Figure 12, according to the real-time image information the operator operates the robot to avoid obstacles. The head of the robot is fixed directly to the forward direction and gets the information about front condition. Which can reduce shaking in the image caused by the phenomenon of periodic motion. It can allow the operator to easily adjust the offset and avoid obstacles due to various causes robot orientation. The amplitude modulation method is adopted for turning motion in the experiment.

In Figure 13, the operator uses the smartphone to carry out the task of in-situ searching the ruins.

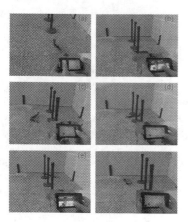

**Fig. 12.** Obstacles avoiding

**Fig. 13.** Ruins exploration

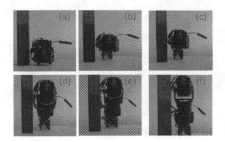

**Fig. 14.** Head up and head down experiment

Figure 14 illustrates the ability of raising its head from the bottom of the snake-like robot.In this experiment,in order to ensure the stability of motion and control,   only two modules are used to raise its head in a limited height during movement.

## 6     Conclusions

In this paper, a snake-like robot with a portable controller has beed developed for search and rescue. In order to meet the requirement in the application, 1) a smartphone based control solution is adopted to simplify complex operations; 2) several motion planning methods for fixing the head in a line direction are also discussed; 3) Head rising during serpentine locomotion is well planned. The experiment results proved that   the improved snake-like robot   and control solutions proposed·in this paper are more efficient   for the search and rescue task.

In the future,our research is to 1) optimize the mechanical structure and increase the degree of freedom of the robot so as to carry out the task in search and rescue more flexible; 2) install a manipulator on the head of the robot to manipulate necessary objects.

## References

1. Liu, J., Wang, Y., Ii, B., et al.: Current research, key performance, and future development of search and rescue robot. Frontiers of Mechanical Engineering in China 2(4), 404–416 (2007)
2. Arai, M., Takayama, T., Hirose, S.: Development of "Souryu-III": connected crawler vehicle for inspection inside narrow and winding spaces. In: IEEE/RSJ International Conference on, vol. 1, pp. 52–57. IEEE (2004)
3. Murai, R., Ito, K., Nakamichi K.: Proposal of a snake-like rescue robot designed for ease of use-Improvement of operability for non-professional operator. In: 34th Annual Conference of IEEE, pp. 1662–1667. IEEE (2008)
4. Paap, K.L., Christaller, T., Kirchner, F.: A robot snake to inspect broken buildings. In: IEEE/RSJ International Conference on, vol. 3, pp. 2079–2082. IEEE (2000)
5. Ma, S., Tadokoro, N.: Analysis of creeping locomotion of a snake-like robot on a slope. Autonomous Robots 20(1), 15–23 (2006)
6. Hirose, S., Matsuno, F.: Development of snake robots for rescue operation/design of the shape and its control. Journal of Japan Society of Mechanical Engineers 106(1019), 769–773 (2003)

7. Miyama, S., Imai, M., Anzai, Y.: Rescue robot under disaster situation: position acquisition with omni-directional sensor. In: IEEE/RSJ International Conference on, vol. 4, pp. 3132–3137. IEEE (2003)
8. Nagatani, K., Ishida, H., Yamanaka, S., et al.: Three-dimensional localization and mapping for mobile robot in disaster environments. In: IEEE/RSJ International Conference on, vol. 4, pp. 3112–3117. IEEE (2003)
9. Guarnieri, M., Debenest, P., Inoh, T., Fukushima, E.F., Hirose, S.: Development of helios vii: an arm-equipped tracked vehicle for search and rescue operations. In: Conference on Intelligent Robots (2004)
10. Chang, C., Brando, A.: Semi-autonomous victim search. Proceedings of the 3rd IEEE International Workshop on Safety, Security and Rescue Robotics (2004)
11. Ito, K., Yang, Z., Saijo, K., et al.: A rescue robot system for collecting information designed for ease of use—a proposal of a rescue systems concept. Advanced Robotics 19(3), 249–272 (2005)
12. Yokokohji, Y.: Interface design for rescue robot operation-Introduction of research outcomes from the human-interface group of the DDT project. Journal-Robotics Society of Japan 22(5), 24–27 (2004)
13. Osuka, K., Kitajima, H.: Development of mobile inspection robot for rescue activities: MOIRA. In: Proceedings of the IEEE/RSJ International Conference on, vol. 4, pp. 3373–3377. IEEE (2003)
14. Hirose, S.: Biologically Inspired Robots (Snake-like Locomotor and Manipulator). Oxford Unibersity Press (1993)
15. Ito, K., Fukumori, Y.: Autonomous control of a snake-like robot utilizing passive mechanism. In: Proceedings of the 2006 IEEE International Conference on Robotics and Automation, ICRA, pp. 381–386. IEEE (2006)
16. Liu, J., Wang, Y., Li, B., et al.: Path planning of a snake-like robot based on serpenoid curve and genetic algorithms. In: Fifth World Congress on Intelligent Control and Automation, WCICA 2004, vol. 6, pp. 4860–4864. IEEE (2004)
17. Liu, J., Wang, Y., Li, B., et al.: Bionic research on concertina motion of a snake-like robot. Chinese Journal of Mechanical Engineering, 2005 5, 022 (2005)
18. Matsuno, F., Suenaga, K.: Control of redundant 3D snake robot based on kinematic model. In: Proceedings of the IEEE International Conference on Robotics and Automation, ICRA 2003, vol. 2, pp. 2061–2066 (2003)
19. Gao, Y., Liu, J.: China's robotics successes abound. Science 345(6196), 523 (2014)
20. Er, M.J., Gao, Y.: Robust adaptive control of robot manipulators using generalized fuzzy neural networks. IEEE Transactions on Industrial Electronics 50(3), 620–628 (2003)
21. Ye, C., Ma, S., Li, B., et al.: Turning and side motion of snake-like robot. In: Proceedings of the 2004 IEEE International Conference on Robotics and Automation, ICRA 2004, vol. 5, pp. 5075–5080. IEEE (2004)

# Design of a Continuum Wearable Robot for Shoulder Rehabilitation

Kai Xu, You Wang, and Zhixiong Yang

RII Lab (Lab of Robotics Innovation and Intervention), UM-SJTU Joint Institute,
Shanghai Jiao Tong University, Shanghai, 200240, China
{k.xu,youwang,yangzhixiong}@sjtu.edu.cn

**Abstract.** A wearable robot for rehabilitation therapy is often shared by a group of patients in a clinic. If the wearable robot only consists of rigid links, the link dimensions usually need to be adjusted from time to time to fit different patients. It is then difficult to make sure these on-site adjustments could introduce the desired kinematic compatibility between the robot and each individual patient. A previous investigation shows it is possible to construct a compliant wearable robot that can provide Anatomy Adaptive Assistances (AAA), which means the robot passively adapts to different patient anatomies while providing consistent motion assistances. However, the previous design also possesses drawbacks such as limited motion ranges and limited payload capabilities. This paper presents a kinematics-based type synthesis for the construction of a new continuum wearable shoulder robot, aiming at overcoming these drawbacks as well as maintaining the capabilities of providing AAA. Three structural concepts of such a continuum wearable shoulder robot are studied through kinematic modeling. One concept is eventually selected based on the comparison results. Preliminary experiments are also presented to demonstrate the feasibility of the selected design.

**Keywords:** Wearable robot, continuum mechanism, type synthesis, kinematics, AAA (Anatomy Adaptive Assistances).

## 1    Introduction

Research on wearable robots and exoskeletons has been quite active in the past a few decades. Many exoskeleton systems or wearable robots were developed for upper and/or lower limbs (e.g. [1, 2]). Some exoskeletons aim at augmenting a healthy wearer's physical performance with robotic actuation (e.g., the BLEEX system [3], the MIT load-carrying exoskeleton [4], etc.), whereas the others aim at delivering rehabilitation therapies to patients with neuromuscular defects after injury or stroke, such as the rehabilitation wearable robots for lower limbs [5-8], and the ones for upper limbs [9-18].

Most of the existing wearable robots and exoskeletons share one similar design approach: a rigid kinematic chain (serial or parallel) is actuated to mobilize an attached user. The use of rigid links in an exoskeleton system might be suitable for

X. Zhang et al. (Eds.): ICIRA 2014, Part I, LNAI 8917, pp. 364–375, 2014.

strength-augmenting applications to shield the wearer so that excessive external loads can be undertaken by the rigid structure. But these rigid links introduce drawbacks such as system bulkiness, high inertia, and the difficulty of maintaining kinematic compatibility between the wearable robot and a wearer. In a clinic for rehabilitation therapy, one wearable robot is often shared by many patients. If the wearable robot has a rigid structure, a physician needs to adjust the dimensions of the wearable robot from time to time to match the robot to different patients. It is difficult to guarantee these on-site adjustments could introduce the desired kinematic compatibility. For this reason, some design alternatives that use compliant components have been investigated [14, 15, 19, 20]. Particularly, a continuum shoulder wearable robot was constructed to demonstrate the capability of providing Anatomy Adaptive Assistances (AAA) [16-18, 21]. Without requiring any hardware adjustments, the continuum wearable robot passively deforms and adapts to different wearer anatomies while providing consistent motion assistances. Such a feature avoids the challenging studies on how to maintain ergonomics as in [22-24].

Besides the identified characteristics of providing AAA, the wearable robot was also found to have some drawbacks, such as limited motion ranges, limited payload capabilities, and the difficulty to wear on an impaired subject. This paper hence presents the descriptions and selection of three structural concepts of a new continuum wearable robot for shoulder rehabilitation, aiming at overcoming the aforementioned drawbacks. The contribution of this paper mainly lies on the proposal and kinematic analysis of the three design variations. The model-based comparison suggests that the selected design is promising in terms of maintaining the capability of providing AAA, enabling big motion ranges, allowing large payload, as well as easing the process of wearing the robot.

The paper is organized as follows. Section 2 summarizes the drawbacks of the existing continuum wearable shoulder robot and presents three design concepts. Section 3 presents nomenclature and a basic kinematics model so that the kinematic analysis and simulation comparisons of the three design variations can be presented in Section 4. Preliminary designs and experiments are reported in Section 5 for the selected design concept. Conclusions and future work are summarized in Section 6.

## 2    Existing Drawbacks and New Design Concepts

The constructed continuum wearable shoulder robot as in [17, 18] is shown in Fig. 1. It consists of i) a rigid armguard, ii) an upper arm sleeve, iii) a flexible continuum shoulder brace, iv) a body vest, v) a set of guiding cannulae, and vi) an actuation unit. Actuation of the continuum shoulder brace orients a patient's arm accordingly.

Structure of the continuum shoulder brace is also depicted in Fig. 3. The brace consists of an end ring, a base ring, a few spacer rings and several backbones. All the backbones are made from thin super-elastic nitinol (Nickel-Titanium alloy) rods. The backbones are only attached to the end ring and can slide in holes of the spacer rings and the base ring. Miniature springs are used to keep the spacer rings distributed evenly. Simultaneous pulling and pushing of these backbones are achieved by the actuation unit as the backbones are routed through the set of guiding cannulae. Bending of the continuum brace orients a patient's upper arm.

Advantages of the continuum wearable robot include: i) safety and comfort introduced by the inherent compliance, ii) passive adaptation to different patient anatomies, iii) a redundant backbone arrangement for load redistribution and reduced buckling risks, and iv) design compactness achieved by dual roles of these backbones as both the motion output members and the structural components.

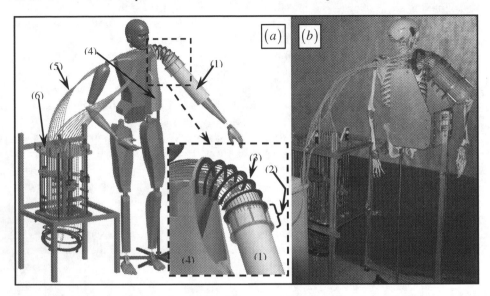

**Fig. 1.** The continuum wearable shoulder robot in [18]: (a.1) a rigid armguard, (a.2) an upper arm sleeve, (a.3) a flexible continuum shoulder brace, (a.4) a body vest, (a.5) a set of guiding cannulae, and (a.6) an actuation unit; (b) the constructed prototype

Although this wearable robot's characteristic of providing AAA (Anatomy Adaptive Assistances) is particularly advantageous in rehabilitation applications, some drawbacks were also identified. Firstly, it can be seen from Fig. 1 and Fig. 3 that the brace has a sparse structure and the payload capability is limited. Secondly, the brace has to be designed big enough to fit a group of patients. When the brace has a big diameter and a relative short length, the available motion range is limited. What's more, the shoulder brace can't be conveniently worn by an impaired subject. Design modifications are hence desired to overcome these drawbacks.

Three design concepts are considered in this paper, attempting to overcome the aforementioned drawbacks, as shown in Fig. 2. These design concepts are all proposed to modify the structure of the shoulder brace for possible improvements. The design concept in Fig. 2(a) has a non-uniform routing intended for an improved payload capability. The design concepts in Fig. 2(b and c) are proposed for enlarged motion ranges and the reduced obstruction of wearing. Nomenclature and basic kinematics are presented in Section 3 so that these three design concepts could be carefully studied in Section 4.

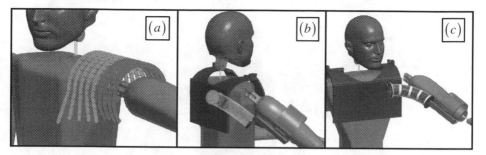

**Fig. 2.** Design concepts for improvements: (a) the non-uniform-routing brace, (b) the back-mounting brace, and (c) the front-mounting brace

# 3    Nomenclature and Kinematics

The nomenclature and the kinematics assume that the continuum brace bends into a planar shape within the bending plane as shown in Fig. 3. Shapes of the backbones are assumed by a sweeping motion of the structure's cross section along the imaginary primary backbone. This work doesn't assume the imaginary primary backbone's shape to be circular, which has been experimentally verified in [17, 18].

## 3.1    Nomenclature and Coordinate Systems

To describe the shoulder brace, nomenclatures are defined in Table I, while coordinate systems of the continuum brace are defined as below

- *Base Ring Coordinate System* (BRS) is designated as $\{b\} \equiv \{\hat{\mathbf{x}}_b, \hat{\mathbf{y}}_b, \hat{\mathbf{z}}_b\}$. It is attached to the base ring, whose XY plane coincides with the base ring and its origin is at the center of the base disk. $\hat{\mathbf{x}}_b$ points from the center of the base disk to the first backbone while $\hat{\mathbf{z}}_b$ is perpendicular to the base ring. The backbones are numbered according to the definition of $\delta_i$.

- *Bending Plane Coordinate System 1* (BPS1) is designated as $\{I\} \equiv \{\hat{\mathbf{x}}_I, \hat{\mathbf{y}}_I, \hat{\mathbf{z}}_I\}$ which shares its origin with $\{b\}$ and has the brace bending in its XZ plane.

- *Bending Plane Coordinate System 2* (BPS2) is designated as $\{2\} \equiv \{\hat{\mathbf{x}}_2, \hat{\mathbf{y}}_2, \hat{\mathbf{z}}_2\}$ obtained from $\{I\}$ by a rotation about $\hat{\mathbf{y}}_I$ such that $\hat{\mathbf{z}}_I$ becomes backbone tangent at the end ring. Origin of $\{2\}$ is at center of the end ring.

- *End Ring Coordinate System* (ERS) $\{e\} \equiv \{\hat{\mathbf{x}}_e, \hat{\mathbf{y}}_e, \hat{\mathbf{z}}_e\}$ is fixed to the end ring. $\hat{\mathbf{x}}_e$ points from center of the end ring to the first secondary backbone and $\hat{\mathbf{z}}_e$ is normal to the end ring. $\{e\}$ is obtained from $\{2\}$ by a rotation about $\hat{\mathbf{z}}_2$.

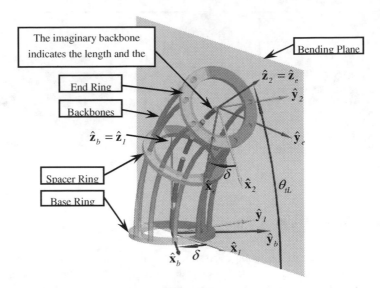

**Fig. 3.** Nomenclature and coordinates of the continuum brace

**Table 1.** Nomenclature used for kinematics modeling

| | |
|---|---|
| $m$ | Index of the backbones, $i = 1,2,\cdots,m$ |
| $r_i$ | The distance from the imaginary backbone to the $i$th backbone in the brace. $r_i$ can be different for different $i$. |
| $\beta_i$ | $\beta_i$ characterizes the division angle from the $i$th backbone to the first backbone. $\beta_1 \equiv 0$ and $\beta_i$ remain constant once the braces are built. |
| $L, L_i$ | Lengths of the imaginary backbone and the $i$th backbone measured from the base ring to the end ring. |
| $\rho(s), \rho_i(s_i)$ | Radius of curvature of the imaginary backbone and the $i$th backbones. |
| $\mathbf{q}_t$ | $\mathbf{q}_t = \begin{bmatrix} q_1 & q_2 & \cdots & q_m \end{bmatrix}^T$ is the actuation lengths of the backbones and $q_i \equiv L_i - L$. |
| $\theta(s)$ | Angle of the tangent to the imaginary backbone in the bending plane. $\theta(L)$ and $\theta(0)$ are denoted by $\theta_L$ and $\theta_0$, respectively. $\theta_0 = \pi/2$ |
| $\delta_i$ | A right-handed rotation about $\hat{\mathbf{z}}_1$ from $\hat{\mathbf{x}}_1$ to a ray passing through the imaginary backbone and the $i$th backbone. |
| $\delta$ | $\delta \equiv \delta_1$ and $\delta_i = \delta + \beta_i$ |
| $\psi$ | $\psi \equiv \begin{bmatrix} \theta_L & \delta \end{bmatrix}^T$ defines the configuration of the shoulder brace. |
| $^b\mathbf{p}(s)$ | Position vector of a point along the imaginary backbone in $\{b\}$. $^b\mathbf{p}(L)$ is the tip position and is designated as $^b\mathbf{p}_L$. |

## 3.2    Kinematics

Thorough kinematics analysis of such a continuum brace can be found in [25, 26]. The model was extended to the cases that the imaginary backbone has a non-circular shape and the backbones are arbitrarily arranged [17, 18]. Basic entities are summarized here so that the kinematic analysis of the three concepts can be elaborated in Section 4.

Configuration of the continuum brace is parameterized by $\psi = [\theta_L \ \delta]^T$. The length of the imaginary backbone is related to the length of the $i$th backbone as in Eq. (1). The integral can be derived to give Eq. (2), as detailed in [18]. Referring to the definition of $q_i$ in Table 1, Eq. (3) is obtained from Eq. (2).

$$L_{ti} = \int ds_{ti} = \int (ds_{ti} - ds_t + ds_t) = \int (ds_{ti} - ds_t) + L_t \ . \tag{1}$$

$$L_{ti} = L_t - r_{ti} \cos \delta_{ti} (\theta_0 - \theta_{tL}) = L_t + r_{ti} \cos \delta_{ti} (\theta_{tL} - \theta_0) \ . \tag{2}$$

$$q_{ti} = r_{ti} \cos \delta_{ti} (\theta_{tL} - \theta_0), \ \ i = 1,2,\cdots,m \ . \tag{3}$$

Equation (3) suggests that actuation of the continuum brace only concerns the values of $\theta_L$ and $\delta$. The actuation doesn't depend on the actual shape of the shoulder brace. This feature provides a particular advantage for rehabilitation in a clinic: when the wearable robot is put on different patients, different anatomies form different shapes of the brace, but the actuation remains the same while orienting a patient's upper arm to the same direction that is characterized by $\theta_L$ and $\delta$.

Rotation matrix $^b\mathbf{R}_e$ that associates $\{e\}$ and $\{b\}$ is as follows:

$$^b\mathbf{R}_e = R(\hat{\mathbf{z}}_b,-\delta)R(\hat{\mathbf{y}}_I,\theta_0 - \theta_L)R(\hat{\mathbf{z}}_2,\delta) \ . \tag{4}$$

Where $R(\hat{\mathbf{n}},\gamma)$ is a rotation about $\hat{\mathbf{n}}$ by an angle $\gamma$.

Tip position of the continuum brace is governed by Eq. (5). When the imaginary backbone has a circular shape, the tip position is then given by Eq. (6).

$$^b\mathbf{p}_L = {}^b\mathbf{R}_I \left[ \int_0^L \cos(\theta(s)) ds \ \ 0 \ \ \int_0^L \sin(\theta(s)) ds \right]^T \ . \tag{5}$$

Where $^b\mathbf{R}_I = \mathrm{R}(\hat{\mathbf{z}}_b, -\delta)$ and the integrals depend on the actual shape of the imaginary backbone.

$$^b\mathbf{p}_L = \frac{L}{\theta_L - \theta_0}\left[\cos\delta(\sin\theta_L - 1) \quad \sin\delta(1 - \sin\theta_L) \quad -\cos\theta_L\right]^T .$$

(6)

## 4    Kinematic Analysis and Simulation Verifications

Three design concepts are proposed to overcome the identified drawbacks of the previous design of the continuum wearable shoulder robot. Kinematic analysis and simulations were conducted to verify the feasibility of the designs. One design was eventually selected and finalized as shown in Section 5.1.

### 4.1    Design Concept #1

The first concept shown in Fig. 2(a) has the backbones with a non-uniform routing. This design was intended for enhanced payload capabilities. The existing design with a uniform routing of the backbones cannot well resist an external twist about $\hat{\mathbf{z}}_e$. The experimental results in [27] showed that a weak torsional rigidity affects the payload capabilities greatly. With a non-uniform routing of the backbones, external twists might be resisted by the helically arranged backbones. Thus the payload capabilities might also be enhanced.

For the non-uniform routing of the backbones, $r_i$ and $\beta_i$ vary along the imaginary backbone. The length of the i*th* backbone shall be calculated as in Eq. (7).

$$L_i = L - \int_0^{\theta_0 - \theta_L} r_i \cos(\delta + \beta_i) d\theta, \quad i = 1, 2, \cdots, m .$$

(7)

A Matlab simulation of this concept is shown in Fig. 4. After several backbone routing patterns were attempted, it was concluded that the length change of the backbones cannot be realized by the existing actuation unit. This design concept was then abandoned.

### 4.2    Design Concepts #2 and #3

From Section 4.1 it was concluded that the continuum brace shall have a uniform routing for its backbones in order to realize the actuation. When the backbones are arranged around a patient's upper arm, the brace's base ring has to be located around a wearer's chest. Then the brace will have a large diameter and a stumpy appearance.

Motion ranges are hence limited when the brace's length is not long enough compared with the brace's diameter. What's more, it is also not very convenient for an impaired subject to wear this brace.

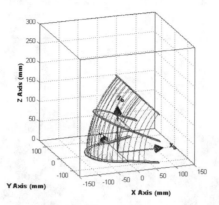

**Fig. 4.** Kinematic simulation of the design concept #1

The design concepts #2 and #3 were then conceived to allow a bigger motion range and ease the difficulty of wearing. A continuum structure was mounted on the back of a wearer and was connected an upper-arm sleeve in the design concept #2. Bending of this continuum structure orients the wearer's upper arm. The continuum structure was moved to the front of the wearer in the design concept #3. The reason will be elaborated below.

Exact shape of the continuum structure in the design concepts #2 and #3 depends on a minimal of the sum of the elastic potential energy of the continuum structure and the gravitational potential energy of the arm and the robot. The exact shape would also be altered by different anatomical parameters (e.g. shoulder widths) of the wearers.

Demonstrated experimentally as in [17], the shape of such a continuum structure differed from a circular arc. In fact the actual shape of the continuum structure would keep changing during the assisted motions for a patient. In order to verify the design concept #2, the simulations were conducted with an approximation that the shape of the continuum structure could be characterized as one circular arc plus a straight line.

The shoulder joint is represented by a spherical joint and the upper arm is represented by a cylinder in Fig. 5. The axis of the upper arm is parallel to $\hat{\mathbf{z}}_e$ so that bending of the continuum structure orients the upper arm.

Main design parameters of the continuum structure in the design concept #2 include i) the length, and ii) the offset of the structure with respect to the shoulder joint. Possible combinations of these parameters were numerated. And it was found the tip of the continuum structure (namely, the origin of $\{e\}$) would translate along the upper arm for a relatively large distance. What's more, the length of the structure also needs to be quite long to allow the desired motion range (the flexion has a wider motion range than the extension in the sagittal plane). Such a prolonged length could reduce the payload capability of the continuum structure.

The design concept #3 is hence obtained as the continuum structure is mounted in front of the wearer. Under this configuration, the wearable robot could be easily put on a user and it entirely allows the desired assistive motion ranges. A structure feature was incorporated into the design to enhance the payload capability, which is presented in Section 5.2.

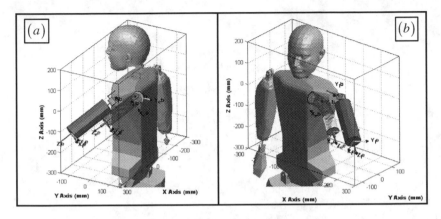

**Fig. 5.** Kinematic simulation of the design concept (a) #2 and (b) #3

# 5     Preliminary System Design and Experimentation

Following the design concept #3, a tentative design of the continuum wearable shoulder robot is reported in Section 5.1. Then a mockup system is constructed as in Section 5.2 for preliminary experimentations.

## 5.1     Tentative System Design

The tentative design shares some similarity to the existing design. The system in Fig. 6 also consists of i) a rigid armguard, ii) an armguard guide, iii) a flexible continuum brace, iv) a body vest, v) a set of guiding cannulae, and vi) an actuation unit.

The armguard can slide on the armguard guide. This will accommodate the translation of the guide with respect to the wearer's upper arm during assisted motions. The actuation unit pushes and pulls the backbones in the shoulder brace to bend the brace so as to orient a wearer's arm. Braided stainless steel overtube as shown in Fig. 6 is attached to the brace's surface to enhance the torsional rigidity.

## 5.2     Preliminary Experimentations

A mockup system is constructed as in Fig. 7 for preliminary experimentations to verify the intended characteristics of the wearable shoulder robot.

The shoulder brace was easily bent for 90° in arbitrary directions. Motion range of the brace is big enough for many intended rehabilitation exercises.

A loading experiment was also carried out as in Fig. 7. Weights of 3 kg and 5 kg were hung to the tip of the flexible brace. Deflections of the brace were acceptable. The brace is hence strong enough for the future use in such a wearable shoulder robot.

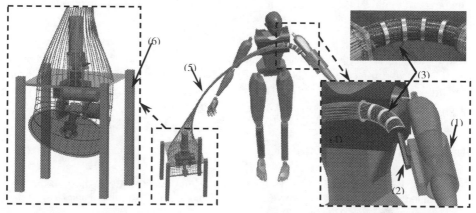

**Fig. 6.** Tentative system design of the new wearable shoulder robot: (1) a rigid armguard, (2) an armguard guide, (3) a flexible continuum brace, (4) a body vest, (5) a set of guiding cannulae, and (6) an actuation unit

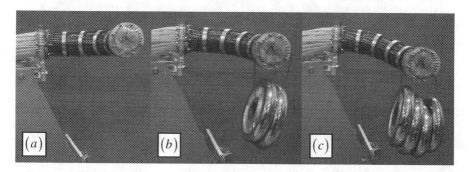

**Fig. 7.** Preliminary payload test of a mockup system: (a) no load, (b) 3 kg, and (c) 5 kg

## 6    Conclusions and Future Work

This paper presents the motivation, kinematics, design concept comparison, preliminary system design and experimentation of a continuum wearable robot for shoulder rehabilitation, aiming at improving the performance of an existing continuum wearable shoulder robot.

The existing continuum wearable shoulder robot could passively adapt to different anatomies while providing assistances. This feature is hence referred to as AAA (Anatomy Adaptive Assistances). However several drawbacks were also identified and three design concepts were hence proposed to overcome these drawbacks.

Based on kinematic analysis, the three concepts are compared and one is selected. The selected design concept is expected to possess a large payload capability, allow desired motion ranges and ease the difficulty for wearing.

Detailed system design of the selected concept was pursued and some preliminary experiments suggested the promising potential of the selected design concept.

Future work mainly includes the finalization and fabrication of the tentative design. Then experimental studies could be carried out for the new continuum wearable robot for shoulder rehabilitation.

**Acknowledgments.** This work was supported in part by the National Program on Key Basic Research Projects (Grant No. 2011CB013300), in part by the Shanghai Rising-Star Program (Grant No. 14QA1402100), and in part by the State Key Laboratory of Mechanical Systems and Vibration (Grant No. MSVZD201406).

# References

1. Brewer, B.R., McDowell, S.K., Worthen-Chaudhari, L.C.: Poststroke Upper Extremity Rehabilitation: A Review of Robotic Systems and Clinical Results. Topics in Stroke Rehabilitation 14(6), 22–44 (2007)
2. Dollar, A.M., Herr, H.: Lower Extremity Exoskeletons and Active Orthoses: Challenges and State-of-the-Art. IEEE Transactions on Robotics 24(1), 144–158 (2008)
3. Zoss, A.B., Kazerooni, H., Chu, A.: Biomechanical Design of the Berkeley Extremity Exoskeleton (BLEEX). IEEE/ASME Transaction on Mechatronics 11(2), 128–138 (2006)
4. Walsh, C.J., Paluska, D., Pasch, K., Grand, W., Valiente, A., Herr, H.: Development of a Lightweight, Underactuated Exoskeleton for Load-Carrying Augmentation. In: IEEE International Conference on Robotics and Automation (ICRA), Orlando, Florida, USA, pp. 3485–3491 (2006)
5. Durfee, W.K., Rivard, A.: Preliminary Design and Simulation of a Pneumatic, Stored-Energy, Hybrid Orthosis for Gait Restoration. In: ASME International Mechanical Engineering Congress, Anaheim, California, USA, pp. 1–7 (2004)
6. Banala, S.K., Agrawal, S.K., Fattah, A., Krishnamoorthy, V., Hsu, W.-L., Scholz, J., Rudolph, K.: Gravity-Balancing Leg Orthosis and Its Performance Evaluation. IEEE Transactions on Robotics 22(6), 1228–1239 (2006)
7. Saglia, J.A., Tsagarakis, N.G., Dai, J.S., Caldwell, D.G.: A High Performance 2-dof Over-Actuated Parallel Mechanism for Ankle Rehabilitation. In: IEEE International Conference on Robotics and Automation (ICRA), Kobe, Japan, pp. 2180–2186 (2009)
8. Farris, R.J., Quintero, H.A., Goldfarb, M.: Preliminary Evaluation of a Powered Lower Limb Orthosis to Aid Walking in Paraplegic Individuals. IEEE Transactions on Neural Systems and Rehabilitation Engineering 19(6), 652–659 (2011)
9. Tsagarakis, N.G., Caldwell, D.G.: Development and Control of a 'Soft-Actuated' Exoskeleton for Use in Physiotherapy and Training. Autonomous Robots 15(1), 21–33 (2003)
10. Perry, J.C., Rosen, J., Burns, S.: Upper-Limb Powered Exoskeleton Design. IEEE/ASME Transaction on Mechatronics 12(4), 408–417 (2007)
11. Gupta, A., O'Malley, M.K., Patoglu, V., Burgar, C.: Design, Control and Performance of RiceWrist: A Force Feedback Wrist Exoskeleton for Rehabilitation and Training. International Journal of Robotics Research 27(2), 233–251 (2008)
12. Stienen, A.H.A., Hekman, E.E.G., Prange, G.B., Jannink, M.J.A., Aalsma, A.M.M., van der Helm, F.C.T., van der Kooij, H.: Dampace: Design of an Exoskeleton for Force-

Coordination Training in Upper-Extremity Rehabilitation. Journal of Medical Devices 3(031003), 1–10 (2009)

13. Klein, J., Spencer, S., Allington, J., Bobrow, J.E., Reinkensmeyer, D.J.: Optimization of a Parallel Shoulder Mechanism to Achieve a High-Force, Low-Mass, Robotic-Arm Exoskeleton. IEEE Transactions on Robotics 26(4), 710–715 (2010)

14. Mao, Y., Agrawal, S.K.: Design of a Cable-Driven Arm Exoskeleton (CAREX) for Neural Rehabilitation. IEEE Transactions on Robotics 28(4), 922–931 (2012)

15. Galiana, I., Hammond, F.L., Howe, R.D., Popovic, M.B.: Wearable Soft Robotic Device for Post-Stroke Shoulder Rehabilitation: Identifying Misalignments. In: IEEE/RSJ International Conference on Intelligent Robots and Systems (IROS), Vilamoura, Portugal, pp. 317–322 (2012)

16. Xu, K., Qiu, D., Simaan, N.: A Pilot Investigation of Continuum Robots as a Design Alternative for Upper Extremity Exoskeletons. In: IEEE International Conference on Robotics and Biomimetics (ROBIO), Phuket, Thailand, pp. 656–662 (2011)

17. Xu, K., Qiu, D.: Experimental Design Verification of a Compliant Shoulder Exoskeleton. In: IEEE International Conference on Robotics and Automation (ICRA), Karlsruhe, Germany, pp. 3894–3901 (2013)

18. Xu, K., Zhao, J., Qiu, D., Wang, Y.: A Pilot Study of a Continuum Shoulder Exoskeleton for Anatomy Adaptive Assistances. ASME Journal of Mechanisms and Robotics 6(4), 041011 (2014)

19. Kobayashi, H., Hiramatsu, K.: Development of Muscle Suit for Upper Limb. In: IEEE International Conference on Robotics and Automation (ICRA), New Orleans, LA, USA, pp. 2480–2485 (2004)

20. Agrawal, S.K., Dubey, V.N., Gangloff, J.J., Brackbill, E., Mao, Y., Sangwan, V.: Design and Optimization of a Cable Driven Upper Arm Exoskeleton. Journal of Medical Devices 3(031004), 1–8 (2009)

21. Xu, K., Wang, Y., Qiu, D.: Design Simulations of the SJTU Continuum Arm Exoskeleton (SCAX). In: International Conference on Intelligent Robotics and Applications (ICIRA), Busan, Korea, pp. 351–362 (2013)

22. Schiele, A., van der Helm, F.C.T.: Kinematic Design to Improve Ergonomics in Human Machine Interaction. IEEE Transactions on Neural Systems and Rehabilitation Engineering 14(4), 456–469 (2006)

23. Kim, H., Miller, L.M., Byl, N., Abrams, G.M., Rosen, J.: Redundancy Resolution of the Human Arm and an Upper Limb Exoskeleton. IEEE Transactions on Biomedical Engineering 59(6), 1770–1779 (2012)

24. Jarrassé, N., Morel, G.: Connecting a Human Limb to an Exoskeleton. IEEE Transactions on Robotics 28(3), 697–709 (2012)

25. Xu, K., Simaan, N.: An Investigation of the Intrinsic Force Sensing Capabilities of Continuum Robots. IEEE Transactions on Robotics 24(3), 576–587 (2008)

26. Xu, K., Simaan, N.: Analytic Formulation for the Kinematics, Statics and Shape Restoration of Multibackbone Continuum Robots via Elliptic Integrals. Journal of Mechanisms and Robotics 2(011006), 1–13 (2010)

27. Xu, K., Fu, M., Zhao, J.: An Experimental Kinestatic Comparison between Continuum Manipulators with Structural Variations. In: IEEE International Conference on Robotics and Automation (ICRA), Hong Kong, China (2014) (accepted for presentation)

# Robust Control with Dynamic Compensation for Human-Wheelchair System

Víctor H. Andaluz[1], Paúl Canseco[1], José Varela[1], Jessica S. Ortiz[1], María G. Pérez[1], Flavio Roberti[2], and Ricardo Carelli[2]

[1] FISEI, Universidad Técnica de Ambato, Ambato-Ecuador
victorhandaluz@uta.edu.ec
[2] Instituto de Automática, Universidad Nacional de San Juan, San Juan-Argentina
vandaluz@inaut.unsj.edu.ar

**Abstract.** This work presents the kinematic and dynamic modeling of a human-wheelchair system, and dynamic control to solve the path following problem. First it is proposed a dynamic modeling of the human-wheelchair system where it is considered that its mass center is not located at the center the wheels' axle of the wheelchair. Then, the design of the control algorithm is presented. This controller design is based on two cascaded subsystems: a kinematic controller with command saturation, and a dynamic controller that compensates the dynamics of the robot. Stability and robustness are proved by using Lyapunov's method. Experimental results show a good performance of the proposed controller as proved by the theoretical design.

**Keywords:** Human-wheelchair system, dynamic control and dynamic modeling.

## 1    Introduction

In recent years, robotics research has experienced a significant change. The research interests are moving from the development of robots for structured industrial environments to the development of autonomous mobile robots operating in unstructured and natural environments. These autonomous robots are applicable in a number of challenging tasks such as cleaning of hazardous material, surveillance, rescue and reconnaissance in unstructured environments which humans are kept away from. Since it is foreseen that this new class of mobile robots will have extensive applications in activities where human capabilities are needed, they have attracted the attention of robotics researchers [1-3].

Therefore, the necessity of technological development in the field of medical and welfare equipment is cried out. By responding to this issue, different types of technologies have been developing by Engineering for the assistance of human [4-7]. Electrical wheelchair is an important means of transport for handicapped and aged people, who do not have the capability of walking, normally can move around using a commercially available wheelchair. However there are many people suffering from

X. Zhang et al. (Eds.): ICIRA 2014, Part I, LNAI 8917, pp. 376–389, 2014.

severe loss of motor function due to variety of accidents or diseases such as a Spinal Cord Injury (SCI) or Amyotrophic Lateral Sclerosis (ALS), in this cases is necessary to provide a new way to command such an electrical vehicle.

Human Machine Interface (HMI) based on electro-biological signal are present in [1,4,8,9]. Such interfaces allow commanding wheelchair robot governed by a computer. The literature shows that an autonomous wheelchair can be successfully driven by persons using only electrical signals generated by eye-blinks, voice, and others [1,2,10,11]. Then, a trajectory will be automatically generated and a trajectory tracking control will guide the wheelchair to the desired target. As indicated, the fundamental problems of motion control of wheelchair robots can be roughly classified in three groups [12]: 1) *point stabilization*: the goal is to stabilize the wheelchair at a given target point, with a desired orientation; 2) *trajectory tracking*: the wheelchair is required to track a time parameterized reference; and 3) *path following*: the wheelchair is required to converge to a path and follow it, without any time specifications; this work is focused to resolve the path following problem.

The path following problem has been well studied and many solutions have been proposed and applied in a wide range of applications. Let $\mathcal{P}_d(s) \in \mathfrak{R}^2$ be a desired geometric path parameterized by the curvilinear abscissa $s \in \mathfrak{R}$. In the literature is common to find different control algorithms for path following where is conseder $s(t)$ as an additional control input. In [13-16], the rate of progression ($\dot{s}$) of a virtual vehicle has been controlled explicitly. Another method for path following of wheelchair robot is the image-based control. The main objective of this method is to detect and follow the desired path through vision sensors [3]. Furthermore, it is important to consider the wheelchair's dynamics in addition to its kinematics because wheelchairs carry relatively heavy loads. As an example, the trajectory tracking task can be severely affected by the change imposed to the wheelchair dynamics when it is carrying a person, as shown in [17]. Hence, some path following control architectures already proposed in the literature have considered the dynamics of the wheelchair robots [11,19].

In such context, this work proposes a new method to solve the path following problem for a wheelchair robot to assist persons with severe motor diseases. Additionally, it is proposed a dynamic modeling of the human-wheelchair system which, which has reference velocities as input signals to the wheelchair, as it is common in commercial robots, and it also has adequate structure for control law designing [11]. The proposed control scheme is divided into two subsystems, each one being a controller itself: 1) the firts on is a kinematic controller with saturation of velocity commands, which is based on the wheelchair robot's kinematic. The path following problem is addressed in this subsystem. It is worth noting that the proposed controller does not consider $s(t)$ as an additional control input as it is frequent in literature; and 2) an dynamic compensation controller that considered the human-wheelchair system dynamic model, which are directly related to physical parameters of the system. In addition, both stability and robustness properties to parametric uncertainties in the dynamic model are proven through Lyapunov's method. To validate the proposed control algorithm, experimental results are included and discussed.

The paper is organized as follows: Section 2 shown the complete dynamic modeling of the human-wheelchair system, while Section 3 describes the path following's formulation problem and it also presents the controllers design. Furthermore, the analysis of the system's stability is developed. Next, experimental results are presented and discussed in Section 4, and finally the conclusions are given in Section 5.

## 2      Human-Wheelchair System Modeling

The wheelchair robot used in this work presents similar characteristics to that of a unicycle-like mobile robot, because it has two driven wheels which are controlled independently by two D.C. motors and four caster wheel to maintain balance. The unicycle-type mobile robots have a caster wheel to maintain stability, but in this work the wheelchair has four caster wheels around the central axis conferring greater stability to the human-wheelchair system. The dynamic modeling of the human-wheelchair system is developed in this section, considering a horizontal work plane where the wheelchair moves. The wheelchair type unicycle-like mobile robot presents the advantages of high mobility, high traction with pneumatic tires, and a simple wheel configuration.

### 2.1      Kinematic Model

It is assumed that the mobile platform moves on a planar horizontal surface. Let $R(X, Y, Z)$ be any fixed frame with $Z$ vertical, as the Fig.1 shows.

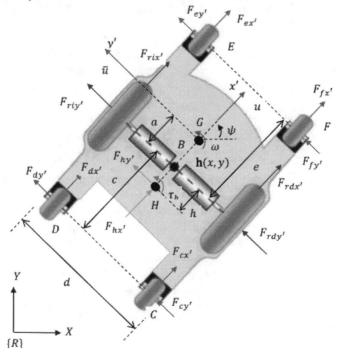

**Fig. 1.** Schematic of the wheelchair

This work is based on Unicycle-like wheelchair. A unicycle wheelchair is a driving robot that can rotate freely around its axis. The term unicycle is often used in robotics to mean a generalized cart or car moving in a two-dimensional world; these are also often called unicycle-like or unicycle-type vehicles. On the other hand, the non-holonomic velocity constraint of the wheelchair determines that it can only move perpendicular to the wheels axis,

$$\dot{x}sen(\psi) - \dot{y}\cos(\psi) + a\dot{\psi} = 0 \tag{1}$$

Therefore, the configuration instantaneous kinematic model of the holonomic mobile platform is defined as,

$$\begin{cases} \dot{x} = u\cos\psi - a\omega sen\psi \\ \dot{y} = u sen\psi + a\omega\cos\psi \\ \dot{\psi} = \omega \end{cases} \tag{2}$$

Equation (2) can be written in compact form as

$$\mathbf{\dot{h}} = \mathbf{J}(\psi)\mathbf{v} \tag{3}$$

$$\dot{\psi} = \omega$$

where $\mathbf{\dot{h}} = [\dot{x} \quad \dot{y}]^T \in \mathcal{R}^2$ represents the vector of axis velocity; $\mathbf{J}(\psi) \in \mathcal{R}^{2\times2}$ is a singular matrix; and the control (*of manoeuvrability*) of the wheelchair is defined $\mathbf{v} = [u \quad \omega]^T \in \mathcal{R}^2$ in which $u$ and $\omega$ which represent the linear and angular velocities of the wheelchair, respectively.

## 2.2 Dynamic Model

Fig.1 illustrates the wheelchair considering in this work; the position of the human-wheelchair system, is given by point $G$, representing the center of mass; $\mathbf{h} = [x \quad y]^T$ represents the point that is required to track a path in $\mathcal{R}$ ; $\psi$ is the orientation of the wheelchair. On the other hand, $u'$ and $\bar{u}$ are the longitudinal and lateral velocities of the center of mass; $\omega$ is the angular velocity; $a, c, e, d,$ and $h$ are distances; $F_{rdx'}$ and $F_{rdy'}$ are the longitudinal and lateral tire forces of the right wheel; $F_{rix'}$ and $F_{riy'}$ are the longitudinal and lateral tire forces of the left wheel; $F_{cx'}, F_{cy'}, F_{dx'}, F_{dy'}, F_{ex'},$ $F_{ey'}, F_{fx'}$ and $F_{fy'}$ are the longitudinal and lateral tire forces exerted on $C, D, E$ and $F$ by the castor wheels; $F_{hx'}$ and $F_{hy'}$ are the longitudinal and lateral force exerted on $E$ by the human; and $\tau_h \, \tau_e$ is the moment exerted by the human.

The force and moment equations for the mobile robot are:

$$\sum F_{x'} = m(\dot{u} - \bar{u}\omega) = F_{rdx'} + F_{rix'} + F_{hx'} + F_{cx'} + F_{dx'} + F_{ex'} + F_{fx'} \tag{4}$$

$$\sum F_{y'} = m(\dot{\bar{u}} - u\omega) = F_{rdy'} + F_{riy'} + F_{hy'} + F_{cy'} + F_{dy'} + F_{ey'} + F_{fy'} \tag{5}$$

$$\Sigma M_z = I_z \dot{\omega} = \frac{d}{2}(F_{rdx'} - F_{rix'}) - a(F_{rdy'} + F_{riy'}) - (h+a)F_{hx'} - (c +$$

$$a)(F_{dy'} + F_{cy'}) + (e-a)(F_{fy'} + F_{ey'}) + \frac{d}{2}(F_{cx'} + F_{fx'} - F_{dx'} - F_{ex'}) + \tau_h \quad (6)$$

where $m = m_h + m_r$ is the human-wheelchair system mass in which $m_h$ is the human mass and $m_r$ is the wheelchair mass; and $I_z$ is the robot moment of inertia about the vertical axis located in $G$. According to [20], velocities $u, \omega$ and $\bar{u}$, including the slip speeds, are given by:

$$u = \frac{r}{2}(\omega_d + \omega_i) \quad (7)$$

$$\omega = \frac{r}{d}(\omega_d - \omega_i) \quad (8)$$

$$\bar{u} = \frac{ar}{d}(\omega_d - \omega_i) \quad (9)$$

where $r$ is the right and left wheel radius; $\omega_d$ and $\omega_i$ are the angular velocities of the right and left wheels, respectively.

The motor models attained by neglecting the voltage on the inductances are:

$$\tau_d = \frac{k_a(v_d - k_b \omega_d)}{R_a} \quad ; \qquad \tau_i = \frac{k_a(v_i - k_b \omega_i)}{R_a} \quad (10)$$

where $v_d$ and $v_i$ are the input voltages applied to the right and left motors; $k_a$ is the torque constant multiplied by the gear ratio; $k_b$ is equal to the voltage constant multiplied by the gear ratio; $R_a$ is the electric resistance constant; $\tau_d$ and $\tau_i$ are the right and left motor torques multiplied by the gear ratio. The dynamic equations of the motor-wheels are:

$$I_e \dot{\omega}_d + B_e \omega_d = \tau_d - F_{rdx'} R_t \quad (11)$$

$$I_e \dot{\omega}_i + B_e \omega_i = \tau_i - F_{rix'} R_t \quad (12)$$

where $I_e$ and $B_e$ are the moment of inertia and the viscous friction coefficient of the combined motor rotor, gearbox, and wheel, and $R_t$ is the nominal radius of the tire.

In general, most market-available robots have low level PID velocity controllers to track input reference velocities and do not allow the motor voltage to be driven directly. Therefore, it is useful to express the mobile robot model in a suitable way by considering rotational and translational reference velocities as input signals. For this purpose, the velocity controllers are included into the model. To simplify the model, a PD velocity controller has been considered which is described by the following equations:

$$v_u = k_{PT}(u_{ref} - u) - \dot{u}k_{DT} \quad (13)$$

$$v_\omega = k_{PR}(\omega_{ref} - \omega) - \dot{\omega}k_{DR} \quad (14)$$

where $k_{PT}, k_{DT}, k_{PR}$ and $k_{DR}$ are gain positive constants of the PD controllers.

From (4 – 14) the following dynamic model of the human-wheelchair system is obtained:

$$\begin{bmatrix} \varsigma_1 & 0 \\ 0 & \varsigma_2 \end{bmatrix} \begin{bmatrix} \dot{u} \\ \dot{\omega} \end{bmatrix} + \begin{bmatrix} \varsigma_4 & \varsigma_3\omega \\ \varsigma_5\omega & \varsigma_6 \end{bmatrix} \begin{bmatrix} u \\ \omega \end{bmatrix} = \begin{bmatrix} u_{ref} \\ \omega_{ref} \end{bmatrix} \tag{15}$$

also, the dynamic equation (15) can be represented as follows,

$$\mathbf{M}(\varsigma)\dot{\mathbf{v}} + \mathbf{C}(\varsigma, \mathbf{v})\mathbf{v} = \mathbf{v_{ref}} \tag{16}$$

where $\mathbf{M}(\varsigma) \in \mathcal{R}^{2x2}$ represents the system's inertia, $\mathbf{C}(\varsigma) \in \mathcal{R}^{2x2}$ represents the components of the centripetal forces, $\mathbf{v} \in \mathcal{R}^2$ is the vector of system's velocity; $\mathbf{v_{ref}} \in \mathcal{R}^2$ is the vector of velocity control signals for the wheelchair; and $\varsigma = [\varsigma_1 \; \varsigma_2 \; ... \; \varsigma_6]^T \in \mathcal{R}^6$ is the vector of sistem's dynamic parameters. Hence the properties for the dynamic model with reference velocities as control signals as:

*Property 1.* Matrix $\mathbf{M}(\varsigma)$ is positive definite, additionally it is known that $\|\mathbf{M}(\varsigma)\| < k_M$, where, $k_M$ i a positive constant;

*Property 2.* Furthermore, the following inequalities are also satisfied $\|\mathbf{C}(\varsigma, \mathbf{v})\mathbf{v}\| < k_C\|\mathbf{v}\|$, where, $k_C$ denote a positive constants.

*Property 4.* The dynamic model of the mobile manipulator can be represented by $\mathbf{M}(\varsigma)\dot{\mathbf{v}} + \mathbf{C}(\varsigma, \mathbf{v})\mathbf{v} = \mathbf{\Phi}(\mathbf{v})\varsigma$, where, $\mathbf{\Phi}(\mathbf{v}) \in \mathcal{R}^{2xl}$ and $\varsigma = [\varsigma_1 \; \varsigma_2 \; ... \; \varsigma_l]^T \in \mathcal{R}^l$ is the vector of $l$ unknown parameters of the human-wheelchair system, *i.e.*, mass of the human, mass of the wheelchair, physical parameters of the wheelchair, motors, velocity, and others.

In order to show the performance of the proposed model dynamic, the identification and validation is shown. The experimental test was implemented on a wheelchair, which admits linear and angular velocities as input reference signals. The identification of the mobile robot was performed by using least squares estimation [21] applied to a filtered regression model [22]. Fig. 2, shown the validation the proposed dynamic model, where it can be seen the good performance of the obtained dynamic model.

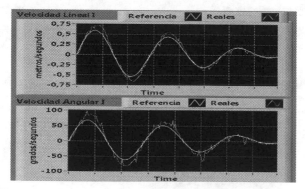

**Fig. 2.** Validation of the proposed dynamic model

Hence, the full mathematical model of the unicycle-like human-wheelchair system is represented by: (3) the kinematic model and (16) the dynamic model, taking the reference velocities of the robot as input signals.

## 3      Formulation Problem and Controllers Design

The solution of the path following problem for mobile robots derived in [12] admits an intuitive explanation. A path following controller should aim to reduce to zero both: i) the distance from the vehicle to a point on the path, and ii) the angle between the vehicle velocity vector and the tangent to the path at this point.

### 3.1     Formulation Problem

As represented in Fig. 3, the path to be followed is denoted as $\mathcal{P}$ . The actual desired location $\mathbf{P}_d = [P_{xd} \quad P_{yd}]^T$ is defined as the closest point on $\mathcal{P}$ to the mobile robot, with a desired orientation $\psi_d$ . In Fig. 3, $\rho$ represents the distance between the robot position $\mathbf{h}$ and $\mathbf{P}_d$ , and $\tilde{\psi}$ is the error orientation between $\psi_d$ and $\psi$ . Given a path $\mathcal{P}$ in the operational space of the mobile robot and the desired velocity module $v$ for the robot, the path following problem for mobile robot consists in finding a feedback control law $\mathbf{v}_{ref}(t) = (s, v, \rho, \tilde{\psi})$, such that $\lim_{t \to \infty} \rho(t) = 0$ and

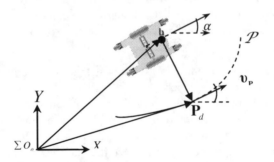

**Fig. 3.** Schematic of the wheelchair

$\lim_{t \to \infty} \tilde{\psi}(t) = 0$ . The error vector of position and orientation between the robot and the point $\mathbf{P}_d$ can be represented as, $\tilde{\mathbf{h}} = \mathbf{P}_d - \mathbf{h}$ and $\tilde{\psi} = \psi_d - \psi$ . Therefore, if $\lim_{t \to \infty} \tilde{\mathbf{h}}(t) = \mathbf{0}$ then $\lim_{t \to \infty} \rho(t) = 0$ and $\lim_{t \to \infty} \tilde{\psi}(t) = 0$ .

Hence, the desired position and desired velocity of the mobile robot on the path $\mathcal{P}$ , are defined as $\mathbf{h}_d(s, h) = \mathbf{P}_d(s, h)$ and $\mathbf{v}_{hd}(s, h) = \mathbf{v}_p(s, h)$ . Where $\mathbf{v}_p$ is the desired velocity of the robot at location $\mathbf{P}_d$ . Note that the component of $\mathbf{v}_p$ has to be tangent to the trajectory due to kinematics compatibility.

Also, The proposed control scheme to solve the path following problem is shown in Fig. 4, the design of the controller is based mainly on two cascaded subsystems: 1) *Kinematic Controller* with saturation of velocity commands, where the control errors $\rho(t)$ and $\tilde{\psi}(t)$ may be calculated at every measurement time and used to drive the mobile robot in a direction which decreases the errors. Therefore, the control aim is to ensure that $\lim_{t \to \infty} \rho(t) = 0$ and $\lim_{t \to \infty} \tilde{\psi}(t) = 0$; and 2) *Dynamic Compensation Controller*, which main objective is to compensate the dynamics of the human-wheelchair system, thus reducing the velocity tracking error. The velocity control error is defined as $\tilde{\mathbf{v}} = \mathbf{v}_c - \mathbf{v}$ Hence, the control aim is to ensure that $\lim_{t\to\infty} \tilde{\mathbf{v}}(t) = \mathbf{0}$.

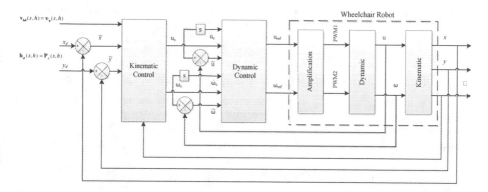

**Fig. 4.** Dynamic compensation controller: block diagram

## 3.2 Kinematic and Dynamic Controllers

The proposed kinematic controller is based on the kinematic model of the unicycle-like mobile robot (3), *i.e.*, $\dot{\mathbf{h}} = f(\mathbf{h})\mathbf{v}$. Hence following control law is proposed,

$$
\begin{bmatrix} u_c \\ \omega_c \end{bmatrix} = \begin{bmatrix} cos\psi & sen\psi \\ -\frac{1}{a}sen\psi & \frac{1}{a}cos\psi \end{bmatrix} \left( \begin{bmatrix} cos\psi_d \\ sen\psi_d \end{bmatrix} + \begin{bmatrix} l_x tanh\left(\frac{k_x}{l_x}\tilde{h}_x\right) \\ l_y tanh\left(\frac{k_y}{l_y}\tilde{h}_y\right) \end{bmatrix} \right) \tag{17}
$$

this equation can be written in compact form as,

$$
\mathbf{v}_c = \mathbf{J}^{-1}\left(\mathbf{v}_P + \mathbf{L}\tanh\left(\mathbf{L}^{-1}\mathbf{K}\,\tilde{\mathbf{h}}\right)\right) \tag{18}
$$

where $\tilde{\mathbf{h}} = [\tilde{h}_x \quad \tilde{h}_y]^T$ represents the position error of the wheelchair defined as $\tilde{h}_x = P_{xd} - x$ and $\tilde{h}_y = P_{yd} - y$; $\mathbf{v}_P = [v\cos\psi_d \quad v\sin\psi_d]^T$ is the desired velocity vector on the path; $\mathbf{L}$ and $\mathbf{K}$ are definite positive diagonal matrices that weigh the control error. In order to include an analytical saturation of velocities in the mobile robot, the **tanh**(.) function, which limits the error $\tilde{\mathbf{h}}$, is proposed. The expression tanh($\mathbf{L}^{-1}\mathbf{K}\,\tilde{\mathbf{h}}$) denote a component by component operation.

Worth noting that the reference desired velocity $(t)$ of the wheelchair during the tracking path need not be constant, with is common in the literature [1,3,13-16],

$$
(t) = f(k, \rho(t), \omega(t), \ldots) 
$$

the wheelchair's desired velocity can be expressed as: constant function, position error function, angular velocities function of the wheelchair; and the others consideration.

Now, the behaviour of the control position error of the robot is now analyzed assuming -by now- perfect velocity tracking *i.e.*, $\mathbf{v}(t) \equiv \mathbf{v}_c(t)$. By substituting (18) in (3), it is obtained, $\left(\mathbf{v}_{hd} - \dot{\mathbf{h}}\right) + \mathbf{L}\tanh\left(\mathbf{L}^{-1}\mathbf{K}\,\tilde{\mathbf{h}}\right) = \mathbf{0}$. Now defining difference signal $\Upsilon$ between $\dot{\mathbf{h}}_d$ and $\mathbf{v}_{hd}$, *i.e.*, $\Upsilon = \dot{\mathbf{h}}_d - \mathbf{v}_{hd}$ and remembering that $\dot{\tilde{\mathbf{h}}} = \dot{\mathbf{h}}_d - \dot{\mathbf{h}}$, it's can be written as

$$\dot{\tilde{\mathbf{h}}} + \mathbf{L}\tanh\left(\mathbf{L}^{-1}\mathbf{K}\,\tilde{\mathbf{h}}\right) = \Upsilon \tag{19}$$

For the stability analysis the following Lyapunov candidate function is considered $V\left(\tilde{\mathbf{h}}\right) = \frac{1}{2}\tilde{\mathbf{h}}^{\mathsf{T}}\tilde{\mathbf{h}} > 0$ Its time derivative on the trajectories of the system is, $\dot{V}\left(\tilde{\mathbf{h}}\right) = \tilde{\mathbf{h}}^{\mathsf{T}}\Upsilon - \tilde{\mathbf{h}}^{\mathsf{T}}\mathbf{L}\tanh\left(\mathbf{L}^{-1}\mathbf{K}\,\tilde{\mathbf{h}}\right)$. Then, a sufficient condition for $\dot{V}\left(\tilde{\mathbf{h}}\right)$ to be negative definite is,

$$\left|\tilde{\mathbf{h}}^{\mathsf{T}}\mathbf{L}\tanh\left(\mathbf{L}^{-1}\mathbf{K}\,\tilde{\mathbf{h}}\right)\right| > \left|\tilde{\mathbf{h}}^{\mathsf{T}}\Upsilon\right| \tag{20}$$

*Remark 1.* $\mathbf{v}_{hd}$ is collinear to $\dot{\mathbf{h}}_d$, then $\Upsilon$ is also a collinear vector to $\mathbf{v}_{hd}$ and $\dot{\mathbf{h}}_d$.

*Remark 2.* For large values of $\tilde{\mathbf{h}}$, it can be considered that: $\mathbf{L}\tanh\left(\mathbf{L}^{-1}\mathbf{K}\,\tilde{\mathbf{h}}\right) \approx \mathbf{L}$. $\dot{V}$ will be negative definite only if $\|\mathbf{L}\| > \|\Upsilon\|$; establishing a design condition which makes path following errors $\tilde{\mathbf{h}}$ to decrease

*Remark 3.* As aforementioned, the desired path velocity can be written as $\mathbf{v}_{hd} = \dot{\mathbf{h}}_d - \Upsilon$. So, for small values of $\tilde{\mathbf{h}}$, $\mathbf{L}\tanh\left(\mathbf{L}^{-1}\mathbf{K}\,\tilde{\mathbf{h}}\right) \approx \mathbf{K}\tilde{\mathbf{h}}$. Thus, the closed loop equation of the system can now written as $\dot{\tilde{\mathbf{h}}} + \mathbf{K}\,\tilde{\mathbf{h}} = \Upsilon$. Applying Laplace representation, one get

$$\tilde{\mathbf{h}}(s) = \frac{1}{s\mathbf{I} + \mathbf{K}}\Upsilon(s) \tag{21}$$

Hence, the direction of the vector of control errors $\tilde{\mathbf{h}}(s)$ tends to the direction of the error velocity vector $\Upsilon(s)$. Therefore, since for finite values $\tilde{\mathbf{h}}(s)$, this location error is normal to $\mathbf{v}_{hd}$ (criterion of minimum distance between the robot and the path), and thus to $\Upsilon$ (see *Remark 1*). Then $\tilde{\mathbf{h}}$ has to be zero. It can now be concluded that $\rho(t) \to 0$ and $\tilde{\psi}(t) \to 0$ for $t \to \infty$ asymptotically.

On the other hand, If there is not considering perfect velocity tracking in kinematic controller design, *i.e.*, $\mathbf{v}(t) \neq \mathbf{v}_c(t)$, the velocity error is defined as, $\tilde{\mathbf{v}}(t) = \mathbf{v}_c(t) - \mathbf{v}(t)$. This velocity error motivates to design of an dynamic compensation controller; the objective of this controller is to compensate the dynamic of the human-wheelchair system, thus reducing the velocity tracking error, hence the following control law dynamic model based (16) is proposed,

$$\begin{bmatrix} u_{ref} \\ \omega_{ref} \end{bmatrix} = \begin{bmatrix} \varsigma_1 & 0 \\ 0 & \varsigma_2 \end{bmatrix} \left( \begin{bmatrix} \dot{u}_c \\ \dot{\omega}_c \end{bmatrix} + \begin{bmatrix} l_u \tanh\left(\frac{k_u}{l_u}\tilde{u}\right) \\ l_\omega \tanh\left(\frac{k_\omega}{l_\omega}\tilde{\omega}\right) \end{bmatrix} \right) + \begin{bmatrix} \varsigma_4 & \varsigma_3\omega \\ \varsigma_5\omega & \varsigma_6 \end{bmatrix} \begin{bmatrix} u \\ \omega \end{bmatrix} \qquad (22)$$

Equation (22) can be written in compact form as

$$\mathbf{v_{ref}} = \mathbf{M}(\varsigma)\boldsymbol{\sigma}(\tilde{\mathbf{v}}) + \mathbf{C}(\varsigma, \mathbf{v})\mathbf{v} \qquad (23)$$

where $\tilde{u} = u_c - u$, $\tilde{\omega} = \omega_c - \omega$ are the linear and angular velocity errors, respectively; $k_u > 0$, $k_\omega > 0$, $l_u > 0$ and $l_\omega > 0$ are positive gain constants.

Next, a Lyapunov candidate function and its time derivative on the system trajectories are introduced in order to consider the corresponding stability analysis $V(\tilde{\mathbf{v}}) = \frac{1}{2}\tilde{\mathbf{v}}^T\tilde{\mathbf{v}}$; the time derivative of the Lyapunov candidate function is,

$$\dot{V}(\tilde{\mathbf{v}}) = \tilde{\mathbf{v}}^T\dot{\tilde{\mathbf{v}}} \qquad (24)$$

After introducing the control laws (16) and (23) in (24), the time derivative $\dot{V}(\tilde{\mathbf{v}})$ is now

$$\dot{V}(\tilde{\mathbf{v}}) = -\tilde{\mathbf{v}}^T\boldsymbol{\Gamma}(\tilde{\mathbf{v}}) < 0 \qquad (25)$$

$$\dot{V}(\tilde{\mathbf{v}}) = -[\tilde{u} \quad \tilde{\omega}]\begin{bmatrix} l_u & 0 \\ 0 & l_\omega \end{bmatrix}\begin{bmatrix} \tanh\left(\frac{k_u}{l_u}\tilde{u}\right) \\ \tanh\left(\frac{k_\omega}{l_\omega}\tilde{\omega}\right) \end{bmatrix} < 0$$

Hence, it can now be concluded that $\tilde{\mathbf{v}}(t) \to 0$, i.e., $\tilde{u} \to 0$ and $\tilde{\omega} \to 0$ with $t \to \infty$ asymptotically.

## 4    Experimental Results

In order to show the performance of the proposed controller and dynamic model several experiments were executed. Some of the results are presented in this section. Fig. 5 presents the wheelchair robot used in this work, which has two independently driven wheels by two D.C. motors (in the center part), and four caster wheel to better balance (two in the rear part and two in the front part). Encoders installed on each one of the motor shafts allow knowing the relative position and orientation of the wheelchair.

Information provided by the encoders is used by the PID controllers responsible for getting an independent velocity control of the left and the right wheel.

The experiment corresponds to the control structure shown on Fig. 4. It was implemented on the wheelchair presented on Fig. 5, using the control laws in (18) and (23). Note that for the path following problem, the desired velocity of the wheelchair will depend on the task, the control error, the angular velocity, etc. For this experiment, it is consider that the reference velocity module depends on the control errors. Then, reference velocity in this experiment is expressed as

$$|\mathbf{v_{hd}}| = v_P / (1 + k\rho)$$

**Fig. 5.** Autonomous wheelchair

where, $k$ is a positive constant that weigh the control error module. Also, the desired location is defined as the closest point on the path to the wheelchair.

Figures 6-8 show the results of the experiment. Fig. 6 shows the movement on the $X$-$Y$ space. It can be seen that the proposed controller works correctly.

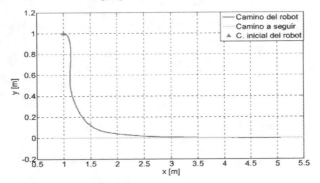

**Fig. 6.** Movement of the mobile robot based on the experimental data

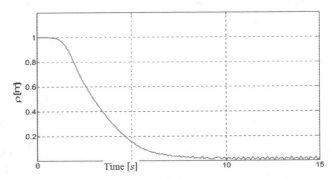

**Fig. 7.** Distance between the wheelchair and the closest point on the path

Figure 7 shows that $\rho(t)$ remains close to zero, while the Fig. 8 shows the relation between the evolution linear velocity commands and the wheelchair's desired velocity.

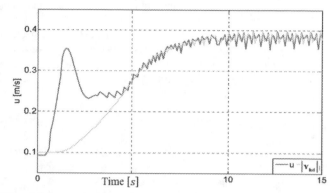

**Fig. 8.** Linear velocity commands and wheelchair's desired velocity

## 5    Conclusion

In this work a compensation dynamic controller for solving the path following problem of the human-wheelchair system was proposed. In addition we proposed a dynamic model for the unicycle-like wheelchair, which has reference velocities as control signals to the robot. The design of the whole controller was based on two cascaded subsystems: a kinematic controller which complies with the task objective (path following), and a dynamic controller that compensates the dynamics of the human-wheelchair system. Finally, the stability and robustness are proved by considering the Lyapunov's method, and the performance of the proposed controller is shown through real experiments.

## Appendix

The following shows the dynamic parameters of the robotic wheelchair model.

$$\varsigma_1 = \frac{\frac{Ra}{ka}(2I_e + mrR_t) + 2rk_{DT}}{2rk_{PT}} \; ; \varsigma_2 = \frac{\frac{Ra}{ka}\left(I_e d^2 + 2rR_t(I_z + a^2 m)\right) + 2rdk_{DR}}{2rdk_{PR}}$$

$$\varsigma_3 = \frac{\frac{Ra}{ka}(amR_t)}{2k_{PT}} \quad ; \quad \varsigma_4 = \frac{\frac{Ra}{ka}\left(\frac{ka k_b}{Ra} + B_e\right)}{rk_{PT}} + 1; \quad \varsigma_5 = \frac{\frac{Ra}{ka}(amR_t)}{dk_{PR}} \quad ;$$

$$\varsigma_6 = \frac{\frac{Ra}{ka}\left(\frac{ka k_b}{Ra} + B_e\right)d}{2rk_{PR}} + 1.$$

# References

1. Bastos-Filho, T., et al.: Towards a New Modality-Independent Interface for a Robotic Wheelchair. IEEE Transactions on Neural Systems and Rehabilitation Engineering 22(3) (2014)
2. Wang, Y., Chen, W.: Hybrid Map-based Navigation for Intelligent Wheelchair. In: IEEE International Conference on Robotics and Automation, China, pp. 637–642 (2011)
3. Cheng, W.-C., Chiang, C.-C.: The Development of the Automatic Lane Following Navigation System for the Intelligent Robotic Wheelchair. In: IEEE International Conference on Fuzzy Systems, Taiwan, pp. 1946–1952 (2011)
4. Abeygunawardhana, P.K.W., Toshiyuki, M.: Self Sustaining Wheelchair Robot on a Curved Trajectory. In: IEEE/ ICIT International Conference on Industrial Technology, pp. 1636–1641 (2006)
5. Biswas, K., Mazumder, O., Kundu, A.S.: Multichannel Fused EMG based Biofeedback System with Virtual reality for Gait Rehabilitation. In: IEEE Proceedings in International Conference on Intelligent Human Computer Interaction, India (2012)
6. Munakata, Y., Tanaka, A., Wada, M.: A Five-wheel Wheelchair with an Active-caster Drive System. In: IEEE International Conference on Rehabilitation Robotics, USA, pp. 1–6 (2013)
7. Yuan, J.: Stability Analyses of Wheelchair Robot Based on Human-in-the-Loop Control Theory. In: IEEE International Conference on Robotics and Biomimetics, Thailand, pp. 419–424 (2009)
8. Soh, H., Demiris, Y.: Involving young children in the development of a safe, smart paediatric wheelchair. Presented at the ACM/IEEE HRI-2011 Pioneers Workshop, Lausanne, Switzerland (2011)
9. Nam, Y., Zhao, Q., Cichocki, A., Choi, S.: Tongue-Rudder: A Glossokinetic-Potential-Based Tongue–Machine Interface. IEEE Transactions on Biomedical Engineering, 290–299 (2012)
10. Munakata, Y., Tanaka, A., Wada, M.: A five-wheel wheelchair with an active-caster drive system. In: IEEE International Conference on Rehabilitation Robotics, pp. 1–6 (2013)
11. De La Cruz, C., Bastos, T.F., Carelli, R.: Adaptive motion control law of a robotic wheelchair. Control Engineering Practice, 113–125 (2011)
12. Soeanto, D., Lapierre, L., Pascoal, A.: Adaptive non-singular path-following, control of dynamic wheeled robots. In: Proceedings of 42nd IEEE Conference on Decision and Control, Hawaii, USA, December 9-12, pp. 1765–1770 (2003)
13. Xu, Y., Zhang, C., Bao, W., Tong, L.: Dynamic Sliding Mode Controller Based on Particle Swarm Optimization for Mobile Robot's Path Following. International Forum on Information Technology and Applications, 257–260 (2009)
14. Wangmanaopituk, S., Voos, H., Kongprawechnon, W.: Collaborative Nonlinear Model-Predictive Collision Avoidance and Path Following of Mobile Robots. In: ICROS-SICE International Joint Conference 2009, Japan, pp. 3205–3210 (2009)
15. Tanimoto, Y., Yamamoto, H., Namba, K., Tokuhiro, A., Furusawa, K., Ukida, H.: Imaging of the turn space and path of movement of a wheelchair for remodeling houses of individuals with SCI. In: IEEE International Conference on Imaging Systems and Techniques (2012)
16. Chen, S.-H., Chou, J.-J.: Motion control of the electric wheelchair powered by rim motors based on event-based cross-coupling control strategy. In: IEEE/SICE International Symposium on System Integration (2011)

17. Martins, F.N., Celeste, W., Carelli, R., Sarcinelli-Filho, M., Bastos-Filho, T.: An Adaptive Dynamic Controller for Autonomous Mobile Robot Trajectory Tracking. Control Engineering Practice 16, 1354–1363 (2008)
18. Chénier, F., Bigras, P., Aissaoui, R.: A new dynamic model of the manual wheelchair for straight and curvilinear propulsion. In: IEEE International Conference on Rehabilitation Robotics (2011)
19. Yahaya, S.Z., Boudville, R., Taib, M.N., Hussain, Z.: Dynamic modeling and control of wheel-chaired elliptical stepping exercise. In: IEEE International Conference on Control System, Computing and Engineering, pp. 204–209 (2012)
20. Zhang, Y., Hong, D., Chung, J.H., Velinsky, S.A.: Dynamic Model Based Robust Tracking Control of a Differentially Steered Wheeled Mobile Robot. In: Proceedings of the American Control Conference, Philadelphia, Pennsylvania, pp. 850–855 (1998)
21. Aström, K.J., Wittenmark, B.: Adaptive Control. Addison-Wesley (1995)
22. Reyes, F., Kelly, R.: On parameter identification of robot manipulator. In: IEEE International Conference on Robotics and Automation, pp. 1910–1915 (1997)

# A Patient-Specific EMG-Driven Musculoskeletal Model for Improving the Effectiveness of Robotic Neurorehabilitation

Ye Ma[1], Sheng Quan Xie[1], and Yanxin Zhang[2]

[1] Department of Mechanical Engineering, Faculty of Engineering, the University of Auckland
[2] Department of Sport and Exercise Science, Faculty of Science, the University of Auckland
yma698@aucklanduni.ac.nz,
{s.xie,yanxin.zhang}@auckland.ac.nz

**Abstract.** An EMG-driven musculoskeletal model for controlling the human-inspired robotic neurorehabilitation is proposed in this paper. This model is built upon the state-of-the-art computer generated musculoskeletal framework which provides patient-specific muscular-tendon physiological, muscular-tendon kinematics parameters. Muscle forces and joint moment during locomotion are predicted through activation dynamics and contraction dynamics based on the hill-type muscle mechanics model. A hybrid Simulink-M simulated anneal algorithm is used for parameters optimization. The preliminary result showed that based on only a few EMG channels, the proposed model could efficiently predict joint moment and muscle forces. The proposed model has the potential to control the rehabilitation robot based only on a few of EMG channels from extensor and flexor muscle.

**Keywords:** musculoskeletal model, neurorehabilitation, patient-specific, EMG-driven model.

## 1    Introduction

Motor impairment and disability are common outcomes of neurological injury and diseases such as stroke and spinal cord injury [1]. It is imperative to investigate effective ways to deliver motor rehabilitation therapy other than physiotherapy because of the conflict between the continuing growing of patients requiring and the limited human resources[2].

Many clinical evidences show that task-oriented repetitive intensive movement training [3] can improve motor performance for patients with motor impairment and disability. Compared with traditional motor rehabilitation solution such as physiotherapy, robotic solutions are accurate, tireless and even can quantitatively assess the rehabilitation effectiveness with high accuracy [4].

Many control strategies have been developed to control rehabilitation robots. Most of the rehabilitation robots controllers are based on kinematic, kinetic information, impedance or admittance [5-8], which have many limitations. First of all,

X. Zhang et al. (Eds.): ICIRA 2014, Part I, LNAI 8917, pp. 390–401, 2014.
© Springer International Publishing Switzerland 2014

appropriately recreating the neural pathway is critical to the development of effective robotic control algorithms for neurological disorder patients [9]. Unfortunately, the existing controllers failed to fulfil this requirement. In addition, many clinical evidence demonstrated that the robotic training paradigms that enforce a fixed kinematic control was suboptimal for rehabilitative training because they did not consider patient's own intention and intrinsic property of neuromuscular control [10, 11]. Moreover, few controllers of these robots are based on the musculoskeletal model of the specific patient. Thus, an optimal controller should be physiologically-based, specific-subject-based and patient's-own-voluntary-movement-and-intention-based in order to generate optimal neurological rehabilitation and safety. The possible way for the physiology based controller is controlling voluntary muscle forces generated by the patient. Using electromyography (EMG) in conjunction with an appropriate musculoskeletal model to estimate the forces produced in each muscle is a suitable way to represent patient's own neuromuscular effort because of the patients' neuroplasticity after neurological injury [12]. Since EMG based models relay on measured muscle activity to estimate muscle force, these models implicitly account for a subject's individual activation patterns without the need to satisfy any constraints imposed by an objective function and these muscle forces are able to be reviewed as patient's intention. To the best knowledge of the authors, such kind of controllers has the potential to optimize the rehabilitation outcome for neurological disorder patients.

For the purpose of developing a physiologically based subject-specific rehabilitation controller, a patient-specific EMG-driven musculoskeletal model will be proposed firstly in this paper. Besides the inherit requirement for a EMG-driven method, which is generating reasonable joint kinematics and kinetics, the implication of robotic controlling puts forward more challenges to the proposed patient-specific EMG-driven musculoskeletal model: real-time application, accuracy and robustness. These requirements mean that the proposed technique should generate muscle forces from EMG signals accurately enough in real-time and also be easily adjusted to different training trials and subjects without further off-line optimization. Kinematic data, kinetic data as well as EMG signals of knee joint from one healthy subject [13] is used as the case study to demonstrate the methodology of proposed patient-specific EMG-driven musculoskeletal model for the purpose of improving the effectiveness of robotic rehabilitation.

## 2    Patient-Specific EMG-Driven Musculoskeletal Model

The patient-specific EMG-driven neuromuscular model is developed for controlling motor rehabilitation robots. According to the robotic application requirements, the proposed model will be improved in the aspects of subject-specific parameters, muscle mechanics, calculation strategies, the amount of muscle involved and the optimization algorithm implemented in this model. The general procedures of the EMG-driven musculoskeletal model is showed in Fig. 1.

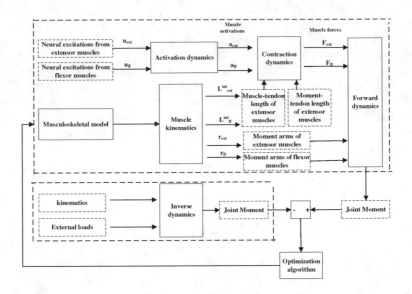

**Fig. 1.** Patient-specific EMG driven musculoskeletal model for biological command based rehabilitation robot controller

As depicted in Fig. 1, the proposed subject-specific EMG-driven algorithm for controlling rehabilitation robot is mainly consists of two parts: the EMG-force modelling and parameters optimization algorithm. The EMG-force modelling is the major work which covers modelling from neural excitations of muscles to muscular tendon force and joint moment generated; the inverse dynamics modelling serves as the reference moment generator and the optimization algorithm is used to determine a set of subject-specific parameters implemented that ensure good predicting performance to the utmost. Inverse dynamics modelling aims to estimate joint moment based on the kinematic data and external loads which is the reference moment implemented in optimization algorithm. This will be carried out use Opensim [14]. In the EMG-driven model, the central nerve system excitations of muscle from extensors and flexors, $u_{ext}$ and $u_{fl}$, which are represented by rectified raw EMG signal acting through activation dynamics to generate muscle activation, $a_{ext}$ and $a_{fl}$. Through muscle contraction dynamics, these activations energize the cross-bridge and develop muscle forces, $F_{ext}$ and $F_{fl}$ [15]. The general model and all of the sub-models will be introduced in the following sections clearly.

## 2.1    Patient-Specific Musculoskeletal Model

In order to develop a patient-specific musculoskeletal model based EMG-driven algorithm for motor disorder recovery, a patient-specific musculoskeletal model is needed. To the best of the authors' knowledge, the state-of-the-art musculoskeletal model is the subject-specific interactive graphic-based geometrical model [16] which is usually scaled from a generic model. The scaled musculoskeletal model is regarded as the patient-specific musculoskeletal model. In this study, the patient-specific model is simplified to meet the real-time calculation requirement. The developed patient-specific model has one degree of freedom. It consists of the femur and tibia representing thigh and shank, treated as rigid bodies with revolute joints which move only in the sagittal plane. The attached muscle rectus femoris muscle (RF) and biceps femoris muscle (BF) represent the main actuator of the knee extension and flexion. The anthropometric properties and muscle-tendon properties are obtained from the scaled patient-specific musculoskeletal model built in Opensim [14] which is demonstrated in Table 1.

**Table 1.** Extensor muscle and flexor muscle properties

|  | rectus femoris | biceps femoris |
| --- | --- | --- |
| Maximum isometric force | 1169 | 896 |
| Optimal muscle fiber length | 0.10889 | 0.10296 |
| Tendon slack length | 0.2961 | 0.30793 |
| Pennation angle at optimal muscle force | 0.0873 | 0 |
| Maximum contraction velocity | 10 | 10 |
| Activation time constant | 0.01 | 0.01 |
| Deactivation time constant | 0.04 | 0.04 |

## 2.2 Musculotendon Kinematics

Musculoskeletal modelling and simulation require accurate estimates of musculotendon kinematics including musculotendon length ($l^{mt}$) and moment arms (r) to accurately predict musculotendon forces ( $F^{mt}$ ) and joint moments. Musculotendon kinematics estimates can be produced by a software that models musculotendon paths wrapping around points and/or surfaces [16, 17]. However, this is based on obstacle detection and may cause discontinuities in the predicted musculotendon kinematics [18, 19]. In order to obtain the continuously musculotendon kinematics, fourth order Gaussian equations are used to represent musculotendon length and moment arms for one entire gait cycle.

## 2.3 EMG to Activation Modelling

The purpose of EMG to activation modelling is to determine each muscle's activation profile from raw EMG signals [20]. The general procedures are as follows.

### EMG Signal Processing
In this study, after full wave rectification, raw EMG signals from RF and BF were high-pass filtered (second order Butterworth filter with a cut-off frequency at 20 Hz). Then the signals were low-pass filtered (second order Butterworth filter with a cut-off frequency at 6 Hz). After that the signals were normalized against maximum voluntary contraction [21].

**Activation Dynamics and Nonlinearity**

Van Ruijven and Weijs showed that including the muscle's twitch response in the EMG-to-activation model can give better predictions of muscle forces [22]. According to [23], the muscle twitch response is able to be represented by a critically damped linear second-order differential system. Buchanan et al. [20] used a formula to represent the nonlinear relationship between EMG and forces:

$$a(t) = \frac{e^{Au(t)}-1}{e^A-1},\tag{1}$$

where A $(-3 < A < 0)$ is the nonlinear shape factor. Fig. 3 showed the results of raw EMG signals after full-wave rectification, high-pass filter, low-pass filter, activation dynamics and nonlinearity.

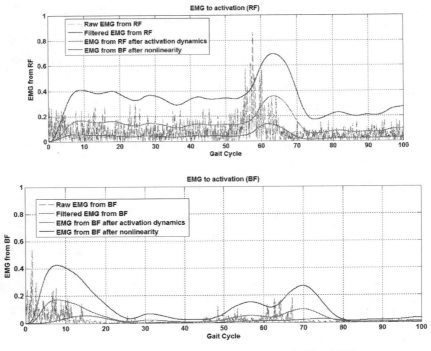

**Fig. 2.** The EMG to activation procedure of RF and BF

## 2.4    The Muscle-Tendon Mechanics Modelling

A model of muscle contraction dynamics is required to determine muscle forces based on the muscle activation levels. Based on requirement of the real-time calculation for controlling rehabilitation robots, muscle mechanical model such as the hill-type models [24-26] rather than physiological model [27] is used to model muscle-tendon mechanics.

## Properties and of Muscle Fiber

In the Hill-type model, the contractile properties of muscle tissue can be represented by a force-length-velocity relationship controlled by muscle activation [31]. The total muscle force $F^m$ is the sum of passive force $F^{PE}$ and active force $F^{CE}$. Force $F^{CE}$ depends on muscle fiber length $L^m$, velocity $V^m$, and the state of activation of the muscle fibers $a(t)$; while force $F^{PE}$ depends on muscle fiber length $L^m$. The relationships are as follows:

$$F^m = F^{CE} + F^{PE},$$
$$F^{CE} = F(L^m, V^m, a(t), F_0^m),$$
$$F^{PE} = F(L^m).$$

(2)

## Passive Force-Length Relationship

The passive force-length relationship of muscle is represented by [28]:

$$\bar{F}^{PE} = \frac{e^{\frac{K^{PE}(\bar{L}^M - 1)}{\varepsilon_0^M} - 1}}{e^{K^{PE}}},$$

(3)

where $\bar{F}^{PE}$ is the normalized passive muscle force, $K^{PE}$ is an exponential shape factor, and $\varepsilon_0^M$ is the passive muscle strain due to maximum isometric force.

## Muscle Contraction Dynamics

In most of the hill-type muscle mechanics models, the force-length-velocity (FLV) relationship is usually considered as a combination of separate force-length (FL) and force-velocity (FV) relationships [29-34]. But it is not accurate and non-physiological based when separately FL and FV relationships are used because the FV equation only describes the FV relationship when the muscle is at its optimal length $(l_m^0)$ [35, 36]. Thus a combined FLV relationship is used in this research which supported by the new experimental result [35].

The muscle fiber velocity $(V^m)$ is assumed to be a function of the muscle fiber length $(\bar{L}^m)$, muscle activation and active muscle force $\bar{F}^{CE}$ :

$$V^m = (0.25 + 0.75a)V_{max}^m \frac{\bar{F}^{CE} - af_l}{b},$$

(4)

where $V_{max}^m$ is the maximum contraction velocity and the parameter $b$ is computed differently depending on whether the muscle fiber is shortening ($\bar{F}^{CE} \leq af_l$) or lengthening ($\bar{F}^{CE} > af_l$) [28].

## Properties of Tendon

Tendons are passive elements like rubber bands. Above the tendon slack length, $l_s^t$, it generates force proportional to the distance stretched; while below $l_s^t$, tendon could not carry any load. Zajac [15] modelled tendon using the relationship between tendon strain $\varepsilon^t$, $\varepsilon^s = \frac{l_t - l_s^t}{l_s^t}$ and normalized tendon force $f^t$. The detailed expressions of $f^t$ could be found in [15].The final tendon force can be calculated by multiplying $f_t$ by the maximum isometric muscle force.

## 2.5    Calculation of Musculotendon Forces

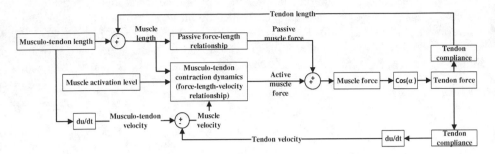

**Fig. 3.** Musculotendon force calculation

Fig. 3 shows the general procedure starts from tendon length [37] to predict muscle forces and joint moment. The model calculation starts from muscle fiber length. Given that the muscle-tendon length could be obtained through the patient-specific musculoskeletal model, the muscle fiber length and muscle fiber velocity would be calculated. Based on the calculated muscle activation level, passive muscle force and active muscle force are calculated which means the muscle force is able to be estimate. By using the relationships between muscle force and tendon force[15], the tendon force is determined. Through the tendon force-strain relationship, tendon length would be calculated which serves the input of the first step. A Simulink model is developed based on the above procedures.

## 2.6    Parameters Optimization

The optimization procedure aims to identify a set of subject-specific parameters that ensure the optimal performance which the muscle forces and joint moment predicted are optimal comparing with the experiment results. Therefore, the objective function of the EMG-to-force global optimization procedure is

$$G = \sum_{i=1}^{n} \sqrt{(M_{sim} - M_{exp})^2} \,, \tag{5}$$

where i is the different experimental trials, $M_{sim}$ is the joint moment calculated by the proposed patient-specific EMG-driven method and the $M_{exp}$ is the so-called experimental joint moment which is calculated by inverse dynamics. EMG-driven models are typically validated to external joint moments using inverse dynamic methods because measuring muscle forces in vivo are difficult [29]. The objective function is the root-mean-square difference between the predicted and experimental joint moments calculated over all calibration trials. By introducing the hybrid Simulink-M technology, the global optimization of the proposed patient-specific EMG-driven method is able to be achieved. A simulated annealing optimization algorithm is used to find the optimal parameters.

## 3    Preliminary Result

Gait data of knee extension/flexion from one healthy adolescent [13] was used to evaluate proposed patient-specific EMG-driven musculoskeletal model. The experimental data included the markers' position data, ground reaction force data and EMG recordings. Model scaling and inverse kinematics were applied to obtain the joint angles. Scaling the musculoskeletal model is to modify the anthropometry, or physical dimensions, of the generic model so that it matched the anthropometry of the particular subject. Based on the measured EMG signals and musculoskeletal model, joint moment calculated by proposed patient-specific EMG-driven musculoskeletal model will be estimated; while with all kinematic data, the inverse dynamic analysis procedure was used to calculate the reference joint moment.

In this section, the simulated joint moment by using the proposed patient-specific EMG-driven musculoskeletal model is compared with the reference moment. Fig. 4 shows the preliminary result and Table 2 showed the statistical analysis of the preliminary result. From Fig. 4 and Table 2, we found that the proposed patient-specific EMG-driven musculoskeletal model could estimate knee joint movement by only the EMG signals from one knee extensor and one knee flexor muscle. Joint moment calculated by inverse dynamics could be viewed as reference joint moment [29] for validation of the proposed method. The statistical analysis shows that the proposed algorithm has good performance which the maximum joint moment deviation is 6.7 Nm, the correlation coefficient value between joint moment calculated by the proposed algorithm and inverse dynamics is 0.91, and the root-mean-square-error is 3.5 Nm. Furthermore, the calculation time of proposed algorithm is within 0.025 seconds which showed big advantage of realizing real-time calculation for the rehabilitation robot controller design.

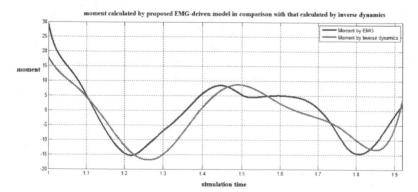

**Fig. 4.** Knee joint moment calculated by proposed patient-specific musculoskeletal model in comparison with that calculated by inverse dynamics method

**Table 2.** Statistical analysis of preliminary result at 95% confident interval

|  | Maximum moment deviation $\Delta M$ | The correlation coefficient | RMSE |
|---|---|---|---|
| Knee joint moment calculated by proposed EMG-driven model in comparison with that calculated by inverse dynamics | 6.7 | 0.91 | 3.5 |

## 4    Discussion

This patient-specific EMG-driven musculoskeletal model aims to provide patient-specific biological command for improving the effectiveness of robotic rehabilitation. More specifically, the proposed model is used for generating accurate joint moment based on patient's own biological signal, i.e., surface EMG signals. Thus, build upon patient's own musculoskeletal model, the patient-specific physiological parameter, muscle kinematics are able to be obtained. Then, muscle forces are able to be calculated after the activation dynamics and contraction dynamics procedures. Afterwards, the optimization algorithm implemented which gives optimal patient-specific parameters set ensures optimal joint moment performances. Based on EMG signals from one knee extensor and flexor muscle, proposed patient-specific EMG-driven musculoskeletal model is able to predict knee joint moment with an acceptable accuracy. Comparing with results from other researchers [29, 30], the preliminary result could be viewed as good performance but also are needed to be improved.

The human-inspired robotic rehabilitation application puts forward three requirements: accuracy, real-time calculation and robustness. Based on these requirements, the proposed patient-specific musculoskeletal model based EMG-driven algorithm made the following improvement: employing the state-of-the-art computed musculoskeletal model to generate accurate physiological parameters, anthropometric parameters, time varying muscle-tendon length and time vary muscle moment arms in Opensim [14, 16]; implementing combined force-length-velocity relationship; using Simulink as a tool instead of the numerically integration algorithm for calculation; the amount of muscles involved are simplified to two. But the effectiveness of this method should be tested in real-world.

Although there are many positive evidences of the separately FL and FV relationship to represent hill-type muscle mechanics, there are also many limitations. The most important issue is that after the FL property is defined, the FV equation only describes the FV relationship when the muscle is at its optimal length ($l_m^0$) [35]. Thus, it is not accurate and non-physiological when you use separately FL and FV relationships. Yeo et al.'s experiment and analysis on hill-type, phenomenological models support that the force scaling with parallel spring and "f-maxing scaling" model are better representations of the FLV relationship [35].

In terms of the numbers of muscle included in the previous EMG-driven model, 13 muscle-tendon units across the knee are incorporated to simulate knee joint moment and muscle forces and34 muscle-tendon units are used to estimate joint moment and forces in the lower extremity [38]. Taken into account the numerically integration algorithm they used for each step and sometimes the tuning algorithm running for every trial, it is impossible to accomplish real-time. The proposed patient-specific musculoskeletal model used two muscle and the calculation time is within 0.025 seconds which showed great potential for the real-time calculation.

There are also some limitations of the proposed patient-specific EMG-driven musculoskeletal model. Firstly, the accuracy of joint moment prediction is highly relay on the quality of measured surface EMG signal and the signal processing algorithm. Thus, a more advanced signal processing algorithm is needed. Secondly, the optimization process which determines the optimal parameters set is calculated off-line, which is inconvenient for catering the changed environment; thirdly, the prediction performance could be improved by incorporating the important muscle vasti and muscle synergy; fourthly, the result is only from one healthy subject, more subject and even patients are needed for validation this method.

The future work is to validate this method based on the data from more subjects, improve the accuracy of the model in the aspects of digital signal processing as well as muscle mechanics model improvement, and develop more effective optimization algorithm which could cater the change situation from both patient and environment.

## 5     Conclusion

A patient-specific EMG-driven musculoskeletal model is proposed and simulated in this paper. Built upon the patient-specific musculoskeletal model, the accurate muscular-tendon physiological parameters, muscle kinematics are determined. Then, joint moment is predicted through the EMG to activation and contraction dynamics process based on the hill-type model. The improved combined force-length-velocity relationship is employed in this model. A set of optimal patient-specific parameters are chosen by the off-line tuned optimization algorithm. The simulation result showed that the proposed patient-specific EMG-driven musculoskeletal model have the potential to predict accurate joint moment and muscle forces in real-time based only on a few of EMG channels from extensor and flexor muscle thus control the rehabilitation robot.

**Acknowledgement.** The support by Chinese Scholarship Council and Faculty research development Fund from the University of Auckland is greatly appreciated. The authors also very appreciate the help from James Pau.

## References

1. Stroke Foundation of New Zealand and New Zealand Guidelines Group. Clinical Guidelines for Stroke Management. Stroke Foundation of New Zealand, Willington (2010)
2. Hogan, N., Krebs, H.I.: Interactive robots for neuro-rehabilitation. Restorative Neurology and Neuroscience 22(3), 349–358 (2004)

3. Pohl, M., et al.: Repetitive locomotor training and physiotherapy improve walking and basic activities of daily living after stroke: a single-blind, randomized multicentre trial (DEutsche GAngtrainerStudie, DEGAS). Clinical rehabilitation 21(1), 17–27 (2007)

4. Patton, J.L., Kovic, M., Mussa-Ivaldi, F.A.: Custom-designed haptic training for restoring reaching ability to individuals with poststroke hemiparesis. Journal of Rehabilitation Research and Development 43(5), 643 (2006)

5. Beyl, P., et al.: Safe and compliant guidance by a powered knee exoskeleton for robot-assisted rehabilitation of gait. Advanced Robotics 25(5), 513–535 (2011)

6. Kong, K., et al.: Mechanical design and impedance compensation of SUBAR (Sogang University's Biomedical Assist Robot). In: 2008 IEEE/ASME International Conference on Advanced Intelligent Mechatronics, AIM 2008, August 2-August 5. Institute of Electrical and Electronics Engineers Inc., Xi'an (2008)

7. Gupta, A., et al.: Design, control and performance of rice wrist: a force feedback wrist exoskeleton for rehabilitation and training. International Journal of Robotics Research 27(2), 233–251 (2008)

8. Stauffer, Y., et al.: The WalkTrainer-A New Generation of Walking Reeducation Device Combining Orthoses and Muscle Stimulation. IEEE Transactions on Neural Systems and Rehabilitation Engineering 17(1), 38–45 (2009)

9. Hidler, J.M., Wall, A.E.: Alterations in muscle activation patterns during robotic-assisted walking. Clinical Biomechanics 20(2), 184–193 (2005)

10. Cai, L.L., et al.: Implications of assist-as-needed robotic step training after a complete spinal cord injury on intrinsic strategies of motor learning. The Journal of Neuroscience 26(41), 10564–10568 (2006)

11. Jezernik, S., et al.: Adaptive robotic rehabilitation of locomotion: a clinical study in spinally injured individuals. Spinal Cord 41(12), 657–666 (2003)

12. Enoka, R.M.: Neuromechanics of human movement. Human Kinetics (2008)

13. Liu, M.Q., et al.: Muscle contributions to support and progression over a range of walking speeds. Journal of Biomechanics 41(15), 3243–3252 (2008)

14. http://opensim.stanford.edu/

15. Zajac, F.E.: Muscle and tendon: properties, models, scaling, and application to biomechanics and motor control. Critical Reviews in Biomedical Engineering 17(4), 359–411 (1989)

16. Delp, S.L., et al.: An interactive graphics-based model of the lower extremity to study orthopaedic surgical procedures. IEEE Transactions on Biomedical Engineering 37(8), 757–767 (1990)

17. Seth, A., et al.: OpenSim: a musculoskeletal modeling and simulation framework for<i> in silico</i> investigations and exchange. Procedia IUTAM 2, 212–232 (2011)

18. Gao, F., et al.: Computational method for muscle-path representation in musculoskeletal models. Biological Cybernetics 87(3), 199–210 (2002)

19. Garner, B.A., Pandy, M.G.: The obstacle-set method for representing muscle paths in musculoskeletal models. Computer Methods in Biomechanics and Biomedical Engineering 3(1), 1–30 (2000)

20. Buchanan, T.S., et al.: Neuromusculoskeletal modeling: estimation of muscle forces and joint moments and movements from measurements of neural command. Journal of Applied Biomechanics 20(4), 367 (2004)

21. Koo, T.K.K., Mak, A.F.T.: Feasibility of using EMG driven neuromusculoskeletal model for prediction of dynamic movement of the elbow. Journal of Electromyography and Kinesiology 15(1), 12–26 (2005)

22. Koo, K.-K.T.: Neuromusculoskeletal modeling of the elbow joint in subjects with and without spasticity (2002)
23. Sartori, M., et al.: An EMG-driven Musculoskeletal Model of the Human Lower Limb for the Estimation of Muscle Forces and Moments at the Hip, Knee and Ankle Joints in vivo. In: Proc. of Int. Conf. on Simulation, Modeling and Programming for Autonomous Robots (2010)
24. Hill, A.V.: First and Last Experiments in Muscle Mechanics. Cambridge University Press, Cambridge (1970)
25. Hill, A.V.: The heat of shortening and the dynamic constants of muscle. Proceedings of the Royal Society of London. Series B, Biological Sciences 126(843), 136–195 (1938)
26. Hill, A.V.: The abrupt transition from rest to activity in muscle. Proceedings of the Royal Society of London. Series B-Biological Sciences 136(884), 399–420 (1949)
27. Huxley, A.: Muscular contraction. The Journal of Physiology 243(1), 1 (1974)
28. Thelen, D.G.: Adjustment of muscle mechanics model parameters to simulate dynamic contractions in older adults. Journal of Biomechanical Engineering 125(1), 70–77 (2003)
29. Lloyd, D.G., Besier, T.F.: An EMG-driven musculoskeletal model to estimate muscle forces and knee joint moments in vivo. Journal of Biomechanics 36(6), 765–776 (2003)
30. Pau, J.W.L., Xie, S.S.Q., Pullan, A.J.: Neuromuscular interfacing: establishing an EMG-driven model for the human elbow joint. IEEE Transactions on Biomedical Engineering 59(9), 2586–2593 (2012)
31. Bogey, R.A., Perry, J., Gitter, A.J.: An EMG-to-force processing approach for determining ankle muscle forces during normal human gait. IEEE Transactions on Neural Systems and Rehabilitation Engineering 13(3), 302–310 (2005)
32. Pandy, M.G., Hull, D.G., Anderson, F.C.: A parameter optimization approach for the optimal control of large-scale musculoskeletal systems. Journal of Biomechanical Engineering 114(4), 450–460 (1992)
33. Pau, J.W.L., et al.: An EMG-driven neuromuscular interface for human elbow joint. In: 2010 3rd IEEE RAS and EMBS International Conference on Biomedical Robotics and Biomechatronics (BioRob). IEEE (2010)
34. Pau, J.W.L., Xie, S.S.Q., Xu, W.L.: Neuromuscular interfacing: A novel approach to EMG-driven multiple DOF physiological models. In: 2013 35th Annual International Conference of the IEEE Engineering in Medicine and Biology Society (EMBC). IEEE (2013)
35. Yeo, S.H., et al.: Phenomenological models of the dynamics of muscle during isotonic shortening. Journal of Biomechanics 46(14), 2419–2425 (2013)
36. Winters, J.: Hill-Based Muscle Models: A Systems Engineering Perspective. In: Winters, J., Woo, S.-Y. (eds.) Multiple Muscle Systems, pp. 69–93. Springer, New York (1990)
37. Shao, Q., et al.: An EMG-driven model to estimate muscle forces and joint moments in stroke patients. Computers in Biology and Medicine 39(12), 1083–1088 (2009)
38. Sartori, M., et al.: EMG-driven forward-dynamic estimation of muscle force and joint moment about multiple degrees of freedom in the human lower extremity. PloS One 7(12), e52618 (2012)

# Performance Analysis of Passenger Vehicle Featuring ER Damper with Different Tire Pressure

Kum-Gil Sung

School of Mechanical and Automotive Engineering Technology,
Yeungnam College of Science and Technology, Daegu, Korea
kgsung@ync.ac.kr

**Abstract.** This paper presents performance analysis of a controllable electro-rheological (ER) damper for a passenger vehicle with different tire pressure. In order to achieve this research goal, an ER damper, which satisfies design specifications for a mid-sized passenger vehicle, is designed and manufactured. After experimentally evaluating the field-dependent characteristics of the manufactured ER damper, the quarter-vehicle suspension system consisting of sprung mass, spring, tire and the ER damper is constructed in order to investigate the ride comfort with different tire pressure. After deriving the equations of the motion for the proposed quarter-vehicle ER suspension system, vertical tire stiffness with respect to different tire pressure is experimentally identified. Ride comfort characteristics such as vertical acceleration RMS (root mean square) of sprung mass are evaluated under bump road conditions using a quarter-vehicle test facility.

**Keywords:** Electro-rheological (ER) Fluid, ER Damper, Quarter-vehicle, Tire Pressure, Ride Comfort.

## 1 Introduction

The recently, the research work on ride comfort of a vehicle using advanced suspension system has been significantly increased. The passive suspension system featuring conventional oil damper provides design simplicity and cost-effectiveness, while performance limitations are inevitable due to the uncontrollable damping force. On the other hand, the active suspension system can provide high control performance in wide frequency range. However, the active suspension requires design complexity high power consumption and actuators such as servo-valve. Consequently, one way to resolve these problems is to adopt the semi-active suspension system such as electro-rheological (ER) fluid. The semi-active suspension system offers a desirable performance generally enhanced in the active mode without complicated hardware.

Nakano proposed several semi-active control algorithms for ER damper and showed that the proportional feedback control using the information of absolute unsprung mass velocity is the most effective control strategy [1]. Sims et al. described a closed-loop control strategy which is capable of linearizing the response of an ER long-stroke damper under experimental conditions [2]. Petek et al. constructed a semi-active full suspension system consisting of four ER dampers and evaluated its

X. Zhang et al. (Eds.): ICIRA 2014, Part I, LNAI 8917, pp. 402–409, 2014.
© Springer International Publishing Switzerland 2014

effectiveness for vibration isolation [3]. They demonstrated experimentally that unwanted pitch, heave and roll motions of vehicle body can be favorably suppressed using the simple skyhook control algorithm. Gordaninejad et al. experimentally evaluated the performance of cylindrical, multi-electrode ER dampers under force vibration [4]. They proposed simple control algorithms such as bang-bang and linear proportional controller, and experimentally demonstrated the successful implementation of the control schemes to a closed-loop system.

As is evident from the previous research work, all of the studies on ride comfort of the ER suspension system have been performed without considering the different tire pressure such as under/over inflation. However, the tire stiffness of the vehicle system has been treated constant value. Consequently, the main contribution of this work is to present performance evaluation of a quarter-vehicle suspension system equipped with continuously controllable ER damper with respect to different tire pressure. As a first step, controllable ER damper is designed and manufactured. After experimentally evaluating dynamic characteristics of the manufactured ER damper, equation of the motion for the proposed quarter-vehicle suspension system consisting of sprung mass, spring, tire and the ER damper is derived in order to investigate the ride comfort. Subsequently, vertical tire stiffness with respect to different tire pressure is experimentally identified. Ride comfort characteristics such as vertical acceleration RMS (root mean square) of sprung mass are evaluated under bump road conditions.

## 2     ER Damper

In this work, an ER damper shown in Fig. 1 is designed and manufactured. The ER damper is divided into the upper and lower chambers by the piston, and it is fully filled with the ER fluid. By the motion of the piston, the ER fluid flows through the duct between inner and outer cylinders from one chamber to the other. Thus, the operating mode of the proposed ER damper is flow mode in which two electrodes are fixed. The control voltage generated by a high voltage supply unit is connected to the inner cylinder and the ground voltage is connected to the outer cylinder. On the other hand, the floating piston between the cylinder and the gas chamber is also used in order to compensate the volume induced by the motion of the piston. The proposed ER damper is applicable to a middle-sized passenger vehicle.

In the absence of the electric field, the ER damper produces damping force only caused by the fluid viscous resistance. However, if a certain level of the electric field is supplied to the ER damper, the ER damper produces additional damping force owing to the yield stress of the ER fluid. This damping force of the ER damper can be continuously tuned by controlling the intensity of the electric field. Therefore, the damping force of the proposed ER damper can be obtained as follows [4].

$$F = k_e x_p + c_e \dot{x}_p + F_{ER} \tag{1}$$

where $k_e$ is the effective stiffness due to the gas pressure, $c_e$ is the effective damping due to the fluid viscosity, $x_p$ is the excitation displacement, and $F_{ER}$ is the field-dependent damping force which is tunable as a function of applied electric field. The controllable damping force $F_{ER}$ can be expressed by

$$F_{ER} = (A_p - A_r) 2 \frac{L}{h} \alpha E^\beta \, \text{sgn}(\dot{x}_p) \qquad (2)$$

where $A_p$ and $A_r$ represent the piston head and piston rod areas, respectively. $\text{sgn}(\cdot)$ is a sign function, $L$ is the electrode length, $h$ is the electrode gap, and $E$ is the electric field. The $\alpha$ and $\beta$ are intrinsic values of the ER fluid to be experimentally determined.

Fig. 2 presents the measured damping force characteristics of the proposed ER damper with respect to the piston velocity at various electric fields. This is obtained by exciting the ER damper with the frequency of 3.0Hz and the amplitude of ±20mm. It is clearly observed that the damping force is increased as the electric field increases.

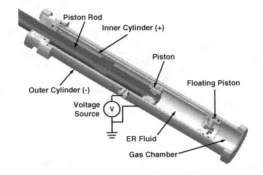

**Fig. 1.** The proposed ER damper for passenger vehicle

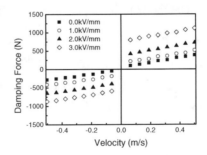

**Fig. 2.** The field-dependent damping force

The static expression of the damping force and input field given by Eq. (2) can be modified to dynamic expression by considering time constant as follows.

$$\tau \frac{d}{dt} F_{ER} + F_{ER} = (A_p - A_r) 2 \frac{L}{h} \alpha E^\beta \, \text{sgn}(\dot{x}_p) \qquad (3)$$

where $\tau$ is the time constant of the ER damper.

## 3    ER Suspension System

In this study, a quarter-vehicle is established to evaluate ride comfort performances of the proposed ER suspension system. Fig. 3 shows the quarter-vehicle model of the semi-active ER suspension system which has two degree-of-freedom. $m_s$ and $m_u$ represent the sprung mass and unsprung mass, respectively. The spring for the suspension is assumed to be linear and the tire is also modeled as linear spring component. Now, by considering the dynamic relationship, the state space control model is expressed for the quarter-vehicle ER suspension system as follows:

$$\dot{x} = Ax + Bu + Lz_r$$
$$y = Cx$$

(4)

where

$$x = \begin{bmatrix} z_s & \dot{z}_s & z_u & \dot{z}_u & F_{ER} \end{bmatrix}^T = \begin{bmatrix} x_1 & x_2 & x_3 & x_4 & x_5 \end{bmatrix}^T$$

$$A = \begin{bmatrix} 0 & 1 & 0 & 0 & 0 \\ -\dfrac{k_s}{m_s} & -\dfrac{c_s}{m_s} & \dfrac{k_s}{m_s} & \dfrac{c_s}{m_s} & -\dfrac{1}{m_s} \\ 0 & 0 & 0 & 1 & 0 \\ \dfrac{k_s}{m_u} & \dfrac{c_s}{m_u} & -\dfrac{k_t + k_s}{m_u} & -\dfrac{c_s}{m_u} & \dfrac{1}{m_u} \\ 0 & 0 & 0 & 0 & -\dfrac{1}{\tau} \end{bmatrix}$$

$$B = \begin{bmatrix} 0 & 0 & 0 & 0 & \dfrac{1}{\tau} \end{bmatrix}^T, \quad C = \begin{bmatrix} 1 & 0 & 0 & 0 & 0 \end{bmatrix},$$

$$L = \begin{bmatrix} 0 & 0 & 0 & \dfrac{k_t}{m_u} & 0 \end{bmatrix}^T, \quad u = (A_p - A_r)2\dfrac{L}{h}\alpha E^{\beta} \, \mathrm{sgn}(\dot{x})$$

where $k_s$ is the total stiffness coefficient of the suspension including the effective stiffness of the ER damper in Eq. (1). $c_s$ is the damping coefficient of the suspension and it is assumed to be equal to $c_e$. In addition, $k_t$ is the stiffness coefficient of the tire. $z_s$, $z_u$ and $z_r$ are the vertical displacement of sprung mass, unsprung mass, and excitation, respectively. Fig. 4 shows the photograph of vertical stiffness test for the tire with different pressure. The tire used in this work is 205-65-R15 for a mid-sized passenger vehicle, and its recommended inflation pressure from manufacturer is 207kPa (30psi). Fig. 5 presents the vertical stiffness characteristics of the tire with respect to different tire pressure. It is clearly observed that the vertical stiffness is increased as tire pressure increases. The vertical stiffness of 153kN/m at tire pressure 124kPa (18psi, under-inflation) is increased up to 211kN/m at 207kPa (30psi, recommended inflation) and 225kN/m at 227kPa (33psi, over-inflation).

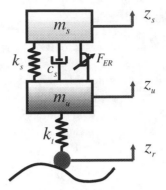

**Fig. 3.** Mechanical model of the quarter-vehicle ER suspension system

**Fig. 4.** Vertical stiffness test of a tire

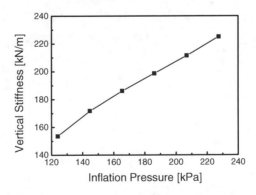

**Fig. 5.** Vertical stiffness of a tire with respect to tire pressure of the tire (205-65-R15)

## 4    Performance Analysis

An experimental apparatus for the quarter-vehicle test is established as shown in Fig. 6 to evaluate performance analysis of the ER suspension system. The ER damper, sprung mass, spring and tire are installed on the hydraulic vibrating system. Ride comfort characteristics of the quarter-vehicle ER suspension system with respect to different tire pressure are evaluated. Fig. 7 and 8 presents ride comfort comparison of the quarter-vehicle ER suspension system under bump road excitation with constant vehicle velocity of 3.08km/h. As shown in Fig. 7, it is obvious that unwanted vibrations induced from the bump excitation have been well reduced by adopting the ER damper associated with the skyhook controller. It is also observed from the vertical displacement of sprung mass that the control performance of the controlled case is better than that of the uncontrolled case independent of tire pressure. In order to evaluate ride comfort characteristics, the RMS (root mean square) of vertical acceleration $\ddot{z}_s$ are implemented as follows [5]

$$RMS = \left[ \sum_{i=1}^{N} \ddot{z}_s(i)^2 \right]^{1/2} \tag{5}$$

As shown in Fig. 8, the RMS values indicate that ride comfort characteristics of a vehicle system can be substantially improved by employing the ER damper and skyhook controller even though tire pressure is changed.

**Fig. 6.** Experimental apparatus for the ER suspension system

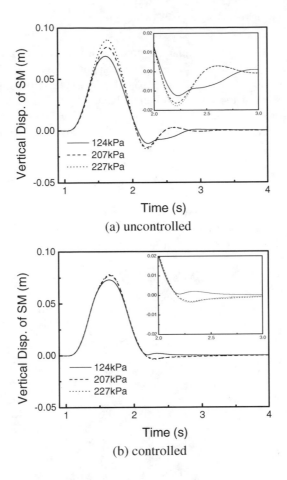

**Fig. 7.** Bump road responses of the quarter-vehicle ER suspension system

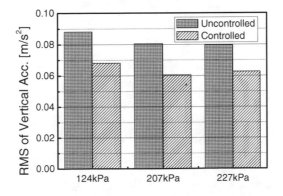

**Fig. 8.** Ride comfort comparison of vertical acceleration RMS of the ER suspension

# 5     Conclusion

In this work, performance analyses of a quarter-vehicle ER suspension system with respect to different tire pressure were investigated. The ER damper was designed and manufactured based on mechanical dimensions required for a mid-sized passenger vehicle. After experimentally evaluating dynamic characteristics of the manufactured ER damper, the quarter-vehicle suspension system consisting of sprung mass, spring, tire and the ER damper was constructed in order to investigate the ride comfort. After deriving the equations of the motion for the proposed quarter-vehicle ER suspension system, vertical tire stiffness with respect to different tire pressure was experimentally identified. Ride comfort characteristics such as vertical acceleration RMS of sprung mass were evaluated under bump road condition. The results presented in this work are quite self-explanatory justifying that the ER damper can provide improved ride comfort for a passenger vehicle even though tire pressure is changed during driving.

# References

1. Nakano, M.: A novel semi-active control of automotive suspension using an electrorheological shock absorber. In: Proceedings of the 5th International Conference on ER Fluid, MR Suspensions and Associated Technology (1995)
2. Sims, N.D., Stanway, R., Beck, S.B.M.: Proportional feedback control of an electrorheological vibration damper. J. Intell. Mater. Syst. Struct. 8, 426 (1997)
3. Petek, N.K., Romstadt, D.J., Lizell, M.B., Weyenberg, T.R.: Demonstration of an automotive semi-active suspension using electrorheological fluid. SAE Tech. Pap. Ser. 950586 (1995)
4. Gordaninejad, F., Ray, A., Wang, H.: Control of forced vibration using multi-electrode electro-rheological fluid dampers. J. Vib. Acoust.-Trnas. ASME 119, 527 (1997)
5. Lin, K.Y., Hwang, J.R., Chang, S.H., Fung, C.P., Chang, J.M.: System Dynamics and Ride Quality Assessment of Automobile. Society Automotive Engineering 2006-01-12 25 (2006)

# Development of Targeting Detection Module and Driving Platform at a Moving Target System

Ki-Ho Yu[1], Seok-Jo Go[2,*], Min- Kyu Park[3], Tae-Hoon Kim[4], and Min-Cheol Lee[5]

[1] Department of Intelligent Mechanical Engineering, Pusan National University, Busan, 609-735, Korea
khyu@nbcore.com
[2] Division of Mechanical Engineering, Dongeui Institute of Technology, Busan, 614-715, Korea
sjgo@dit.ac.kr
[3] School of Mechanical and Automotive Engineering Technology, Yeungnam College of Science and Technology, Daegu, 705-703, Korea
mk_park@ync.ac.kr
[4] Department of Electronic Engineering, Dongeui Institute of Technology, Busan, 614-715, Korea
kth@dit.ac.kr
[5] School of Mechanical Engineering, Pusan National University, Busan, 609-735, Korea
mclee@pusan.ac.kr

**Abstract.** Recently, there are increasing demands for a sort of improved combat simulator. In case of rifle shooting simulator with fixed target, it is useful to improve a basic rifle marksmanship. However, there is not vivid because a target in combat or leisure situation moves. Therefore, we develop a moving target system. The moving target system consists of a driving platform and targeting detection module. The driving platform of the moving target system is developed to mount various targets and to move in slope and unpaved terrain. The targeting detection module is developed to check a hit portion of target by using acceleration sensor. A sensing performance is confirmed by the real bullet fired tests.

**Keywords:** Combat simulator, Moving target system, Targeting detective module, Driving platform, Acceleration sensor, Bullet fired test.

## 1 Introduction

Korea military and other armed forces around the world are much effort to improve a combat capability by training combat simulator. MILES (Multiple Integrated Laser Engagement System) is a prime example in improved combat simulator. It uses lasers and blank cartridges to simulate actual combat. Combat simulator like MILES is widely used in many countries. Nowadays, this system gradually spread to game and leisure industries [1-6].

---

* Corresponding author.

X. Zhang et al. (Eds.): ICIRA 2014, Part I, LNAI 8917, pp. 410–419, 2014.

Fig. 1 shows a rifle shooting simulator. A rifle mounted optical sensor receives its input from the light emitted by on-screen targets. A sort of this simulator is widely used in military, game, and leisure because it is needed a small space and can support various kinds of scenarios [1, 2]. However this simulator has a significant difference from actual combat environment. The main difference is that the target is fixed. Therefore, a moving target system based on Segway is used in Australia military and U.S. military as shown in Fig. 2(a). And, a commercial product has come in leisure and game as shown in Fig. 2(b) by modifying a military version. However, the conventional moving target system detects only 'hit' or not. In real combat, it is important to know what part of body is targeted.

Therefore, this study is focus on developing a target detection module which is able to check a hit portion by using variation of vibration from acceleration sensor. The developed target detection module is installed at the top of the driving platform. We also design and fabricate the platform which is able to move a slope terrain as well as an unpaved terrain. Finally, we carry out experiment on bullet fired test for evaluating a performance.

**Fig. 1.** Shot training simulator (EINTECH Co. Ltd)

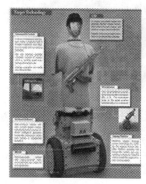

(a) Marathon targets Co. Ltd                    (b) Target Tracker Co. Ltd

**Fig. 2.** Moving target system

## 2    Driving Platform of Moving Target System

Fig. 3 shows the overall concept of a proposed moving target system. Driving platform should be designed and controlled to move quickly and flexibly to anywhere for implementing a similar level of actual bullet fired situation. In this study, 4-wheel steering control [7, 8] is applied to the driving platform for improving mobility of the platform. For this study, Two driving platform (30 kg and 60 kg) was developed as shown in Fig. 4 and Table 1.

**Fig. 3.** Design of a moving target system

(a) 30 kg                                    (b) 60 kg

**Fig. 4.** Driving platform of a moving target system

**Table 1.** Specifications of the developed moving platforms

|                 | Specification   |
| --------------- | --------------- |
| Weight          | 30 kg / 60 kg   |
| Moving velocity | 18 km/h         |
| Driving time    | 3.5 h           |

| (a) 5% slope road | (b) 10% slope road | (c) dirt road |

**Fig. 5.** Driving tests of the developed driving platform

The driving platform of the moving target system is developed to mount various targets and to move in slope and unpaved terrain.

In order to evaluate performance of the developed driving platform, driving experiments was carried out on slope road and unpaved terrain. The experiment results show that the platform can stably operate on a 5% slope road, a 10% slope road and a dirt road as shown in Fig. 5.

## 3     Targeting Detection Module of Moving Target System

It is important to know a hitting portion because it can estimate the extent of injury by a shot. There are three ways to detect the hit such as pressure detection, sound-wave detection, and vibration detection. In this study, we choose a vibration detection method by using an acceleration sensor. The sensors are installed for covering the upper body as shown in Fig. 6.

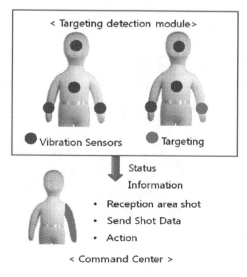

**Fig. 6.** Sensors for targeting detection

## 3.1     Design of the Targeting Detection Module

Figure 7 shows the developed targeting detection module by using acceleration sensors. The sensor used in this study can be measured up to 12 G and the resolution is 3,200 Hz.

A performance evaluation is carried out about detecting on hitting portion in the Lab. Figure 8 shows a target (300 mm × 300 mm × 5 mm in size, Formed PVC). The targeting detection module is attached on the top of the target. We set 5 hitting point as 50 mm interval on the target, iron ball (45g) from a height of 10 cm fall down to the set position of the target. When the ball hit the target, acceleration is measured from the sensor in Z-axis as shown in Fig. 9.

Table 2 is result data from acceleration sensors. As the results, the acceleration value is lowered as increasing the distance from the module to the hitting position. However, the test 4 (150mm) is measured as higher value than we expect. It can be seen that the only one module bring an error on determining a hitting portion as vibration characteristics of target. The characteristics is depends on the material.

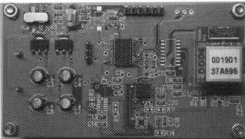

**Fig. 7.** Developed the Acceleration sensor module

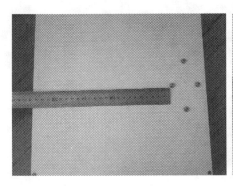

**Fig. 8.** Targeting test of the acceleration sensor module

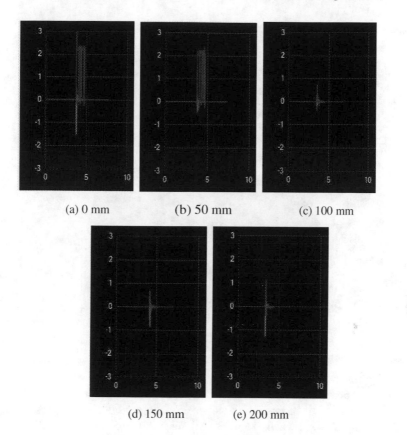

(a) 0 mm              (b) 50 mm              (c) 100 mm

(d) 150 mm              (e) 200 mm

**Fig. 9.** Acceleration data measured at targeting point

**Table 2.** Acceleration data (Unit: G)

| Axis | 0 mm | 50 mm | 100 mm | 150 mm | 200 mm |
|------|------|-------|--------|--------|--------|
| Z (Blue) | 2.9 | 2.1 | 0.7 | 1.2 | 0.5 |

## 3.2    Selection of Pad Attached to the Targeting Detection Module

In this section, we deal with selection of pad which is attached on the target for increasing the sensitivity and the stability of detection. We choose 6 kinds of different pad such as PP (poly propylene), PE (poly ethylene), PC (poly carbonate), PTFE (poly tetra fluoro ethylene), Poly Urethane, and Foamex. Figure 10 shows a 10 meter - bullet fired test.

**Fig. 10.** Shot test with Smith & Wesson SW1911 .45ACP

Figure 11-16 shows the shooting results on 6 kinds of different pads. We shoot 5 bullets per a pad and the acceleration values are measured in Z-axis by using targeting detection module as shown in Table 3. All materials are acceptable and stable, but heat distortion is occurred in the poly propylene and polyethylene. In case of PTFE, polycarbonate, and Foamex, there is severe damage in bullet pierced area. Therefore, Poly Urethane is finally selected.

## 4    Conclusion

This study was dealt with a realistic moving target system. For this study, we developed the driving platform and the target detection module. The driving platform was designed to be able to move quickly even a slope and unpaved terrain. And the target detection module is able to check a hit portion by using variation of vibration by using acceleration sensors. To evaluate the performance, experiment were conducted on driving platform and target detection module. As the results, drivability and detectability are confirmed. In the future work, we will integrate a whole moving target system and test real bullet fired test in various field.

**Acknowledgments.** This work (Grants No. C0123068) was supported by Business for Cooperative R&D between Industry, Academy, and Research Institute funded Korea Small and Medium Business Administration in 2013.

**Fig. 11.** Results of shot test with PP(poly propylene) pad

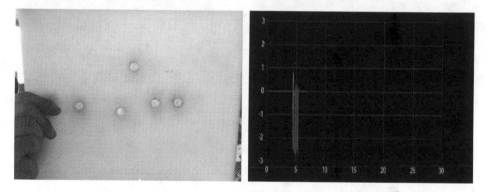

**Fig. 12.** Results of shot test with PE(poly ethylene) pad

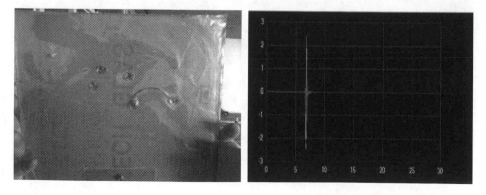

**Fig. 13.** Results of shot test with PC(poly carbonate) pad

**Fig. 14.** Results of shot test with PTFE pad

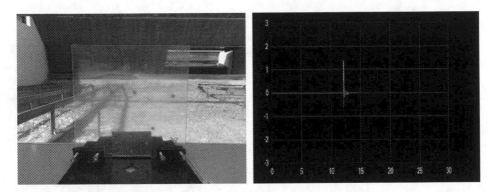

**Fig. 15.** Results of shot test with Poly Urethane pad

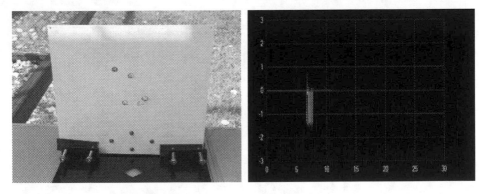

**Fig. 16.** Results of shot test with Foamex pad

**Table 3.** Acceleration data measured by the targeting detection module (Unit: G)

| Material of pad | Measured acceleration data of Z-axis |
|---|---|
| Poly Propylene | 1 |
| Poly Ethylene | 1.8 – 2.8 |
| Poly Carbonate | 2.1 – 3.1 |
| PTFE | 0.9 – 3.2 |
| Poly Urethane | 0.6 – 1.5 |
| Foamex | 0.9 – 1.3 |

# References

1. EINTECH Co. Ltd, http://www.eintech.net
2. FTS Co. Ltd, http://www.ftssystem.co.kr
3. Sports Marketing Surveys, MSR Consumer Report 2010, National Shooting Sports Foundation (2010)
4. Responsive Management/National Shooting Sports Foundation, The Future of Hunting and the Shooting Sports, the U.S. Fish and Wildlife Service (2008)
5. Marathon targets Co. Ltd, http://www.marathon-targets.com
6. Target Tracker Co. Ltd., http://www.targettracker.net
7. Campion, G., Bastin, G., D'Andrea-Novel, B.: Structural properties and classification of kinematic and dynamic models of wheeled mobile robots. IEEE Trans. Robot. Automat. 12(1), 47–62 (1996)
8. Caracciolo, L., De Luca, A., Iannitti, S.: Trajectory tracking control of a four-wheel differentially driven mobile robot. In: IEEE Int. Conf. Robotics and Automation, Detroit, MI, pp. 2632–2638 (1999)

# A Nonlinear Control of 2-D UAVs Formation Keeping via Virtual Structures

Zhuang Shao[1], Xiaoping Zhu[2], Zhou Zhou[1], and Yanxiong Wang[1]

[1] Science and Technology on UAV Laboratory, Northwestern Polytechnical University,
710065 Xi'an, China
[2] UAV Research Institute, Northwestern Polytechnical University,
710065 Xi'an, China
{shaozhuang233,wangyanxiongli}@163.com,
zhouzhou@nwpu.edu.cn

**Abstract.** In this paper, a nonlinear controller for UAVs formation keeping in 2-D situation is designed. In this design, virtual structure approach is applied for formation definition, and Lyapunov stability theory is applied to obtain the original nonlinear controller. And, a nonlinear function is used to make the controller more efficient combined with the differential and integral techniques. Simulation results show that UAVs formation keeping is well achieved with the controllers designed and which present better performance than the existing linear PI controllers.

**Keywords:** UAV, formation keeping, virtual structure, nonlinear control.

## 1 Introduction

Formation keeping control of unmanned aerial vehicles (UAVs) has got a spurt of interest in recent years and has many applications in both civilian and military domains such as search and rescue, surveillance, and reconnaissance [1] and [2].

In the literature, commonly used formation control schemes are mainly grouped into three kinds, namely, leader-follower [3] and [4], virtual structure [5] and [6], and behavior-based approach [7]. In the leader-follower approach, one or more vehicles are designed as leaders which determine formation trajectory and communicate with the ground station, while others are designed as followers which keep following the leader so that the formation can be maintained. This scheme is easy to understand and implement, but not robust with respect to leader's failure, and has a poor disturbance rejection property because of its error propagation. In the virtual structure approach, the entire formation is treated as a single virtual rigid structure, and each vehicle follows a moving point on virtual rigid. Since all vehicles are retreated as a rigid, the guidance of a group is easier than other approaches, while it's difficult to consider obstacle avoidance and lack of flexibility. In the behavior-based approach, the main idea is to prescribe several desired behaviors for each vehicle, and then to make the control action of each vehicle a weighted average of the control for each behavior. Possible behaviors commonly include collision avoidance, obstacle avoidance, goal seeking and formation keeping. It's suitable for uncertain environment, but lack of a

X. Zhang et al. (Eds.): ICIRA 2014, Part I, LNAI 8917, pp. 420–431, 2014.

rigorous theoretic analysis [8]. Each of these approaches has its own advantages and disadvantages, and this paper is based on virtual structure approach.

Virtual structures are first proposed in [5] to maintain the geometric configuration of a group of cooperative robots during movement. In [8], UAVs formation keeping is achieved by nonlinear model predictive control combined with virtual structure, this approach takes more computation time. In [1] and [9], a flexible virtual structure is proposed for formation keeping, while the trajectory controller designed has a low performance during formation turning with constant rate.

In this paper, a nonlinear controller is designed for formation keeping, and a nonlinear function is adopted to modify the controller for better performance, differential and integral techniques are also added.

## 2    Problem Formulation

### 2.1    2-D UAV Dynamics Model

Generally, UAVs formation flight control is implemented by a two-loop system. The inner-loop controller is allowing tracking of commanded velocity, altitude, and heading, which is the so-called autopilot. The outer-loop controller or formation controller generates a reference path command for the inner loop to maintain a certain formation configuration [4]. Since only the 2-D formation flight in horizontal plane is considered here, the UAV kinematic equations of motion are modeled as:

$$\begin{cases} \dot{x}_i = V_i \cos \chi_i \\ \dot{y}_i = V_i \sin \chi_i \end{cases} \tag{1}$$

where $(x_i, y_i)$ is UAV $i$'s inertial position in horizontal plane, $V_i$ and $\chi_i$ are its velocity and heading angle. And the inner loop dynamic of UAV $i$ is modeled as a first-order model [8]:

$$\begin{cases} \dot{V}_i = \dfrac{1}{\tau_V}(V_i^c - V_i) \\ \dot{\chi}_i = \dfrac{1}{\tau_\chi}(\chi_i^c - \chi_i) \end{cases} \tag{2}$$

where $V_i^c$ and $\chi_i^c$ are the control commands to UAV $i$'s autopilots, $\tau_V$ and $\tau_\chi$ are the time constants of the autopilots. Assume that all UAVs have equipped with the same autopilots and each UAV has an airspeed envelop of $0 < V_{min} \leq V_i \leq V_{max}$, and a turning rate envelop of $|\omega_i| \leq \omega_{max}, \omega_i = \dot{\chi}_i$.

### 2.2    2-D Formation Relative Dynamics Model

In this paper, virtual structure approach is used to define the UAVs formation configuration. Imagine that there is a virtual rigid moving along the reference formation trajectory exactly, as in Fig.1. Thus, the trajectory of virtual structure is the reference formation trajectory, and we assume that it is generated by a formation flight trajectory generator and satisfies the following kinematic equations:

$$\begin{cases} \dot{x}_f = V_f \cos \chi_f \\ \dot{y}_f = V_f \sin \chi_f \\ \dot{\chi}_f = \omega_f \end{cases} \tag{3}$$

where $(x_f, y_f)$ denote the instantaneous position of the virtual structure or reference trajectory, $\chi_f$ is the instantaneous reference heading angle, and $(V_f, \omega_f)$ are the reference velocity and reference turning rate.

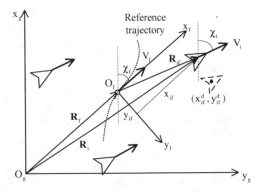

**Fig. 1.** Relative movement analysis on UAVs formation

The formation reference coordinate system $O_f x_f y_f$ is defined as in Fig.1, and $O_g x_g y_g$ is the inertial coordinate system. We call the origin $O_f$ of the virtual rigid as virtual point, and the desired formation configuration is defined by a set of relative coordinates $\{(x_{if}^d, y_{if}^d), i = 1, ..., m\}$ in $O_f x_f y_f$, where m is the total number of UAVs. According to Fig.1, we have the following equation satisfied:

$$\mathbf{R}_{if} = \mathbf{R}_i - \mathbf{R}_f \tag{4}$$

where $\mathbf{R}_i$, $\mathbf{R}_f$ and $\mathbf{R}_{if}$ are UAV i's inertial position vector, $O_f$'s inertial position vector and their relative position vector. By differentiating equation (4) and applying the rotation matrix between $O_f x_f y_f$ and $O_g x_g y_g$, the following relative kinematic equations are obtained:

$$\begin{cases} \dot{x}_{if} = V_i \cos \chi_{ei} - V_f + y_{if} \omega_f \\ \dot{y}_{if} = V_i \sin \chi_{ei} - x_{if} \omega_f \end{cases} \tag{5}$$

where heading error $\chi_{ei} = \chi_i - \chi_f$ . Let relative position error in $O_f x_f y_f$ as $x_{eif} = x_{if} - x_{if}^d$ , $y_{eif} = y_{if} - y_{if}^d$ , and combined with equation (5) we obtain the formation keeping error equations:

$$\begin{cases} \dot{x}_{eif} = V_i \cos\chi_{ei} - V_f + (y_{eif} + y_{if}^d)\omega_f \\ \dot{y}_{eif} = V_i \sin\chi_{ei} - (x_{eif} + x_{if}^d)\omega_f \end{cases} \tag{6}$$

Clearly, our control objective is to determine the control law $(V_i^c, \chi_i^c)$ so that the formation keeping errors $(x_{eif}, y_{eif})$ converge stably to a neighborhood about zero.

# 3    Control Design

In this paper, Lyapunov stability theory is first used to derive the control laws $(V_i^c, \chi_i^c)$, and then, we modify them for better performance. First, let following positive definite function:

$$V_{Li} = \frac{1}{2}(x_{eif}^2 + y_{eif}^2) \tag{7}$$

Its derivative along equation (6) is

$$\dot{V}_{Li} = (V_i \cos\chi_{ei} - V_f + y_{if}^d \cdot \omega_f) \cdot x_{eif} + (V_i \sin\chi_{ei} - x_{if}^d \cdot \omega_f) \cdot y_{eif} \tag{8}$$

According to Lyapunov stability theory, $(x_{eif}, y_{eif})$ converge stably to zero when $\dot{V}_{Li}$ is negative definite. Thus, we choose

$$V_i \cos\chi_{ei} = V_f - y_{if}^d \cdot \omega_f - k_{xi} \cdot x_{eif} \tag{9}$$

$$V_i \sin\chi_{ei} = x_{if}^d \cdot \omega_f - k_{yi} \cdot y_{eif} \tag{10}$$

where $k_{xi} > 0, k_{yi} > 0$. $\dot{V}_{Li}$ becomes

$$\dot{V}_{Li} = -k_{xi} \cdot x_{eif}^2 - k_{yi} \cdot y_{eif}^2 \tag{11}$$

which is negative definite. Then, according to equation (9) and (10), we can obtain the following nonlinear control laws:

$$V_i^c = \sqrt{(V_f - y_{if}^d \cdot \omega_f - k_{xi} \cdot x_{eif})^2 + (x_{if}^d \cdot \omega_f - k_{yi} \cdot y_{eif})^2} \tag{12}$$

$$\chi_i^c = \arctan(\frac{x_{if}^d \cdot \omega_f - k_{yi} \cdot y_{eif}}{V_f - y_{if}^d \cdot \omega_f - k_{xi} \cdot x_{eif}}) + \chi_f \tag{13}$$

According to [10], dealing with the feedback error of a feedback control law by a suitable nonlinear function can make the controller more efficient. Here we apply following function to deal with $(x_{eif}, y_{eif})$ [10]:

$$\text{fal}(e, \alpha, \delta) = \begin{cases} \dfrac{e}{\delta^{1-\alpha}}, & |e| \le \delta \\ |e|^{\alpha} \cdot \text{sign}(e), & |e| > \delta \end{cases} \tag{14}$$

where e denotes $x_{eif}$ or $y_{eif}$, constants $0 < \alpha < 1$, and $\delta > 0$. It's easy to prove that this process does not change $\dot{V}_{Li}$'s negative definite, because only $(x_{eif}, y_{eif})$ are replaced with $(\text{fal}(x_{eif}, \alpha, \delta), \text{fal}(y_{eif}, \alpha, \delta))$. In order to eliminate the steady state error and improve the system dynamic response properties, differential and integral techniques are applied. Thus, the modified control laws based on above ones are:

$$V_i^c = \sqrt{(V_f - y_{if}^d \cdot \omega_f - k_{xi} \cdot \text{fal}(x_{eif}, \alpha, \delta))^2 + (x_{if}^d \cdot \omega_f - k_{yi} \cdot \text{fal}(y_{eif}, \alpha, \delta))^2} -$$
$$k_{Ixi} \int x_{eif} dt - k_{Dxi} \frac{dx_{eif}}{dt} \tag{15}$$

$$\chi_i^c = \arctan(\frac{x_{if}^d \cdot \omega_f - k_{yi} \cdot \text{fal}(y_{eif}, \alpha, \delta)}{V_f - y_{if}^d \cdot \omega_f - k_{xi} \cdot \text{fal}(x_{eif}, \alpha, \delta)}) + \chi_f - k_{Iyi} \int y_{eif} dt - k_{Dyi} \frac{dy_{eif}}{dt} \tag{16}$$

where $k_{Ixi} > 0, k_{Dxi} > 0, k_{Iyi} > 0, k_{Dyi} > 0, i = 1, ..., m$.

## 4      Simulation Results

In order to illustrate the effectiveness of the designed control laws and the better performance compared with the existing PI controllers proposed in [3], a two-UAV formation flying in a diamond configuration is considered via Matlab Simulink.

The time constants of autopilots are $\tau_V = 10/3 s^{-1}$, $\tau_\chi = 5 s^{-1}$, and $18 m/s \le V_i \le 35 m/s$, $|\omega_i| \le 0.05 rad/s$, $i = 1, 2$. The desired formation configuration is given by $(x_{1f}^d, y_{1f}^d) = (25m, 25m)$, $(x_{2f}^d, y_{2f}^d) = (-25m, -25m)$. Initial formation states are given in Table 1. The total simulation time is 200s, and the reference trajectory is given as:

$$(x_{f0}, y_{f0}) = (0m, 0m), \chi_{f0} = 0rad, V_f = 25m/s, t \in [0, 200]s$$

$$\omega_f = \begin{cases} 0(rad/s), t \in [0, 80]s \\ 0.03(rad/s), t \in [80, 130]s \\ 0(rad/s), t \in [130, 200]s \end{cases}$$

As comparation, simulation results with existing PI controllers are given in Fig.2(a)-Fig.6(a). For control laws (12) and (13), simulation results are shown in Fig.2(b)-Fig.6(b), with $k_{xi} = 1, k_{yi} = 1.5, i = 1, 2$. For modified control laws (15) and (16), $\alpha = 0.2$, $\delta = 0.5, k_{xi} = 1, k_{yi} = 3.75, k_{Ixi} = 0.1, k_{Dxi} = 0.5, k_{Iyi} = 0.01, k_{Dyi} = 0.1, i = 1, 2$, and simulation results are shown in Fig.2(c)-Fig.6(c). Fig.7 shows the formation flight trajectory and the position of each UAV at snapshots of time at 0s, 40s, 100s and 140s.

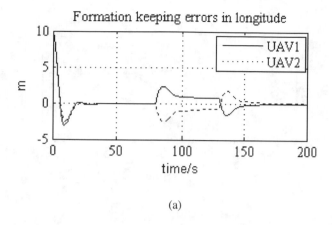

(a)

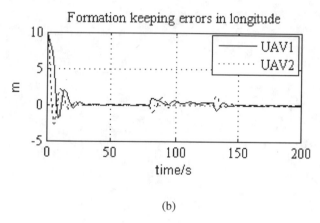

(b)

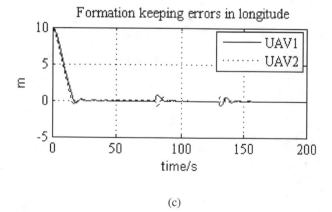

(c)

**Fig. 2.** Formation keeping errors in longitude by using (a) the existing PI controllers, (b) the original nonlinear controllers designed, and (c) the modified nonlinear controllers designed

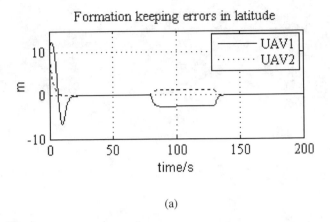

(a)

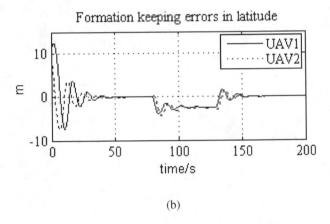

(b)

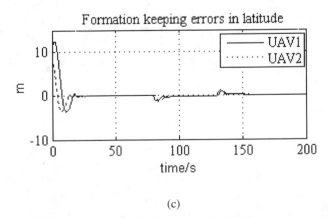

(c)

**Fig. 3.** Formation keeping errors in latitude by using (a) the existing PI controllers, (b) the original nonlinear controllers designed, and (c) the modified nonlinear controllers designed

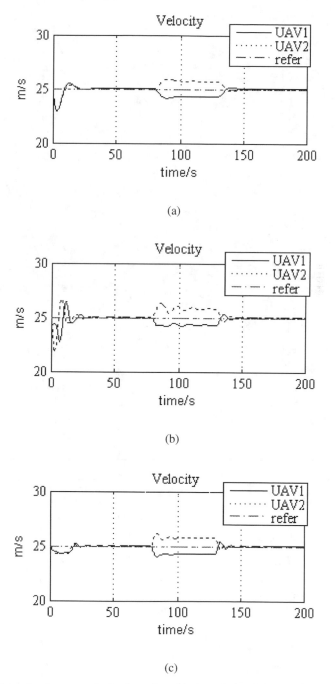

**Fig. 4.** UAVs and reference velocity by using (a) the existing PI controllers, (b) the original nonlinear controllers designed, and (c) the modified nonlinear controllers designed

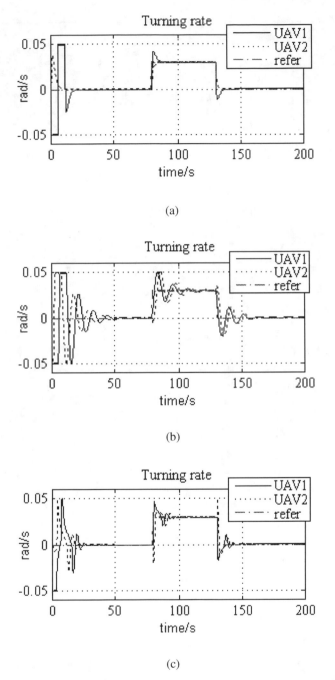

(a)

(b)

(c)

**Fig. 5.** UAVs and reference turning rate by using (a) the existing PI controllers, (b) the original nonlinear controllers designed, and (c) the modified nonlinear controllers designed

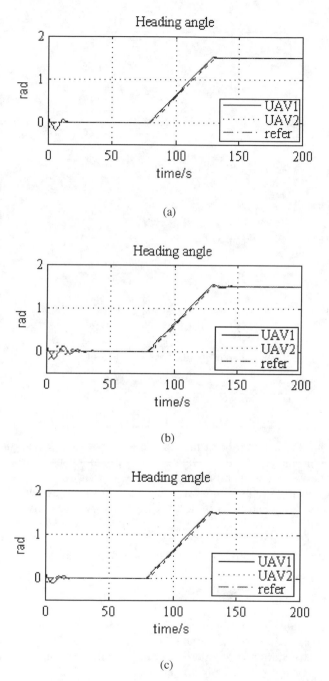

**Fig. 6.** UAVs and reference heading angle by using (a) the existing PI controllers, (b) the original nonlinear controllers designed, and (c) the modified nonlinear controllers designed

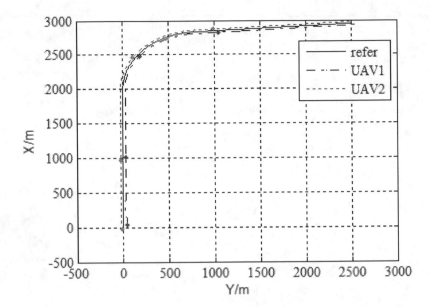

**Fig. 7.** Formation flight trajectory

**Table 1.** Initial formation states

| UAV | $V_i$ (m/s) | $\chi_i$ (rad) | $(x_{eif} / m, y_{eif} / m)$ |
|-----|-----|-----|-----|
| 1 | 25 | 0.1 | (10,10) |
| 2 | 25 | - 0.1 | (10,10) |

According to Fig.2 and Fig.3, the existing PI controllers and the nonlinear controllers we designed can both eliminate the initial relative errors stably and maintain the formation configuration steady during turning, while the existing linear PI controllers and the original ones we designed both make a stable errors during formation turning. Fig.4 and Fig.5 show that the both UAVs' velocity and turning rate are within the envelops. By comparing Fig 2(a)-Fig 6(a) with Fig 2(b)-Fig 6(b) and Fig 2(c)-Fig 6(c), it's clearly to see that the modified control laws present a much better performance both than the original ones and the existing linear PI controllers.

## 5    Summary

In this paper, a nonlinear formation keeping controller is designed based on virtual structure, which can eliminate the initial relative formation errors stably and maintain the formation configuration steady during turning. In order to improve the efficiency of the controller designed, we apply a nonlinear function combined with differential and integral techniques to modify the controller. Simulation results of a two-UAV formation show that the modified controllers make a much better performance both than the original ones and the existing linear PI controllers. A more precise 6DOF UAV model will be used in future work.

# References

1. Low, C.B., Quee, S.N.: A flexible virtual structure formation keeping control for fixed-wing UAVs. In: IEEE International Conference on Control and Automation, pp. 621–626. IEEE Press, Santiago (2011)
2. Fabrizio, G., Maro, I., Marcello, N., Lorenzo, P.: Dynamic and control issues of formation flight. In: Aerospace Science and Technology, pp. 65–71 (2005)
3. Pachter, M., D'Azzo, J.J., Dargan, J.L.: Automatic formation flight control. Journal of Guidance, Control, and Dynamics, 1380–1383 (1994)
4. Giulietti, F., Pollini, L., Innocenti, M.: Autonomous formation flight. IEEE Control System Magazine, 566–572 (2000)
5. Anthony Lewis, M., Tan, K.-H.: High precision formation control of mobile robots using virtual structures. Autonomous Robots, 387–403 (1997)
6. Brett, J.Y., Randal, W.B., Jed, M.K.: A control scheme for improving multi-vehicle formation maneuvers. In: Proceedings of the American Control Conference, pp. 704–709. IEEE press, Arlington (2001)
7. Mark, R.A., Andrew, C.R.: Formation flight as a cooperative game. In: Proceedings of the AIAA Guidance, Navigation and Control Conference, pp. 24–251. American Institute of Aeronautics and Astronautics (1998)
8. Zhou, C., Shao, L.Z., Lei, M., Wen, G.Z.: UAV Formation Flight Based on Nonlinear Model Predictive Control. In: Mathematical Problems in Engineering, Hindawi Publishing Corporation (2012)
9. Low, C.B.: A trajectory tracking control design for fixed-wing unmanned aerial vehicles. In: IEEE Conference on Control Applications, pp. 2118–2123. IEEE Press, Yokohama (2010)
10. Han, J.Q.: Active Disturbance Rejection Control Technique, pp. 119–206. National Defense Industry Press, Beijing (2009) (in Chinese)

# Research of Cooperative Control Method for Mobile Robots

Chang-Jun Woo[1], Jae-Hoon Jung[2], and Jang-Myung Lee[3]

[1] Department of Electrical and Computer Engineering, Pusan National University,
Jangjeon 2-dong, Geumjeong-gu, Busan, Korea
changjun1696@pusan.ac.kr
[2] Department of Electrical and Computer Engineering, Pusan National University,
Jangjeon 2-dong, Geumjeong-gu, Busan, Korea
jaehoon1696@pusan.ac.kr
[3] Department of Electrical Engineering, Pusan National University,
Jangjeon 2-dong, Geumjeong-gu, Busan, Korea
jmlee@pusan.ac.kr

**Abstract.** This paper proposes a control method for multiple mobile robots to move one object. It will show how each other mobile robot, with a hand and manipulator with three degrees of freedom and the two wheels, can cooperate with other. Firstly, via the kinematics interpretation, the position of the end-effector is calculated. Then, by using inverse kinematics, the end-effector is moved to the object's position. This cooperation of these two robots seems like a closed-chain form. Additionally, the angles of the both robot's manipulators are the same as each other. Utilizing the distance between robots, the constant force can be added to the object. When lifting the object, there are pushing forces from both sides. It is a study of how to control the angle of the manipulator and the position of the robot when know the exact position of the object with the assuming that the robot keeps perfect balance. The result is also simulated by using MSRDS 3D simulation tool.

**Keywords:** cooperative control, closed-chine, manipulator.

## 1 Introduction

Robot's manipulators are widely used in such industries as automobile manufacture to perform simple repetitive tasks, or in industries where work must be performed in environments hazardous to humans. Recently, there has been an increasing attention on cooperation of two robot's manipulators. As in human cases, where the use of two arms presents an advantage over the use of only one arm in several circumstances, with the cooperation of two hands, we can have more flexible in task, such as: process, maintenance and assembly. The cooperation among robots is not only essential for carrying the heavy or bulky objects, which are not in constant size, but also more efficient than using only one robot's hand because of stronger form.

In this paper, instead of lifting the object by using gripping power, it is to provide a method of research to lift the object with giving the power of both sides

X. Zhang et al. (Eds.): ICIRA 2014, Part I, LNAI 8917, pp. 432–437, 2014.

**Fig. 1.** Lifting the object with the both sides forces

## 2    Body

In this research, we experienced some following steps. Firstly, the permanent position of each robot's end-effectors are identified by solving kinematics equation. Secondly, with the knowing value of object's coodinates, it is necessary to identify the desired position of the end-effector. In this step, after using the inverse kinematics equation, we can know exactly the desired values of angles of each links, which is required for controlling. Finally, the result of controlling process is described by simulation via MSRDS 3D simulation program.

The following diagram is the simple description of our research's object.

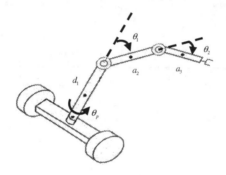

**Fig. 2.** Schematic diagram of a mobile manipulator

As the first step, it is possible by using the robot kinematics modeling to determine the coordinates of the end-effector. Using the D-H method for kinematics modeling, it is possible to obtain the following table.

**Table 1.** D-H table of mobile manipulator

| link | $\theta$ | d | a | $\alpha$ |
|------|----------|-----|------|----------|
| 1 | $\theta_p$ | $d_1$ | 0 | 0 |
| 2 | $\theta_1$ | 0 | $a_2$ | +90 |
| 3 | $\theta_2$ | 0 | $a_3$ | 0 |

We can know exactly the point where the robots can catch the object. Using inverse kinematics, it is possible to find end-effector position which coincides with the target's position.

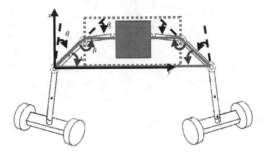

**Fig. 3.** Schematic diagram of a closed-chine form mobile manipulator

The operation of lifting the object is transformed from trapezoid form, as the solid red line, to rectangle form, as the dotted line, by approaching two robots and lifting the arm. It is opposite to the lifting operation that put down the object. In this process, it does not take into account the rotation of link 1. Therefore, the position of end-effector can be found by simplifying as follows, with the kinematics model and added height of the vehicle body and length of the link1 to x-axis. The following are simplified D-H table

**Table 2.** Simplified D-H table

| link | θ | d | a | α |
|------|-----------|---|-------|---|
| 1 | $\theta_1$ | 0 | $a_1$ | 0 |
| 2 | $\theta_2$ | 0 | $a_2$ | 0 |

Base on the simplified D-H table, we can obtain the homogeneous matrix of each link as follows:

$$^0H_1 = \begin{bmatrix} \cos\theta_1 & -\sin\theta_1 & 0 & -a_1\sin\theta_1 \\ \sin\theta_1 & \cos\theta_1 & 0 & a_1\cos\theta_1 \\ 0 & 0 & 1 & 0 \\ 0 & 0 & 0 & 1 \end{bmatrix} \tag{1}$$

$$^1H_2 = \begin{bmatrix} \cos\theta_2 & -\sin\theta_2 & 0 & -a_2\sin\theta_2 \\ \sin\theta_2 & \cos\theta_2 & 0 & a_2\cos\theta_2 \\ 0 & 0 & 1 & 0 \\ 0 & 0 & 0 & 1 \end{bmatrix} \tag{2}$$

$$^0H_2 = {}^0H_1\,{}^1H_2$$
$$= \begin{bmatrix} \cos(\theta_1+\theta_2) & -\sin(\theta_1+\theta_2) & 0 & x \\ \sin(\theta_1+\theta_2) & \cos(\theta_1+\theta_2) & 0 & y \\ 0 & 0 & 1 & 0 \\ 0 & 0 & 0 & 1 \end{bmatrix} \tag{3}$$

*with* :

$$x = -a_1 \sin\theta_1 - a_2 \sin(\theta_1 + \theta_2)$$
$$y = a_1 \cos\theta_1 + a_2 \cos(\theta_1 + \theta_2)$$

Since the sum of the Cabinet of the rectangle is 360 degree. Robot forms a perfect balance, so the sum $\theta_1 + A$ is 90 degree. Namely, we can find Eqs. (4) ~ (6).

$$\theta_1 + A = 90 \tag{4}$$

$$\theta_2 + B = 180 \tag{5}$$

$$A + B = 180 \tag{6}$$

It is possible to make simultaneous Eq. (7) by using Eqs. (4) ~ (6).

$$\theta_1 + \theta_2 + A + B = 180 + 90 \tag{7}$$

As a result, it is possible to obtain the following equation.

$$\theta_1 + \theta_2 = 90 \tag{8}$$

Substituting Eq. (8) to x and y of Eq. (3), we can find follow equation.

$$\theta_1 = \sin^{-1}\left(-\frac{x + a_2}{a_1}\right) \ or \ \theta_1 = \cos^{-1}\left(\frac{y}{a_1}\right) \tag{9}$$

Finally, $\theta_1$ and $\theta_2$ can be obtained.

With the knowing value of $\theta_1$ and $\theta_2$, the result of controlling process can be illustrated by using MSRDS 3D simulation program. The following figures are results of MSRDS 3D simulation.

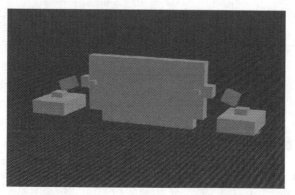

**Fig. 4.** The simulation result before lift object

When the two robots lift the object, both robots push force from both sides rather than the power of the robot's hands to grasp the object. It is possible to interpret the trapezoid which is based on distance of between robots. In addition, when two robots are taking one object, it seems like a closed-chain form. Make sure that both robots captured one object from both side, it is to follow the contact force by using the

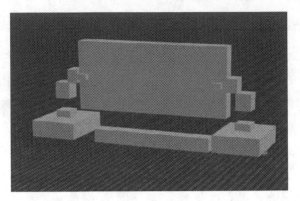

**Fig. 5.** The simulation result after lift object

impedance control. By narrowing the distance between the robots, the contact force can follow the desired value. The goal of lifting process the object is the A and B angle of the robot arm to be 90 degrees each. To maintain a constant contact force by two robots are closed, the arms are controlled to rise at the same time.

## 3    Conclusion

Normally, the object is lifted by using one robot's hand with the appropriate gripping force, or applying the force at the bottom of the object. In this paper, a method to lift an object by applying a constant force on both sides is introduced. The result can be stimulated by using MSRDS 3D simulation with the notice that in this simulation, it is responsible for using the hand, to prevent the object from shaking. The study also shows the use the dynamics modeling, specific method of lifting the object.

Further, in this study, only the state in which robot makes a perfect balance, and is assumed to know the exact position of the object is considered. In later studies, the balance control of the robot and the control of the manipulator according to the running will be evaluated. It is also a plan to study how can take advantage of not only transport but also assembly and fixing.

**Acknowledgments.** "This work was supported by the National Research Foundation of Korea(NRF) Grant funded by the Korean Government(MSIP) ( NRF-2013R1A1A2021174)."

"This research was supported by the MOTIE (Ministry of Trade, Industry & Energy), Korea, under the Industry Convergence Liaison Robotics Creative Graduates Education Program supervised by the KIAT (N0001126)."

## References

1. Sun, D., Liut, Y., Mills, J.K.: Cooperative Control of a Two-Manipulator System Handling a General Flexible Object. In: Proc. ICROS, Intelligent Robots and Systems, vol. 1, pp. 5–10 (1997)

2. Kim, B.H., Oh, S.R., Suh, I.H., Yi, B.J.: A Compliance Control Method for Robot Manipulators Using Nonlinear Stiffness Adaptation. Journal of Control, Automation and System Engineering 6(8), 703–709 (2000)
3. Choi, H.S.: A Study on the Control of Two-Cooperating Robot Manipulators for Fixtureless Assembly. Trans. Korean Soc. Mech. Eng (A) 21(8), 1209–1217 (1997)
4. Yeo, H.J.: A Coordination Control Methodology for Two cooperating Arms Handling a Single Object. Journal of Control, Automation and System Engineering 6(2) (2000)
5. Jang, J.Y., Seo, K.: Evolutionary Generation of the Motions for Cooperative Work between Humanoid and Mobile Robot. Journal of Institute of Control, Robotics and Systems 16(2), 107–113 (2010)
6. Kim, K., Ko, N.Y.: Trends Survey in Force/Torque Control of Robot Manipulator by Patent Analysis. J. of Advanced Engineering and Technology 3(3), 295–299 (2010)

# A Locomotion Driving of the Capsule Robot in Intestinal Tract

Zhou Hongfu[1,*]

School of Mechanical and Automotive Engineering, South China University of Technology, Guangzhou, Guangdong Province, China
mezzhang@scut.edu.cn

**Abstract.** For the fact of most capsule robots traveling in the intestinal tract is controlled by the natural peristalsis of gastrointestinal tract, which can't be controlled effectively with the peristalsis method, for the peristalsis capsule endoscope it can't be observed the suspicious lesions in detail and repeatable in the intestinal tract and also it can't be touched in some points, such as diverticulum in the intestinal tract. The paper presents a capsule robot locomotion driving movement controller method in the intestinal tract, which is by spiral lead-screw mechanism to propeller the capsule robot forward and backward movement.

**Keywords:** Capsule robot, Locomotive capsule, MEMS, Capsule endoscope.

## 1 Introduction

Gastrointestinal endoscopies are widely used in intestinal diseases test, but there are two drawbacks to face with the endoscope. One is the intestinal tract is narrow and winding, which is difficult for endoscopies pipe to insert into the intestinal tract for a long distance medic test and reach some narrow points. For the other drawback, the contact friction between the endoscope pipe and the wall of intestinal tract makes the intestinal tract lesion and arouse serious discomfort and pain to subjects. To solve the problems mentioned above, physicians and experts in digestive disease area, devote their efforts into the research and design, such as China, Israel, USA, Japan, and South Korea developed their capsule robots. Such as Omom capsule endoscopy in China, it's a capsule robot which applies in clinic more years for replacement of gastrointestinal endoscope[1]. However, most of them for traveling in the intestinal tract by the peristalsis of gastrointestinal tract and can't be controlled effectively. The research presents a driving method of the capsule robot in intestine tract, and makes the capsule robot movement controllable. Also in fabricating the capsule robot, the micro electro mechanical system (MEMS) method is a key technology to fabricate the robot in the research.

---

[*] Correspondent author.

X. Zhang et al. (Eds.): ICIRA 2014, Part I, LNAI 8917, pp. 438–445, 2014.

## 2    The Driving of the Capsule Robot

At present, there are a few driving method for capsule robot movement in the intestinal tract, they are earthworm-like type, multi-legged type, rotating magnetic field type, hydrodynamic pressure spiral type, leg locomotive drive type[2] and mechanical shock type. Following it introduces the capsule robot driving in earthworm-like type, rotating magnetic field type, and hydrodynamic pressure spiral type.

### 2.1    Earthworm-Like Drive

The earthworm-like micro-driving [3, 4] is designed based on the moving principle of earthworm. The robot consists of a few pairs of bidirectional linear drives, the principle is shown in Fig.1, where it is consists of 3 components, unit 1, unit 2 and unit 3.

The earthworm peristaltic working processing in Fig 1:

Stage 0: during a work cycle, the spring remains the original length at first,

Stage 1: next, the spring in tail, unit 3, contracts and the unit 3 moves forward for a step; at this moment, other components stay unchanged, and the robot reaches state 1.

Stage2: then the spring in mid part, unit2, contracts and the unit 2 moves a step forward; while other components remain in same position, the robot reaches state 2.

Stage 3: at last, the spring in head, unit 1, extends, and the robot reaches state 3(namely state 0). During the cycle, the robot forward for a step. Repeated processing the above steps, it makes the robot continuous to forward or backward movement.

Reverse the control sequence would make the robot move a step backward.

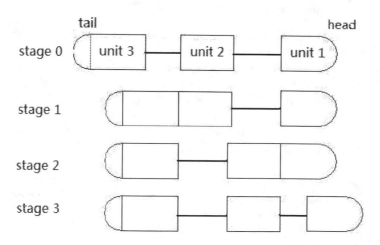

**Fig. 1.** The principle of earthworm type capsule robot

## 2.2    Rotating Magnetic Field Driving [5, 6]

Rotating magnetic field drive capsule robot is proposed by Ishiyama K, etc [7],   from Japan. They invented an in-tube moving micro-robot driven by the external magnetic field, which is based on the magnetic interactivities of the external executive magnetic field with inside capsule magnetic field to produce rotating magnetic force for spinning the capsule inside NdFeB permanent magnet. The driving principal shows in Fig 2[5,6,7].

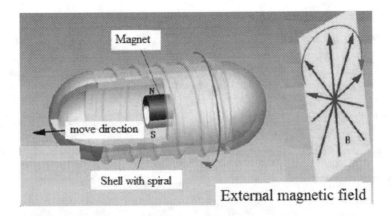

**Fig. 2.** Principle of the capsule robote drived by the external magnetic field

## 2.3    Hydrodynamic Pressure Spiral Driving [8, 9,10]

Zhou Yinsheng, the professor of Zhejiang University, invented a non-invasive intestinal micro-robot based on the spiral driving principle. This kind of robot can be run in the intestine track with full of liquid environment. It consists of a micro-motor with a dextral slot, a cylinder with a sinistral slot and a flexible coupling. When the micro-motor is switched to the positively power, the cylinder with a slot and the micro-motor with a slot rotates oppositely. The axial force between them opposites in direction drives the robot to move forward. In same way, when the micro-motor switches to the power negatively, the robot moves backward. In Fig 3 [11,12] it shows the spiral movement.

The spinning of the robot produces a hydrodynamic pressuring membrane on the surface. And the helical structure of the capsule robot makes the thickness of mucus membranes perpendicular to the spiral line varied according to different positions, which produces a difference liquid pressure perpendicular to the spiral line, and drives the robot to move forward or backward.

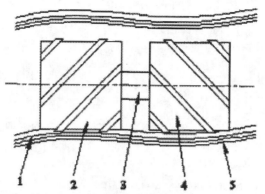

1) liquid   2) cylinder   3) elastic coupling
4) micro-motor   5) inlet wall

**Fig. 3.** Principle of the spiral robot driven by micro-motor

# 3   The Main Elements Part in the Capsule Robot Design

For making the robot traveling pass through the intestinal tract, the research designs the compact size robot. In size, the capsule robot diameter is about 10mm, and length in 28mm.

## 3.1   Image Sensor

The common image sensor for capsule robot choice is the CMOS and CCD sensitive sensors. The main performance of the CMOS and CCD image sensor devices lists in the Table 1, where it is shown that the CMOS integrated circuits have more application functions than CCD device, such as CMOS with AD converter, sequential control signal processing, etc. Besides the CMOS sensor has advantages, such as quicker response, smaller size, lower power consumption and cost, while the CCD owns stronger light sensitivity.

This research design selects CMOS sensitive image sensor chip, OV6920, which can output analogue color image signals. The tiny package size of OV6920 is 2.1mm*2.3mm. The image sensor has great stability and low power consumption for capsule robot. For cut the PCB board size to a fit in the tiny size, the chip can be designed again with MEMS method.

**Table 1.** The comparison of the main performance of CMOS and CCD image sensor

| Project | CMOS | CCD |
|---|---|---|
| AD | Included | Not included |
| Time sequence and control circuit | Included | Not included |
| Automatic gain control | Included | Not included |

**Table 1.** (*Continued*)

| | | |
|---|---|---|
| Signal processing | Included | Not included- |
| System consumption | <150mW | >1.5W |
| Sensitivity | 21X | 11X |
| Signal to noise ratio | 46dB | 50dB |
| Dynamic range | >70dB | >70dB |
| Integrated situation | Monolithic highly centralized | Multi    disc,    multi component |

### 3.2    Micro RF Signal Emitting Chip

The camera itself, it broadcasts image via digital modulated RF (Radio Frequency) using a custom protocol. ASIC RF emitting chip is highly integrating in a tiny chip which makes it be appropriate for the capsule robot. ASIC RF signal emitting chip consists of the low pass filter(LPF), frequency synthesizer(FS), voltage-controlled oscillator(VCO) and power amplifier(PA).

This design selects RF emitting component DX-2, the size in 7.62mm*7.62mm*6.55mm, the power voltage in 5-12V, current consumption in 12mA/9V, and emission frequency is 2.4GHz.

## 4    Capsule Robot Driving Design

### 4.1    Basic Capsule Robot Design

The basic capsule robot is a non-locomotion driving capsule endoscope, which is designed with camera for sending the intestine tract image out. This robot design mainly consists of image sensor, battery, camera lens, wireless transmission module, and control circuit PCB board. The design size of it is about Φ11mm×27mm. The robot moves by the intestinal peristalsis, and the power supplies for camera and wireless board is by the enclosed batteries in the capsule. The basic capsule robot is a capsule endoscope without locomotive driving. The capsule robot structure shows in Fig4 and Fig 5.

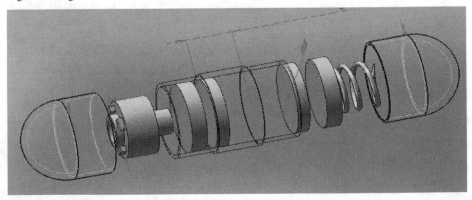

**Fig. 4.** The exploded view of the non-driven capsule robot

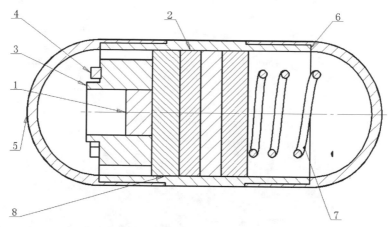

1.Image sensor    2.Battery    3.Camera lens support    4.Light-emitting diode(LED)  5.Transparent enclosure    6.Intermediate connecting shell 7.Transmission 9. Controlled board

**Fig. 5.** Basic capsule endoscope

## 4.2    Design the Spiral Lead-Screw Driven Capsule Robot

The spiral lead screw driven capsule robot is driving along the spiral curve, moves forward or backward inside the intestine tract, and send the tract image out meanwhile. It mainly consists of shell, camera, control module, wireless, motor stator,

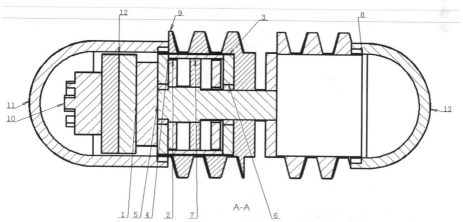

1.Control module    2.Motor stator   3. Shell cover    4.Motor shell  5.Motor shaft  6.Motor bearings  7. Motor rotor  8. Spiral screw  9. Fix spiral screw 10.Camera  11.Transparent cover  12.Battery   13.Transparent tail cap

**Fig. 6.** Spiral lead screw driving capsule robot

motor rotor, left screw and right screw. The design size is about Φ12mm×37mm. In PCB circuit, the drive circuit with DSP, MSP430F1232, power boost circuit with 1f17451-5, and wireless circuit with CC2500 RF circuit. In Fig 8, the capsule robot move locomotion is by a motor, and it drives the spiral screw to rotate, which touch the internal tract and generate an axis force to makes the robot axis movement. The whole system power consumption is about to 2mW.

The capsule robot drives via spiral lead-screw force at capsule tail. In the Fig 7, it shows the exploded drawing with Solidworks, and in the Fig.8, it is one drive design of the spiral lead-screw driven capsule robot.

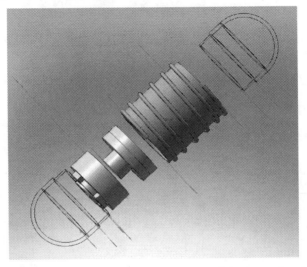

**Fig. 7.** The exploded view of spiral lead-screw driven capsule robot

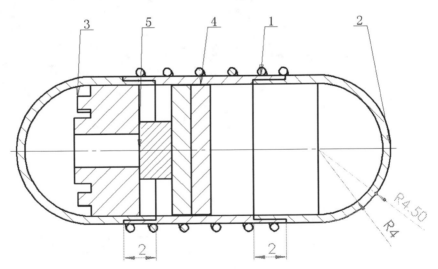

1.Threaded shell   2.Transparent shell   3.Camera   4.Battery   5.Image sensor

**Fig. 8.** Spiral lead-screw driven capsule robot

# 5     Conclusion

In the research it studies a capsule robot locomotion driving method for capsule driving in the intestinal tract. In the research, it designs a robot endoscope without locomotive driving as beginning, following it design a lead-screw for generating force to drive the capsule robot, where the spiral lead screw mechanism guides the capsule robot to move in the intestinal tract forward and backward movement. This design can be used as a method in the capsule robot locomotion movement.

# References

1. Liao, Z., Li, F., Li, Z.-S.: Clinical application of omom capsule endoscopy in china: a review of 1,068 cases, Gastrointest. Endosc. 67(5), AB265 (2008)
2. Quaglia, C., Buselli, E., Webster III, R.J., Valdastri, P., Menciassi, A., Dario, P.: An endoscopic capsule robot: a meso-scale engineering case study, J. Micromech. Microeng. 19, 105007 (11pp) (2009)
3. Ueno, S., Takemura, K., Yokota, S., Edamura, K.: Micro inchworm robot using electro-conjugate fluid. Sensors and Actuators A PP 216, 36–42 (2014)
4. Kim, H.M., Yang, S., Kim, J., Park, S., Cho, J.H., Park, J.Y., Kim, T.S., Yoon, E.-S., Song, S.Y., Bang, S.: Active locomotion of a paddling-based capsule endoscope in an in vitro and in vivo experiment. Tribology International 70, 11–17 (2014)
5. 张永顺,王娜,杜春雨,孙颖,王殿龙,胶囊机器人弯曲环境内万向旋转磁矢量控制原理, 技术科学,2013年,第43卷,第3 期, 274–282
6. Ciuti, G., Valdastri, P., Menciassi, A., et al.: Robotic magnetic steering and locomotion of capsule endoscope for diagnostic and surgical endoluminal procedures. Robotica 28(2), 199–207 (2010)
7. Sendoh, M., Ishiyama, K.: Fabrication of magnetic actuator for use in a capsule endoscope. IEEE Transactions on Magnetics 39(5), 232–233 (2003)
8. Liang, H., Guan, Y., Xiao, Z., Hu, C., Liu, Z.: A Screw Propelling Capsule Robot. In: Proceeding of the IEEE International Conference on Information and Automation, Shenzhen, China, pp. 786–791 (June 2011)
9. Sendoh, M., Ishiyama, K.: Fabrication of magnetic actuator for use in a capsule endoscope. IEEE Transactions on Magnetics 39(5), 3232–3234 (2003)
10. Wang, X.: Meng, M.Q.H.: Guided Magnetic Actuator for Active Capsule Endoscope. In: Proc. IEEE Int. Conf. on Nano/Micro Engineered and Molecular Systems, January 16-19, pp. 1153–1158 (2007)
11. Menciassi, Stefanini, C., et al.: Locomotion of a Legged Capsule in the Gastrointestinal Tract: Theoretical Study and Preliminary Technological Results. In: Proc. of EMBC 2004, San Francisco, USA (September 2004)
12. Lkeuchi, K., Yoshinaka, K., Hashimoto, S., Tomita, N.: Locomotion of Medical Micro Robot with Spiral Ribs Using Mucus. In: Proc. of the Seventh International Symposium on Micro Machine and Human Science (1996)

# Multi-agent Coordination for Resource Management of Netted Jamming

Wang Xu, Jindan Chang, Xiaowei Shi, and Guangjie Wu

Science and Technology on Electronic Information Control Laboratory,
610036 Chengdu, China
xuwang@mail.ustc.edu.cn, xiao_shou_@163.com,
shixiaowei@126.com, 348604891@qq.com

**Abstract.** Recently, as jammers have to face to a large number of enemy networking radars, the method of netted jamming is proposed, and become an active area of research. However, because of the data transfer rate and time limitation, how to manage the finite jamming resource still remains a tough problem. Two multi-agent coordination mechanisms including the coordination based on Center Control and the coordination based on Commitments and Conventions are taken into this resource allocation modeling. Finally, a confrontation scene is designed to check these two allocation methods, and the results shows the coordination based on Commitments and Conventions is of great feasibility and more effective than the other mechanism.

**Keywords:** jamming resource management, multi-agent coordination, netted jamming, networking radars.

## 1    Introduction

With the worldwide development and dispositions of networking radar system, it is more and more difficult to penetrate into the radar system effectively. In this context, a method of using jammers to screen the radars is proposed. Compared to the enemy networking radars, however, jamming resource is so finite, and one jammer could not jam so many radars. So, the jammers should be netted and work cooperatively to face these threats as well.

Because of the limitation of time and data transfer rate, few means of coordination among networking jammers are put forward at present. As a result, the public reports on netted or cooperating jamming rarely appear at present.

Essentially, these networking jammers construct an intelligent system, since they must work highly cooperatively with some common goals. And in artificial intelligent field, multi-agent joint mission planning has got a great achievement during the past several decades ([Cai 2004], [Russell 2008]). These techniques could be applied to this jamming resource management as well. With this enlightenment, [1]we cast the

---

[1] This research was financially supported by the fund of Science and Technology on Electronic Information Control Laboratory.

X. Zhang et al. (Eds.): ICIRA 2014, Part I, LNAI 8917, pp. 446–456, 2014.
© Springer International Publishing Switzerland 2014

jammer cooperating process during the confrontation with the networking radars as a process of multi-agent coordination for a team goal, and try to optimize the resource allocation among these agents. The paper is structured as follows, the confrontation process of the networking jammers and networking radars is analyzed, and the main problems and limitations of network jamming is presented in next section. And then, two resource management models, based on multi-agent coordination, are put forward in section 3. During section 4, an experiment of the confrontation between 4 jammers and 8 radars is designed and simulated to check the feasibility of model proposed. And finally, some conclusions and future researches are discussed in section 5.

## 2    The Process of Netted Jamming

In order to finish some tactical missions, our air force and equipments have to face to enemy networking radars as shown in Fig. 1. However, because of the quantitative advantage of the enemy and the shortcoming of the time division of jamming and observing, one jammer cannot deal with so many enemy networking radars.

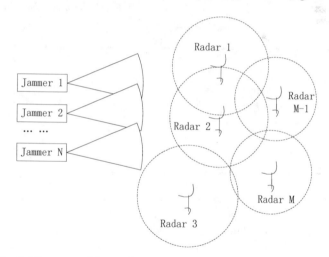

**Fig. 1.** The scene of the confrontation of jammers and networking radars

In this context, these jammers should work cooperatively, and a straightforward way may be taken is showed as follows:

(1) Set one of the jammers as the role of observer and coordinator, and note it as center-jammer and the others as sub-jammer.

(2) When a pulse comes, the center jammer distributes the corresponding jamming task (such as threat parameters and jamming measures) to these sub-jammers by some rules and methods, such as GA and Linear Programming.

(3) When a task arrives, the corresponding sub-jammer executes it, and sends a feedback to center-jammer at once.

These radar attributes included in the jamming tasks sent by center-jammer are composed of Radar Frequency (RF), Pulse Width (PW), Pulse Repetition Interval (PRI), ..., Radar Polar (RPolar) as shown in [Xie 2011], and the size of these parameters is defined in Table 1.

**Table 1.** Description of enemy radar

| Attribute | Length/bit | Description |
|-----------|------------|-------------|
| RF/MHz | 18 | Frequency range: X, C, Ku |
| PW/μs | 13 | Average value, invalid value for Continuous-wave radar |
| AOA/° | 12 | Angle of signal arrival |
| PRI/μs | 20 | Average value, 0: Continuous wave radar |
| Min_RF/MHz | 18 | Min value of RF |
| ...... | ...... | ...... |
| Min-PRI/μs | 20 | Min value of PRI |
| Max_PRI/μs | 20 | Max value of PRI |
| Radar Polar | 3 | Left-Circular, Right-Circular, Horizontal, Vertical, or Slant. |
| Total | 578 | |

**Table 2.** Description of jamming measure

| Attribute | Length/bit | Description |
|-----------|------------|-------------|
| Jamming Method | 20 | Noise, Deception Jamming, etc. |
| Interference Time | 20 | Interference Time, unit: μs |
| Total | 40 | |

If time and resource permit, this cooperative way can meet our interference demand, whereas several MB per second is needed at least if there are thousands of pulses of target radars during one second, and some other messages are not included in it, such as the message head and the feedback from the sub-jammer to center-jammer. And to ensure the reliability of communication, actually far more than ten MB/s are needed. So the data transfer speed is one of the most limitations. Meanwhile, center-jammer cannot switch promptly, while two pulses of different radars arrived with a very small time interval, as indicated in Fig. 2. That is to say, a new coordination model is needed to construct under the limitation of time and data transfer rate.

Besides time and data transfer rate, we also need to optimize the task distributing mechanism since both the jamming time resource and the number of the networking jammers is finite. During next part, an optimization model based on multi-agent coordination is constructed to solve the resource management problem mentioned above.

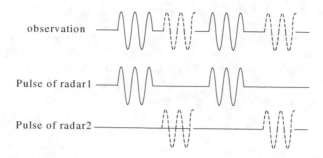

**Fig. 2.** Pulses arrives in a small interval

## 3 Multi-agent Coordination for Jamming Resource Management

The jamming resource management is equal to joint mission planning of multi-agent and could use multi-agent coordination to manage the jamming resource, if we regard these networking jammers as agents with the common goal of jamming. Since observing and jamming are of time division for one jammer, and one jammer cannot deal with two or more radars simultaneously, one jammer is needed to play the role of observer as mentioned in section 2. Meanwhile, because of the limitation of data transfer rate, real-time dynamic control for jamming resource management is impossible. In order to economize resource, the observer jammer should act as a planning role, too. Note the observing and planning jammer as center-agent, and the others as sub-agents. From the analysis in section 2, two fundamental rules should be obeyed during the joint mission planning is showed as follows:

- Because of the limitation of time and data transfer rate, these agents cannot exchange messages with each other real-time;
- As jamming resource is finite and the number of enemy networking radars is relatively large, these sub-agents should work highly cooperatively.

These two rules indicated that center-agent should identify enemy radar's signals, plan for some goal, and distribute the planning results, including threat parameters and corresponding jamming measures, to these sub-agents dynamically and respectively, but the distributing behavior cannot be too frequently to exceed the maximum value of data transfer rate. Meanwhile, these sub-agents should work cooperatively with few information exchanges.

Two mechanisms are usually used in multi-agent coordination of joint task planning: coordination based on Center Control, and coordination based on Commitments and Conventions. To achieve synergic jamming, these two methods are taken into jamming resource management.

### 3.1    Coordination Based on Center Control

The frame of the coordination based on center control is showed in Fig. 3, simply. Similar to the method in section 2, Center-agent keeps observing the world, identifying the threat signals, and storing them in its Data Base (DB). However, because of time limitation and data transfer rate, it plans for task and distributes the results, including threat parameters and corresponding counter measures, to these sub-agents with some mechanisms in Knowledge Base (KB), only when the changes of threat signals exceed a threshold or the time interval after the last distribution exceeds a time threshold. Meanwhile, when new jamming strategy arrive, these sub-agents update their jamming parameters, such as jamming targets and jamming measures in their DB, instantly. While a pulse arrives, the sub-agent identifies whether the pulse is a threat signal, which sub-agent is responsible for it, and what action it should take based on the mechanism in its KB/DB.

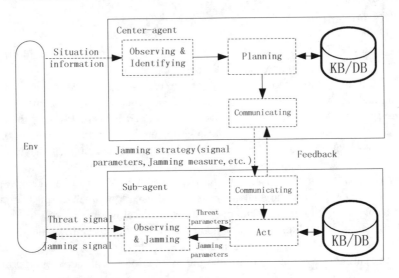

**Fig. 3.** Multi-agent coordination of center control for jamming resource management

As indicated in Fig. 3, three main steps should be taken to design the coordination mechanism of center control for jamming resource management:

Step 1 (Initiation): Add threat signals to the threat signal base, and set parameters for the threat assessment, planning, and strategy distribution for center-agent as follows:

$K$, the maximum num of jamming objects;

$CurJamObj$, the table of signal objects jammed currently;

$MinThD$, the minimum threat degree of jamming objects;

$Dec\_Tim$, the last planning time of center-agent;

$Dec\_Tim\_Interl$, the maximum time interval of twice planning for center-agent.

*Deta_time*, the minimum time interval that allows sub-agents to change the jamming object

*Maxchg_Num*, the maximum tolerated number of jamming object. If the number of the change of jamming objects is smaller than *Maxchg_Num*, center-agent does not have to plan.

*N*, the number of sub-agents (sub-jammers).

Step 2 (Center-agent Planning): As mentioned above, Center-agent planning process could be divided into several parts: Observing, Identifying threat signal, Planning, and Policy Distributing. And this process is described in Algorithm 1.

It should be mentioned here that plan(*JamObj, N*) in Algorithm 1 is the planning function, and many methods could be used here, such as Genetic Algorithms (GA) in [Qiu 2012], Linear Programming (LP) in [Guo 2006], and Artificial Immune System in [Xue 2012].

Step 3(Sub_agents Jamming): As the role of jamming executor, the function of Sub_agents is mapping (strategy, threat signal) to interference signals. So, the jamming process of them could be described as Algorithm 2.

---

*Algorithm 1*

**Function center_agent(Cur_Percept, Dec_Time, Cur_Time) return Ture or False**
// *True*: Plan and send strategy to sub_agents successfully; *False*: otherwise)

       **Input**: *Cur_Percept*, percept of the signal of current state;
                   *Dec_Time*, last planning time of center-agent;
                   *Cur_Time*, current time.
       **Static**:*K, MinThD, Maxchg_Num, N*, const variable as mentioned above;
              *JamObj*, the table of current signal objects of interference, initially empty;
              *NewObj*, new threat signal objects, initially empty;
              *Address*, the table of address of sub_agent.
       **for** $i \leftarrow 1$ **to** sizeof(*Cur_Percept*) **do**
         **if** threat_assess(*Cur_Percept*(*i*)) $\geqslant MinThD$ add(*NewObj, Cur_Percep*(*i*),*K*);
       **if** difference(*NewObj, JamObj*) $\geqslant Maxchg\_Num \parallel Cur\_Time\text{-}Dec\_Time \geqslant Dec\_Tim\_Inter$
        *Dec_Time* = *Cur_Time*;
        *JamObj* = *NewObj*;
        *strategy* = plan(*JamObj, N*); // plan for the current state.
        *note* = strategysend(*strategy, Address*); //distribute strategy to sub_agents, respectively.
              //*Note=True, Plan and send strategy to sub_agents successfully;*
              //*note=False*: otherwise
        **return** note;
       **return** False;

---

Compared to the method in section 2, Center control for jamming resource allocation is not so real time but of great feasibility, since its data transfer behavior is less frequent. However, it is also static after the distribution of interference strategy of center-agent.

```
Algorithm 2
Function sub_agent(Cur_Signal,strategy) return True or False
//True: deal successfully; False: otherwise
        Input: Cur_Signal, percept of the threat signal;
                strategy, strategy received from center_agent;
        Static:SubJamObj, the table of interference objects of this sub_agent;
                JamMeas, the table of interference measures corresponding to the
objects;
                update(strategy, SubJamObj);//update the interference object of sub_agent
                for i← 1 to sizeof(SubJamObj) do
                    if Cur_Signal == SubJamObj(i)
                        action = jam(SubJamObj(i), JamMeas);
                        //action = True, send interference signal successfully;False, otherwise.
                    return action;
                return True;
```

## 3.2    Coordination Based on Commitments and Conventions

During the mechanism of center control, every sub-agent may be allotted more than one interference signal object. And in this context, some sub-agent may be very "busy" and the others may be very "free". As showed in Fig. 2, if both radar 1 and radar 2 are the interference objects of sub-agent 1, it almost cannot deal with that, while the other sub-agents have nothing to do since no corresponding signal object arriving currently.

Compared to coordination based on Center Control, coordination based on Commitments and Conventions add some privities which are some rules obeyed by all sub-agents to sub-agents after the distribution of strategy of center-agent. Considering the limitation of data transfer rate, we reform these two algorithms in 3.1.1 as follows:

(1) Center-agent sends the interference strategy and the table of objects to every sub-agent by Broadcast Communication. And the table of objects is defined as Table 3 in which $S_{ij}$ are defined in Table 1.

**Table 3.** Table of Interference objects

| Sub-agent No. | Interference objects | | | |
|---|---|---|---|---|
| Sub-agent 1 | $S_{11}$ | $S_{12}$ | ...... | $S_{1n_1}$ |
| Sub-agent 2 | $S_{21}$ | $S_{22}$ | ...... | $S_{2n_2}$ |
| ... ... | ... ... | ... ... | ... ... | ... ... |
| Sub-agent 3 | $S_{N1}$ | $S_{N2}$ | ...... | $S_{Nn_3}$ |

(2) To achieve sub-agent coordination, each sub-agent should note a state sheet of all the sub-agents, as showed in Table 4, in which the value of every row is equal to the subscript of the interference object in Table 3 if the corresponding jammer is in jamming state and equal 0 otherwise.

**Table 4.** State sheet of all sub-agents

| Sub-agent No. | Interference objects | | Start time |
|---|---|---|---|
| Sub-agent 1 | $i_1$ | $j_1$ | $t_1$ |
| Sub-agent 2 | $i_2$ | $j_2$ | $t_2$ |
| … … | … … | … … | … … |
| Sub-agent N | $i_N$ | $j_N$ | $t_N$ |

When a threat signal arrives, each sub-agent identifies it, searches from interference table (Table 3), and confirms the corresponding sub-agent it belong to, firstly. Contrary to the coordination based on Center Control (*Algorithm 1* and *Algorithm 2*), if the corresponding sub-agent is in interference state with other threat object, other sub-agents in free state will try to give it a hand. Note the threat signal as $S_{ij}$ (that is, it is the *j*th interference object of Sub-agent *i*), each of these sub-agents coordinates with others with the privities as follows:

- If sub-agent *i* is free, it jams $S_{ij}$ directly, and all these sub-agents update the state of sub-agent *i*.
- If sub-agent *i* is in interference state and some of other sub-agents is free, then:
  a) Search from sub-agent $mod(i+1,N)$ to $mod(i+2,N)$, … , $mod(i+N-1,N)$, until the state of sub-agent $mod(i+k,N)$ is free, then sub-agent $mod(i+k,N)$ jams $S_{ij}$ and stops search. Here, $mod(a,b)$ is the integer remainder, when *a* is divided by *b*.
  b) Update state stored in each sub-agent.
- If all sub-agents are in interference state, then:
  a) If some of these sub-agents will finish their current interference behavior in some threshold ($\Delta t$), then the sub-agent with the least remain time of interference will response to $S_{ij}$. And update state stored in each sub-agent.
  b) Ignore $S_{ij}$, otherwise.

It is noteworthy that when current time surpasses the end time of some sub-agents' interference, all sub-agents will update the states stored in their DB correspondingly and automatically. In this way, these free or not so busy sub-agents could try to help the busy sub-agent with little message. And in next section, a confrontational scene is designed to check the feasibility of these two coordination methods and the difference between them during the process of jamming resource allocation.

**Table 5.** Parameters of enemy networking radar

| Radar No. | RF/MHz | PRI/ms | Pw/μs | Scan Period/s | Initial Scan Azimuth /° |
|---|---|---|---|---|---|
| Radar 1 | 1600 | 4 | 30 | 10 | 0 |
| Radar 2 | 1720 | 4 | 30 | 9 | 45 |
| Radar 3 | 1840 | 3 | 30 | 10 | 90 |

**Table 5.** (*continued*)

| Radar 4 | 1960 | 4 | 20 | 9 | 135 |
|---------|------|---|----|----|-----|
| Radar 5 | 2080 | 3 | 20 | 10 | 180 |
| Radar 6 | 2200 | 2 | 10 | 9 | 225 |
| Radar 7 | 2320 | 2 | 10 | 10 | 270 |
| Radar 8 | 2440 | 2 | 10 | 9 | 315 |

## 4    Experiment and Simulation

In order to illustrate these two coordination mechanisms for jamming resource management, a simulation example of the confrontation of 4 jammers and 8 networking radars is designed with the parameters as shown in Table 5 and Table 6, in which the initial scan azimuths of these radars are of uniform distribution of [0, 2π]. Besides that, setting these jamming parameters in

Table 2, we could get the sequence of interference objects of sub-agent 1 allocated by center-agent as shown in Fig. 4, in which real line and dashed line represent two adjacent allocation. And the sequences of interference objects of sub-agent 2 and sub-agent 3 are similar to that of sub-agent 1, as well.

As shown in Fig. 4 and Fig. 5, center-agent updates the interference object per 1 second, and the sequence of actual interference of Coordination based on Center-agent Control and Coordination based on Commitments and Conventions are shown in Fig. 5.

As shown in Fig. 5, under the coordination mechanism of the Center Control, it may be sometimes so busy and sometimes so free for every sub-agent. Meanwhile, it may be very busy while others are very free. However, on the contrary, during the process of coordination based on Commitments and Conventions, all the sub-agents' interferences are uniform on the time dimension, and it is impossible to appear that someone is very busy and someone has nothing to do.

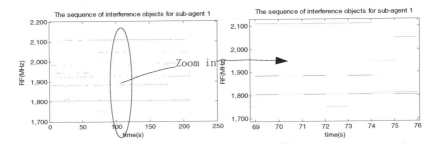

**Fig. 4.** The sequence of interference objects of sub-agent 1

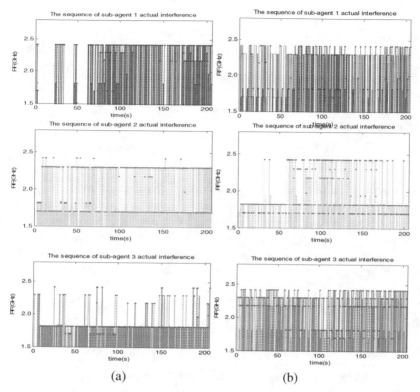

**Fig. 5.** The sequences of sub-agents actually jamming ((a) shows the coordination based on Center Control, while (b) shows the coordination based on Commitments and Conventions)

**Table 6.** Parameters for jammers

| Parameter | Value |
|---|---|
| MinThD/rank | 2 |
| Dec_Time/s | 0 |
| Dec_Tim_Interl/s | 1 |
| Deta_time/ms | 10 |
| Maxchg_Num | 3 |
| N | 3 |

**Table 7.** The signal number lost by these two mechanisms

| | Missing rate of threat pulses | |
|---|---|---|
| Scene No. | Coordination based Center Control | Coordination based on Commitments and Conventions |
| Scene 1 | 8.40% | 1.00% |
| Scene 2 | 18.75% | 2.30% |
| Scene 3 | 11.35% | 1.10% |
| Scene 4 | 9.40% | 1.60% |

Finally, besides the scene (note it as Scene 1) described in Table 5, we also choose another 3 scenes (note them as Scene 2, 3 and 4) stochastically to check the rate of signal number lost by these two methods, and the results is shown as Table 7 in which the coordination mechanism of based on Commitments and Conventions, is also more effective than that of Center Control.

## 5    Conclusion

Resource management problem of netted jamming is analyzed, and two multi-agent coordination mechanisms including Center Control, and Commitments and Conventions are taken into this problem modeling. And a confrontational example is designed to check these two mechanisms. During the future study, we'll try to consider the individual character. For example, for jammer 1, noise jamming is the best interference measure of it to jam radar 1, while Deception Jamming is that for jammer 2. So, when jammer 2 tries to help jammer 1 jam radar 1, it should take Deception Jamming.

## References

1. Cai, Z., Xu, G.: Artificial Intelligence: Principles and Applications, pp. 332–333. Tinghua University Press (2004)
2. Russell, S.J., Norvig, P.: Artificial Intelligence: A Modern Approach, 3rd edn., pp. 428–430. Prentice Hall (2009)
3. Zhengqiao, X.: Radar signal Real-time Sorting of Technology, Nanjing University of Science and Technology Master's Thesis (2011)
4. Qiu, W., Li, M.: Research on Target Assignment for Cooperative Jamming with a Variety of Interfere Styles. Journal of CAEIT 7(6) (2012)
5. Xiaoyi, G.: Study on Methods of Radar Jamming Resource Assignment from Multitactics. Graduate School of National University of Defense Technology Master's Thesis (2006)
6. Yu, X., Yi, Z., et al.: Efficiently Immune Genetic Algorithm for Solving Cooperative Jamming Problem. Journal of University of Electronic Science and Technology of China 42(3), 453–458 (2012)

# Simulation of Mobile Robot Navigation Using the Distributed Control Command Based Fuzzy Inference

Mingoo Kim and Taeseok Jin[*]

Dept. of Mechatronics Engineering, DongSeo University,
San 69-1 Churye-dong, Sasang-ku, Busan 617-716, Korea
jints@dongseo.ac.kr

**Abstract.** This paper propose simulation results of navigation for a mobile robot with an active camera, which is intelligently searching the goal location in unknown dynamic environments using sensor fusion, data that is usually required in real-time mobile robotics or simulation. Instead of using "physical sensor fusion" method which generates the trajectory of a robot based upon the environment model and sensory data. In this paper, "command fusion" method is used to govern the robot motions. The navigation strategy is based on the combination of fuzzy rules tuned for both goal-approach and obstacle-avoidance. To identify the environments, a distributed control command technique is introduced, where the sensory data of ultrasonic sensors and a vision sensor are fused into the identification process.

**Keywords:** Fuzzy, Control, Navigation, Mobile robot, Obstacle, avoidance.

## 1 Introduction

Autonomous mobile robot is intelligent robot that performs a given work with sensors by identifying the surrounded environment and reacts on the state of condition by itself instead of human. Unlike general manipulator in a fixed working environment[1,2], it is required intelligent processing in a flexible and variable working environment. And studies on a fuzzy-rule based control are attractive in the field of autonomous mobile robot. Robust behavior in autonomous robots requires that uncertainty be accommodated by the robot control system. Fuzzy logic is particularly well suited for implementing such controllers due to its capabilities of inference and approximate reasoning under uncertainty [3,4,5].

Many fuzzy controllers proposed in the literature utilize a monolithic rule-base structure. That is, the precepts that govern desired system behavior are encapsulated as a single collection of *if-then* rules. In most instances, the rule-base is designed to carry out a single control policy or goal. In order to achieve autonomy, mobile robots must be capable of achieving multiple goals whose priorities may change with time. Thus, controllers should be designed to realize a number of task-achieving behaviors that can be integrated to achieve different control objectives.

---

[*] Corresponding author.

X. Zhang et al. (Eds.): ICIRA 2014, Part I, LNAI 8917, pp. 457–466, 2014.

This requires formulation of a large and complex set of fuzzy rules. In this situation a potential limitation to the utility of the monolithic fuzzy controller becomes apparent. Since the size of complete monolithic rule-bases increases exponentially with the number of input variables [6,7,8,9], multi-input systems can potentially suffer degradations in real-time response. This is a critical issue for mobile robots operating in dynamic surroundings. *Hierarchical* rule structures can be employed to overcome this limitation by reducing the rate of increase to linear [16, 17].

First, this paper briefly introduces the operation of each command and the fuzzy controller for navigation system in chapter 2. Chapter 3 explains about behavior hierarchy based on fuzzy logic. In chapter 4, experimental results to verify efficiency of system are shown. Finally, Section 5 concludes this research work and mentions possible future related work.

## 2     Fuzzy Controller Design

The proposed fuzzy controller is shown as follows. We define three major navigation goals, i.e., target orientation, obstacle avoidance and rotation movement; represent each goal as a cost function. Note that the fusion process has a structure of forming a cost function by combining several cost functions using weights. In this fusion process, we infer each weight of command by the fuzzy algorithm that is a typical artificial intelligent scheme. With the proposed method, the mobile robot navigates intelligently by varying the weights depending on the environment, and selects a final command to keep the minimum variation of orientation and velocity according to the cost function[11-15].

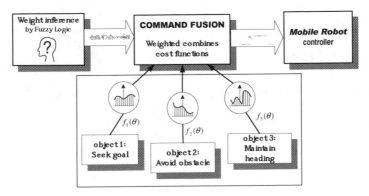

**Fig. 1.** Overall structure of navigation algorithm

### 2.1     Commands for Navigation

*Seeking Goal* command of mobile robot is generated as the nearest direction to the target point. The command is defined as the distance to the target point when the robot moves present with the orientation, θ and the velocity, $v$. Therefore, a cost function is defined as Eq. (1).

$$E_d(\theta) = \{x_d - x_c + v \cdot \Delta t \cdot \cos\theta)\}^2$$
$$+ \{y_d - (y_c + v \cdot \Delta t \cdot \sin\theta)\}^2 \tag{1}$$

where, $v$ is $v_{max} - k \cdot |\theta_c - \theta|$ and $k$ represents the reduction ratio of rotational movement.

*Avoiding obstacle* command is represented as the shortest distance to an obstacle based upon the sensor data in the form of histogram. The distance information is represented as a form of second order energy, and represented as a cost function by inspecting it about all $\theta$ as shown in Eq. (2).

$$E_0(\theta) = d^2_{sensor}(\theta) \tag{2}$$

To navigate in a dynamic environment to the goal, the mobile robot should re cognize the dynamic variation and react to it.

*Maintain heading* command is minimizing rotational movement aims to rotate wheels smoothly by restraining the rapid motion. The cost function is defined as minimum at the present orientation and is defined as a second order function in terms of the rotation angle, θ as Eq. (3).

$$E_r(\theta) = (\theta_c - \theta)^2 \quad \theta_c : present\ angle \tag{3}$$

The command represented as the cost function has three different goals to be s atisfied at the same time. Each goal differently contributes to the command by a different weight, as shown in Eq. (4).

$$E(\theta) = w_1 \cdot E_d(\theta) + w_2 \cdot E_o(\theta) + w_3 \cdot E_r(\theta) \tag{4}$$

## 2.2    Inference of Cost Function

We infer the weights of the usual fuzzy if-then rule by means of fuzzy algorithm. The main reason of using fuzzy algorithm is that it is easy to reflect the human's intelligence into the robot control. Fuzzy inference system is developed through the process of setting each situation, developing fuzzy logic with proper weights, and calculating weights for the commands.

Fig. 2 shows the structure of a fuzzy inference system. We define the circumstance and state of a mobile robot as the inputs of fuzzy inference system, and infer the weights of cost functions. The inferred weights determine a cost function to direct the robot and decide the velocity of rotation. For the navigation control of the mobile robot, the results are transformed into the variation of orientation and angular velocities by the inverse kinematics of the robot.

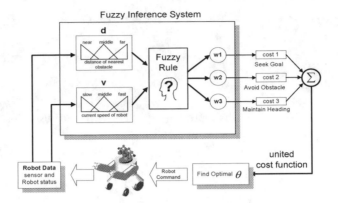

**Fig. 2.** Structure of Fuzzy Inference System

Fig. 3 shows the output surface of the fuzzy inference system for each weight fuzzy subsets using the inputs and the output. The control surface is $\omega_1$ fuzzy logic controller of seeing goal (a), $\omega_2$ fuzzy logic controller of avoiding obstacle (b) and $\omega_3$ fuzzy logic controller of minimizing rotation (c).

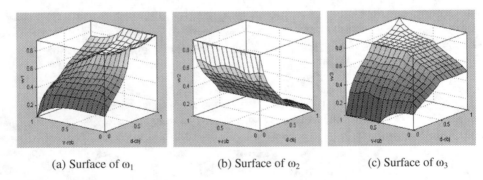

(a) Surface of $\omega_1$          (b) Surface of $\omega_2$          (c) Surface of $\omega_3$

**Fig. 3.** Input-output Surface of Weight Inference System

## 3      Building an Local Map

A mobile robot moves by selecting a more secure path after recognizing the environment to naviagte using image information. When we estimate the environments from the conditions of camera's actuator, there are many uncertainties. That is, environment informations estimated from the errors of camera angle $\alpha$, $\beta$, link parameters $l_1$, $l_2$ and $l_3$ of camera actuator, and calculating process, all of them have uncertain informations. In this paper, therefore, we propose the method of making probability map from image information, considering these uncertainties. Let j, k resulted in image processing have uncertainties ( $\Delta j$, $\Delta k$ ) of Eq. 5.

$$j = j_m + \Delta j, \quad k = k_m + \Delta k \tag{5}$$

where $\Delta j$, $\Delta k$ are 2 jointly caussian random variables with PDF as shown in Eq. 7. and $\sigma_{\Delta j} = \sigma_{\Delta k} = 0$, $m_{\Delta j} = m_{\Delta k} = 0$, $\rho_{\Delta j, \Delta k} = 0$

$$f_{\Delta j, \Delta k} = \frac{1}{2\pi\sigma^2} \exp\left\{-\frac{1}{2\sigma^2}\left(\Delta j^2 + \Delta k^2\right)\right\} \tag{6}$$

Image information j, k mapped real distance information x, y respectively by translation equation as follows

$$x = f_X(j,k), \quad y = f_Y(j,k) \tag{7}$$

Uncertainty of images, therefore, represented that of real distance information x ,y. PDF which presented uneertainty of x,y about a trasformation of Eq. 7 is given by

$$f_{X,Y} = \frac{1}{2\pi\sigma^2} \exp\left[-\frac{1}{2\sigma^2}\left\{(f_J(x,y) - j_m)^2 + (f_K(x,y) - k_m)^2\right\}\right] \cdot |J_{JK}(x,y)| \tag{8}$$

whrere $f_J(x,y)$, $f_K(x,y)$ are the relation of transformation that represented mapping relation between real coordinate x, y and camera image. $J_{JK}(x,y)$ is the jacobian matrix as follows

$$J_{JK}(x,y) = \begin{bmatrix} \dfrac{\partial f_J(x,y)}{\partial x} & \dfrac{\partial f_J(x,y)}{\partial y} \\ \dfrac{\partial f_k(x,y)}{\partial x} & \dfrac{\partial f_k(x,y)}{\partial y} \end{bmatrix} \tag{9}$$

and $|J_{JK}(x,y)|$ is the determinant of a jacobian matrix and is in the following form

$$|J_{JK}(x,y)| = \frac{\partial f_J(x,y)}{\partial x} \cdot \frac{\partial f_K(x,y)}{\partial y} - \frac{\partial f_K(x,y)}{\partial x} \cdot \frac{\partial f_J(x,y)}{\partial y}$$

$$= \frac{P_x P_y}{\theta_{ry}\theta_{rx}} \cdot \left(\frac{(l_1 + l_2)^2}{(l_1 + l_2)^2 + x^2 + y^2}\right) \cdot \sqrt{x^2 + y^2} \tag{10}$$

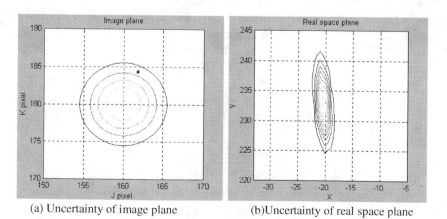

(a) Uncertainty of image plane          (b)Uncertainty of real space plane

**Fig. 4.** Uncertainties between image plane and space plane

Fig. 4 shows the simulation results about uncertainties between image plane and real space plane on condition that $\alpha = 15°$, $\beta = 0°$, j=160 and k=180. The longer X, Y image distance, PDF of x, y are lower. It meant that Long distance information has lower trust.

## 4 Experimental Results

This navigation method that includes the proposed algorithm is applied for mobile robot named as *AmigoBot* that has been developed in the laboratory for Intelligent Robotics as shown in Fig, 5.

**Fig. 5.** AmigoBot mobile robot

This modified *AmigoBot* robot had to be fast, flexible and offer real time image processing capabilities for navigation, so we applied Controller Area Network(CAN) to Pioneer-DX. CAN is a serial bus system especially suited for networking "intelligent" devices as well as sensors and actuators within a system or sub-system. CAN nodes can request the bus simultaneously and the maximum transmission rate is specified as 1M bit/s.[9].

With the proposed method, we make an experiment on building environmental map. Parameter values used for experiment are shown in Table 1.

**Table 1.** Parameter values used for experiment

| $l_1$ : 55 cm , $l_2$ : 7.5 cm , $l_3$ : 4 cm | | | |
|---|---|---|---|
| $P_x$ | 320 pixel | $P_y$ | 240 pixel |
| $\theta_x$ | 50° | $\theta_y$ | 40° |

Fig. 6(a) is the image used on the experiment; Width of corridor is 2m and Joint angle parameter α, β of active camera are 11° and 0° respectively. After capturing the image, The 'LOG' operator is utilized to extract the edge elements. It is suitable for detecting edge element at corridor that appear noises (e.g.: Patterns of the bottom, wall) sensitively because it is difficult for edge detection in case of other edge operators which has the characteristic of high-pass filter.

The essential information to map building is edge information that meets with the bottom in the edge information extracted from LOG operation. Fig. 6(b) shows the image that the points meet with the bottom through the thining after LOG operation. We construct probability map by this informations.

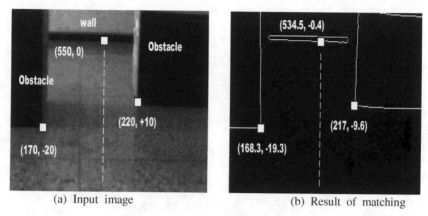

(a) Input image                    (b) Result of matching

**Fig. 6.** Experimental result of the vision system

Fig. 7(a) shows the map including the experimental environment. We exclude the information over 6m because that has low probability. Because we can estimate the reliability through the probability approach, the better map can be acquired.

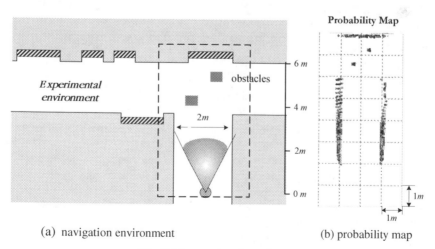

(a) navigation environment          (b) probability map

**Fig. 7.** The results of matching

Fig. 7(b) is the values resulted from matching after image processing which shows the estimated map over front 6m. The brightness presents the probability, and through the transformation, in case of having the area of the same distance   in the image, the farther the point is, the smaller the probability is. Therefore the information which extracted image has low truthness bacause it has wide proability density.

Fig. 7(b) shows that maximum matching error is within 4% of the dash-line area in Fig 7(a). Therefore, it can be seen that above vision system is proper to apply to navigation. The mobile robot navigates along a corridor with 2m widths and without obstacles and with some obstacles, respectively, as shown in Fig. 8. The real trace of the mobile robot is shown in Fig. 8(b). It demonstrates that the mobile robot avoids the obstacles intelligently and follows the corridor to the goal.

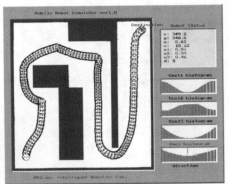 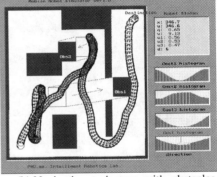

(a)   Navigation  trajectory without  obstacle      (b) Navigation  trajectory  with  obstacles

**Fig. 8.** Navigation of robot in corridor environment

## 5    Conclusion

A fuzzy control algorithm for both obstacle avoidance and path planning has been implemented in experiment so that it enables the mobile robot to reach to goal point under the unknown environments safely and autonomously.

And also, we showed an architecture for intelligent navigation of mobile robot which determine robot's behavior by arbitrating distributed control commands, seek goal, avoid obstacles, and maintain heading. Commands are arbitrated by endowing with weight value and combining them, and weight values are given by fuzzy inference method. Arbitrating command allows multiple goals and constraints to be considered simultaneously. To show the efficiency of proposed method, real experiments are performed.

To show the efficiency of proposed method, real experiments are performed. The experimental results show that the mobile robot can navigate to the goal point safely under unknown environments and also can avoid moving obstacles autonomously.

Our ongoing research endeavors include the validation of the more complex sets of behaviors, both in simulation and on an actual mobile robot. Further researches on the prediction algorithm of the obstacles and on the robustness of performance are required.

**Acknowledgment.** This research was supported by the Basic Science Research Program through the National Research Foundation of Korea(NRF) funded by the Ministry of Education, Science and Technology(No. 2010-0021054).

# References

1. Er, M., Tan, T.P., Loh, S.Y.: Control of a mobile robot using generalized dynamic fuzzy neural networks. Microprocessors and Microsystems 28, 491–498 (2004)
2. Zadeh, L.A.: Outline of a New Approach to the Analysis of Complex Systems and Decision Processes. IEEE Transactions on Systems, Man, and Cybernetics 3(1), 28–44 (1973)
3. Nair, D., Aggarwal, J.K.: Moving Obstacle Detection from a Navigation Robot. IEEE Trans. on robotics and automation 14(3), 404–416 (1998)
4. Bentalba, S., El Hajjaji, A., Tachid, A.: Fuzzy Control of a Mobile Robot: a New Approach. In: IEEE International Conference on Control Applications, pp. 69–72 (1997)
5. Furuhashi, T., Nakaoka, K., Morikawa, K., Maeda, H., Uchikawa, Y.: A study on knowledge finding using fuzzy classifier system. Journal of Japan Society for Fuzzy Theory and Systems, vol 7(4), 839–848 (1995)
6. Itani, H., Furuhashi, T.: A study on teaching information understanding by autonomous mobile robot. Trans. of SICE 38(11), 966–973 (2002)
7. Beom, H.R., Cho, H.S.: A sensor-Based Navigation for a Mobile Robot Using Fuzzy Logic and Reinforcement Learning. IEEE Trans. on system, man, and cybernetics 25(3), 464–477 (1995)
8. Ohya, A., Kosaka, A., Kak, A.: Vision-Based Navigation by a Mobile Robot with Obstacle Avoidance Using Single-Camera Vision and Ultrasonic Sensing. IEEE Transactions on Robotics and Automation 14(6), 969–978 (1998)
9. Tunstel, E.: Fuzzy-behavior synthesis, coordination, and evolution in an adaptive behavior hierarchy. In: Saffiotti, A., Driankov, D. (eds.) Fuzzy Logic Techniques for Autonomous 470 TUNSTEL, de OLIVEIRA, AND BERMAN Vehicle Navigation. STUDFUZZ, vol. 61, pp. 205–234. Springer, Heidelberg (2000)
10. Tunstel, E.: Fuzzy behavior modulation with threshold activation for autonomous vehicle navigation. In: 18th International Conference of the North American Fuzzy Information Processing Society, pp. 776–780 (1999)
11. Er, M., Tan, T.P., Loh, S.Y.: Control of a mobile robot using generalized dynamic fuzzy neural networks. Microprocessors and Microsystems 28, 491–498 (2004)
12. Jouffe, L.: Fuzzy inference system learning by reinforcement method. IEEE Trans. Syst., Man, Cybern., Part C 28(3), 338–355 (1998)
13. Leng, G., McGinnity, T.M., Prasad, G.: An approach for on-line extraction of fuzzy rules using a self-organising fuzzy neural network. Fuzzy Sets and Systems 150, 211–243 (2005)
14. Nishina, T., Hagiwara, M.: Fuzzy inference neural network. Neurocomputing 14, 223–239 (1997)
15. Takahama, T., Sakai, S., Ogura, H., Nakamura, M.: Learning fuzzy rules for bang-bang control by reinforcement learning method. Journal of Japan Society for Fuzzy Theory and Systems 8(1), 115–122 (1996)

16. Mehrjerdi, H., Saad, M., Ghommam, J.: Hierarchical Fuzzy Cooperative Control and Path Following for a Team of Mobile Robots. IEEE/ASME Transactions on Mechatronics 16(5), 907–917 (2011)
17. Wang, D.S., Zhang, Y.S., Si, W.J.: Behavior-based hierarchical fuzzy control for mobile robot navigation in dynamic environment. In: 2011 Chinese Control and Decision Conference(CCDC), pp. 2419–2424 (2011)

# Robotics and Road Transportation: A Review

José A. Romero[1,*], Alejandro A. Lozano-Guzmán[2],
Eduardo Betanzo-Quezada[3], and Carlos S. López-Cajún[4]

[1] Querétaro Autonomous University - SJR, Río Moctezuma 249, San Juan del Río,
Querétaro, México 76806
jaromero@uaq.mx
[2] CICATA – IPN, Querétaro Unit. Cerro Blanco 141, Querétaro, México 76090
alozano@ipn.mx
[3] Querétaro Autonomous University, Centro Universitario, Querétaro, Querétaro,
México 76010
betanzoe@uaq.mx
[4] Querétaro Autonomous University, - SJR, Río Moctezuma 249, San Juan del Río,
Querétaro, México 76806
cajun@uaq.mx

**Abstract.** In this paper a critical review of road infrastructure and vehicle robotic technologies is presented. It is found that many infrastructure-related robotic technologies have not reached the implementation stage, which is attributed to reliability concerns as such technologies involve high risk operations such as crack sealing. However, that the greatest effort to robotize operations in road transportation has been aimed at getting driver assisted and autonomous vehicles. The use of a crash avoidance system to prevent the impact of a double tractor-semitrailer truck onto a scholar bus is further analyzed, finding that a longitudinal crash-avoidance robotic system might have saved as many as seven lives. It is found that the main limitation of autonomous vehicles has to do with their ability to recognize atypical road irregularities that might endanger driving.

**Keywords:** Robotics, road transportation, road crashes, lateral stability, directional stability, autonomous vehicles.

## 1 Introduction

Road transport externalities include the social and economic consequences of crashes and the multidimensional effects of pollutants emitted by motor vehicles. On the road safety side, each year approximately 1.3 million people are killed on the roads while 20 to 50 million individuals result injured [1]. On the other hand, emissions of toxic gases due to transportation represent a real threat to sustainability [2] [3], with congestion and driving style representing prominent influential factors [4].

---

* Corresponding author.

X. Zhang et al. (Eds.): ICIRA 2014, Part I, LNAI 8917, pp. 467–478, 2014.

In this context, since the very beginning of motorized road transportation, electromechanical systems have been designed and implemented to increase driver and passenger comfort, road safety and fuel efficiency of engines. Examples of such early developments include electric windshield wipers (1939), antilock brakes (1971) and computer-controlled fuel injection engines (1976) [5]. Further improvements in electronics and mechanical devices resulted in systems with a broader range of functions representing different levels of intervention and automation, aiming to increase road safety and fuel efficiency. In this context and in a wider sense, robotics has been considered among the space-time adjusting technologies, as a result of improvements in automation and efficiency in the transportation infrastructure [6]. In this paper, a critical literature review of road transportation robotic technologies is presented, focusing on road safety, fuel efficiency and prevailing challenges.

## 2    Infrastructure-Related Robotics

Robots have been designed to perform operations for building and maintaining infrastructures, such as excavation works [7] and unmanned construction [8]. Robotic devices for road construction and maintenance have evolved since the first description made by professor Dah-Cheng Woo in 1995 [9]. At that time, a wide scope about the use of robots for construction and maintenance operations was conceived, focusing on areas of automated pavement inspection and crack sealing, automated bridge inspection and maintenance, automated bridge construction, and site integration. This last concept included, for example, the optimal earth moving operations. Future developments at that time included "accurate means to detect pavement distress at the earliest stage", robotic aids for working zones, underwater inspection of abutments and pier scours, and "sound and continuous operations for bridge painting and paint removal" [9]. In 1992, the Strategic Highway Research Program also recognized robotic operations for maintenance operations to identify, map, track and fill pavement cracks, involving sensors to specify crack length, size and depth [10].

Automation and robotics in construction in general, and road construction in particular, has represented a crucial interest to the construction industry as indicated by the creation of specialized associations such as the International Association for Automation and Robotics in Construction (IAARC) which publishes the Automation in Construction Journal and organizes the Annual International Symposium on Automation and Robotics in Construction. As a result of these activities, a diversity of estimations has been made on the future of robotics in the construction industry. In particular, Elattar lists the following developments [11]: automatic asphalt operations (reception, conveyance, spreading, paving, longitudinal crack sealing, roadside cleaning endeavors). In this respect, it is claimed that by robotizing these operations a better quality of the work will result while the workers will be less exposed to dangerous operations. Additionally, improved efficiencies are also thought to be the potential result of robotic earth moving operations, involving a variety of sensors to minimize the number of earth moving operations for a certain volume of material. Nevertheless, the main technological limitation for these robotic earth-moving operations derives from the complexity to model machine – soil interaction as a result of the plurality of factors that affect soil properties, including moisture content, stress history, time and environmental conditions [12]. Failed infrastructures have also been the subject of robotic approaches as in the case of rescue and surveillance operations [13].

Robotic total stations have been used for scanning and producing 3D models for construction and upgrade projects [14], or to inspect bridges [15]. To prevent road workers exposure, robotic highway safety markers have been proposed and tested [16] [17] [18]. Under different principles of operation (Lasser, global hearthbeat), these robotic systems provide accurate means to position barrels along the road work areas without exposing workers to accidents. Constraints for these designs include stability under wind loads, climb slopes and low cost [17].

Unfortunately, in spite of the numerous studies and prototypes there is no evidence that these work zones technologies have been actually deployed.

## 3    Computer Aided Driving

Robotics in vehicles involves different levels of intervention of the robotic system for the operation of the vehicle with the purpose of preventing crashes, increase comfort or to save fuel. The maximum level of intervention of the robotic system is represented by autonomous vehicles, which do not need a driver to circulate even under normal traffic conditions, with assisted driving providing some aid to the driver during specific situations.

The maneuvers that have been robotized and that have been incorporated into commercial vehicles include the following: Parallel Parking (PP), Automatic Cruise Control (ACC), Crash or Collision Avoidance (CA), Overtaking or Passing Maneuvers (OPM), rollover prevention, and lateral vehicle guidance. While such systems have been incorporated into cars and light trucks, heavy trucks represent in general different conditions leading to a crash. Particularly, articulated vehicles can get into lost control situations such as the jackknifing or rollover. For trucks, the following vehicular stability systems have been developed [19]: (i) the Roll Stability Control (RSC), which is a system that automatically intervenes to assist the driver to avoid a rollover through reducing the throttle and potentially applying the engine brakes; and (ii) the Electronic Stability Program (ESP) which controls vehicle´s oversteer or understeer through automatically controlling the throttle and selectively activating brakes to eliminate such instability condition. Safety systems to control the vehicle under emergency situations thus include rollover and steering stability considerations [20]. Active safety systems comprise the Electronic Stability Control (ESC) and the Active Front Steering (AFS), under an integrated scheme. However, brake control can also be used to prevent rollover risk situations, through the combination of real time data with simulation information about the precise moment at which the rollover might occur [21]. Such a system is based on a Linear Parameter Varying model (LPV) of yaw-roll dynamics of heavy vehicles that include the prediction of critical values while monitoring the lateral load transfer.

Advanced robotic applications include the Fully Adaptive Cruise Control system (FACC), which takes into account even the driver´s preferences that define his/her driving style [22], involving a learning process of driver´s attitudes and preferences. The purpose of a FACC system is to maintain a certain pre-calibrated distance with respect to the vehicle traveling ahead, regardless of the speed changes of that vehicle. Other advanced robotic devices consist of the CAS CWS systems (Automatic Cruise Control, Collision Avoidance Systems and Collision Warning Systems), for which it is necessary to have reliable information about the kinematics of the vehicle [23].

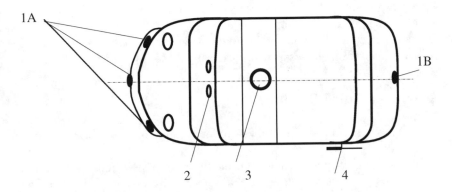

Parts:

    1A, 1B   Radar sensors, determine the position of distance objects;

    2          A camera near the rear-view mirror to detect traffic lights and helps the car´s onboard computer recognize moving obstacles like pedestrians and bicyclist;

    3          Rotating sensor on the roof scans more than 200 feet in all directions to generate precise three-dimensional map of the car´s surroundings;

    4          Sensor on the left wheel measures small movements made by the car and helps to accurately locate its position on the map.

**Fig. 1.** Autonomous car. Figure made using information from Markoff (2011) [25]

An assisted driving vehicle can become an autonomous, robotic vehicle once the different detection, cognition and acting systems govern the steering, braking and acceleration controls of the vehicle. Autonomous systems can also be used for the purpose of controlling the path of a vehicle when following another vehicle [24]. Figure 1 illustrates an autonomous vehicle that has been recently tested at a prototype level, listing the set of sensors needed to drive the vehicle under normal traffic conditions [25]. As it can be seen in this figure, this design combines different technologies to locate the vehicle within the space and to identify the objects in such space. It includes a geographic positioning system as well as cameras, radars and inertial references. Developers of this vehicle do not point out whether the vehicle is able to recognize road profile to detect bumps, potholes and other pavement disturbances which might affect vehicle´s stability and integrity. While the elements and sensors in this vehicle recalled those used by other car manufacturers that have participated in sponsored events by the Defense Advanced Research Projects Agency (DARPA) it seems that recent developments are more compact and less cumbersome.

Efforts to provide autonomy to trucks have been limited to some functions and operations such as longitudinal speed control, lane position detection and control, and turning at low speed (40 km/h) [26]. As far as the fuel consumption of heavy trucks is concerned, optimized geographic information algorithms have been considered to control the throttle position during changes of road geometry [27]. In particular, an intelligent system has been proposed to minimize fuel consumption during flat to uphill transitions, through the estimation of power demand and by controlling the throttle position [28].

In this context, the technological complexities associated to the creation of autonomous vehicles derive from the change of environments, including "unengineered" environments subject to sudden changes [29].

## 3.1   Robotic Vehicles and Professional Associations

The Society of Automotive Engineers (SAE) classifies the level of robotic intervention as follows [30]: autonomous vehicles, collision avoidance, electronic control systems, intelligent vehicles, and total vehicle integration. In 2009, SAE created the Standards Technical Committee AS-4 named Unmanned Systems, including four subcommittees focusing on architectural framework, network environment, model and performance measures. There is also the ITS (Intelligent Transport Systems) safety and human factors technical committee [31]. The Transport Research Board (TRB) includes the AHB30 committee to cover topics dealing with the Highway Automation, Intelligent Transportation Systems (AHB15), and Vehicle User Characteristics. There is the TRB´s Committee on Artificial Intelligence and Advanced Computing, Emerging technology law, Unmanned ground vehicles, Autonomous vehicles, Vehicle platooning, and vehicle platoons [32]. In a worldwide context, there is the Association for Unmanned Vehicle Systems International (AUVSI), which locates different State of the Art centers for autonomous road vehicles in different countries: four in The United States, two in Germany, and one in each of the following countries: Sweden, Italy, United Arabic Emirates, Australia, Japan and China [33].

While these organizations seem to cover all of the technical aspects related to driverless cars and robotic applications in road transportation, an important topic seems to be needed to be addressed, which is the circulation on damaged roads, as driverless vehicles do not have the ability to detect and to response to potholes and other pavement defects [34].

## 3.2    Autonomous Vehicles and the Law

In the United States of America three States, New Jersey, Nevada and California, have recently introduced the term "autonomous vehicle" to promote the use of such technologies. In the case of the State of New Jersey, the respective standard defines the following [35]: ""Autonomous vehicle" means a motor vehicle that uses artificial intelligence, sensors, global positioning system coordinates, or any other technology to carry out the mechanical operations of driving without the active control and continuous monitoring of a human operator". The State of Nevada legislation, however, recognizes such autonomy as an operational mode, allowing self-driving automobiles provided the use of a special red license plate and the payment of an extra insurance bond [36]. In the State of California a law will take effect on January 1, 2013, allowing driverless cars to be operated on public roads for testing purposes [37]. While these regulations refer to "autonomous vehicle", commercial publications use the term "Driverless cars" [36], which is not exact in the sense that autonomy of the vehicle is a non-permanent mode of operation, that is, legislation assumes a potential driver in the vehicle who can take control of it under special circumstances.

Regarding assisted driving, in May 2012 the National Highway Traffic Safety Administration (NHTSA) proposed a Federal Motor Vehicle Safety Standard (FMVSS) to mandate the preferential use of the Electronic Stability Control (ESC) over the Roll Stability Control systems (RSC), on all new trucks with a weight greater than 26000 pounds [38]. Although such preferential criteria was the result of several studies and statistical analyses, the American Transportation Research Institute (ATRI), has pointed out that ESC systems are less effective and more expensive than RSC systems [39]. While there is no doubt about the positive effect of using any of these technologies, apparently there is still discussion about which system represents the greater cost-benefit ratio.

## 3.3    Advantages and Disadvantages of Autonomous Vehicles

It has been argued that autonomous technology could enhance road safety and fuel efficiency, in addition to a potential economic development. However, the main issue here is the needed enhanced liability [25]. In spite of the open interests that different institutions and organisms have demonstrated toward the development of autonomous vehicular systems, it has been argued that such systems do not necessarily represent a road safety improvement, as drivers that might get used to automatic driving will be less aware and responsive than if they were in permanent control of the vehicle [36].

On the other hand, in addition to the potential safety improvements due to robotized vehicles, it has been argued that driverless systems could become so reliable that other activities could be performed while traveling so that such activity would not be a waste of time anymore [40]. In this respect, users of autonomous vehicles could communicate through cell phones, whether through SMS or call [41].

Potentially, repercussions of having autonomous vehicles include the capacity of roads and even vehicles' design. While larger road traffic capacities are expected as shorter gaps between cars could be possible, a diminished crash probability would signify to build lighter cars that consume less fuel. Less accidents with autonomous vehicles is expected as such vehicles would not get fatigued nor fell asleep or get intoxicated, in addition to having a faster reaction time than humans and a better, 360 degree perception [42].

However, autonomous vehicles face difficulties to avoid crashes as there are multiple scenarios for crashes. In the case of bus transit, Dunn et al. [43] recognize at least 60 different collision scenarios, involving seven collision warning for object detection systems.

# 4    Crash Prevention Potentials

A scenario is described to analyze what would be needed from the autonomous vehicle technologies perspective to avoid a tragic road crash. On the morning of April 12, 2012, a double tractor – semitrailer combination (DTSC) crashed onto a bus carrying 36 university students while negotiating a turn on a 2.2% downgrade three-lane Mexican highway. 7 people died at the spot while as many as 10 suffered grave disability (e.g., no arm). It was reported that the DTSC truck had lost its brakes, accelerating the truck out of control and causing the failure of the trailer's double-axle dolly, centrifuging it onto the scholar bus. Part (a) of Figure 2 describes a 2.5 s potential kinematics for such a crash, assuming a constant speed for the vehicles. According to the timeline shown in this figure, it took less than two seconds for the DTSC to hit the bus from a starting position 20 m behind it. To prevent this crash, actions could have been taken on the DTSC or on the bus. For the DTSC, the use of positive engine-linked braking systems could have functioned to stop the full vehicle once the brake malfunction was detected. On the bus side, evasive and accelerating maneuvers could have been executed, whether separately or in conjunction. Apparently, the bus driver was not aware of the coming out-of-control DTSC, and no reaction from him took place. On the one hand, in order to perform such a crash-avoidance maneuver through a robotic vehicle, the bus should have been equipped with 360° radar sensors in order to detect the coming vehicle, further assessing the available space for the crash avoidance steering maneuver. Additionally, the other action that might have been taken by the robotic system on the bus could have included an energetic acceleration, aiming to diminish the relative DTSC-bus speed and, if possible, get out of the trailer's way. The time available for these two crash-avoidance maneuvers is less than 2.5 s. In this regard, it is taken into account that accelerating is the faster response to avoid a crash as it represents shorter processing time [44]. Consequently, part (b) of Figure 2 illustrates the possible acceleration maneuvers that could have avoided this fatal crash, assuming an acceleration of 1.5 $m/s^2$ for the bus once the radar system detects the truck. Such an acceleration maneuver could have been feasible for the bus, as starting accelerations on zero slope roads can be up to 2.5 $m/s^2$, according to graphs presented by Rakha et al. [45]. For such maximum attainable acceleration, an additional safety gap could have been even gained in this case.

## 5    Discussion

The impetuous institutional impulse given to the creation and developing of autonomous cars, represented by multiple institutional and academic endorsements, suggests that such vehicles will become a technological reality in the near future. However, the cost of such equipment can be a problem, resembling a parallel situation with another development such as the hybrid traction systems, which are still not affordable for the gross of the population. A big difference between these developments, however, which might impulse autonomous vehicles technologies, resides in the fact that such technologies cross many critical transport issues such as safety and fuel economy. Another impulse to these technologies can be gained from the need to assist an ageing population that would wish the mobility independence provided by cars. In this respect, robotics in cars takes a new dimension, being part of the overall picture of population´s future mobility.

## 6    Conclusions

Robotics in road transportation has encompassed many critical safety-related areas from construction and maintenance of infrastructures (road and bridges) to vehicles operation. Although infrastructure operations were the subject of many early academic and government endeavors, this review shows that no significant technological deployment has occurred. On the other hand, autonomous vehicles have gained industrial momentum recently as a result of emerging technologies such as the precise and reliable global positioning systems, and rapid image processing. In this respect, incorporation of robotic systems into road transportation has evolved from relatively simple operations to fully autonomous vehicles in which the vehicle becomes the robot itself. The social acceptance of such technological developments is demonstrated by the issuing of standards and laws for such advanced vehicles. However, unresolved technological issues are related to the detection of atypical infrastructure defects such as open manholes and other road perturbations, which might represent major road safety hazards. In this context, autonomous vehicles could also be a mobility alternative to a growing aged population. The analysis of a crash involving a scholar bus revealed that a robotic longitudinal crash avoidance system might have saved many lives or diminished the gravity of its effects.

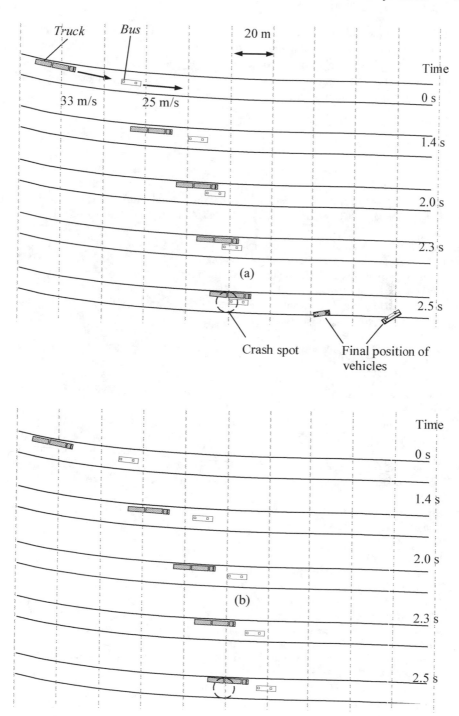

**Fig. 2.** Potential crash sequences: (a) Crash occurrence situation; (b) Crash-avoidance situation

Robotic longitudinal crash avoidance systems might thus represent a meaningful means to increase road safety at a short term. While cars have been the main focus of robotic technological developments so far, the implementation of autonomous heavy trucks would represent a major challenge as far as the liability is concerned, but with potentially enhanced benefits.

# References

1. Smith, WHO, Burden of disease from environmental noise. Quantification of healthy life years lost in Europe. World Health Organization. Geneva. Switzerland (2011)
2. Stanley, J.K., Hensher, D.A., Loader, C.: Road transport and climate change: Stepping off the greenhouse gas. Transportation Research Part A: Policy and Practice 45, 1020–1030 (2011)
3. Uherek, E., Halenka, T., Borken-Kleefeld, J., Balkanski, Y., Berntsen, T., Borrego, C., Gauss, M., Hoor, P., Juda-Rezler, K., Lelieveld, J., Melas, D., Rypdal, K., Schmid, S.: Transport impacts on atmosphere and climate: Land transport. Atmospheric Environment 44, 4772–4816 (2010)
4. Santos, G., Behrendt, H., Maconi, L., Shirvani, T., Teytelboym, A.: Part I: Externalities and economic policies in road transport. Research in Transportation Economics 28, 2–45 (2010)
5. Glancey, J.: The Car: A History of the Automobile. Carlton Publishing Group, London (2008)
6. Janelle, D.G., Gillespie, A.: Space-time constructs for linking information and communication technologies with issues in sustainable transportation. Transport Reviews 24(6), 65–677 (2004)
7. Ha, Q., Santos, M., Nguyen, Q., Rye, D., Durrant-whyte, H.: Robotic excavation in construction automation. IEEE Robotics & Automation Magazine (March 2002)
8. Arai, T.: Advanced Robotics & Mechatronics And Their Applications In Construction Automation. In: Proceedings of the 28th ISARC, Seoul, Korea (2011)
9. Woo, D.C.: Robotics in highway construction and maintenance. Public Roads 58(3), 26–34 (1995)
10. SHRP, Investigation of a pavement crack-filling robot. National Research Council. Strategic Highway Research Program Report SHRP-ID/UFR-92-616. Washington. D.C. USA (1992)
11. Elattar, S.M.S.: Automation and robotics in construction: opportunities and challenges. Emirates Journal for Engineering Research 13(2), 21–26 (2008)
12. Halbach, E.: Development of a simulator for modeling robotic earth-moving tasks. Helsinki University of Technology, Finland (2007)
13. TRB, TCRP Report 86: Public Transportation Security, Robotic Devices: A Guide for the Transit Environment. Federal Transit Operation, Washington, D.C. USA (2003)
14. Griffin, R., Navon, R., Brecher, A., Livingston, D., Haas, C., Bullock, D.: Emerging Technologies for Transportation Construction. TRB A2F09. Committee Report. Washington. D.C (2009)
15. DeVault, J.E., Hudson, W.B., Hossain, M.: Robotic system for underwater bridge inspection and scour evaluation. NCHRP-ID043 Final Report. The IDEA Program. TRB, Washington.D.C. (1998)
16. Mukhopadhyay, S., Shane, J.S., Strong, K.C.: Safety analysis and proposing risk mitigation strategies for operations and maintenance activities in highways: A qualitative

method. In: Proceedings, Construction Research Congress 2012, ASCE, West Lafayette, Indiana, USA, May 21-23 (2012)

17. Shen, X., Dumpert, J., Farritor, S.: Design and control of robotic highway safety markers. IEEE/ASME Transactions on Mechatronics 10(5), 513–520 (2005)

18. Bennet, D.A., Feng, X., Velinsky, S.A.: Robotic machine for highway crack sealing. Transportation Research Record 1827, 18–26 (2003)

19. USDOT, Concept of Operations and Voluntary Operational Requirements for Vehicular Stability Systems (VSS) On-board Commercial Motor Vehicles. FMCSA-MCRR-05-006. Federal Motor Carrier Safety Administration, Washington D.C. USA (2005)

20. Ghoneim, Y.A.: Control strategy for integrating the active front steering and the electronic stability control system: analysis and simulation. International Journal of Vehicle Autonomous Systems 8(2/3/4), 106–125 (2010)

21. Gaspar, P., Szabo, Z., Bokor, J.: Brake control using a prediction method to reduce rollover risk. International Journal of Vehicle Autonomous Systems 8(2/3/4), 126–145 (2010)

22. Bifulco, G.N., Pariota, L., Simonelli, F., Di Pace, R.: Development and testing of a fully Adaptive Cruise Control system. Transportation Research Part C: Emerging Technologies (2011A) (in Press)

23. Bifulco, G.N., Pariota, L., Simonelli, F., Di Pace, R.: Real-time smoothing of car-following data through sensor-fusion techniques. Procedia Social and Behavioral Sciences 20, 524–535 (2011B)

24. Travis, W., Martin, S., Bevly, D.M.: Automated short distance vehicle following using a dynamic base RTK system. International Journal of Vehicle Autonomous Systems 9(1/2), 126–141 (2011)

25. Markoff: Google lobbies Nevada to Allow Self-Driving cars. New York Times Newspaper–A18 (May 11, 2011)

26. Ukawa, H., Idonuma, H., Fujimura, T.: A study on the autonomous driving system of heavy duty vehicle. International Journal of Vehicle Autonomous Systems 1(1), 45–62 (2002)

27. Huang, W., Bevly, D.M.: Evaluation of 3D road geometry based heavy truck fuel optimization. Int. J. Vehicle Autonomous Systems 8(1), 39–55 (2010)

28. Krahwinke, W.: Robustness Analysis of Look-ahead Control for Heavy Trucks, Thesis work. TU-Braunschweig, Department of electrical engineering, Division of vehicular systems. Braunschweig, Germany (2009)

29. Newman, P.: C4B – Mobile robots. E-book. Mobile Robotics Group. Oxford University, England (2003)

30. SAE: Taxonomy and definitions for terms to on-road autonomous vehicles. SAE document J3016 (2012A)

31. SAE, Committees and Forums. SAE International, http://committees.sae.org/ (2012B)

32. TRB, Committees and Panels. Transportation Research Board of the National Academies (2012), http://www.trb.org/CommitteeandPanels/CommitteesAndPanels.aspx

33. Lucey, D.: Amazing race: blind, but now able to drive. AUVSI's Unmanned Systems Mission Critical 1, 13–22 (2011)

34. Winston, C.: Opinion: Paving the way for driverless cars. Driver's seat (2012), http://blogs.wsj.com/drivers-seat/2012/07/18/opinion-paving-the-way-for-driverless-cars/ (retrieved December 10, 2012)

35. SNJ. Assembly No. 2757. State of New Jersey. 215th legislature (2012) , `http://www.njleg.state.nj.us/2012/Bills/A3000/2757_I1.HTM` (May 10, 2012)
36. Marks, P.: Hands off the wheel. New Scientist 31 March 2012 (2012)
37. ENS. Driverless cars allowed on California roads. Environment News Service, September 26, 2012 (2012), `http://ens-newswire.com/2012/09/26/driverless-cars-allowed-on-california-roads/` (retrieved November 10, 2012)
38. NHTSA. FMVSS No. 136. Electronic stability control systems on heavy vehicles. Preliminary regulatory impact analysis. U.S. Department of Transportation. National Highway Traffic Safety Administration. Washington, D.C., USA (2012)
39. ATRI. Roll Stability Systems: Cost Benefit Analysis of Roll Stability Control Versus Electronic Stability Control Using Empirical Crash Data. American Transportation Research Institute. Arlington. Virginia. USA (2012)
40. Tomlin, J.: University to introduce driverless car. The Oxford Student April 6, 2012. Oxford, England (2012)
41. HD, Highlights of robot history (2012), `http://www.historydiary.com/electronics/Highlights-Of-Robot-Car-History.html` (retrieved November 10, 2012)
42. Markoff, J.: Google cars drive themselves, in traffic. New York Times newspapers, A1. New York (October 10, 2010)
43. Dunn, T., Laver, R., Skorupski, D., Zyrowski, D.: Assessing the Business Case for Integrated Collision Avoidance Systems on Transit Buses. Federal Transit Administration, Washington, D.C. (August 2007)
44. Jansson, J., Johansson, J., Gustafsson, F.: Decision Making for Collision Avoidance Systems, SAE paper 2002-01-0403 (2002)
45. Rakha, H., Lucic, I., Demarchi, S., Setti, J., Can Aerde, M.: Vehicle dynamics model for predicting maximum truck accelerations. Journal of Transportation Engineering 127(5), 418–425 (2001)

# Exploration of Unknown Multiply Connected Environments Using Minimal Sensory Data

Reem Nasir and Ashraf Elnagar

Computational Intelligence Center, Department of Computer Science,
University of Sharjah, P.O. Box 27272, Sharjah, UAE
ashraf@sharjah.ac.ae

**Abstract.** In robotics, Bug/Gap algorithms have shown good results as an alternative for traditional roadmap techniques, with a promising future, these results were locally optimal and sufficient to navigate and achieve goals. However, such algorithms have not been applied, or tested, on all types of environments. This work is aiming at improving and adding to this category of algorithms using minimal sensory data. To achieve this objective, we adapt a dynamic data structure called Gap Navigation Trees (GNT) that represents the depth discontinuities (gaps). The final GNT characterizes a roadmap that robots can follow. The basic GNT data structure is reported to model simple environments. In this paper, we extend GNT to unknown multiply connected environments. In addition, we add landmarks to eliminate infinite cycles. The proposed algorithm can be used in a variety of solid applications such as exploration, target finding, and search and rescue operations. The solution is cost effective, which enables the production of affordable robots in order to replace expensive ones in such applications. The simulation results had validated the algorithm and confirmed its potential.

**Keywords:** motion planning, gap-navigation trees, roadmap, robotics, local environments.

## 1 Introduction

The problem of "exploring the environment surrounding a robot" can be divided into different sub-categories based on three parameters. The first parameter describes the environment itself and it is divided into simple and multiply connected. Simple environments are environments that have no holes (i.e., obstacles) or discontinuity in its walls while multiply connected environments include holes. The second parameter is status of the obstacles whether it is static or dynamic. As more information is made available to the robot, the problem becomes more challenging [2], [4], [5]. The third parameter is knowledge about the robot's environment which is classified into known, partially known or unknown to a robot [2], [5].

When a point robot is placed in a given (known) polygonal region, the focus of path planning research would be finding the path a robot would follow. The tasks required of a robot would determine which path should be considered, the fastest,

X. Zhang et al. (Eds.): ICIRA 2014, Part I, LNAI 8917, pp. 479–490, 2014.
© Springer International Publishing Switzerland 2014

shortest, safest, or less mechanical movements from the robot. When the task involves search and rescue, then the shortest path is considered, and in known environments, computing shortest paths is a straightforward task. The most common approach is to compute a visibility graph which is accomplished in $O(n \log n)$ time by a radial sweeping algorithm [9] where n is the number of visible points for the robot.

When the environment is unknown (or partially unknown), sensors are used to explore and map the environment to develop navigation strategies. In this case, the robot is equipped with sensors that measure distances, able to identify walls from objects and track them. The information reported from these sensors, are used by the robot to build a map for the environment that will be used later to navigate through. This approach needs the robot to build an exact model of the environment with all of its walls, edges and objects with their exact coordination and measurements [17]. This leads to raise the question of whether this is practical or not and how accurate these maps are? That leads to many proposed solutions [2], which of course will keep increasing as new issues with the current algorithms arise [3]. On the other hand, some research has been conducted on how probabilistic techniques would be applied to the current algorithms [16] and how that would affect their performance [14].

Our work focuses on a relatively new approach where researchers studied the sensory data of a robot. They studied which of these data are essentially to the robot's ability to explore and navigate an environment, and that led to what they called minimal information called Gap Navigation Trees (GNT).

## 2    Problem Statement

The problem tackled in this paper is of two fold. The first is to explore a local environment cluttered with static obstacles. The second is to navigate such environment for carrying out a specific task.

Given an unknown, multiply connected environment with static obstacles, and a robot; the goal is (for the robot) to explore such environment and build a data structure (roadmap) while using the least amount of sensors possible. This data structure shall to be sufficient to achieve the robot's goals (finding an object, tracking an object or even just learning the environment).

We assume that the robot is modeled as a point moving in static, unknown and multiply connected environment, to achieve its goal. The robot is equipped with depth discontinuities sensor. The robot is expected to use the sensor's feedback to build a data structure which is used along with building local grid map as virtual landmarks, to learn and navigate among obstacles in the environment.

## 3    Proposed Algorithm

### 3.1    Gap-Navigation Trees (GNT)

GNT is a dynamic data structure constructed over direction of depth discontinuities [1]. The robot is modeled as a point moving in an unknown planar environment. The robot is assumed to have an abstract sensor (in the sense of [7]) that reports the order of depth discontinuities of the boundary, from the current position of the robot. These discontinuities are called gaps, and the abstract sensor may be implemented in a number of ways, using a directional camera or a low-cost laser scanner. Once GNT is constructed, it encodes paths from the current position of the robot to any place in the environment. As the robot moves, the GNT is updated to maintain shortest-path information from the current position of the robot. These paths are globally optimal in Euclidean distance. GNT showed that the robot can perform optimal navigation without trying to resolve ambiguities derived from different environments which result in having same GNT [1].

More research have been done on GNT in simple, closed, piecewise-smooth curved environment [15].Then it has been extended to multiply-connected environments (where the environment has holes, obstacles, jagged holes and edges) and new modification been added to handle the new events that have occurred [6], [8].

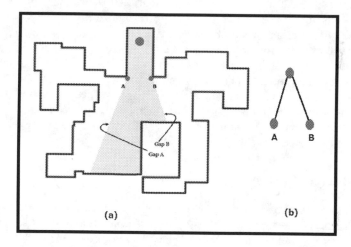

**Fig. 1.** (a) Visibility region of a robot with depth sensor, (b) GNT

We assume that the robot is equipped with a depth sensor. This sensor has the ability to report discontinuities which are referred to as gaps.

Fig. 1 shows how a robot detects gaps, where the nodes A and B represent starting points of the visibility edges of the robot's view. Beyond these edges, the robot has no knowledge of the environment; hence these nodes are added as children of the robot's current position. These nodes are called gaps. By adding children to each current position of the robot, and exploring them, the robot navigate and explore the whole environment until the whole environment been visible from these gaps.

There are three events that can occur while the robot investigates gaps:

- Addition of New Gaps: Each discontinuity in the robot's visibility field from its current position will be added as child gap node of the robot's node. This takes place while preserving the cyclic ordering from the gap sensor.

- Merger of Gaps: Two or more gaps could be merged into one gap, if they are the children of the same parent robot position and cover same area when investigated.
- Deletion of Gaps: If a gap becomes redundant, by being covered in the visibility range of the robot while examining another gap, and not a child for the same robot position parent node, then it will be deleted. When gaps belong to the same parent, but are in different directions (cannot be merged), the current gap will be kept, and the ones been seen (in the visibility range of the current gap) but not visited yet will be deleted.

## 3.2    Landmarks Strategy

GNT works perfectly in simple planner environments and gives optimal results, but when it is used in more complex environments, it is not guaranteed to work. The depth discontinuity sensor is no longer sufficient; at least another piece of data is required; because the robot needs to be able to recognize a gap that was previously visited. In simple environments each gap can be reached from one way only while in multiply connected environments, a gap could be reached from different sides. Therefore, the need to label such gaps becomes necessary. Previous works and suggestions had been made about combining GNT and landmarks, such as simple color landmark models [12], range measurements with respect to distinct landmarks, such as color transitions, corners, junctions, line intersections [13], scale-invariant image features as natural landmarks [11], and matching technology-RANSAC (Random Sample Consensus) [10].

## 4    Results and Analysis

In our simulations we classified the environments into six categories based on the size of obstacles and their distribution. The first four classifications are determined from the obstacles' sizes and the areas they occupy in the environment. The simulation started with sparse environments (i.e., large free space) with uniform-size obstacles followed by sparse environments with variable-size obstacles. We repeated both simulations but in cluttered environments. At last, we simulated two popular problematic environments in the literature of the path planning field. Namely, the Narrow-Passages problem and Indoor Environments. The following observations help appreciate the simulation results:

- Blue rectangles represent the obstacles in the environment.
- Red discs signify the initial robot position ($Rs$) or the current robot's position in the environment ($Ri$).
- Light orange discs indicate already visited gaps.
- Yellow discs represent the newly encountered gaps, which were added as children from the current robot position.
- Green discs indicate the next gap to be inspected.
- Green/black lines demonstrate the roadmap.
- **Maximum Number of Gaps** (MNG): This refers to the theoretical maximum number of gaps, in the given multiply connected environment. The upper

limit of the number of gaps is a function of the number of obstacles and their locations. For example, one rectangular obstacle would produce four gaps. If we consider having two obstacles in the environment, there will be a maximum number of eight gaps because the obstacles could be sharing a gap or more. Therefore we might end up with less than eight gaps, but it will not exceed eight. This info is reported by the simulator.

- **Unique Number of Gaps** (UNG): For validation purposes, this is obtained manually, by counting the number of gaps in a given environment. This number is less or equal to the maximum number of gaps. As the number of obstacles increase, in a given environment, this measure tends to be far less than the MNG.

- **Algorithm-computed Number of Gaps** (ANG): This refers to the total number of gaps encountered while constructing the GNT by the proposed algorithm. In an optimal scenario, ANG is the same as UNG. However, in a worse-case scenario it is close to MNG.

- **Gap Redundancy Rate Reduction** (GRR): This refers to the gap reduction in redundant gaps. The best-case scenario would yield testing UNG gaps only. Therefore, redundancy would be MNG-UNG. Therefore, we compute two reduction rates: optimal gap redundancy rate (OGRR) and algorithm-computed gap redundancy rate (AGRR), which are measured by:
    - OGRR= UNG/MNG.
    - AGRR= ANG/MNG.

Our objective is to minimize (AGRR-OGRR). That is the closer AGRR to OGRR, the less is the redundancy in gaps.

We will start our simulation results with sparse environments which would have obstacles that occupy a small area of the environment. Here is an example of a uniform-size obstacles distribution. The environment has four small obstacles. Each 2 adjacent obstacles share a common gap.

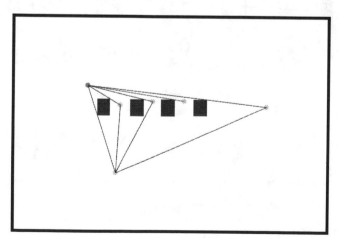

**Fig. 2.** Sparse environment with four obstacles, complete GNT

As shown in Fig. 2, the robot was able to identify each unique gap. Redundant gaps are well dealt with. The red node is the original robot's position and the root of the GNT (Rs). The light orange nodes represent the other gaps in the GNT. In this example, all of the gaps are the children of the root, which are sufficient to cover the whole environment. There were no further gaps to be added from any of its children.

**Table 1.** Observations on Fig. 2

| Performance Criterion | Measurement |
| --- | --- |
| Number of obstacles in the environment | 4 |
| MNG | 16 |
| UNG | 5 |
| ANG | 6 |
| OGRR | 68.7% |
| AGRR | 62.5% |

Table 1 summarizes the observations on Fig. 2. There were four obstacles in the environment. The final GNT consists of 6 nodes. In this example AGRR was 62.5%, which signifies a very good reduction of redundant gaps; it is close to the OGRR as well (68.7%).

Now we show how a robot explores a cluttered environment, in which the obstacles cover a large area of the environment. This example has 20 obstacles of uniform-size and uniform distribution. These obstacles cover a large area of the environment and are aligned as a grid. Rs is represented in red color.

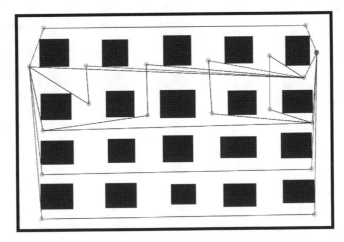

**Fig. 3.** Cluttered environment with 20 obstacles, complete GNT.

Fig. 3 depicts the resulting GNT. The light orange vertices correspond to the visited gaps. The four gaps in the first row of obstacles cannot cover the whole area and therefore another set of gaps has been added among the second row of obstacles.

Although there are many obstacles, the final GNT was small because obstacles share significant part of the free-space, which is inversely proportional to the number of the required gaps in GNT.

**Table 2.** Observations on Fig. 3

| Performance Criterion | Measurement |
| --- | --- |
| Number of obstacles in the environment | 20 |
| MNG | 80 |
| UNG | 11 |
| ANG | 19 |
| OGRR | 86.3% |
| AGRR | 76.3% |

Table 2 summarizes the observations on Fig. 3. There are 20 obstacles in the environment with 80 unique vertices but only 19 gaps were added to the GNT. The final GNT was sufficient for this environment. Although the redundancy reduction rate (76.3%) is high, GNT provides a full coverage of the environment. The measure AGRR is not far from OGRR.

We also tested the performance in environments with variable-size obstacles. The obstacles are few; the free-space is sparse. The next example has only two obstacles.

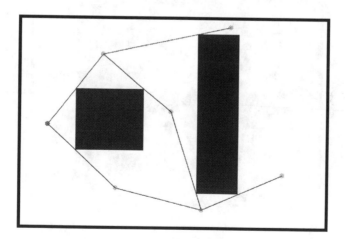

**Fig. 4.** Sparse environment with four obstacles, complete GNT

The performance of the proposed algorithm is expected as when the number of obstacles decreases, the overlapping gaps are few and therefore GNT would yield a higher number of gaps. In practice, such environments are scarce.

**Table 3.** Observations on Fig. 4

| Performance Criterion | Measurement |
| --- | --- |
| Number of obstacles in the environment | 2 |
| MNG | 8 |
| UNG | 3 |
| ANG | 7 |
| OGRR | 62.5% |
| AGRR | 12.5% |

Obstacle sizes and distribution in any environment are key parameters to construct the GNT. The algorithm computed a close number of gaps to the MNG. However, the actual number of gaps that are sufficient for this environment (3 gaps, see Table 3). This is the result of a global-based computation of the environment.

The following examples have environments with different obstacles' sizes that consume a large area of the environments. Some of these obstacles hide part of the environment (such part is not visible/shared with other obstacles) which increases ANG. Of course, it would decrease the AGRR as well.

Fig. 5 represent an example of 5 obstacles that cover a large area of the environment and they differ in their sizes and distribution. One of these obstacles is a polygonal shape of 20 vertices.

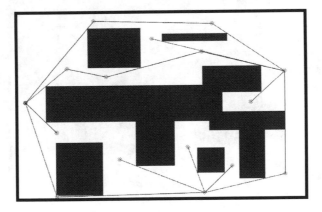

**Fig. 5.** Polygonal-shape obstacle in a cluttered environment

The polygonal-shape obstacle covers most of the middle area of the environment. The GNT structure grew surrounding this obstacle in order to cover the whole environment. The gap reduction rate of 55.5% is an encouraging result (Table 4).

**Table 4.** Observations on Fig. 5

| Performance Criterion | Measurement |
|---|---|
| Number of obstacles in the environment | 5 |
| MNG | 36 |
| UNG | 11 |
| ANG | 16 |
| OGRR | 69.4% |
| AGRR | 55.5% |

The proposed algorithm can handle the popular narrow passages problem [98] as well. It is a challenging environment that has been frequently reported in the literature. Fig. 6 shows an example of such environment with 5 obstacles that cover most of the workspace. There are two narrow passes in this environment.

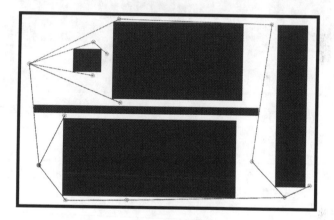

**Fig. 6.** The narrow-passage problem

The resulting GNT completely covers the free space. It is noticeable that the larger the obstacles get, the less gaps they share with other obstacles which leads to a less redundancy rate.

**Table 5.** Observations on Fig. 6

| Performance Criterion | Measurement |
|---|---|
| Number of obstacles in the environment | 5 |
| MNG | 20 |
| UNG | 13 |
| ANG | 14 |
| OGRR | 35% |
| AGRR | 30% |

Fig. 7 combines a narrow-passage problem and a trap-type obstacle. The resulting GNT covers the whole environment's free-space eliminating the effect of such troublesome obstacles. There are eight obstacles. The robot successfully explored the environment and built the GNT. The performance of the algorithm is satisfactory as summarized in Table 6.

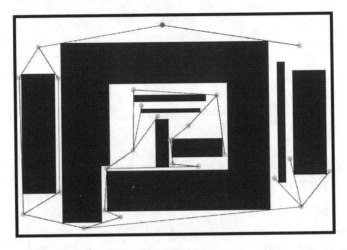

**Fig. 7.** Trap-Shape obstacles in an "Indoor Environment"

**Table 6.** Observations on Fig. 7

| Performance Criterion | Measurement |
|---|---|
| Number of obstacles in the environment | 8 |
| MNG | 40 |
| UNG | 16 |
| ANG | 26 |
| OGRR | 60% |
| AGRR | 35% |

In summary, the vast set of simulations shed light on the performance of the proposed algorithm to construct a roadmap to cover the free-space. The algorithm is complete as it handles all types of environments including the challenging ones such as the narrow-passage or indoor environments. The resulting GNT is a function of Rs, obstacle size and placement. Although, the algorithm reduces redundant gaps, it does not eliminate all redundancy. This is very much attributed to the limited sensory data we use.

The following bar chart (Fig. 8) describes the performance of the proposed algorithm with respect to gap redundancy reduction rate. The x-axis represents 13 simulations where Fig.s 2, 3, 4, 5, 6, 7 are represented as examples 2, 3, 6, 8, 9 and 11respectively while the y-axis demonstrates the redundancy rate. This chart is a comparison between OGRR and AGRR.

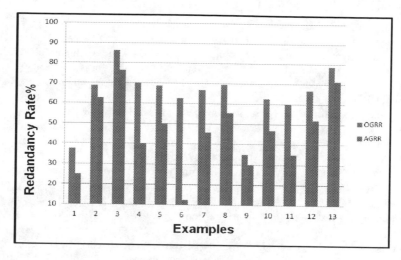

**Fig. 8.** Bar-chart of the gap redundancy reduction rate

Finding the minimum (unique) number of gaps is considered NP-hard problem. However, the performance of the algorithm (which uses local sensory data) is very much satisfactory when compared to the optimal one (NP-hard and based on global data). It should be noted that Example 6 (Fig. 4) is a worst-case scenario where we had two obstacles; a scarce case.

## 5    Conclusions

In this paper, we have proposed using GNT while enabling the robot to count and record the steps it takes (as building its own grid) in order to enable GNT to work in multiply, connected environments. This solution would require using more RAM storage. However, the GNT is a versatile data structure that can serve a variety of applications.

The proposed solution is a cost-effective one which makes it practical to produce it in bigger quantities for specific application that require multiple robots such as search and rescue. It also, relaxed one of the major assumptions made to solve the robot motion planning using GNT, which is obstacles are uniquely identified. Now, the robot is able to explore unknown multiply connected environments on its own with one very affordable sensor. Our solution eliminates the need for a landmark based sensor such as camera and it provides a cost effective prototype of a robot.

## References

1. Tovar, B., Murrieta, R., LaValle, S.M.: Distance-Optimal Navigation in an Unknown Environment without Sensing Distances. IEEE Transactions on Robotics 23(3), 506–518 (2007)

2. Canny, J., Reif, J.: New lower bound techniques for robot motion planning problems. In: Proceedings IEEE Symposium on Foundations of Computer Science, pp. 49–60 (1987)
3. Thrun, S., Burgard, W., Fox, D.: Probabilistic Robotics. MIT Press, Cambridge (2005)
4. Murphy, L., Newman, P.: Using incomplete online metric maps for topological exploration with the gap navigation tree. In: Proceedings IEEE International Conference on Robotics and Automation (2008)
5. Craig, J.: Introduction to robotics: mechanics and control, 3rd edn (2004)
6. Tovar, B., LaValle, S.M., Murrieta, R.: Locally-optimal navigation in multiply-connected environments without geometric maps. In: IEEE/RSJ International Conference on Intelligent Robots and Systems, vol. 4, pp. 3491–3497 (2003)
7. Erdmann, M.: Understanding action and sensing by designing action- based sensors. Int. J. Robot. Res. 14(5), 483–509 (1995)
8. Tovar, B., Guilamo, L., LaValle, S.M.: Gap navigation trees: Minimal representation for visibility-based tasks. In: Proc. Workshop on the Algorithmic Foundations of Robotics (2004)
9. LaValle, S.M., Hinrichsen, J.: Visibility-based pursuit-evasion: The case of curved environments. IEEE Transactions on Robotics and Automation 17(2), 196–201 (2001)
10. Zhao, L., Li, R., Zang, T., Sun, L.-N., Fan, X.: A method of landmark visual tracking for mobile robot. In: Xiong, C.-H., Liu, H., Huang, Y., Xiong, Y.L. (eds.) ICIRA 2008, Part I. LNCS (LNAI), vol. 5314, pp. 901–910. Springer, Heidelberg (2008)
11. Se, S., Lowe, D., Little, J.: Mobile robot localization and mapping with uncertainty using scale-invariant visual landmarks. The International Journal of Robotics Research 21(8), 735–758 (2002)
12. Yoon, K.-J., Kweon, I.: Landmark design and real-time landmark tracking for mobile robot localization. In: Electrical Engineering, vol. 4573, pp. 219–226 (2002)
13. Bais, A., Sablatnig, R.: Landmark based global self-localization of mobile soccer robots. In: Computer Vision–ACCV, pp. 842–851 (2006)
14. Elnagar, A., Lulu, L.: An art gallery-based approach to autonomous robot motion planning in global environments. In: IEEE/RSJ International Conference on Intelligent Robots and Systems, pp. 2079–2084 (2005)
15. Tovar, B., LaValle, S., Murrieta, R.: Optimal navigation and object finding without geometric maps or localization. In: IEEE Int. Conf. Robot and Automation (2003)
16. Thrun, S., Burgard, W., Fox, D.: Probabilistic mapping of an environment by a mobile robot. In: IEEE Int. Conf. Robot and Automation (1998)
17. Batalin, M.A., Sukhatme, G.S.: Coverage, Exploration and Deployment by a Mobile Robot and Communication Network, pp. 376–391 (2003)

# Navigation System Development
# of the Underwater Vehicles
# Using the GPS/INS Sensor Fusion

Won-Suck Choi[1], Nhat-Minh Hoang[1], Jae-Hoon Jung[1], and Jang-Myung Lee[2]

[1] Department of Electrical and Computer Engineering,
Pusan National University, South Korea
{wonsuck1696,nhatminh1696,jaehoon1696}@pusan.ac.kr
[2] Department of Electronic Engineering, Pusan National University, South Korea
jmlee@pusan.ac.kr

**Abstract.** Sensor fusion of GPS/INS using the Kalman filter design is proposed in this paper. GPS/INS data is utilized for estimating the position of AUV (Autonomous Underwater Vehicle) and Kalman filter simplifies the position estimation. The received GPS signals are stable most of the time, because they determine the position vector of the receiver which can receive microwaves transmitted from satellites (more in practice) with 24 hours' orbitation in the GPS real-time Otherwise, A low data rate and the impact of the disturbance. These disadvantages the INS data (gyroscope sensor, accelerometer, magnetic compass) compensatation for the inaccuracy of the GPS data. The noise in the acceleration data from INS data is reduced by Kalman filter in this paper. So, the KF localization system applied to the surface of water is proposed in this paper and the system performance is confirmed by experiments.

**Keywords:** Kalman Filter, GPS/INS, Sensor Fusion, IMU.

## 1 Introduction

Ocean is occupied for 70% of the Earth surface, 97% of the water on earth is present as seawater. also, Recently, as the importance of marine resources's development has been increased, the need for ocean exploration activities which is essential to the development of marine resources has been also increased but still exists as the unknown world that didn't explore. In order to explore the ocean, the development of underwater location estimation technology is needed, in addition to It is carried out construction of underwater structures, underwater research observation, underwater work, water industry, such as military applications, in a variety of fields. To perform these roles, location estimation techniques by fusing various sensors, and determine the position by recognizing the external environment is essential.

In the field of underwater location estimation, GPS and INS are often used to recognize the location of underwater vehicle. INS consists of the accelerometer sensor, the Gyroscope sensor, the magnetic compass and so forth to estimate the absolute location and the relative location at the same time. It provides the precise

X. Zhang et al. (Eds.): ICIRA 2014, Part I, LNAI 8917, pp. 491–497, 2014.
© Springer International Publishing Switzerland 2014

location information in a short time, but there occur accumulated errors when used for long time because of the errors that come from the characteristics of the sensor itself and the disturbance of the external environment. GPS error is large in a short period of time. In the specific environment that isn't able to receive the GPS signals, those are cut off. However, it provides the long-time stable and absolute location information. It helps to correct the location by receiving the signals continuously from the satellite by real time and brings the result of not having a cumulative error. These two navigations complement each other's characteristics and make possible to estimate the precise location from outdoors.

This paper proposes precise localization and control of underwater vehicle that is preferentially required for underwater localization using INS sensor and GPS. I will introduce outdoor precise localization estimation system of the AUVapplying KF for INS/GPS sensor fusion in this paper.

## 2     Summary of Kalman Filter

Many researchers study to solve estimation problem of states variables about the dynamic system. The method based on theory of probability is configured probability space consisting of state variables. It is based on the estimate about state variable using System's dynamic characteristics and the measurement value. Typically based on Bayesian estimation technique, Kalman Filter (KF), a particle filter and so forth have been actively studied in the field of localization

$$\hat{x}_{\bar{k}} = A\hat{x}_{k-1} + Bu_k$$
$$P_{\bar{k}} = AP_{k-1} + A^T + Q$$
(1)

In the System equation (1), $\hat{x}_{\bar{k}}$ is the state variable that want to optimize as the Kalman Filter, $A$ is the transform coefficients that connected between the previous step and the next step, $B$ and $u_k$ are the additional input value regardless of the system. $P$ is the error covariance value, so can be obtained by the system error $Q$ and covariance value in the previous step. In the Kalman Filter, $Q$ is the most important element with the $R$ of the Observation equation.

$$K_k = P_{\bar{k}}H^T (HP_{\bar{k}}H^T + R)^{-1}$$
$$\hat{x}_k = \hat{x}_{\bar{k}} + K_k (z_k - H\hat{x}_{\bar{k}})$$
$$P_k = (I - K_k H)P_{\bar{k}}$$
(2)

In the Observation equation (2), Kalman gain $K_k$ is obtained by the error covariance $P$ and the Observation error $R$. Through the obtained Kalman gain previously, state variable values can be determined by predicted state variable values. This process not end instead of a single calculation and $P$ values of the current state are readjusted by the Kalman gain, affect to the next step.

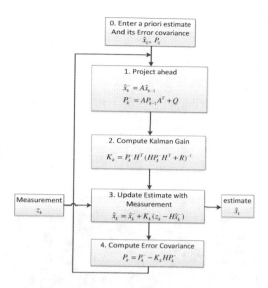

**Fig. 1.** Algorithm of Kalman Filter

## 3    GPS/INS Sensor fusion

The INS used in this experiment is composed of the 3-axis Accelerometer Sensor that measure coordinates and rotational angle of the underwater vehicle, 3-axis Gyroscope Sensor that measure rotational angle by the angular rate, and Magnetic Compass that measure traveling direction of the underwater vehicle.

In order to fuse acceleration value from the accelerometer sensor and angular rate from the gyroscope sensor, attitude $\hat{q}$ that expressed by quaternion is obtained by angular rate. Because x-axis of the acceleration sensors is same as progression direction, the position of the underwater vehicle is gotten as perform double integral for acceleration value $\hat{a}_x$. Using Roll-axis and Pitch-axis value from the gyroscope sensor and $M_x, M_y, M_z$ from the Magnetic Compass, Yaw is obtained.

$$Yaw = \arctan(Yh / Xh)$$
$$Xh = M_x \cos\phi + M_y \sin\theta - M_z \cos\theta \sin\phi \qquad (3)$$
$$Yh = M_y \cos\theta + M_z \sin\theta$$

Origin and destination of the latitude, longitude that obtained from the GPS sensor is converted to distance value. Distance value from the GPS sensor is used to Measurement $z_k$ of the Kalman Filter algorithm, so compensate for the cumulative error of the INS data. Through this process, we can obtain result that closed to the actual traveled distance.

# 4    Experiment and Result

## 4.1    Experiment Environment

GPS and INS(Gyroscope Sensor, Accelerometer Sensor, Magnetic Compass) was modularity for the localization of underwater vehicle. It is made as one-board to be used on ARM3s8962.

In this experiment, The INS sensor was used EBIMU-9DOF, and The GPS sensor was used UIGGUB01R004. In Fig 2, the INS sensor and ARM was used in the experimental underwater vehicle, which was positioned in front of.

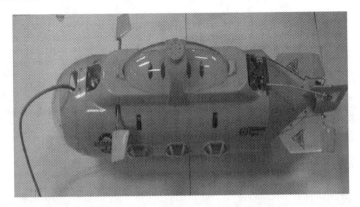

**Fig. 2.** Experimental underwater vehicle

## 4.2    Experiment Result

For Performance evaluation of this system, Standard moving information of AUV is obtained from INS data applying KF. GPS can receive AUV's Position information. And, Position error was compensated through using GPS.

The KF combined Acceleration signals, gyro signals and magnetic compass signals with NMEA(The National Electronic Association) GPRMC code of GPS. By combining, We estimated driving-trajectory of AUV in Fig. 3.

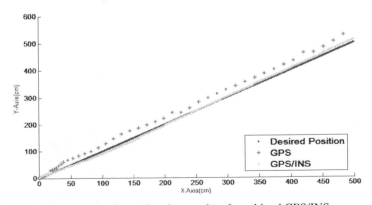

**Fig. 3.** Location estimation results of combined GPS/INS

In this picture, the blue line is the actual moving trajectory, the red line is the GPS data according to location of the AUV and the green line is the graph applying KF. The error of estimated trajectory increased over time for the reason that the increasing error of yaw axis as the input of KF and the influence of the disturbance (sea breeze, wave etc.).

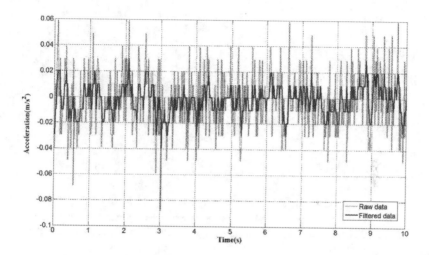

**Fig. 4.** Noise remove using Kalman filter

**Table 1.** Noise removing using Kalman filter

| Data Type | The Average error of Accelation |
|---|---|
| Raw data | $0.0178 \, m/s^2$ |
| Filtered data | $0.0085 \, m/s^2$ |
| Error Improvement[%] | 52.12% |

In this picture, the blue line is the Filtered data, the red line is the raw data according to chart can see Error Improvement up to 52.12% and Noise is removed using Kalman Filter.

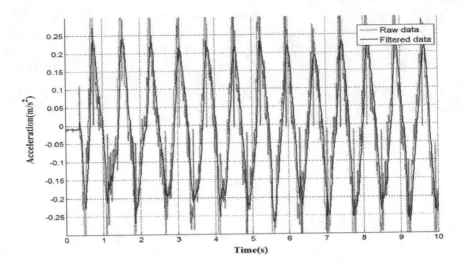

**Fig. 5.** IMU noise and Error removal test using Kalman filter

**Table 2.** IMU noise and Error removal test using Kalman filter

| Data Type | The Average error of Accelation |
|---|---|
| Raw data | $0.0428 \, m \, / \, s^2$ |
| Filtered data | $0.0175 \, m \, / \, s^2$ |
| Error Improvement[%] | 59.12% |

In this picture, the blue line is the Filtered data, the red line is the raw data according to chart can see Error Improvement up to 59.12% and removed noise of IMU using Kalman Filter.

## 5    Conclusion

The combination systems of GPS/INS compensate for noises of GPS data and errors of INS data in an outdoor. So, it generally uses in outdoor precision localization. However, there is a difficulty that it can't effectively remove natural noise used in INS. And, the motion of ship's body is assumed to be a linear. In addition, the noises of GPS/INS are assumed to be regular Distribution noise. Thus, Owing to System Configuration, it is many difficult in outdoor precision localization. In addition, because of the influence of the disturbance (sea breeze, wave etc.) underwater vehicle was affected when it moves along the trace.

In the future, We have plan to use Extended Kalman Filter(EKF) or Unscented Kalman Filter(UKF) suitable for de-noising irregular distribution noise, and more precise location estimation expected to obtained.

**Acknowledgments.** "This research was supported by the MOTIE (Ministry of Trade, Industry & Energy), Korea, under the Industry Convergence Liaison Robotics Creative Graduates Education Program supervised by the KIAT (N0001126)."

"This research was supported by Basic Science Research Program through the National Research Foundation of Korea(NRF) funded by the Ministry of Education( NRF-2010-0024129 )".

# References

1. Kim, K.J., Park, C.G., Yu, M.J., Park, Y.B.: A performance comparison of extended and unscented Kalman filters for INS/GPS tightly coupled approach. Journal of Control, Automation, and Systems Engineering 12(8) (August 2007)
2. Eom, H.S., Kim, J.Y., Baek, J.Y., Lee, M.C.: Reduction of Relative Position Error for DGPS Based Localization of AUV using LSM and Kalman Filter. Journal of the Korean Society for Precision Engineering 27(10), 52–60
3. Schmidt, G.T.: INS/GPS technology trends, NATO Research and Technology Organization, pp. 1–16 (May 2009)
4. Lee, J.H., Kim, H.S.: A Study of High Precision Position Estimator Using GPS/INS Sensor Fusion. Journal of The Institute of Electronics Engineers of Korea 49(11) (November 2012)
5. Hwang, S.Y., Lee, J.M.: Estimation of attitude and position of moving objects using multi-filtered inertial navigation system. The Institute of Electronics Engineers of Korea 60(12), 2383–2396 (2011) (in Korean)
6. Kim, T.G., Choi, H.T., Lee, Y.J., Ko, N.Y.: Localization for pose of an Underwater Robot Using EKF Method. In: Summer scholarship Conference of The institute of Electronics and information Engineers (July 2013)
7. Rigaud, V., March, L., Michel, J.L., Borot, P.: Sensor fusion for AUV localization, The Institute of Electrical and Electronics Engineers (Jun 1990)

# Component-Based System Integration
# Using Proper Augmented Marked Graphs

K.S. Cheung

The Open University of Hong Kong
Good Shepherd Street, Homantin, Kowloon, Hong Kong
kscheung@ouhk.edu.hk

**Abstract.** A key challenge in component-based system integration is to ensure the correctness of the integrated system in terms of liveness and boundedness, especially as there involve distributed components competing for some shared resources. This paper proposes a formal integration method for distributed component-based systems using proper augmented marked graphs. A subclass of Petri nets, proper augmented marked graphs possess a special structure for modelling systems with shared resources and desirable properties pertaining to liveness and boundedness. By composing a set of proper augmented marked graphs via common resource places, liveness and boundedness can be preserved under simple conditions. In this paper, based on the theory of proper augmented marked graphs, the modelling and integration of distributed component-based systems with shared resources are elaborated and illustrated.

**Keywords:** component-based system, system integration, Petri net, augmented marked graph, distributed system, shared-resource system.

## 1    Introduction

Component-based system design emphasizes the compositional synthesis of a system from components, where a system is structurally considered as the integrated whole of a set of interacting components [1, 2]. The system exhibits a collection of behavioural patterns, each delineating a scenario in which the components interact with each other for some functional purposes. In component-based system design, the components are first identified and defined. An integrated system is then obtained by integrating these components.

In component-based system integration, a key challenge is to ensure the integrated system is correct in the sense that it is free from erroneous situations such as deadlock and capacity overflow. For a system involving distributed components which compete for some shared resources, a deadlock would occur when two or more components are each holding some resources and waiting for another component to release some other resources. Capacity overflow is another erroneous situation, where the capacity of a component exceeds its capacity limit. It is therefore necessary to ensure the integrated system is live (implying freeness from deadlock) and bounded (implying freeness from capacity overflow.

X. Zhang et al. (Eds.): ICIRA 2014, Part I, LNAI 8917, pp. 498–509, 2014.

System integration is an essential step in component-based system design. For this integration, liveness and boundedness of the integrated system is not guaranteed even though every component is live and bounded, meaning that these properties of the components may not be preserved after the integration.

A subclass of Petri nets, proper augmented marked graphs are basically a special type of augmented marked graphs [3, 4, 5]. Like augmented marked graphs, proper augmented marked graphs possess a special structure, useful for representing a system involving shared resources. Not only inheriting the properties of augmented marked graphs, proper augmented marked graphs possess a number of desirable properties pertaining to liveness and boundedness [6]. Typically in a proper augmented marked graph, a specific subset of places (called resource places or common resource places) is defined to denote common resources. When composing a set of proper augmented marked graphs via these common resource places, liveness and boundedness as well as other properties such as reversibility and conservativeness can be preserved under some simple conditions.

In this paper, based on proper augmented marked graphs, an integration method for distributed component-based systems with shared resources is proposed. It begins with modelling the individual components as proper augmented marked graphs. An integrated system is then obtained by composing these proper augmented marked graphs via their common resource places. Liveness and boundedness of the integrated system can be effectively derived through the property-preserving composition of proper augmented marked graphs. The rest of this paper is structured as follows. Section 2 is the preliminaries for those who do not have prior knowledge of Petri nets. Section 3 describes proper augmented marked graphs, with a focus on the property-preserving composition. Section 4 presents the integration method for distributed component-based systems. The dining philosopher problem is used for illustration. Section 5 briefly concludes this paper.

# 2    Preliminaries

A Petri net or place-transition net (PT-net) is a bipartite directed graph consisting of two sorts of nodes called places and transitions, such that no arcs connect two nodes of the same sort [7, 8, 9]. In graphical notation, a place is represented by a circle, a transition by a rectangular box, and an arc by a directed line.

**Definition 2.1.** A PT-net is a 4-tuple $N = \langle P, T, F, W \rangle$, where $P$ is a set of places, $T$ is a set of transitions, $F \subseteq (P \times T) \cup (T \times P)$ is a flow relation that represents the arcs, and $W : F \rightarrow \{ 1, 2, ... \}$ is a weight function which assigns a weight to every arc.

**Definition 2.2.** A PT-net $N = \langle P, T, F, W \rangle$ is said to be pure or self-loop free if and only if $\forall x, y \in (P \cup T) : (x, y) \in F \Rightarrow (y, x) \notin F$, and ordinary if and only if the range of $W$ is $\{ 1 \}$. An ordinary PT-net can be written as $\langle P, T, F \rangle$.

**Definition 2.3.** A marked graph is an ordinary PT-net $N = \langle P, T, F \rangle$, such that $\forall p \in P : | {}^{\bullet}p | = | p^{\bullet} | = 1$.

**Definition 2.4.** For a PT-net $N = \langle P, T, F \rangle$, a path is a sequence $\langle x_1, x_2, ..., x_n \rangle$, where $(x_i, x_{i+1}) \in F$ for $i = 1, 2, ..., n-1$. A path is said to be elementary if and only if it does not contain the same place or transition more than once.

**Definition 2.5.** For a PT-net $N = \langle P, T, F \rangle$, a cycle is a set of places $\{ p_1, p_2, ..., p_n \}$, where there exist $t_1, t_2, ..., t_n \in T$ such that $\langle p_1, t_1, p_2, t_2, ..., p_n, t_n \rangle$ forms an elementary path and $(t_n, p_1) \in F$.

**Definition 2.6.** Let $N = \langle P, T, F \rangle$ be a PT-net. For $x \in (P \cup T)$, $^\bullet x = \{ y \mid (y, x) \in F \}$ and $x^\bullet = \{ y \mid (x, y) \in F \}$ are called the pre-set and post-set of x, respectively.

**Definition 2.7.** Let $N = \langle P, T, F \rangle$ be a PT-net, where $P = \{ p_1, p_2, ..., p_n \}$. N is marked if tokens are assigned to its places (graphically denoted by dots). A marking is defined as a function $M : P \rightarrow \{ 0, 1, 2, ... \}$, where $M(p)$ represents the number of tokens assigned to a place $p \in P$. $(N, M_0)$ represents N with an initial marking $M_0$.

**Definition 2.8.** For a PT-net $N = \langle P, T, F, W \rangle$, a transition $t \in T$ is said to be firable at a marking M if and only if $\forall p \in {}^\bullet t : M(p) \geq W(p,t)$. On firing t, M is changed to M' such that $\forall p \in P : M'(p) = M(p) - W(p,t) + W(t,p)$, in notation, $M [t\rangle M'$.

**Definition 2.9.** For a PT-net $(N, M_0)$, a sequence of transitions $\sigma = \langle t_1, t_2, ..., t_n \rangle$ is said to be firable if and only if there exists markings $M_1, M_2, ..., M_n$ such that $M_{i-1} [t_i\rangle M_i$, for $i = 1, 2, ..., n$. On firing $\sigma$, $M_0$ is changed to $M_n$, in notation, $M_0 [\sigma\rangle M_n$.

**Definition 2.10.** For a PT-net $(N, M_0)$, a marking M is said to be reachable, in notation $M_0 [*\rangle M$, if and only if there exists a firable sequence $\sigma$ such that $M_0 [\sigma\rangle M$. $[N, M_0\rangle$ represents the set of all reachable markings of $(N, M_0)$.

**Definition 2.11.** For a PT-net $(N, M_0)$, a set of places S is called a siphon if and only if $^\bullet S \subseteq S^\bullet$. S is said to be minimal if and only if there does not exist another siphon S' in N such that $S' \subset S$.

**Definition 2.12.** For a PT-net $(N, M_0)$, a set of places S is called a trap if and only if $S^\bullet \subseteq {}^\bullet S$. S is said to be a marked trap if and only if $\forall p \in S : M_0(p) > 0$.

**Definition 2.13.** For a PT-net $(N, M_0)$, a transition t is said to be live if and only if $\forall M \in [N, M_0\rangle, \exists M' : M [*\rangle M'$ and t is enabled at M'. $(N, M_0)$ is said to be live if and only if every transition of $(N, M_0)$ is live.

**Definition 2.14.** Let $(N, M_0)$ be a PT-net, where $N = \langle P, T, F \rangle$. $(N, M_0)$ is said to be deadlock-free if and only if $\forall M \in [N, M_0\rangle, \exists t \in T: t$ is enabled at M.

**Definition 2.15.** For a PT-net $(N, M_0)$, a place p is said to be bounded if and only if there exists $k > 0$, such that $\forall M \in [N, M_0\rangle : M(p) \leq k$. $(N, M_0)$ is said to be bounded if and only if every place is bounded.

# 3    Proper Augmented Marked Graphs

Augmented marked graphs were first introduced by Chu for modelling systems with shared resources, and their properties are reported in the literature [3, 10, 11, 12, 13]. Augmented marked graphs possess a special structure for modelling of systems with shared resources, such as manufacturing systems [14, 15]. Proper augmented marked graphs is a special type of augmented marked graphs, recently found by Cheung [6]. This Section summarizes the definition and properties of augmented marked graphs and proper augmented marked graphs. The property-preserving composition of proper augmented marked graphs is discussed.

## 3.1    Augmented Marked Graphs and Proper Augmented Marked Graphs

**Definition 3.1.** An augmented marked graph $(N, M_0; R)$ is a PT-net $(N, M_0)$ with a specific subset of places R satisfying the following conditions : (a) Every place in R is marked by $M_0$. (b) The PT-net $(N', M_0')$ obtained from $(N, M_0; R)$ by removing the places in R and their associated arcs is a marked graph. (c) For each $r \in R$, there exist $k_r > 1$ pairs of transitions $D_r = \{ \langle t_{s1}, t_{h1} \rangle, \langle t_{s2}, t_{h2} \rangle, ..., \langle t_{skr}, t_{hkr} \rangle \}$ such that $r^\bullet = \{ t_{s1}, t_{s2}, ..., t_{skr} \} \subseteq T$, $^\bullet r = \{ t_{h1}, t_{h2}, ..., t_{hkr} \} \subseteq T$ and that, for each $\langle t_{si}, t_{hi} \rangle \in D_r$, there exists in N' an elementary path $\rho_{ri}$ connecting $t_{si}$ to $t_{hi}$. (d) In $(N', M_0')$, every cycle is marked and no $\rho_{ri}$ is marked.

**Definition 3.2.** For an augmented marked graph $(N, M_0; R)$, a minimal siphon is called a R-siphon if and only if it contains at least one place in R.

**Definition 3.3.** Let $(N, M_0; R)$ be an augmented marked graph to be transformed into $(N', M_0')$ as follows. For each place $r \in R$, where $D_r = \{ \langle t_{s1}, t_{h1} \rangle, \langle t_{s2}, t_{h2} \rangle, ..., \langle t_{skr}, t_{hkr} \rangle \}$, r is replaced by a set of places $Q = \{ q_1, q_2, ..., q_{kr} \}$, such that $M_0'[p_i] = M_0[r]$ and $q_i^\bullet = \{ t_{si} \}$ and $^\bullet q_i = \{ t_{hi} \}$. $(N', M_0')$ is called the R-transform of $(N, M_0; R)$.

**Definition 3.4.** Let $(N, M_0; R)$ be an augmented marked graph, and $(N', M_0')$ be the R-transform of $(N, M_0; R)$. $(N, M_0; R)$ is a proper augmented marked graph if and only if the following two conditions are satisfied : (a) For each place $r \in R$, $| ^\bullet r | = | r^\bullet | \geq 1$. (b) Every place in $(N', M_0')$ belongs to a cycle.

**Property 3.1.** A proper augmented marked graph $(N, M_0; R)$ is live if every R-siphon contains a trap marked by $M_0$ [6].

**Property 3.2.** A proper augmented marked graph is bounded [6].

Figure 1 shows a proper augmented marked graph $(N, M_0; R)$. For $(N, M_0; R)$, every R-siphon contains a trap marked by $M_0$. $(N, M_0; R)$ is live and bounded.

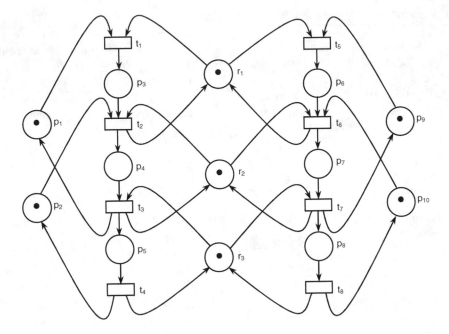

**Fig. 1.** A proper augmented marked graph $(N, M_0; R)$ which is live and bounded

### 3.2    Property-Preserving Composition of Proper Augmented Marked Graphs

Consider two proper augmented marked graphs $(N_1, M_{10}; R_1)$ and $(N_2, M_{20}; R_2)$. Suppose $r_{11} \in R_1$ and $r_{21} \in R_2$ refer to the same common resource. Then, $r_{11}$ and $r_{21}$ are called common resource places in $(N_1, M_{10}; R_1)$ and $(N_2, M_{20}; R_2)$. Suppose there also exist two other common resource places $r_{12} \in R_1$ and $r_{22} \in R_2$. By composing $(N_1, M_{10}; R_1)$ and $(N_2, M_{20}; R_2)$ via these common resource places, in the sense that $r_{11}$ and $r_{21}$ are fused as $r_1$, and $r_{12}$ and $r_{22}$ as $r_2$, the PT-net so obtained is also a proper augmented marked graph, as formally stated below.

**Property 3.3.** Let $(N_1, M_{10}; R_1)$ and $(N_2, M_{20}; R_2)$ be two proper augmented marked graphs. $R_1' = \{ r_{11}, r_{12}, ..., r_{1k} \} \subseteq R_1$ and $R_2' = \{ r_{21}, r_{22}, ..., r_{2k} \} \subseteq R_2$ are the common resource places, where $M_{10}(R_1') = M_{20}(R_2')$. Suppose that $r_{11}$ and $r_{21}$ are to be fused as $r_1$, $r_{12}$ and $r_{22}$ as $r_2$, ..., $r_{1k}$ and $r_{2k}$ as $r_k$. The PT-net so obtained is a proper augmented marked graph $(N, M_0; R)$, where $R = (R_1 \setminus R_1') \cup (R_2 \setminus R_2') \cup \{ r_1, r_2, ..., r_k \}$. $(N, M_0; R)$ is called the integrated proper augmented marked graph [6].

**Definition 3.5.** Let $(N, M_0; R)$ be the integrated proper augmented marked graph obtained by composing two proper augmented marked graphs $(N_1, M_{10}; R_1)$ and $(N_2, M_{20}; R_2)$ via a set of common resource places $\{ (r_{11}, r_{21}), (r_{12}, r_{22}), ..., (r_{1k}, r_{2k}) \}$, where $r_{11} \in R_1$ and $r_{21} \in R_2$ are fused as $r_1$, $r_{12} \in R_1$ and $r_{22} \in R_2$ are fused as $r_2$, $r_{13} \in R_1$ and $r_{23} \in R_2$ are fused as $r_3$, and so on. $R_F = \{ r_1, r_2, ... r_k \} \subseteq R$ is called the set of fused resource places.

**Property 3.4.** Suppose $(N_1, M_{10}; R_1)$ and $(N_2, M_{20}; R_2)$ are two live proper augmented marked graphs, where every $R_1$-siphon in $(N_1, M_{10}; R_1)$ contains a trap marked by $M_{10}$ and every $R_2$-siphon in $(N_2, M_{20}; R_2)$ contains a trap marked by $M_{20}$. Let $(N, M_0; R)$ be the integrated proper augmented marked graph obtained by composing $(N_1, M_{10}; R_1)$ and $(N_2, M_{20}; R_2)$ via their common resource places, where $R_F \subseteq R$ is the set of fused places. $(N, M_0; R)$ is live if every place in $R_F$ contains a marked trap [6].

**Property 3.5.** The integrated proper augmented marked graph $(N, M_0; R)$ obtained by composing two proper augmented marked graphs $(N_1, M_{10}; R_1)$ and $(N_2, M_{20}; R_2)$ via their common resource places is bounded [6].

Figure 2 shows two proper augmented marked graphs $(N_1, M_{10}; R_1)$ and $(N_2, M_{20}; R_2)$. $R_1 = \{ r_{11}, r_{12} \}$ and $R_2 = \{ r_{21}, r_{22} \}$ are the common resource places, where $M_{10}(R_1) = M_{20}(R_2)$. Figure 3 shows the integrated proper augmented marked graph $(N, M_0; R)$ obtained by composing $(N_1, M_{10}; R_1)$ and $(N_2, M_{20}; R_2)$ via $\{ (r_{11}, r_{21}), (r_{12}, r_{22}) \}$, where $R_F = \{ r_1, r_2 \}$.

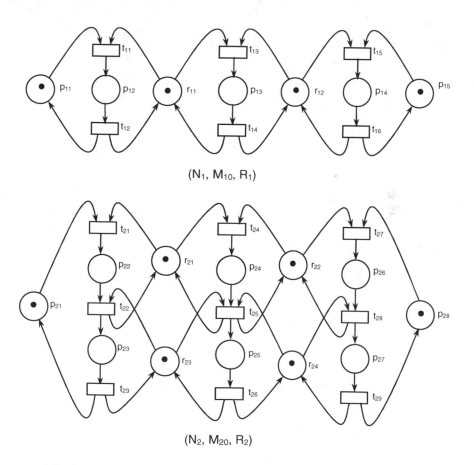

$(N_1, M_{10}, R_1)$

$(N_2, M_{20}, R_2)$

**Fig. 2.** Two proper augmented marked graphs $(N_1, M_{10}; R_1)$ and $(N_2, M_{20}; R_2)$

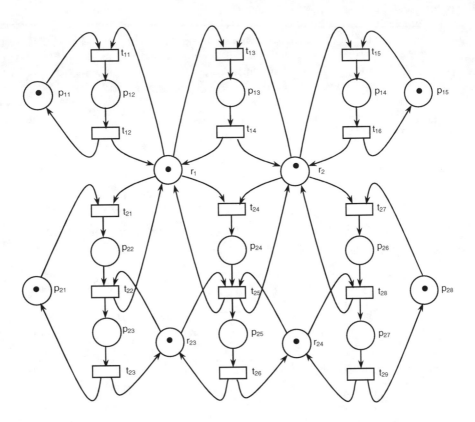

**Fig. 3.** The integrated proper augmented marked graph $(N, M_0; R)$ obtained by composing $(N_1, M_{10}; R_1)$ and $(N_2, M_{20}; R_2)$ via $\{ (r_{11}, r_{21}), (r_{12}, r_{22}) \}$

Both $(N_1, M_{10}; R_1)$ and $(N_2, M_{20}; R_2)$ are live, where every $R_1$-siphon in $(N_1, M_{10}; R_1)$ contains a trap marked by $M_{10}$ and every $R_2$-siphon in $(N_2, M_{20}; R_2)$ contains a trap marked by $M_{20}$. For $(N, M_0; R)$, every in $R_F$ contains a trap marked by $M_0$. Hence, $(N, M_0; R)$ is live. Besides, both $(N_1, M_{10}; R_1)$ and $(N_2, M_{20}; R_2)$ are bounded. $(N, M_0; R)$ is also bounded.

## 4    Component-Based System Integration

This section describes the modelling and integration of distributed component-based systems with shared resources, based on proper augmented marked graphs. We show the modelling of components as proper augmented marked graphs. Then, we show the integration by composing the proper augmented marked graphs via their common resource places. Liveness and boundedness of the integrated system is derived, based on the property-preserving composition of proper augmented marked graphs. The dining philosopher problem is used for illustration.

Consider the dining philosopher problem introduced by Dijkstra [16]. There are 4 meditating philosophers, namely, $H_1$, $H_2$, $H_3$ and $H_4$, sitting around a table for dinner. The foods are placed at the centre, and there are 4 pieces of chopsticks, namely, $C_1$, $C_2$, $C_3$ and $C_4$, shared by the philosophers. Figure 4 shows the set-up of the table. In order for a philosopher to get the foods, both the chopsticks at the left-hand side and right-hand side must be available. The philosopher grasps both chopsticks to take the foods. The chopsticks are then released.

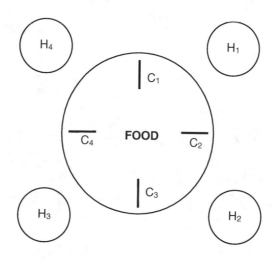

**Fig. 4.** The dining philosopher problem

Taking the component-based design approach, each philosopher is considered as a component. For $H_1$, the process for $H_1$ to grasp the chopsticks to get the foods is modelled as a proper augmented marked graph $(N_1, M_{10}; R_1)$, where $R_1 = \{r_1, r_2\}$. Likewise, the processes for $H_2$, $H_3$ and $H_4$ to grasp the chopsticks to get the foods are modelled as proper augmented marked graphs $(N_2, M_{20}; R_2)$, $(N_3, M_{30}; R_3)$ and $(N_4, M_{40}; R_4)$, where $R_2 = \{r_2, r_3\}$, $R_3 = \{r_3, r_4\}$ and $R_4 = \{r_1, r_4\}$. Figure 5 shows these proper augmented marked graphs.

For $(N_1, M_{10}; R_1)$, $(N_2, M_{20}; R_2)$, $(N_3, M_{30}; R_3)$ and $(N_4, M_{40}; R_4)$, $r_{11} \in R_1$ and $r_{41} \in R_4$ refer to the same resource $C_1$, Likewise, $r_{12} \in R_1$ and $r_{22} \in R_2$ refer to the same resource $C_2$, $r_{23} \in R_2$ and $r_{33} \in R_3$ refer to the same resource $C_3$, and $r_{34} \in R_3$ and $r_{44} \in R_4$ refer to the same resource $C_4$. Figure 6 shows the integrated proper augmented marked graph $(N, M_0; R)$ obtained by composing $(N_1, M_{10}; R_1)$, $(N_2, M_{20}; R_2)$, $(N_3, M_{30}; R_3)$ and $(N_4, M_{40}; R_4)$ via $\{ (r_{12}, r_{22}), (r_{23}, r_{33}), (r_{34}, r_{44}), (r_{11}, r_{41}) \}$.

$(N_1, M_{10}; R_1)$ is live, since every $R_1$-siphon in $(N_1, M_{10}; R_1)$ contains a trap marked by $M_{10}$. Likewise, $(N_2, M_{20}; R_2)$, $(N_3, M_{30}; R_3)$ and $(N_4, M_{40}; R_4)$ are live, where every $R_2$-siphon in $(N_2, M_{20}; R_2)$ contains a trap marked by $M_{20}$, every $R_3$-siphon in $(N_3, M_{30}; R_3)$ contains a trap marked by $M_{30}$, and every $R_4$-siphon in $(N_4, M_{40}; R_4)$ contains a trap marked by $M_{40}$. For $(N, M_0; R)$, every in $R_F$ contains a trap marked by $M_0$. Hence, $(N, M_0; R)$ is live. Besides, $(N_1, M_{10}; R_1)$, $(N_2, M_{20}; R_2)$, $(N_3, M_{30}; R_3)$ and $(N_4, M_{40}; R_4)$ are bounded. $(N, M_0; R)$ is also bounded.

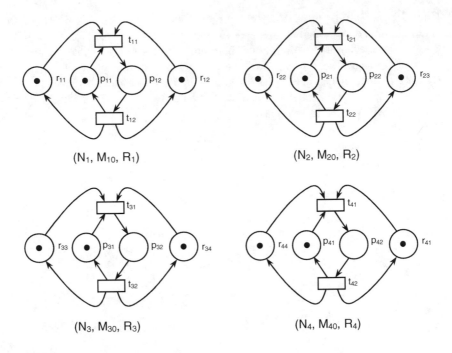

$(N_1, M_{10}, R_1)$    $(N_2, M_{20}, R_2)$

$(N_3, M_{30}, R_3)$    $(N_4, M_{40}, R_4)$

| Place / Transition | Semantic meaning when marked / fired |
|---|---|
| $p_{11}$ | $H_1$ is meditating. |
| $p_{12}$ | $H_1$ has $C_1$ and $C_2$ and takes the food. |
| $p_{21}$ | $H_2$ is meditating. |
| $p_{22}$ | $H_2$ has $C_2$ and $C_3$ and takes the food. |
| $p_{31}$ | $H_3$ is meditating. |
| $p_{32}$ | $H_3$ has $C_3$ and $C_4$ and takes the food. |
| $p_{41}$ | $H_4$ is meditating. |
| $p_{42}$ | $H_4$ has $C_4$ and $C_1$ and takes the food. |
| $r_{11}$ | $C_1$ is available. |
| $r_{12}$ | $C_2$ is available. |
| $r_{22}$ | $C_2$ is available. |
| $r_{23}$ | $C_3$ is available. |
| $r_{33}$ | $C_3$ is available. |
| $r_{34}$ | $C_4$ is available. |
| $r_{41}$ | $C_1$ is available. |
| $r_{44}$ | $C_4$ is available. |
| $t_{11}$ | $H_1$ takes the action to grasp $C_1$ and $C_2$. |
| $t_{12}$ | $H_1$ takes the action to return $C_1$ and $C_2$. |
| $t_{21}$ | $H_1$ takes the action to grasp $C_2$ and $C_3$. |
| $t_{22}$ | $H_1$ takes the action to return $C_2$ and $C_3$. |
| $t_{31}$ | $H_1$ takes the action to grasp $C_3$ and $C_4$. |
| $t_{32}$ | $H_1$ takes the action to return $C_3$ and $C_4$. |
| $t_{41}$ | $H_1$ takes the action to grasp $C_4$ and $C_1$. |
| $t_{42}$ | $H_1$ takes the action to return $C_4$ and $C_1$. |

**Fig. 5.** Components modelled as proper augmented marked graphs

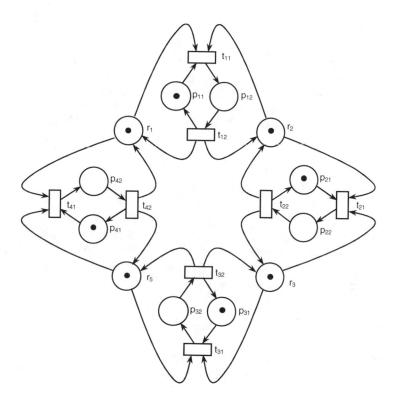

| Place / Transition | Semantic meaning when marked / fired |
|---|---|
| $p_{11}$ | $H_1$ is meditating. |
| $p_{12}$ | $H_1$ has $C_1$ and $C_2$ and takes the food. |
| $p_{21}$ | $H_2$ is meditating. |
| $p_{22}$ | $H_2$ has $C_2$ and $C_3$ and takes the food. |
| $p_{31}$ | $H_3$ is meditating. |
| $p_{32}$ | $H_3$ has $C_3$ and $C_4$ and takes the food. |
| $p_{41}$ | $H_4$ is meditating. |
| $p_{42}$ | $H_4$ has $C_4$ and $C_1$ and takes the food. |
| $r_1$ | $C_1$ is available. |
| $r_2$ | $C_2$ is available. |
| $r_3$ | $C_3$ is available. |
| $r_4$ | $C_4$ is available. |
| $t_{11}$ | $H_1$ takes the action to grasp $C_1$ and $C_2$. |
| $t_{12}$ | $H_1$ takes the action to return $C_1$ and $C_2$. |
| $t_{21}$ | $H_1$ takes the action to grasp $C_2$ and $C_3$. |
| $t_{22}$ | $H_1$ takes the action to return $C_2$ and $C_3$. |
| $t_{31}$ | $H_1$ takes the action to grasp $C_3$ and $C_4$. |
| $t_{32}$ | $H_1$ takes the action to return $C_3$ and $C_4$. |
| $t_{41}$ | $H_1$ takes the action to grasp $C_4$ and $C_1$. |
| $t_{42}$ | $H_1$ takes the action to return $C_4$ and $C_1$. |

**Fig. 6.** The integrated proper augmented marked graph for the integrated system

## 5    Conclusion

Proper augmented marked graphs possess a structure which is especially useful for modelling distributed component-based systems with shared resources. Inheriting the properties of augmented marked graphs, proper augmented marked graphs are live under a simple condition (on a small set of siphons called $R_F$-siphons). Besides, they are always bounded. Liveness and boundedness can be preserved after composition via the common resource places. These are useful for the modelling and integration of distributed components with shared resources.

This paper shows an integration method for distributed component-based systems with shared resources. The dining philosopher example is used for illustration. In designing distributed component-based systems with shared resources, it is necessary but difficult to ensure the correctness of the integrated system in the sense that the integrated system is live and bounded. Based on proper augmented marked graphs and the property-preserving composition, liveness and boundedness of the integrated systems are derived. This theoretically sound integration method can be effectively applied to manufacturing system engineering.

## References

[1] Haskins, C. (ed.): INCOSE Systems Engineering Handbook: A Guide for System Life Cycle Processes and Activities. International Council of System Engineering (2010)

[2] Leavens, G.T., Sitaraman, M. (eds.): Foundations of Component-Based Systems. Cambridge University Press (2000)

[3] Chu, F., Xie, X.: Deadlock Analysis of Petri Nets Using Siphons and Mathematical Programming. IEEE Transactions on robotics and Automation 13(5) (1997)

[4] Cheung, K.S.: Augmented Marked Graphs. Informatica 32(1), 85–94 (2008)

[5] Cheung, K.S.: Augmented Marked Graphs. Springer (2014)

[6] Cheung, K.S.: Proper Augmented Marked Graphs. In: Cheung, K.S. (ed.) Augmented Marked Graphs, pp. 55–66. Springer (2014)

[7] Peterson, J.L.: Petri Net Theory and the Modeling of System. Prentice-Hall (1981)

[8] Reisig, W.: Petri Nets: An Introduction. Springer (1985)

[9] Murata, T.: Petri Nets: Properties, Analysis and Applications. Proceedings of the IEEE 77(4) (1989)

[10] Cheung, K.S.: New Characterisation for Live and Reversible Augmented Marked Graphs. Information Processing Letters 92(5), 239–243 (2004)

[11] Cheung, K.S., Chow, K.O.: Cycle-Inclusion Property of Augmented Marked Graphs. Information Processing Letters 94(6), 271–276 (2005)

[12] Cheung, K.S.: Boundedness and Conservativeness of Augmented Marked Graphs. IMA Journal of Mathematical Control and Information 24(2), 235–244 (2007)

[13] Chen, C.L., Chin, S.C., Yen, H.C.: Reachability Analysis of Augmented Marked Graphs via Integer Linear Programming. The Computer Journal 53(6), 623–633 (2009)

[14] Cheung, K.S., Chow, K.O.: Manufacturing System Design Using Augmented Marked Graph. In: Proceedings of the Chinese Control Conference, South China University of Technology, pp. 1209–1213 (2005)

[15] Cheung, K.S.: Augmented Marked Graphs and the Analysis of Shared Resource Systems. In: Kordic, V. (ed.) Petri Net: Theory and Application, pp. 377–400. I-Tech Publishing (2008)

[16] Dijkstra, E.W.: Cooperating Sequential Processes. In: Genuys, F. (ed.) Programming Languages, Academic Press, London (1965)

# Author Index